ANIMAL PHYSIOLOGY AND BIOCHEMISTRY

ANIMAL PHYSIOLOGY AND BIOCHEMISTRY

For Undergraduate Students of Indian Universities in Accordance with UGC Model Curriculum

Dr. R.A. Agarwal
Ph.D.

Dr. Anil K. Srivastava
Ph.D.

Dr. Kaushal Kumar
Ph.D.

Formerly Professors and Heads
Department of Zoology, DDU Gorakhpur University
Gorakhpur, Uttar Pradesh

S. CHAND
PUBLISHING

S Chand And Company Limited
(ISO 9001 Certified Company)

S Chand And Company Limited

(ISO 9001 Certified Company)

Head Office: D-92, Sector–2, Noida – 201301, U.P. (India), Ph. 91-120-4682700

Registered Office: A-27, 2nd Floor, Mohan Co-operative Industrial Estate, New Delhi – 110 044, Phone: 011-49731800

www.**schandpublishing.com;** e-mail: **info@schandpublishing.com**

Marketing Offices:

Chennai	:	Ph: 23632120; chennai@schandpublishing.com
Guwahati	:	Ph: 2738811, 2735640; guwahati@schandpublishing.com
Hyderabad	:	Ph: 40186018; hyderabad@schandpublishing.com
Jalandhar	:	Ph: 4645630; jalandhar@schandpublishing.com
Kolkata	:	Ph: 23357458, 23353914; kolkata@schandpublishing.com
Lucknow	:	Ph: 4003633; lucknow@schandpublishing.com
Mumbai	:	Ph: 25000297; mumbai@schandpublishing.com
Patna	:	Ph: 2260011; patna@schandpublishing.com

First Edition 1978
Subsequent Editions & Reprints 1979, 80, 81, 82, 83, 84, 85, 86, 90, 94, 95, 97, 98, 99, 2001, 2002, 2005, 2007, 2008, 2009, 2010, 2011, 2013, 2014 (Twice), 2015, 2017, 2018, 2019, 2020, 2021, 2022

Reprint 2023

ISBN: 978-81-219-0447-6 **Product Code:** H6APH68ANMS10ENAE0XO

PRINTED IN INDIA

By Vikas Publishing House Private Limited, Plot 20/4, Site-IV, Industrial Area Sahibabad, Ghaziabad – 201 010 and Published by S Chand And Company Limited, A-27, 2nd Floor, Mohan Co-operative Industrial Estate, New Delhi – 110 044.

PREFACE TO FIFTH REVISED EDITION

Physiology and Biochemistry are the most rapidly advancing disciplines of life sciences. This fifth edition is broader than its name implies, since attempt has been made not only to corrlelate the two fields to related subjects as cell biology, histology and anatomy but also to achieve a better-balanced presentation of the over-all picture of the area with appropriate emphasis on human systems.

This edition has grown considerably. I have added more illustrations in chapters 1, 5, 11, and 13 which have risen from 1 to 8, 27 to 39, 11 to 19, and 18 to 22, respectively. I have updated all these chapters, including chapters 14 and 16, contributed by me, with latest developments to further enhance the usefulness and scope of the book. The remaining twelve chapters dealing with classical physiology and biochemistry have been contributed by my co-authors. They are statisfied with the material present in the earlier four editions with little or no change. Maybe in a future edition, they might discover new information to add.

Many corrections and suggestions have been received from readers during the long history of more than three decades of the earlier four editions. I wish to encourage this reaction from the field, not only to permit corrections to be made in reprinted issues, but also as a basis for preparing future editions. As my co-authors and I have retired, all letters relating to chapters 7, 8, 9, 10, 15 and 17 be mailed to Dr. R.A. Agarwal, 43-A, Jail Road, Shahpur, Gorakhpur; for chapters 2, 3, 4, 6, 12 and 18 to Dr. Kaushal Kumar, 192-Q Shivpuram, Basharatpur East, Gorakhpur or directly to the publishers who will forward these to the authors.

Professor H.S. Chaudhary, to whom we are greatly indebted for writing the Foreword of the first edition and for the values he instilled in his colleagues, has passed away into the caring hands of God at a ripe old age of 94 years. In remembrance, his Foreword is reproduced on the next page.

I take this opportunity to thank the management of S.Chand & Co. who all have been very supportive in making this edition possible and, also, for inviting me to write the preface of this revised Fifth Edition of the book in its new format.

Dr. ANIL K. SRIVASTAVA

G-7, Dronacharya Apts.
Mayur Vihar, Ph. 1 Extn.
New Delhi 110091

FOREWORD

In the present perspective, physiology and biochemistry have attained an important place in the college and university curricula of animal sciences. This book is, therefore, extremely welcome; it is clearly written and should be easily comprehensible to everyone with an elementary background of these sciences.

It is pleasant to recall that, as Professor and Chairman of the Department of Zoology, University of Gorakhpur, I was able to introduce this exciting and promising area of biological sciences in the university cirricula. The authors of this book have been associated with the teaching and growth of the subject in the department for many years. They themselves have made important contributions in the area both here and abroad; a good deal of their enthusiasm for the subject is reflected in the present book. The excellent illustrations given in the book extend the range of usefulness of this effort. I congratulate the authors for their efforts. I am confident the book will give both teachers and students exciting and serious reading.

Gorakphur
December 1975

Late Dr. H.S. CHAUDHARY
Former Vice-Chancellor
University of Gorakhpur

PREFACE TO THE FIRST EDITION

It is our pleasure and privilege to present the first edition of Textbook of Physiology and Biochemistry to all those who elect to use this book. After having taught physiology to a large number of students, both under- and postgraduates, and involved in researches in this area here and abroad, it was felt that there was pressing demand for such a book especially to cater the curricula requirements of undergraduate students of Indian universities and medical colleges. We were, in fact, encouraged in our endeavour by many of our past and present students who also felt that a textbook covering physiology and biochemistry in an adequate manner, and yet not too terse in reading would go a long way in creating interest in the area.

The present knowledge of physiology has already assumed gigantic proportions, and something keeps on being added every day. In determining the size of this book we were compelled to talk into consideration the extent to which any student can be stretched during the time allowed to him for a particular course. Emphasis, therefore, has been placed more on central concepts in depth than details wherever we were confronted with the choice of picking between the two. Since this book was intended to be a textbook, no attempt has been made to cite individual authorities for each statement. We have devoted hundreds of hours in making the language simple and intelligible to our young students. Its length is such that it will be read, and hopefully, eagerly enjoyed both by students and teachers of the subject.

The book has been designed for readers who are being initiated into the mysteries of human body. Although it expects a minimum knowledge of physical, chemical and anatomical terms from the students, every effort has been made to minimize the need of reference books. Although anatomy is outside the scope of this book, we have given a large number of illustrations so that the reader many not have to refer to books on anatomy in relating structure to body function. A list of Selected Readings is given at the end of the book which may be used for supplementary reading or as a guide to more specific literature.

The book starts out with the ultrastructure of the cell as this, we thought, should be the natural choice and the stepping stone for gaining a lasting insight into the succeeding chapters on bio-chemistry and physiology. A regorous physico-chemical treatment has been avoided in the chapters on body function. The chapters (2 to 7) in the beginning are designed in a manner which will give the reader an idea of the various biological melecules; the subsequent chapters deal with the physiology of different organo-systems.

It is indeed a pleasure to thank a large number of persons who have directly or indirectly helped in making this book a reality. Our foremost thanks are due to those physiologists and biochemists who have provided, and are continually providing information to make these disciplines two of the most rapidly advancing branches of biology. We are indebted to Dr. H.S. Chaudhry, formerly Senior Professor and Chair man, Department of Zoology, and now Vice-chancellor, University of Gorakhpur, who has always gone out of his way to encourage productive work in the department, and at large.

A large number of illustrations have been drawn by the authors themselves. However, valuable assistance has also been provided by Mrs. Madhulika Kumar, Onkar Singh, S.P. Srivastava, Deepak Kumar, and others who like their names to be kept anonymous. Of course, any errors in the drawings remain ours alone.

The authors are aware of the fact that shortcomings are bound to creep in any first edition. We shall be grateful to our students and colleagues if they point out these to us with brutal frankness.

Finally, we fondly acknowledge the love, unlimited patience, and understanding of three wonderful wives. They not only waited with hot meals, but also made numerous and valuable suggestions regarding the contents and format of the book itself. Only their modesty prevents us from naming them as co-authors. We dedicate this book to them.

GORAKPHUR **AUTHORS**

CONTENTS

1 Chapter

The Cell

"We shall not obtain the best insight into things until we actually see them growing from the beginning." So said Aristotle, centuries ago, and it prompted us to begin this book by first giving an insight into the cell — *the unit of life*. This approach is rather unconventional for a book on physiology. Nevertheless, it is not possible to explore the intricacies of living organisms unless we understand the basic structure of the cell that is fundamental to all of the biological sciences.

The cells have evolved into two main categories, namely, the *prokaryotic cells* and *eukaryotic cells*. The prokaryotic cell is characteristic of the super kingdom Prokaryotes that include various types of all bacteria which lack a membrane-bound nucleus. Cell division in these involves *binary fission*; centrioles, mitotic spindles and mitochondria are absent. They also lack microtubules, (or the cytoskeleton). The chromosome in a prokaryotic cell is a single circular DNA (*deoxyribonucleic acid*) molecule containing the entire set of genetic instructions essential for the life of the cell. This super-kingdom contains one kingdom the **Monera**.

Eukaryotic cells are characteristic of the superkingdom *Eukaryotes* which contain all organisms that are unicellular (e.g., yeast, amoeba, etc.), multicellular aggregates (e.g., *Volvox*) or consists of cells with true nuclei bounded by nuclear membrane and which undergo meiosis. Cell division occurs by mitosis. Oxidative enzymes are packaged within mitochondria. The chromosomes inside the nucleus consist of chromatin which carry genetic information in *linear* DNA molecules. The eukaryotes contain four kingdoms, namely, **Protoctista, Fungi,** ***Animalia*** and ***Plantae***. In spite of these differences, the same basic properties, such as nutrition, production of energy, biosynthesis, movement, respiration, growth, irritability, and reproduction that govern the lives of both prokaryotes and eukaryotes. This points out that all present day cells are descended from a single primordial ancestor. This has led to the *endosymbiont theory* (or *Serial Symbiosis Theory*) which postulates that eukaryotic cells evolved from bacterial ancestors by a series of symbiotic associations. The theory in its most modern form suggests that the mitochondria and the microtubules organizing systems of present day eukaryotes evolved from bacteria and spirochaetes that lived as symbionts in a line of single-celled eukaryotes that were the ancestors of both fungi and animals.

In the following chapters a detailed account of these vital processes, which underlie the manifestations and maintenance of the many and diversified activities of life, given. However, before, we embark on — are the study of these processes which go on in living organisms, it will be necessary to understand the structure and functions of an *animal cell*.

BASIC CELL DESIGN

A cell, as observed under the light microscope, can ordinarily be differentiated into a dense inner **nucleus** surrounded by a less dense mass of **cytoplasm** bounded by a **plasma membrane**. But in the past four decades or so, since the development of the electron microscope, the study of cell and its components has reached new dimensions. In the following pages, the electron microscopic structure (**ultrastructure**) of the cells has been described.

Figure 1.1 illustrates the basic ultrastructural organization of an individual cell.

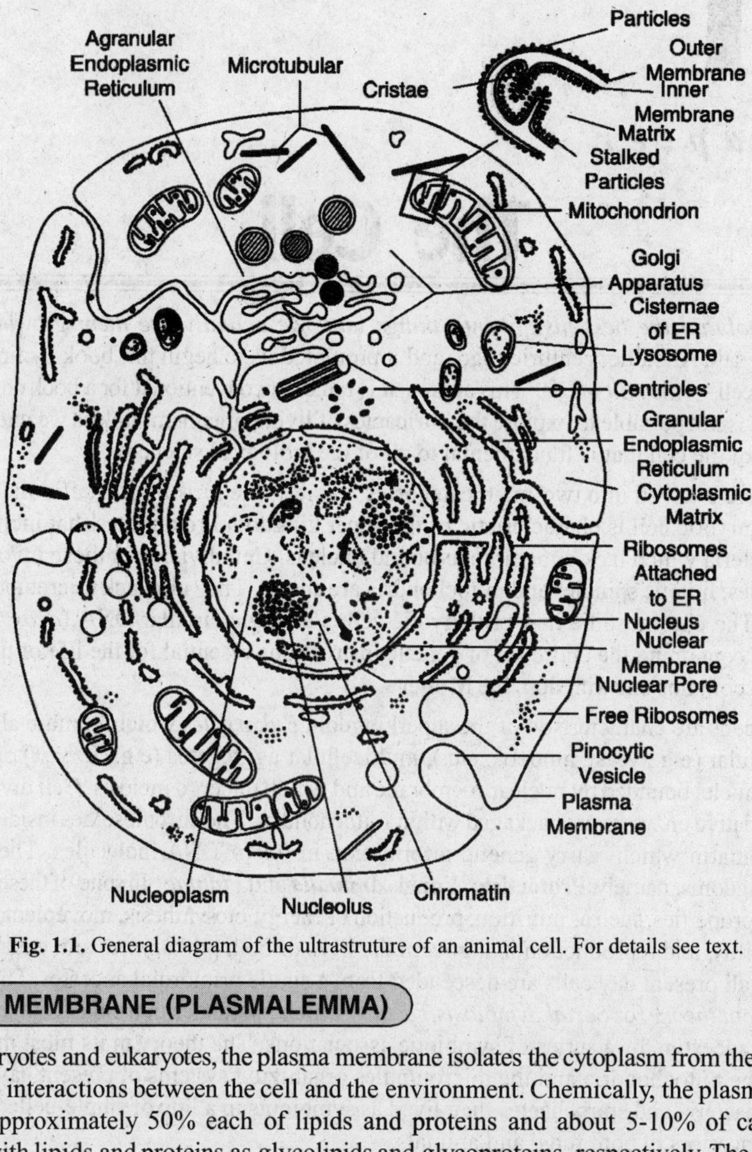

Fig. 1.1. General diagram of the ultrastruture of an animal cell. For details see text.

PLASMA MEMBRANE (PLASMALEMMA)

In both prokaryotes and eukaryotes, the plasma membrane isolates the cytoplasm from the environment and mediates interactions between the cell and the environment. Chemically, the plasma membrane consists of approximately 50% each of lipids and proteins and about 5-10% of carbohydrates, conjugated with lipids and proteins as glycolipids and glycoproteins, respectively. The lipids are the fundamental structural elements of the membrane. A large part of the lipid in phospholipids (namely, phosphatidylcholine, phosphatadilethanolamine, phosphatidylserine, and sphingomyelins). A fifth phospholipid, phosphatidylinositol is also a minor membrane component. These consist of two fatty acids joined to a polar head group. The two fatty acids are bound to glycerol which is then bound to a phosphate group. The latter is attached to polar molecule such as choline, ethanolamine, serine or inositol. The sphingomyelin is a nonglycerol phospholipid in which glycerol is replaced by serine and the phosphate group attached to choline. All phospholipids have *hydrophobic* tails, consisting of two hydrocarbon chains, and *hydrophilic* head groups, consisting of phosphate group and its polar attachments. This means that phospholipids are **amphipathic** molecules, i.e., part of which are water – soluble and part water–insoluble. Besides the phospholipids the plasma membrane also contains glycolipids and cholesterol (cholesterol is absent in bacteria). Both of these are also amphipathic.

The fundamental structure that plasma membrane is a phospholipid bilayer evolved from the study of membrane structure of mammalian erythrocytes by E. Gorton and R. Grendel in 1925. When sectioned plasma membrane is viewed under the electron microsope, it has a total thickness of 75 Å and appears as two dense strata each about 20 Å thick separated by a light interzone about 35Å wide. This structure has been referred to as '*trilaminar*' or '*railroad track*' appearance. The polar head group of the phospholipids which face out of the cell and towards the cytosol (cytosol is the fluid portion of the cytoplasm exclusive of organelles) are hydrophilic and appear as dark lines; the less dense interior portion contains the hydrophobic fatty acid tails. The latter imparts the characteristic property to the membrane that it is impermeable to water–soluble molecules, including ions and most biological molecules. The glycolipids are found exclusively in the other layer of the plasma membrane (Fig. 1.2).

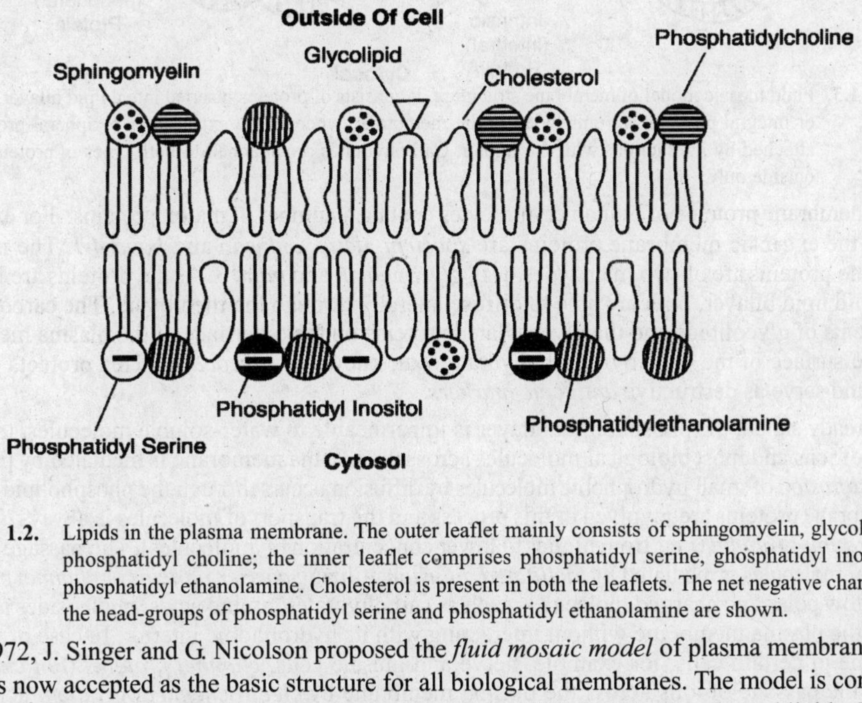

Fig. 1.2. Lipids in the plasma membrane. The outer leaflet mainly consists of sphingomyelin, glycolipids and phosphatidyl choline; the inner leaflet comprises phosphatidyl serine, ghosphatidyl inositol and phosphatidyl ethanolamine. Cholesterol is present in both the leaflets. The net negative charges (-) of the head-groups of phosphatidyl serine and phosphatidyl ethanolamine are shown.

In 1972, J. Singer and G. Nicolson proposed the *fluid mosaic model* of plasma membranes. This model is now accepted as the basic structure for all biological membranes. The model is considered to be a *two-dimensional viscous solution* consisting of a bilayer of highly oriented lipids. The two layers are discontinuous being interrupted by protein molecules that penetrate one or both layers. Two classes of proteins, namely *peripheral* or *extrinsic* and *integral* or *intrinsic* proteins have been identified. The extrinsic proteins do not penetrate the phospholipid bilayer and are associated with the membrane through protein–protein interactions. On the other hand, the intrinsic proteins are embedded in the lipid bilayer, sometimes being exposed on the outer surface, sometimes on the inner surface, and sometimes penetrating both surfaces (transmembrane proteins). Most transmembrane proteins are glycoproteins with their oligosaccharides exposed on the surface of the cell (Fig. 1.3). Transmembrane proteins, known as *porins*, in *Escherichia* coli do not span the lipid bilayer but form aqueous channels in the outer membrane which makes it permeable to ions and small molecules.

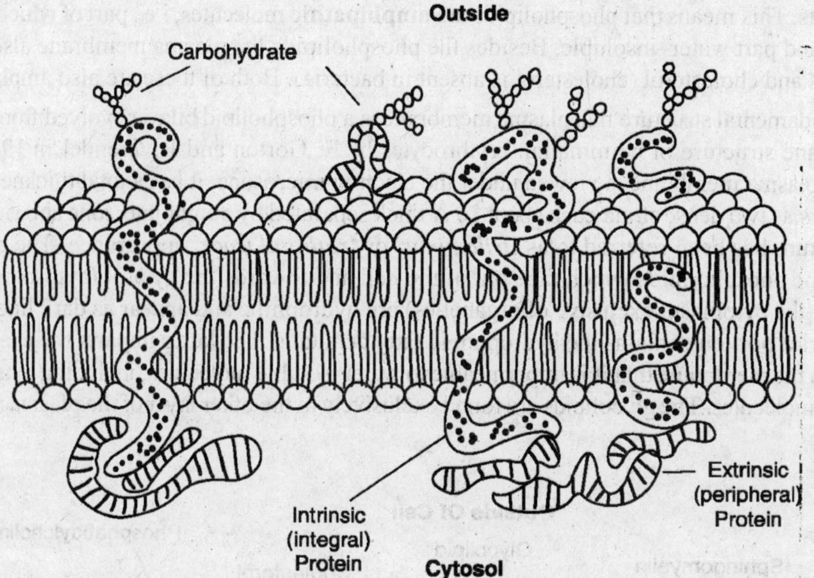

Fig. 1.3. Fluid mosaic model of membrane strucuture. It consists of proteins inserted into a lipid bilayer. Intrinsic or integral proteins are embedded in the membrane, whereas the extrinsic or peripheral proteins are attached by interactions with the former. Carbohydrates are attached to both types of proteins on the outside only.

The membrane proteins of human erythrocytes contain a number of major proteins. For example, some of the extrinsic membrane proteins are *spectrin, actin*, *ankyvin* and *band 4.1*. The intrinsic membrane proteins are of two main types, *viz.*, *glycophorin* and *band 3*. As the proteins are inserted into a fluid lipid bilayer, both are able to diffuse laterally through the membrane. The carbohydrate components of glycolipids and the glycosylated proteins on the outer face of the plasma membrane cover the surface of the cells by a carbohydrate coat, known as *glycocalyx*. This protects the cell surface and serve as destructive *antigenic markers.*

As already stated, the phospholipid bilayer is impermeable to water-soluble molecules, therefore passage of ions and most biological molecules across the plasma membrane is mediated by proteins. *Passive diffusion* of small hydrophobic molecules by diffusion occurs through the phospholipid bilayer. No membrane proteins are involved in this process and the transport of molecules is always *down the concentration gradient* (*i.e.*, from higher to lower concentration of molecules). The passage of most biological molecules is mediated by *facilitated diffusion* utilizing *carrier proteins* or *channel proteins*. These allow polar and charged molecules, such as carbohydrates, amino acids, nucleosides and ions to cross the plasma membrane without interacting with its hydrophobic interior. In case of plasma membrane of certain cells, for example, nerve and muscle cells, *channel proteins* (*ion channels*) mediate the passage of ions across the plasma membrane by *electrochemical gradient* across the membrane. The regulated opening and closing of ion channels is responsible for the *transmission of electric signals* (see chapter 10). *No external source of energy is provided in such transports.*

The cell, however, must transport molecules against their concentration gradients. In such cases, transport takes place by *active transport* in which energy is provided by the hydrolysis of ATP (*adenosine triphosphate*) to drive the molecules against the electrochemical gradients. For example, the concentrations of Na^+ (*higher outside the cell*) and K^+ (*higher inside the cell*) are maintained by Na^+ K^+ pump (or Na^+ - K^+ ATPase). In addition to the ionic pump, *ABC transporters* are utilized to transport a wide variety of molecules (ions, sugars and amino acids) and *MDR transporter* removes foreign toxic compound from the liver, intestine and kidneys. Ion gradients are sometimes coupled with the transport of a second molecule such as sugars, amino acids and ions by either *symport* (e.g. uptake of glucose and Na^+) or *antiport* (such as $Na^+ - Ca^{++}$ antiporter transporting Na^+ into the cells and Ca^{++} out) or $Na^+ - H^+$. The latter forms a mechanism for preventing acidification of the cytoplasm.

These two phenomena are different from the transport of a single molecule, such as glucose, in facilitated diffusion which as an example of *import*.

The plasma membrane is able to take up macromolecules and particles by a process called *endocytosis*. The plasma membrane invaginates in which the material is taken up. The invagination containing the ingested material is then pinched off to form a small vesicle. In this way ingestion of large particles and uptake of fluids take place. The former is known as *phagocytosis* (*cell eating*) and the latter as *pinocytosis* (*cell drinking*). Pinocytosis is eukaryotic cells may also be *receptor-mediated endocytosis*. This involves binding of ligand (such as intellogenin) to a plasma receptor. The *ligand-receptor complex* moves towards a coated pit which is a basket-like network formed by three protein complexes called *triskelions*, each composed of three molecules of *clathrin proteins, a* 185-kd protein and three smaller polypeptides. These clathrin-coated pits following accumulation of a large number of ligand-receptor complexes invaginate into the cytoplasm and a small vesicle pinched off known as *receptosome*. The receptor molecule is then returned to the plasma membrane. The uptake of cholesterol and LDL (low-density lipoprotein) are some common examples of receptor-mediated endocytosis.

Selective cell-cell interactions are characteristics of multi cellular organisms. Such interactions are mediated by four *cell adhesion proteins*, namely, selections, integrins, inmmunoglobulin super family members and *cadherins*. Cadherin-mediated junctions are called *adherens junctions* and desmosomes. In addition to the above cell-cell interactions, *tight junctions* and *gap junctions* are also present. The former prevent free passage of molecules between cells and the latter form open channels connecting the cytoplasma of adjacent cells.

CYTOPLASM

The cytoplasm consists of a colloidal solution called **cytoplasmic matrix** in which various bodies are present. These include *cytoskeleton, centrioles, endoplasmic reticulum, Golgi apparatus, lysosomes, mitochondria, peroxisomes, cytoplasmic inclusions*, and *nucleus*. These are persistent structures with specialized functions. They are analogous to an organ in a whole organism and are, therefore, referred to as **cytoplasmic organelles**. In addition to the cytoplasmic organelles, the cytoplasm also contains secretory products of the cell called **cytoplasmic inclusions**.

Cytoplasmic Matrix (Ground Substance)

It is surrounded on the outside by the plasma membrane and limited on the inside by the nuclear membrane. It appears as a heterogeneous substance containing the cytoskeleton, cytoplasmic organelles and inclusions.

The cytoplasmic matrix contains several inorganic ions, *viz*., Cl^-, Na^+, K^+, Ca^{++}, Mg^{++}, SO_4^{--}, CO_3^-,$^-$ HCO_3^-, $H_2PO_4^-$, HPO_4^{--}, etc., certain unionized mineral components, *viz*., iron, magnesium, copper, cobalt, iodine, selenium, nickel, and molybdenum along with RNA molecules, globular and fibrillar proteins (which include many enzymes and coenzymes such as those needed for energy production by anaerobic glycolysis and those used in the activation of amino acids for protein synthesis, and hormones), lipids, amino acids (forming an amino acid pool), simple sugars, and large quantities of water. The small molecules such as amino acids, simple sugars, ions, etc., are in solution, whereas the macromolecules, for instance, proteins and lipids form a colloid. The cytoplasmic matrix, thus, assumes the form of a **heterogeneous colloid**. Due to its colloidal nature the cytoplasmic matrix can be converted from fluid (*sol*) to semisolid (*gel*) state and *vice versa*. Changes between the *sol* and *gel* states are brought about by changes in temperature, pressure, pH, salt concentration, and ATP, etc.

Cytoskeleton

In the cytoplasmic matrix of eukaryotic cells is found another level of internal organization referred to as the *cytoskeleton*. It is an internal network of protein filaments that provides the structural framework of the cell, give its ability to move, to assume a characteristic shape, to divide, to assist chromosomal movements during cell division, to undergo pinocytosis to arrange its organelles, and to transport the organelles from one location to another. The cytoskeleton contains *microfilaments* or *actin filaments intermediate filaments and microtubules*. These are interconnected by a network of accessory filaments. This complex is called *microtrabecular lattice* and can be visualized with the electron microscope. Microfilaments contain actin monomers which polymerize to form a helix. They function to determine cell shape, anchor the cell at regions of cell-cell and cell-substrate in contact, phagocytosis, and cell locomotion. The intermediate filaments comprise a heterogeneous class of cytoskeletal proteins. A particular class of intermediate filament is characteristic of a particular cell type. For instance, keratin filaments in epithelial cells, *neurofilaments* in neurons, *desmin filaments* in glial cells and vimentin filaments in fibroblasts. The intermediate filaments provide mechanical support to cells and tissues. The microtubules are long, non-branching, thin cylinders composed of 13 strands called *protofilaments*. Each of these contain α and β *tubulin* molecules. Microfilaments play key role in cell division, secretion, intercellular transport, morphogenesis and ciliary and flagellar movements. The microtubules extending the length of the cilium and flagellum are called *axoneme* that contains *dynein*- a protein. The microtubules in the *axoneme* are arranged in a characteristic '9 + 2' pattern (*i.e.*, a pair of central microtubules surrounded by nine outer microtubule-doublets).

The microtubules in most cells extend outward from a *microtubule-organization centre* or *centrosome*, located adjacent to the nucleus of interphase cells. In animal cells the centrosome contains a pair of *centrioles* surrounded by pericentriolar material. The growth of the microtubules is initiated in the pericentriolar material, which serves to anchor their minus ends (Fig. 1.4).

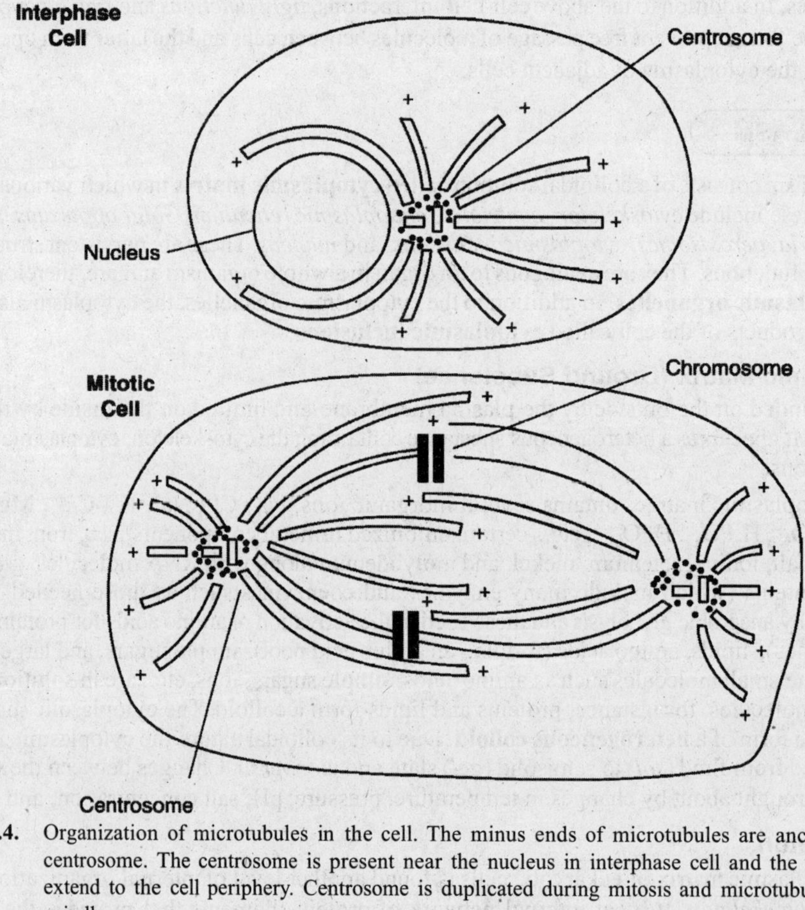

Fig. 1.4. Organization of microtubules in the cell. The minus ends of microtubules are anchored in the centrosome. The centrosome is present near the nucleus in interphase cell and the microtubules extend to the cell periphery. Centrosome is duplicated during mitosis and microtubules form the spindle.

Centrioles

As already discussed, the centrosome of interphase cell contain a pair of centrioles. A centriole is a self-reproducing organelle consisting of short cylinders containing 9 triplets of peripheral microtubules disposed about a central cavity. When a centriole reproduces, it forms a *daughter* centriole at right angles to itself. The 'daughter' centriole grows outward and move from the 'mother' until it reaches the mature size. They are capable of movement and lie at the polar region of the mitotic spindle in dividing animal cells. *Centrioles do not occur in higher plants.*

Endoplasmic Reticulum (ER)

It is a system of membrane-enclosed tubules and sacs (cisternae) in the cytoplasm of most eukaryotic cells. In places the ER is continuous with the plasma membrane or the outer membrane of the nuclear envelope. When the outer surface of the ER membranes is covered with *ribosomes*, the ER is said to be *rough ER* (RER) or *granular ER*; if devoid of ribosomes the ER is said to be *smooth ER (SER)* or non-granular ER. The RER functions in processing of proteins which are translated on RER-bound ribosomes. The SER is the major site of lipid synthesis in *eukaryotic* cells. Both the proteins and lipids are transported in vesicles from RER and SER, respectively, to the Golgi apparatus.

Golgi Apparatus (Golgi Complex or Golgi Substance).

It is a cell organelle identified in electron micrographs as a complex made up of closely packed broad cisternae and small vesicles. The Golgi apparatus is distinguished from the ER by the arrangement of the membranous vesicles and by the lack of ribosomes. The Golgi apparatus consists of three distinct regions, namely, the *cis Golgi network*, *Golgi stack* and the *trans Golgi* network (Fig. 1.5). The organelle functions to collect and sequester substances such as N-linked oligosaccharides to protein (protein glycosylation) synthesized by the rough ER. It is also the site of synthesis of lipids (e.g. glycolipids and sphingomyelin) and polysaccharides.

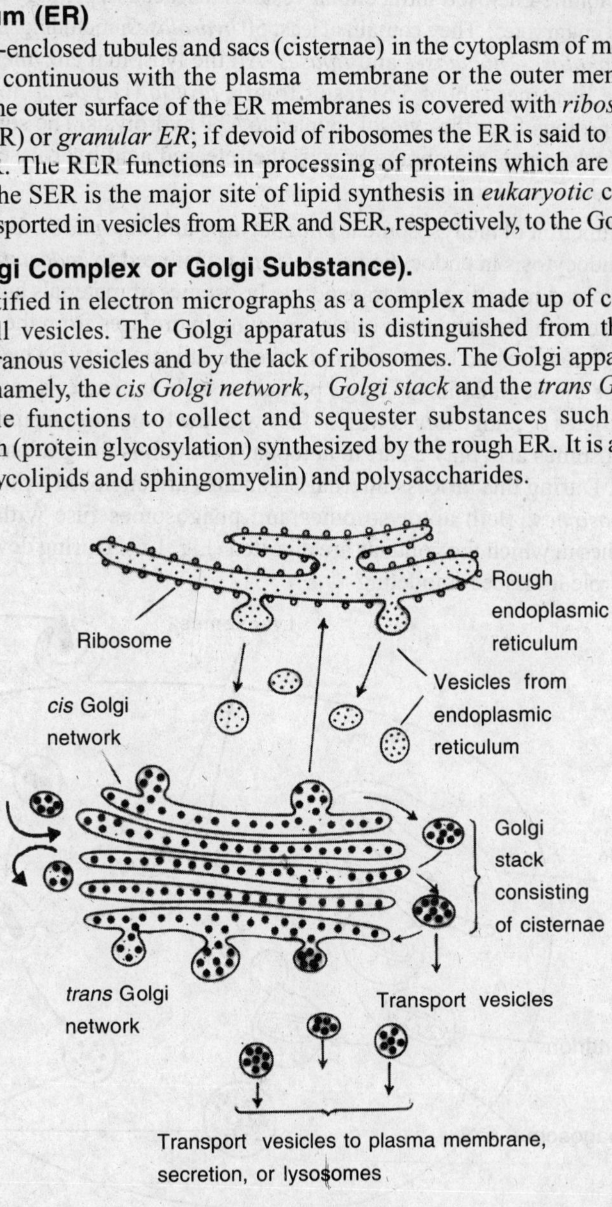

Fig. 1.5. Golgi apparatus consisting of the *cis* and *trans* Golgi network. The former is the site at which ER vesicles enters the Golgi and ER proteins are returned via the recycling pathway. The Golgi stack formed by the cisternae is the site for the modification of proteins. These proteins are then transferred to the *trans* Golgi network for transport to plasma membrane, secretion, or lysosomes.

Proteins packaged in ER vesicles enter the Golgi apparatus of cis Golgi network. These are sorted out here and those meant for return to ER are recognized and recycled back to ER. Other proteins are taken to Golgi stack which is the site of metabolic activity of Golgi apparatus. From here, the vesicles carry the modified proteins, lipids and polysaccharides to the trans Golgi network. This works as a sorting and distributing centre for transport to the plasma membrane, to the cell exterior or the lysosomes

Lysosomes

These are membrane-enclosed intracellular vesicles that act as a primary component for intrecellular digestion in all eukaryotes. They contain at least 50 *hydrolases*, including *phosphatases, glycosidases, proteases*, *sulphatases*, *nucleases* and *lipases*. All the lyosomal enzymes are *acid hydrolases*. The acid pH of the lyosomes (about 5.5) result from a *proton* (H^+) *pump* in the lysosomal membrane which import protons from the cytosol coupled to ATP hydrolysis. The significane of the acidic pH is that even if the lysomal membrane breaks, the released acid hydrolases would be inactive at the neutral pH (about 7.2) of the cytosol.

The main function of the lysosome is *phagocytosis* and *autophagy.* Extracellular materials when taken up by endocytosis in endocytic vesicles, are transferred to *endosomes* (a specialized vesicle in the cytosol involved in sorting and transport to lysosomes of materials taken up by endocytosis). The endosomes fuse with the transport vesicles carrying hydrolases from the *trans* Golgi network. Such endosomes then mature into lysosomes and have the full quota of acid hydrolases. The lysosomes are responsible for the degradation of large particles (such as bacteria, cell debris, etc.) taken up in phagocytic resicles or *phagosomes* which fuse with the lysosomes resulting in the digestion of their contents. Lysosomes also play a part in autophagy (which is the gradual turnover of the cell's own components). During this process internal organelles are enclosed by the membranes of the ER to form *autolysosomes*. Both autolysosomes and phagosomes fuse with lysosomes to form large phagolysosomes in which the contents are digested (Fig. 1.6). During development, autophagy plays an important role in tissue remodeling.

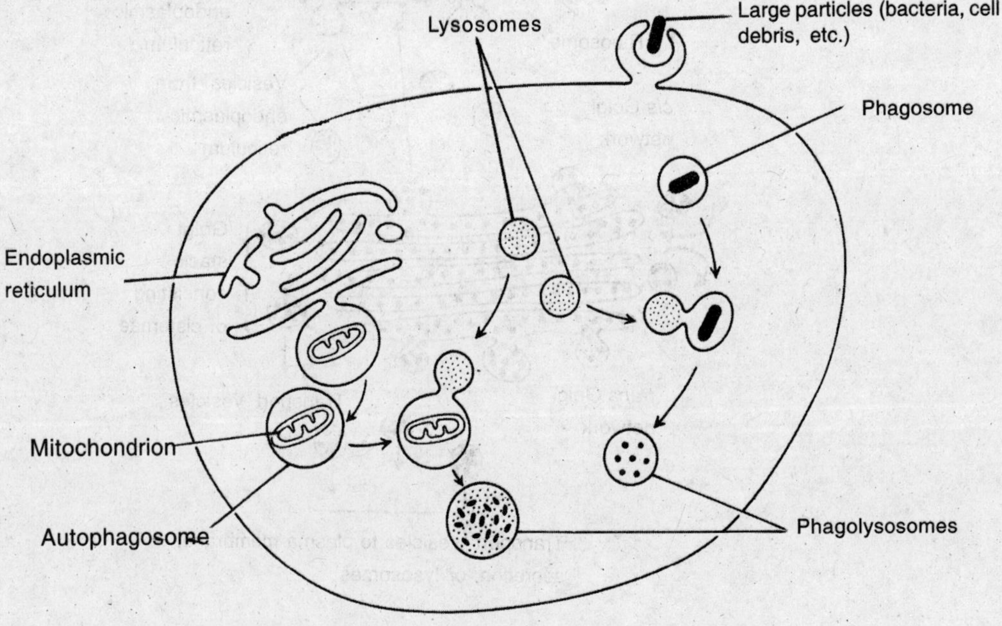

Fig. 1.6. Role of lysosomes in phagolysosomes of large particles by phagosomes and autophagy of cytoplasmic organelles (such as mitochondria) by autophagosomes formed from ER. Lysosomes fuse with both phagosomes and autophagosomes to form phagolysosomes wherein their contents are digested by the lysosomal hydrolytic enzymes.

Mitochondria

These organelles play critical role in energy metabolism. They are *semiautonomous, self-replicating* organelles that occur in the cytoplasm of all cells of most, *but not all*, eukaryotes. For example, *microsporidia* (primitive eukaryotes) lack mitochondria. Mitochondria are believed to have arisen from aerobic bacteria (such as cyanobacteria) that established a symbiotic relationship (*endosymbiosis*) with protoeukaryotes. Each mitochondrion is surrounded by a *double limiting membrane*, consisting of *inner* and *outer* mitochondrial membranes separated by an *intermembrane space*. The inner membrane is highly invaginated and its projections are called *cristae*. Each of these extend into the interior or *matrix* of the organelle. The matrix and the inner membrane represent the major functional components of the mitochondria. The matrix contains the *mitochondrial genetic system* as well as the enzymes required for the central reactions of *oxidative metabolism*.

The mitochondria contain their own circular DNA molecules which encode some of their components. Mitochondrial DNA is replicated each time the organelle divides and the genes they encode are *transcribed* within the organelle and *translated* on organelle ribosomes. They have, thus, their own genetic systems, which are distinct from the nuclear genome of the cell. Mitochondria depend on the genes within the nucleus of the cells in which they reside for many essential mRNA. The proteins translated from the mRNAs on free ribosomes in the cytosol are imported into the organelle. The mitochondria contain distinctive ribosomes, tRNA (transfer RNA), aminoacyl-tRNA synthetase, and elongation and termination factors.

The mitochondria are sites of *oxidative metabolisim* which results in generation of ATP derived from the breakdown of organic molecules (carbohydrates and fatty acids). The initial stages of glucose metabolism (*glycolysis*) occur in the cytosol where glucose is converted to pyruvate. Both the pyruvate and fatty acids are imported from the cytosol and converted to acetyl CoA (*coenzyme A*) in the mitochondrial matrix. Acetyl CoA is then converted to CO_2 via the citric acid cycle which is regarded as the central pathway of oxidative metabolism. The oxidation of acetyl CoA to CO_2 is coupled to the reduction of NAD+ (nicotinamide-adenine-dinucleotide) and FAD (flavin adenine dinucleotide) to NADH and $FADH_2$, respectively. Most of the energy (ATP) derived from oxidative metabolism is then produced by oxidative phosphorylation which takes place in the inner mitochondrial membrane (Fig. 1.7). The electrons of NADH and $FADH_2$ combine with O_2 and the energy released from the process drives the synthesis of 32 to 34 ATP molecules from ADP. This process takes place through a services of *carriers*, organized into four protein complexes I, II, III and IV in the inner mitochondrial membrane, which constitute the *electron transport chain*. A fifth protein complex (complex V or ATP synthetase) then serves to couple the energy yielding reactions of electron transport to ATP synthesis. Complex V is also referred to as '*elementary particles*' or *oxysomes*. This ATP yield when added to the 4 molecules of ATP produced during glycolysis makes the total yield to 36 or 38 ATPs per molecule of glucose (see Chapter 13).

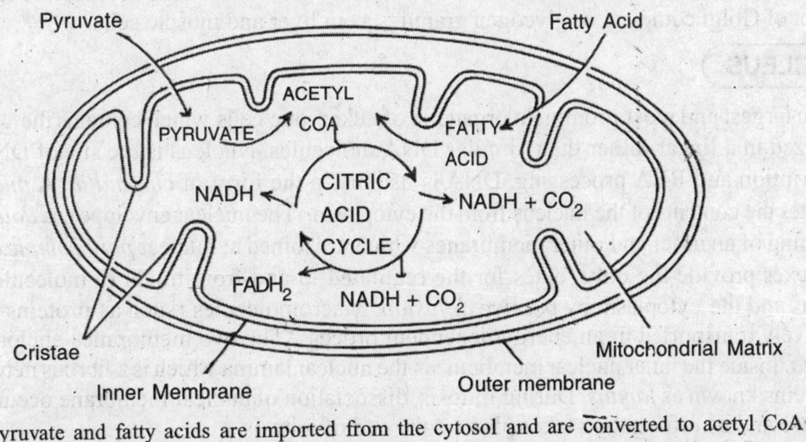

Fig. 1.7. Pyruvate and fatty acids are imported from the cytosol and are converted to acetyl CoA in the mitochondrial matrix. Acetyl CoA is then oxidized to CO_2 via the citric acid cycle.

Peroxisomes (microbodies)

These are membrane-bound organelles which like mitochondria, replicate by division and are assembled from proteins synthesized on free ribosomes. They are morphologically similar to lysosomes but contain at least 50 different enzymes involved in various biochemical pathways in different types of cells. Fatty acid oxidation, a major source of metabolic energy, is carried out in peroxisomes, resulting in the production of hydrogen peroxide (H_2O_2). As H_2O_2 is toxic to the cell, it is decomposed by a peroxisomal enzyme *catalase*, either by conversion to water or by oxidation of other organic compounds such as uric acid, amino acids (designated as XH_2) as under:

$$R-CH_2-CH_2-\overset{\overset{\displaystyle O}{\|}}{C}-S-CoA + O_2 \rightarrow R-CH=CH-\overset{\overset{\displaystyle O}{\|}}{C}-CoA + H_2O_2$$

$$2H_2O_2 \xrightarrow{\text{Catalase}} 2H_2O + O_2$$

or

$$H_2O_2 + XH_2 \xrightarrow{\text{Catalase}} 2H_2O + X$$

Peroxisomes synthesize lipids (such as cholesterol), bile acids, plasmalogens (phospholipid membrane component in heart and brain).

In plants the peroxisomes are called *glyoxisomes*. They are capable of synthesizing carbohydrates from fatty acids by the *glyoxylate cycle*. Peroxisomes are also involved in photorespiration.

Ribosomes

These are small particles of universal occurrence and are usually associated with RER. They also occur free in the cytoplasmic matrix. Ribosomes are ribonucleoprotein particles 10-12 mμ in diameter, that are sites of translation (or protein synthesis). Each ribosome consists of two unequal subunits bound together by magnesium ions. Each subunit is made up of roughly equal parts of *ribosomal RNA* (rRNA) and protein. Each ribosomal unit is assembled from one molecule of rRNA that is non-covalently bonded to 20-30 smaller protein molecules to form a compact tightly coiled particle. In eukaryotes, the rRNA of cytoplasmic ribosomes are formed by *cistrons* (see Chapter 5) localized in the nucleolus organizer of chromosomes. At least four classes of ribosomes (bacterial chloroplasts; plant cytoplasm; animal cytoplasm; mitochondria) exist that can be characterized by their *sedimentation coefficients* (S); for example, 70S and 80S for prokaryotic and eukaryotic ribosomes, respectively (for details, see Chapter 5).

Cytoplamic Inclusion

These are of various kinds; they are either secretory granules which accumulate in the cells as a product of Golgi complex, or glycogen granules as in liver and muscle cells.

NUCLEUS

It is the largest and most prominent organelle of eukaryotic cells which contains the genetic material organized in a linear rather than circular DNA molecules. Nucleus is the site of DNA replication, transcription and RNA processing. DNA is usually in the form of *chromatin*. A *nuclear envelope* separates the contents of the nucleus from the cytoplasm. The nuclear envelope is a *double membrane*, consisting of an inner and outer membranes which are joined at nuclear *pore complexes*. These pore complexes provide the only routes for the regulated to-and-from travel of molecules between the nucleus and the cytoplasm by *passive diffusion*. Macromolecules (such as proteins and RNAs are selectively transported in an energy-dependent process. The two membranes enclose a *prinuclear cisterna*. Inside the inner nuclear membrane is the nuclear lamina which is a fibrous network composed of proteins known as *lamins*. During mitosis, dissociation of nuclear membrane occurs at the end of prophase and its reformation takes place at the end of mitosis.

The chromatin becomes highly condensed during mitosis to form the compact metaphase chromosomes. During interphase some of the chromatin remains highly condensed and is transcriptionally inactive and is known as *heterochromatin*; the remainder of the chromatin remains decondensed and distributed throughout the nucleus and is called *euchromatin*. Chromatin is a complex of nucleic acid (DNA and RNA) and *proteins* (*histones* and *nonhistones*) comprising eukaryotic chromosomes.

The basic structural unit of chromatin is the *nucleosome*. The nucleosome contains 146 base pairs of DNA wrapped 1.75 times around a histone core consisting of two molecules each of H2A, H2B, H3 and H4 (*the core histones*). One molecule of a fifth histone H1 is bound to the DNA as it enters and exits each nucleosome particle. This forms a subunit known as *chromatosome*. Nonhistone proteins bind to the *linker* DNA segments between nucleosome core particles. The linker DNA contains about 80 base pairs that connect adjacent nucleosomes of a chromosome (Figs. 1.8A and B).

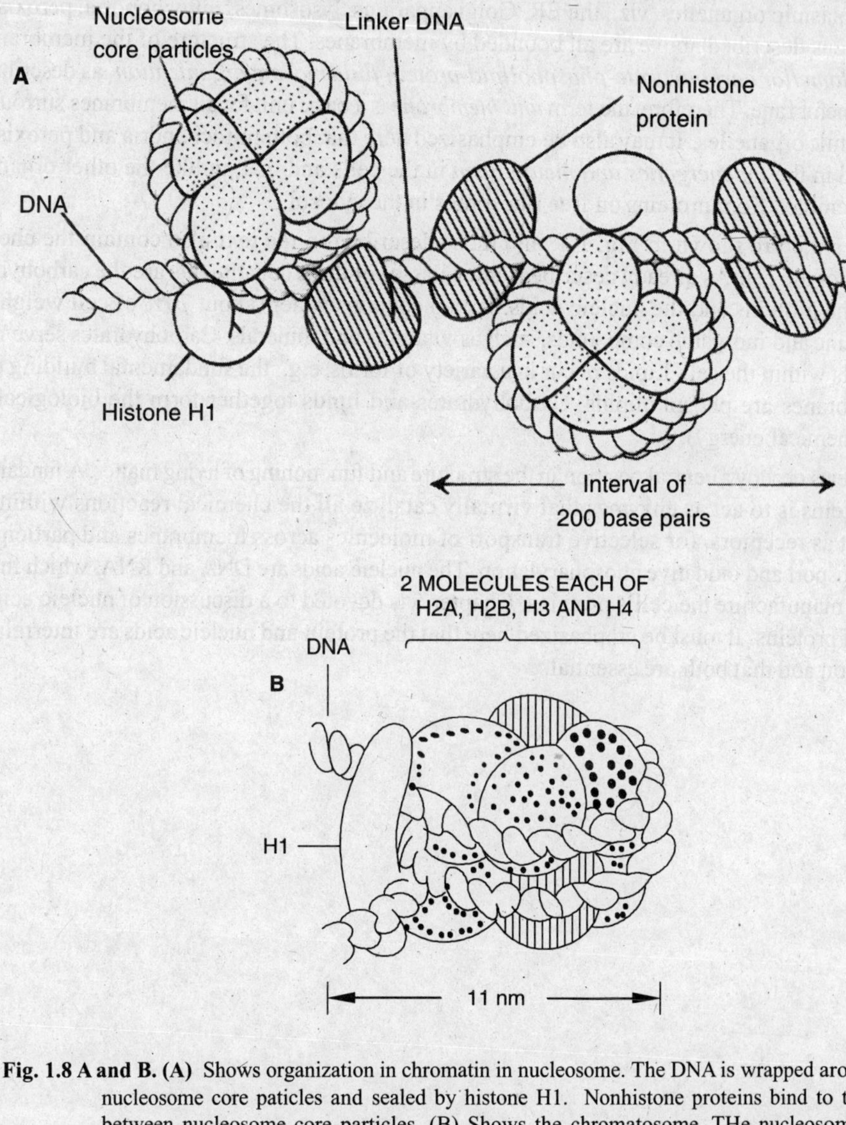

Fig. 1.8 A and B. (A) Shows organization in chromatin in nucleosome. The DNA is wrapped around histones in nucleosome core paticles and sealed by histone H1. Nonhistone proteins bind to the linker DNA between nucleosome core particles. (B) Shows the chromatosome. THe nucleosome core particle consists of 146 base pains of DNA wrapped 1.75 turns around a histone octamer consisting of two molecules each of H2A, H2B, H3 and H4. A chromatosome contains two full turns of DNA (166 base pairs) locked in place by one molecule of H1.

Nucleosomes impart a beaded appearance about 10 nm in diameter, to the chromatin fiber. The chromatin is further condensed by coiling into a 30 nm diameter fibre about six nucleosomes per turn. This is referred to as the *nucleosome model* of R.D. Kornberg (1974).

The nucleolus is the most prominent substructure in the nucleus. It is an RNA-rich spherical body associated with specific chromosomal regions that contain the 5.8 S, 18 S and 28S rRNA genes, which are therefore called *nucleolar organizing regions*. Thus, there may be one or more nucleoli in a nucleus. Consequenly, the nucleolus contains the primary products of these genes, their associated proteins and a variety of enzymes (e.g., RNA polymerase, RNA methylase and RNA endonucleases). Electron micrographs show three distinct regions in the nucleolus, viz., fibrillar centre, dense fibrillar component and granular component. When the nucleus is dividing, the nucleolus disappears.

The cytoplasmic organelles, viz., the ER, Golgi apparatus, lysosomes, mitochondria, peroxisomes and the nucleus described above are all bounded by membranes. The structure of the membrane has the same *trilamellar carbohydrate-phospholipid-protein fluid mosaic organization* as described for the plasma membrane. Therefore, the term *unit membrane* is a basic unit for all membranes surrounding the cytoplasmic organelles. It may also be emphasized here that both mitochondria and peroxisomes are involved in the *bioenergetics and metabolism* in the cells and that, unlike the other organelles, they are assembled from proteins on free ribosomes in the cytosol.

The cell structure shown in Fig. 1.1, and those described in the text also contain the chemical components of the cell. As already described, chief classes of these constituents are the carbohydrates, the lipids, the proteins and the nucleic acids. It also contains water (about 70% of cell weight) and several organic and inorganic compounds, such as vitamins and minerals. Carbohydrates serve mostly as food stuffs within the cell. Lipids occur in a variety of forms, e.g., the fundamental building blocks of all membranes are phospholipids. Carbohydrates and lipids together form the biological fuels (stores of chemical energy).

The proteins occupy a central position in the structure and functioning of living matter. A fundamental task of proteins is to act as enzymes that virtually catalyze all the chemical reactions within cells. Proteins act as receptors, for selective transport of molecules across membranes and participate in electron transport and oxidative phospharylation. The nucleic acids are DNA and RNA, which function together to manufacture the cell's proteins. Chapter 5 is devoted to a discussion of nucleic acids and synthesis of proteins. It must be emphasized here that the protein and nucleic acids are interrelated in their function and that both are essential.

2 Chapter

Proteins

Proteins are polymers formed by *monomeric units* (molecular units) called *amino acids*. They are the principal structural and metabolically active compounds found in the body. Proteins are made up of hundreds or even thousands of amino acid molecules.

Linked together by *peptide linkages* ($\overset{\overset{\displaystyle O}{\|}}{C}$ — NH) into one or more chains which are variously folded.

They are large molecules (macromolecules) having molecular weight from about 5,000 up to several millions—the molecular weight of insulin is 5,733. Proteolytic enzymes break proteins into peptones. Another term which is frequently used in protein chemistry is polypeptide. A polypeptide may be defined as a chain of three or more amino acids joined together by peptide linkages. It may even consist of several hundred amino acids. Proteins consist of polypeptide chains joined together in various ways. Sometimes the term polypeptide is interchangeably used with proteins of low molecular weight.

AMINO ACIDS

Amino acids are carboxylic acids which contain the amino group ($-NH_2$). The carboxyl ($-COOH$) and amino ($-NH_2$) groups are attached to the α -carbon atom. Hence, they are termed as α -amino acids and their general formula is:

$$\begin{array}{c} H \\ | \\ R - C - COOH \\ | \\ NH_3 \end{array}$$

where *R* is the side chain consisting of a variable grouping of atoms fundamentally forming a chain or a ring.

The amino acids, proline and hydroxyproline, however, instead of having the amino group ($-NH_2$) have an *imino group* (NH). The latter is a part of the cyclic structure of the molecule. These two are, therefore, termed as *imino* acids

Essential and Nonessential Amino Acids

Twenty amino acids are commonly. Found in proteins. Some more rate amino acids can be added to this number. *Autotrophic organisms* (such as green plants) synthesize their own amino acids. Animals (heterotrophic organisms), however, must obtain certain amino acids in their diet because they cannot be synthesized in the body. These amino acids are called *essential amino acids. Nonessential amino acids* are those which can be synthesized in the body even through some of these require essential amino acids for their synthesis. *In man, the essential amino acid are* value, leucine, phenylalanine, tryptophan, lysine, isoleucine, methionine threonine, orginine and histidine. The list of essential amino acids in other animal is also similar.

CLASSIFICATION OF AMINO ACIDS

Amino acids may be classified by one of the following ways: (1) on the number of amino and carboxyl group and (2) on the nature of their side chain (R).

According to the first criterion they can be classified as (*a*) *neutral amino acids,* (*b*) *acidic amino acids and* (*c*) basic amino acids.

According to the second criterion, the amino acids can be classified as (a) *aliphatic amino acids*, (b) *hydroxyl amino acids*, (c) aromatic amino acids, (d) sulphur containing amino acids and (e) heterocycle amino acids. The following classification is based on both these criteria.

Neutral Amino Acids

These are those amino acids which contain only a single amino and carboxyl group. They are, therefore, also called mono-amino-monocarbolylic acids

(*a*) *Aliphatic Amino Acids*:

R group α-carbon	*Name*	*Abb.*
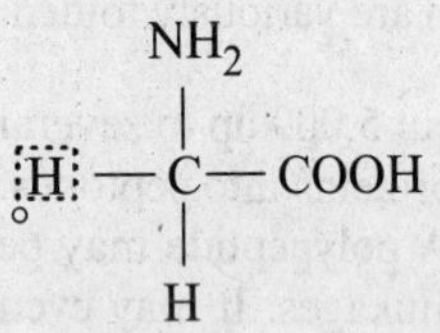	Glycine (Amino accetic acid)	Gly
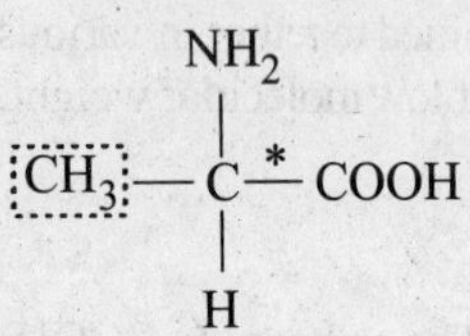	Alanine (α-amino propionic acid)	Ala
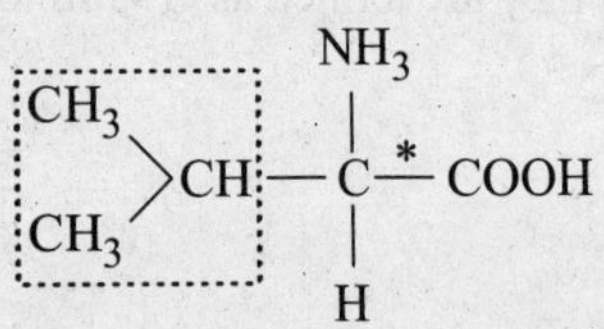	Valine (c-amino-iso valeric acid)	Val
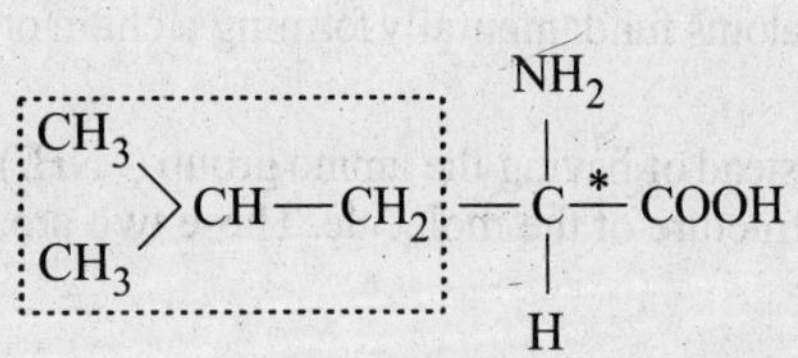	Leucine (α-amino-isocaproic acid)	Leu
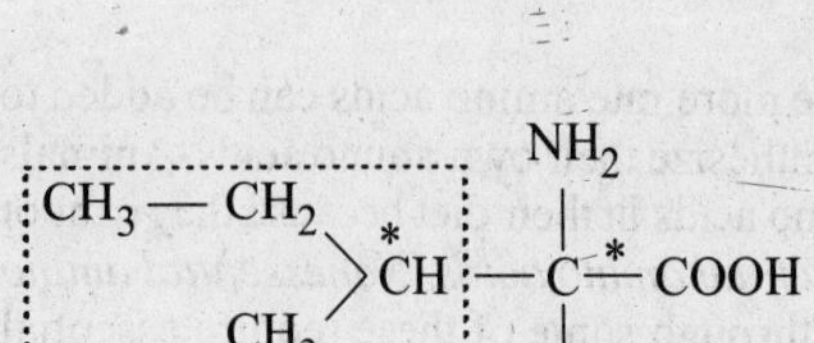	Isoleucine (α-amino b-methyl valeric acid)	Ile

* Indicates asymmetrical carbon atom.

(b) Hydroxy Amino Acids:

R group α-carbon	*Name*	*Abb.*
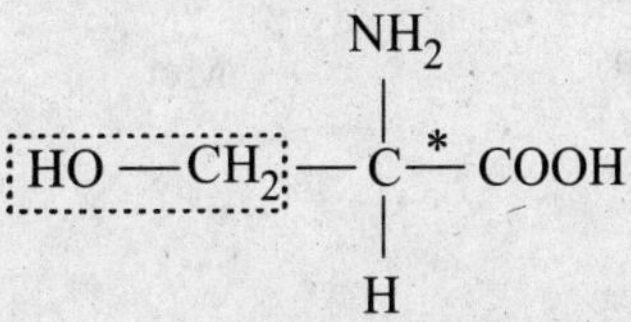	Serine (α-amino-β-hydroxy butyric acid)	Ser
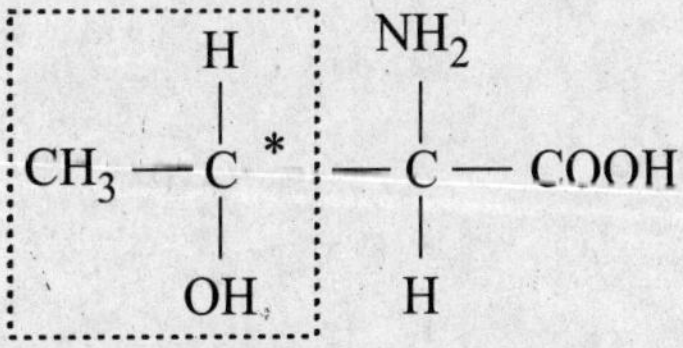	Threonine (α-amino-β-hydroxy butyric acid)	Thr

(c) Aromatic Amino Acids

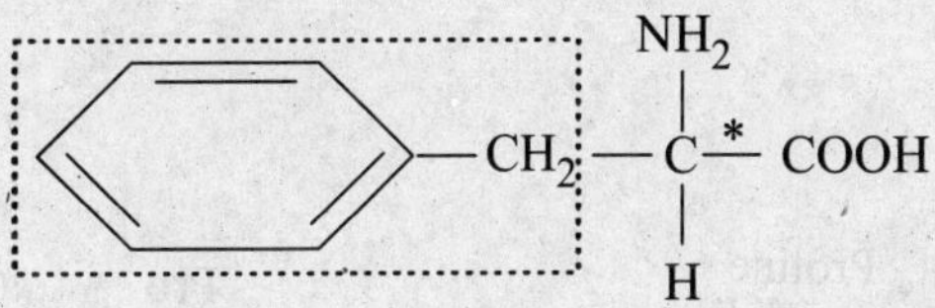	Penylalanine (α-amino-β-phenyl propionic acid)	Phe
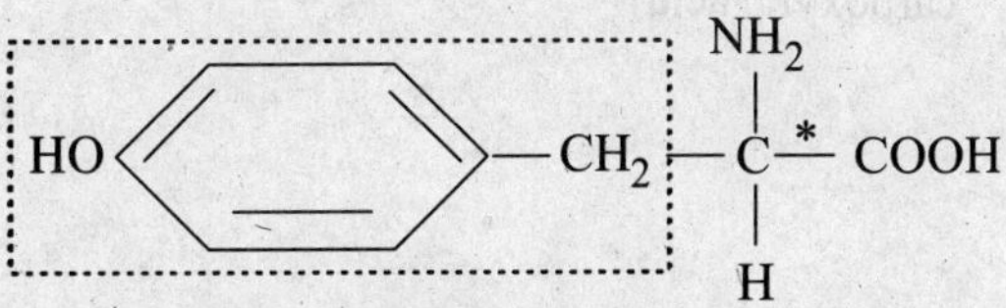	Tyrosine (a-amino-b-hydroxyl phenyl propionic acid)	Tyr

(d) Sulphur-containing Amino Acids:

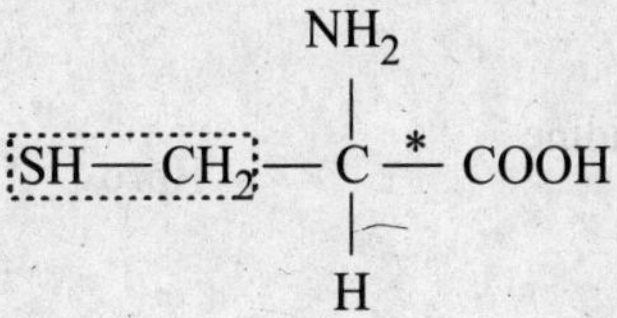	Cysteine (α-amino-β-thiol propionic acid)	Cys
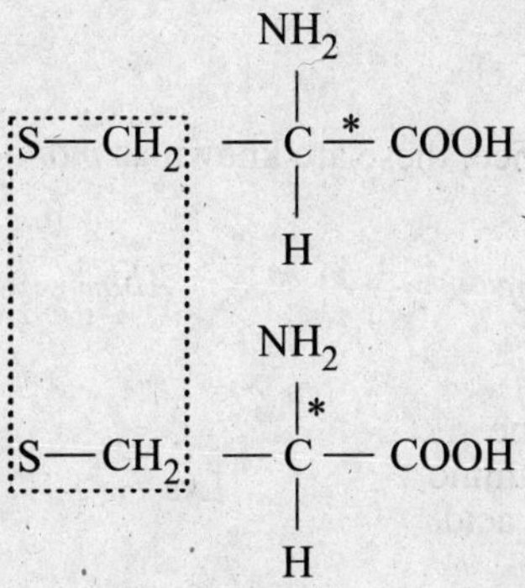	Cysteine (α-amino-β-thio propionic acid)	Cyseey

---

* **Indicates asymmetrical carbon atom.**

R group	*α-carbon*	*Name*	*Abb.*

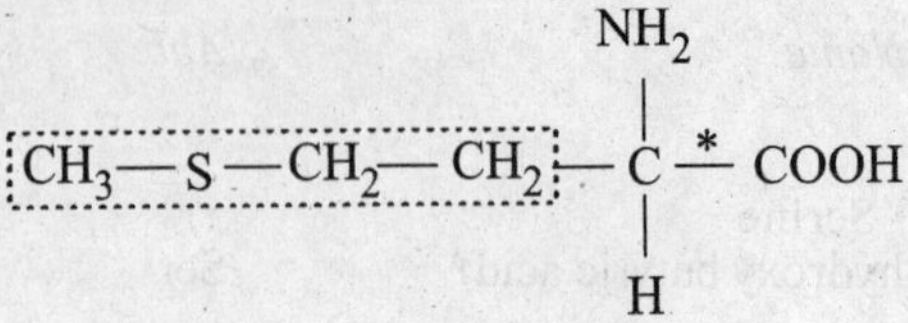

Methionine (α-amino metyl thio butyric acid) — Met

(e) Heterocyclic Amino Acids:

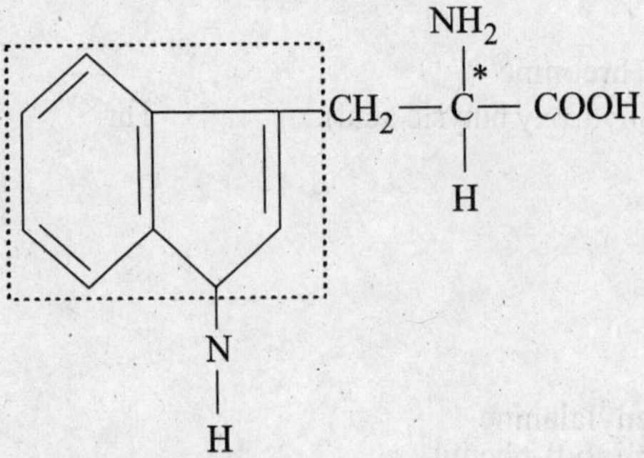

Tryptophan (α-amino metyl thio butyric acid) — Trp

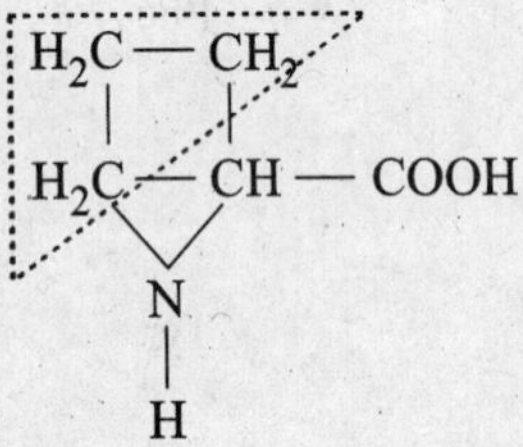

Proline (α-pyrrolidine carboxylic acid) — Pro

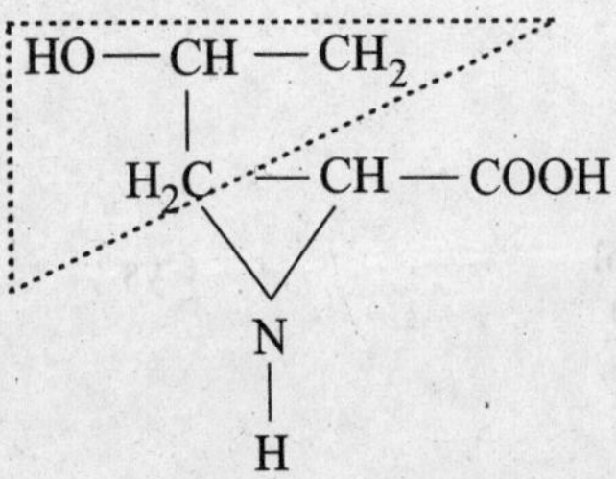

Hydroxyproline (γ – hydroxy-α-pyrrolidine carboxylic acid) IMINO ACID — Hpro

Acidic Amino Acids

These amino acids contain two carboxyl and one amino group. Hence, these are known as *mono-amino-dicarboxylic acids.*

R group	*α-carbon*	*Name*	*Abb.*

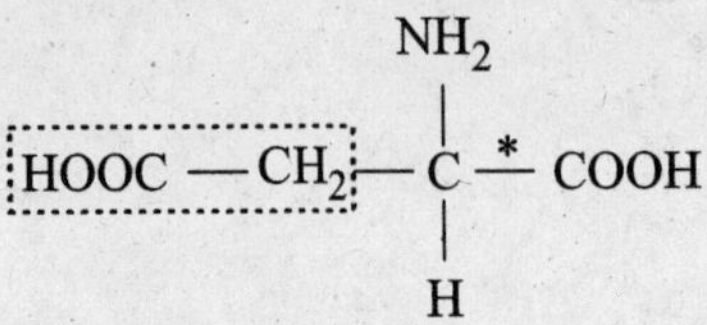

Lysine (α-Σ-diamino capioic acid — Lys

* Indicates asymmetrical carbon atom.

R group / *α-carbon*	*Name*	*Abb.*
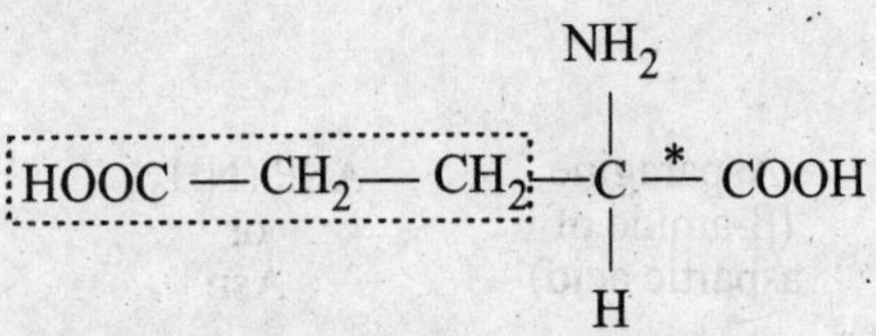	Glutamic acid (α-amino-glutaric acid)	Glu

Basic Amino Acids

These contain two basic and one carboxyl group and hence are known as *diamino-monocarboxylic acids.*

(i) Simple Basic Amino Acids:

R group / *α-carbon*	*Name*	*Abb.*
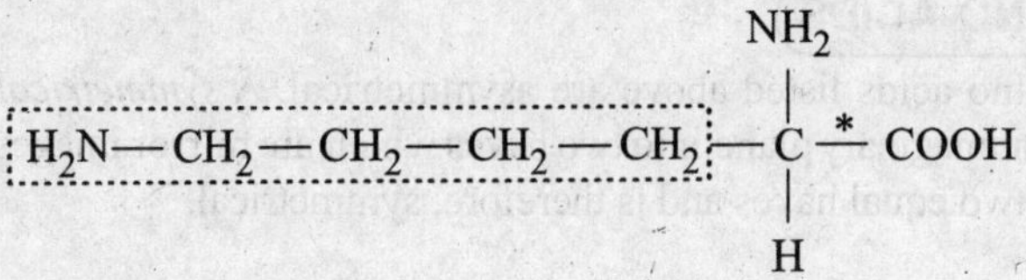	Lysin e (α-Σ-diamino captoic acid)	Lys
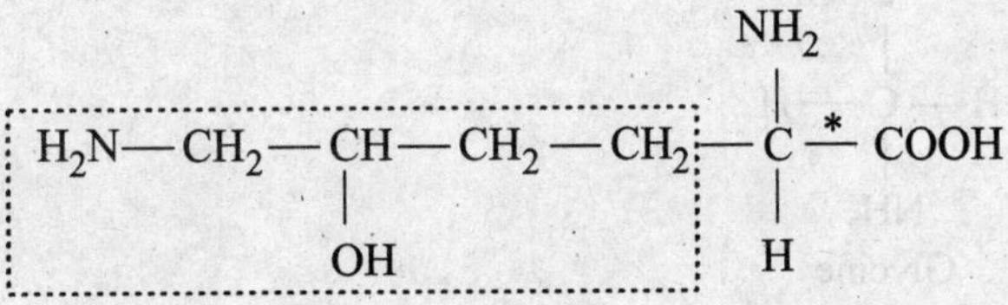	Hydroxylysine (α-Σ-diamino δ-hydro-xycaproic acid)	Hlys
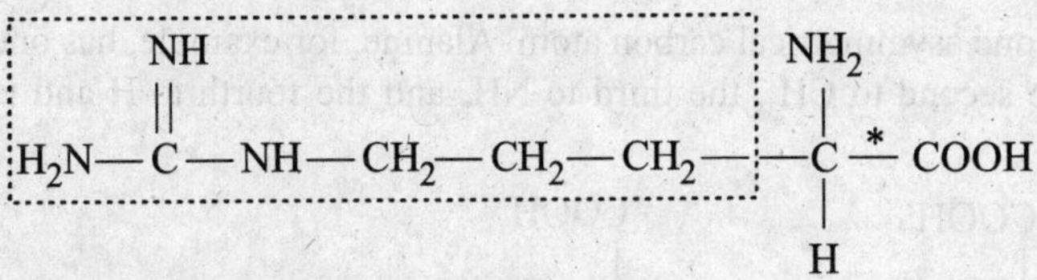	Arginine (αamino-guanidine valeric acid)	Arg

(ii) Simple Basic Amino Acids:

R group / *α-carbon*	*Name*	*Abb.*
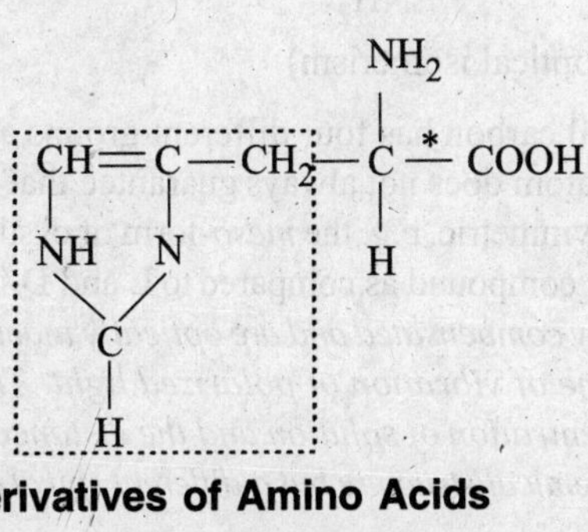	Histidine (α-amino-β-imidazole propionic acid)	His

Amide Derivatives of Amino Acids

In addition to the above amino acids, two amide derivatives of aspartic and glutamic acids are also generally included in the list of naturally occurring amino acids. These are *asparagine* and *glutamine* respectively.

---

* Indicates asymmetrical carbon atom.

R group	*α -carbon*	*Name*	*Abb.*
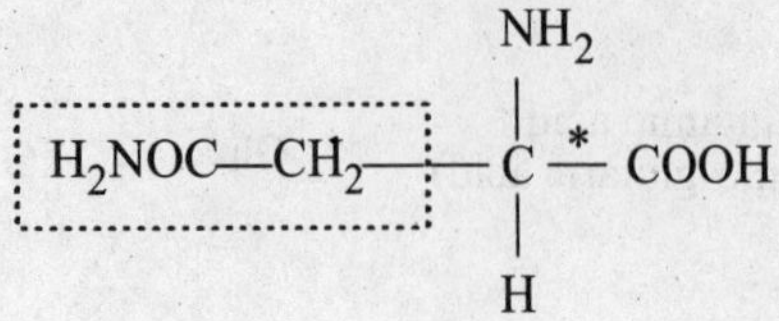		Asparagine (β-amide of aspartic acid)	ASP (NH_2) or Asn
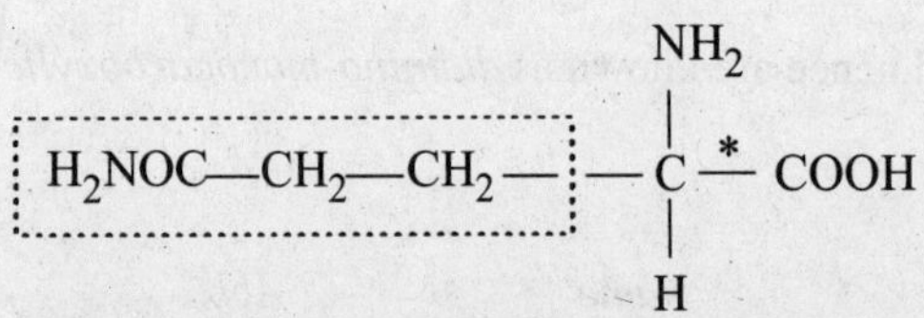		Glutamine	Glu (NH_2) or Glu

MOLECULAR SYMMETRY OF AMINO ACIDS

With the exception of glycine, all the amino acids listed above are asymmetrical. A *symmetrical molecule* is that which can be divided by an imaginary plane into two haves which are mirror images of each other. Glycine can be divided into two equal haves and is therefore, symmetrical.

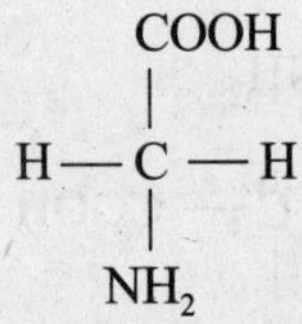

Glycine

Nearly all amino acids have at least one asymmetrical carbon atom. Alanine, for example, has one valency linked to COOH group, the second to CH_3, the third to NH_2 and the fourth to H and is therefore, asymmetrical.

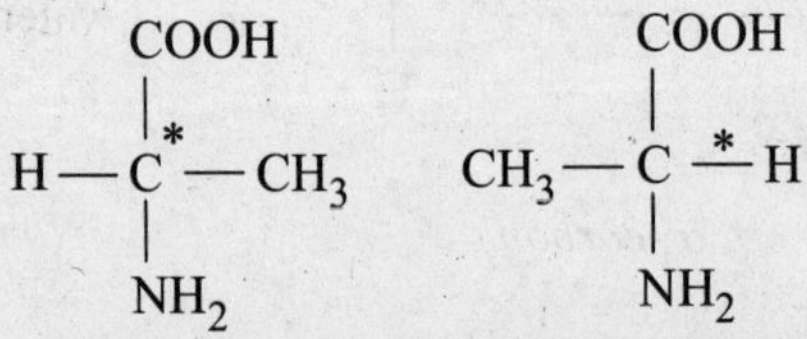

Alanine (showing optical isomerism)

In nearly all the known examples an asymmetrical carbon has four different groups attached to its four valencies. However, an asymmetrical carbon atom does not always guarantee that the molecule containing such as asymmetrical carbon will be asymmetric, *e.g.* the *meso*-form of cystine which has two asymmetrical carbon atoms but is a symmetric compound as compared to L and D-cystine which are asymmetric. Such molecules are called *internally compensated and are optically inactive (optically active substances are those which rotate the plane of vibration of polarized light. The amount of rotation depends on the wavelength of light, concentration of solution and the distance light travels in the medium. Compounds which have the same chemical structure but a different spiral configuration are called isomers).*

---

* Indicates asymmetrical carbon atom.

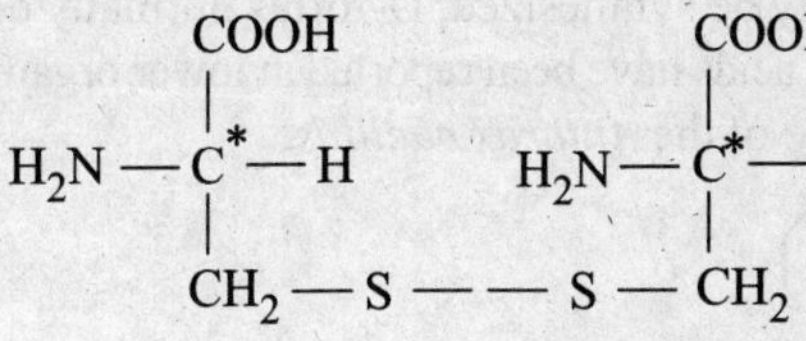

L-cystine

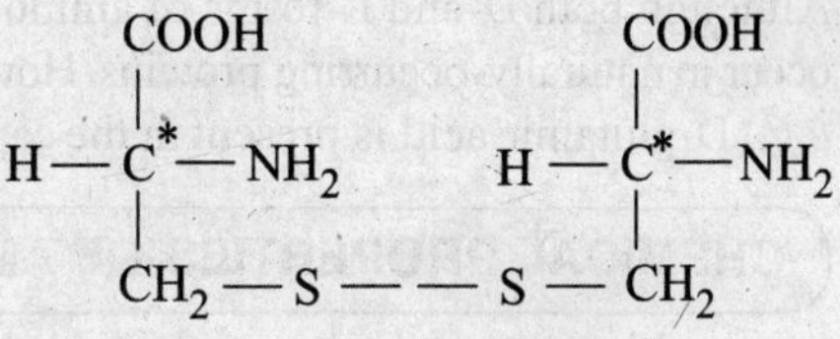

L-cystine

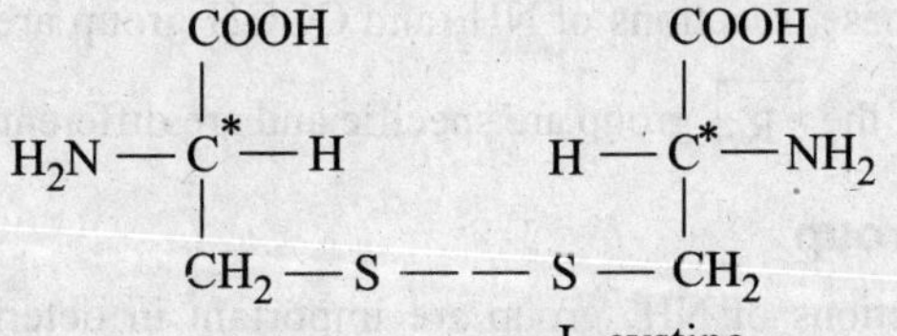

L-cystine

meso-Cystine
(optically inactive)

Asymmetrical amino acids exhibit optical activity. The amino-acids in which there is one asymmetrical carbon atom exist in two optically active forms. Those containing two asymmetrical carbon atoms, exist in four optically active forms. This means the total number of optically active forms follow exactly The *Le bel* and *Van't Hoff's Rule, i.e., 2^n,* where *n* is the number of asymmetrical carbon atoms. For example, valine has one asymmetrical carbon atom, so according to *Le Bel and Van't Hoff's Rule* it has only two ($2^n = 2^1 = 2$ optically active forms.

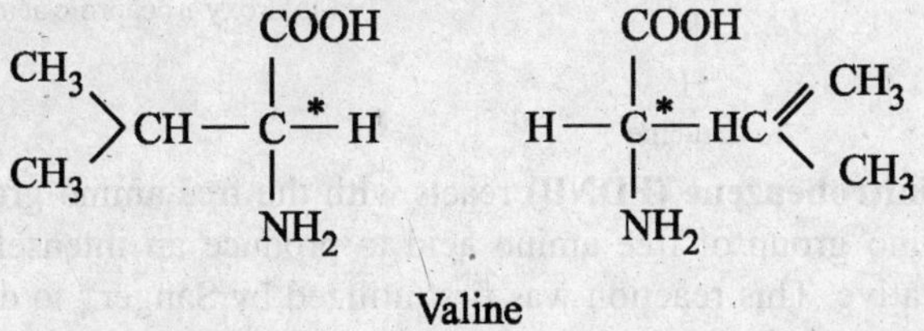

Valine

Similarly threonine having two asymmetrical carbons exists in four optically active forms ($2^n = 2^2 = 4$). These four forms are called diastereomers and are designated as L, D, L-allo and D-allo.

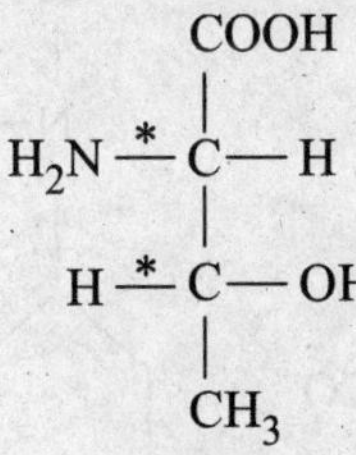

L-Threonine

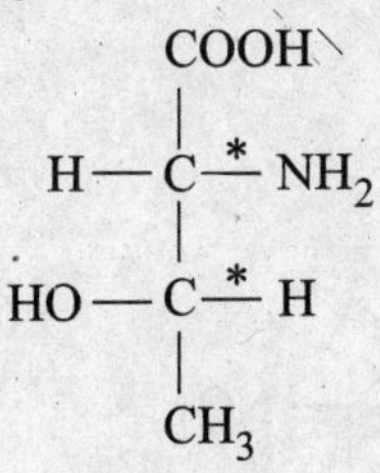

D-Threonine

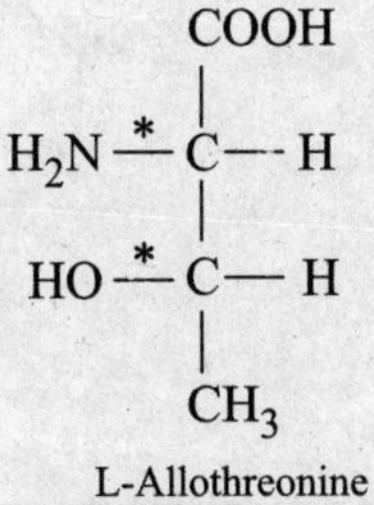

L-Allothreonine

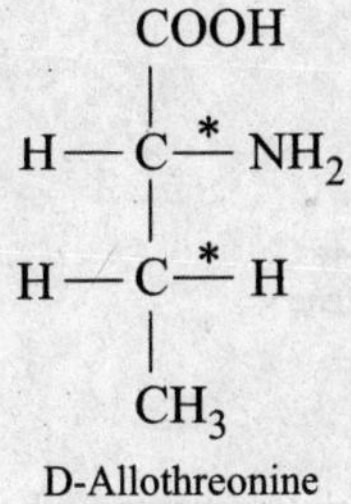

D-Allothreonine

* Indicates asymmetrical carbon atom.

Although both D-and L-forms of amino acids can easily be synthesized, D-forms normally do not occur in naturally-occurring proteins. However, D-amino acids have been reported in lower organisms, *e.g.*, D-glutamic acid is present in the capsular substance of the *Anthrax bacillus.*

CHEMICAL PROPERTIES OF AMINO ACIDS

Amino acids contain three groups, *viz.,* COOH, NH_2 and the R Group. Hence, they show three types of reactions. Of these, reactions of NH_2 and COOH group are common to all amino acids, whereas the reactions of the R group are specific and are different for different amino acids

Properties of NH_2 group

The following two reactions of NH_2 group are important in determining protein structure and composition.

The NH_2 group reacts with **nitrous acid** to liberate gaseous nitrogen. This reaction is utilized for quantitative estimation of NH_2 nitrogen. Nitrous acids does not act with **NH (imino)** group, hence, imino acids cannot be estimated in this way.

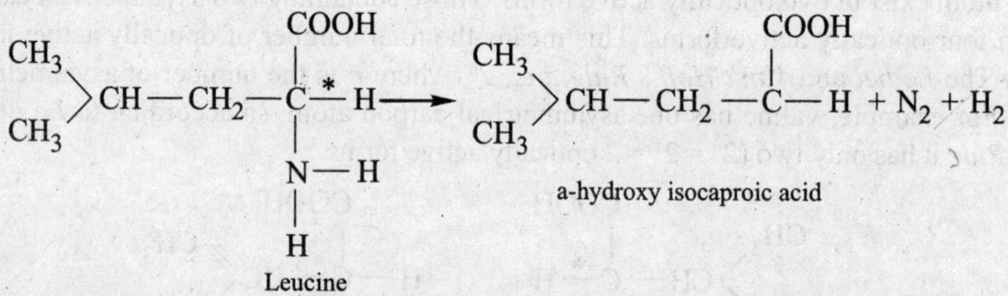

1-Fluro, 2, 4 Dinitrobenzene (FDNB) reacts with the free amino group of the N-terminal of peptides and α-amino group of free amino acid to produce an intensely coloured yellow 2, 4-dinitrophenyl derivative. This reaction was first utilized by Sanger* to determine the N-terminal amino acids in insulin hydrolysate.

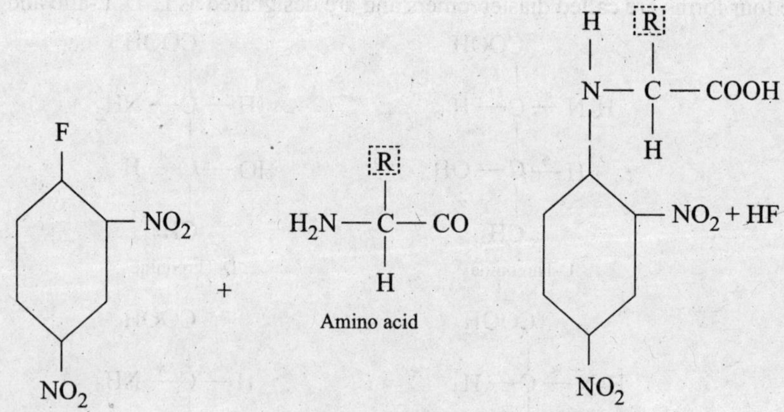

* Sanger was an English Biochemist who got Nobel Prize for determining the amino acid sequence of Insulin.

Properties of COOH group

The group shows all the properties commonly found in organic acids, *e.g.* salt formation, ester formation etc.

Alkalies fact with amino acids to form salts. In case of dicarboxylic amino acids one COOH group is neutralized at a time.

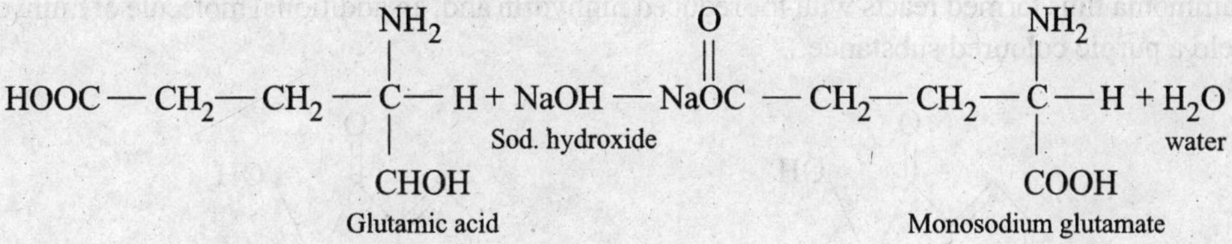

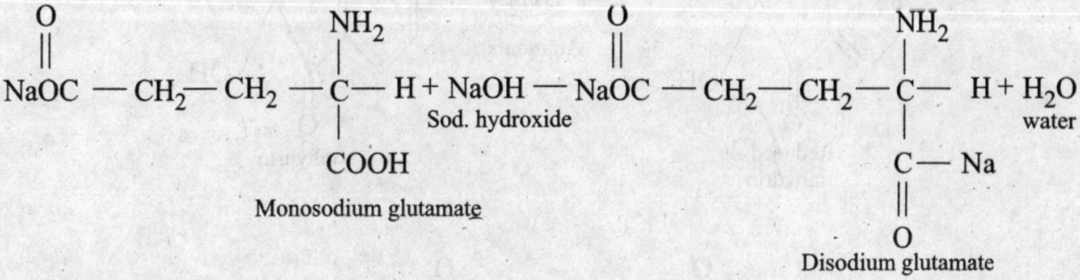

(Monosodium glutamate is extensively used as a flavouring agent in the preparation of oriental dishes).

Alcohols form esters with carboxyl groups of amino acids to form esters.

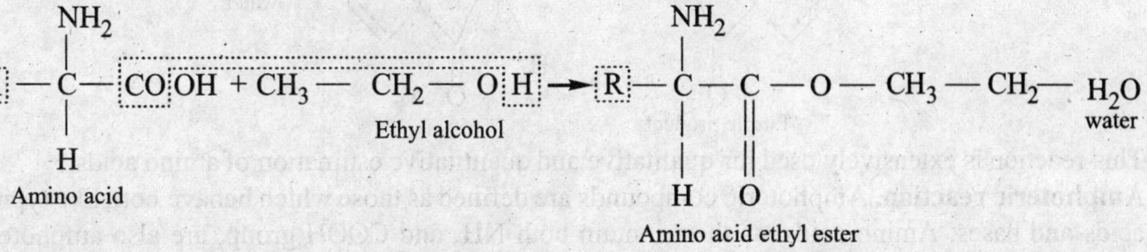

The esters so obtained are volatile and can be easily separated and identified by gas chromatography. Indeed the preparation of methyl esters was the first procedure to separate the naturally occurring amino acids.

Reactions Shared by both Amino and carboxyl groups

Nihydrin ($C_6H_4(CO)_3H_2O$, triketohydrindene hydrate) is a strong oxidizing agent which acts on amino acids in three steps. During the initial reaction ninhydrin is reduced whereas the amino acid is oxidized to form reduced ninhydrin and imino acids respectively.

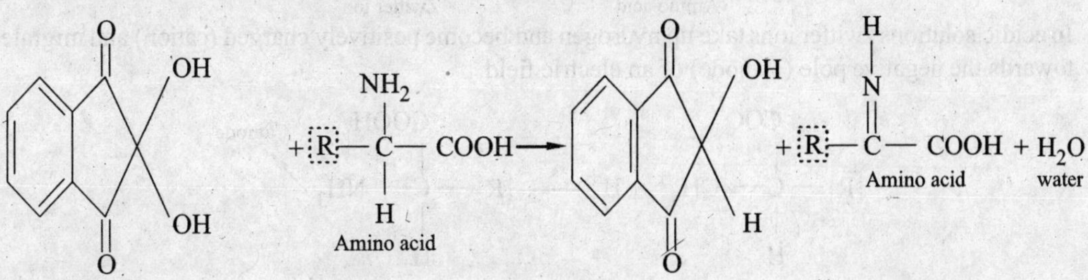

The imino acid is then hydrolysed to form NH_3, keto-acid and CO_2.

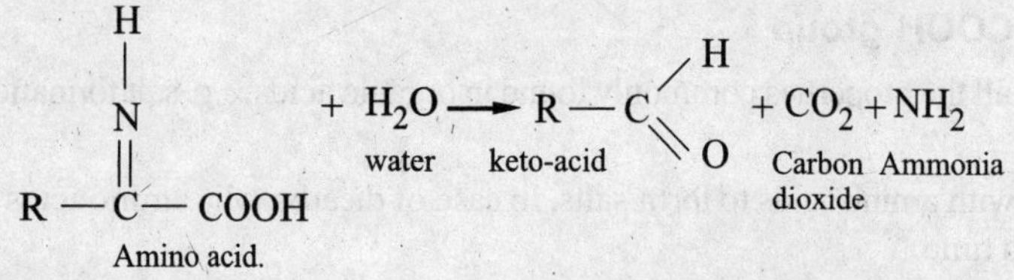

The ammonia thus formed reacts with the reduced ninhydrin and, an additional molecule of ninhydrin to yield a purple coloured substance.

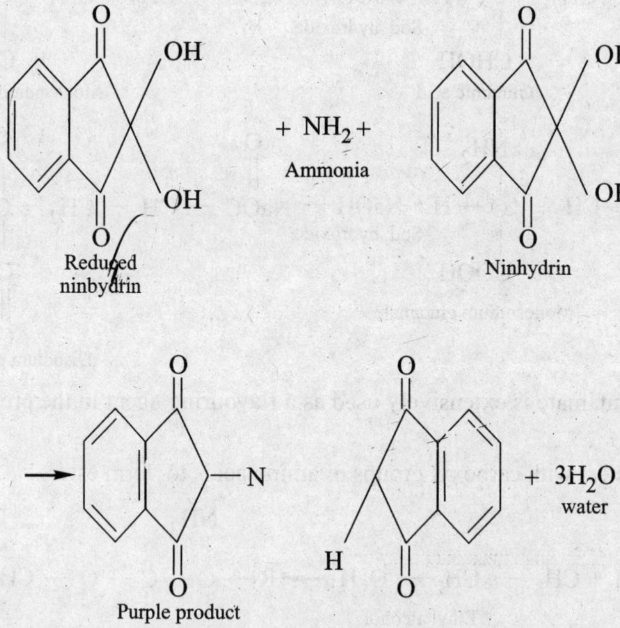

This reaction is extensively used for qualitative and quantitative estimation of amino acids.

Amphoteric reaction. Amphoteric compounds are defined as those which behave both like typical acids and bases. Amino acids, as they contain both NH_2 and COOH group, are also amphoteric compounds forming salts both the acids and base. The NH_2 and COOH group of the amino acids also act with each other to form *inner salts* called *zwitter ions.*

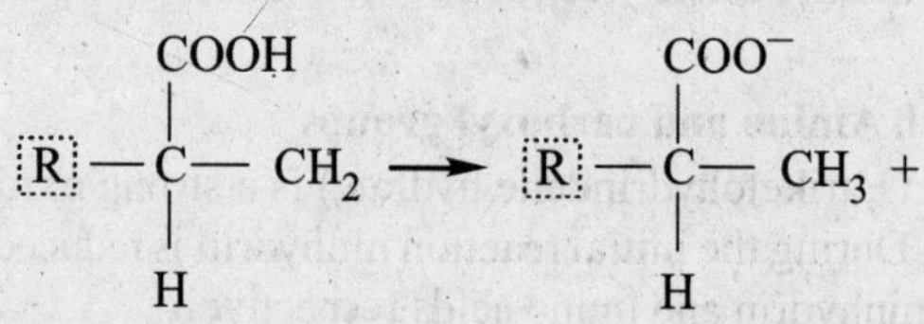

In acidic solutions zwitter ions take up hydrogen and become positively charged (cation) and migrate towards the negative pole (cathode) of an electric field.

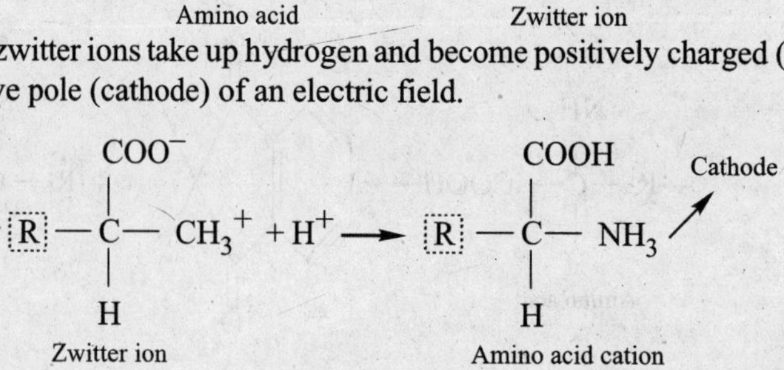

In basic solutions the zwitter ions donate one H+ and become negatively charged (anion) and migrate towards the positive pole (anode).

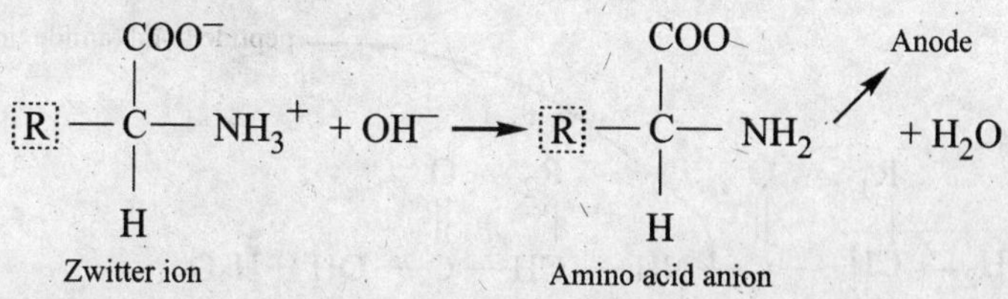

Isoelectric point of Amino Acid

At a certain pH, specific for each amino acid, the number of zwitter ion is maximum and there are no anions or cations. At this pH the net charge is zero so that there is no accumulation of amino acid either on the cathode or on the anode of an electrical field. This particular pH is known as isoelectric point (pI) of an amino acid (or protein). All the aliphatic amino acids are completely converted to zwitter ions at their isoelectirc point whereas in the case of aromatic amino acids only 50% conversion takes place.

Amino acids as buffers. If a strong base such as hydroxide ions (OH–) is added to a solution of an amino acid, the following reaction occurs.

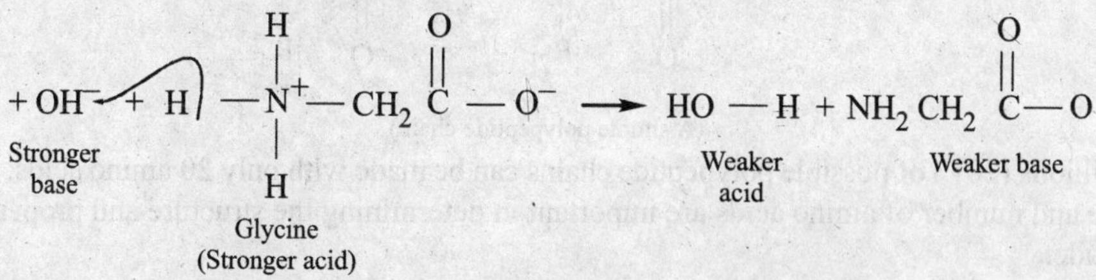

If a strong acid (e.g., H_3O+) is added to a solution of an amino acid; it is neutralized as follows:

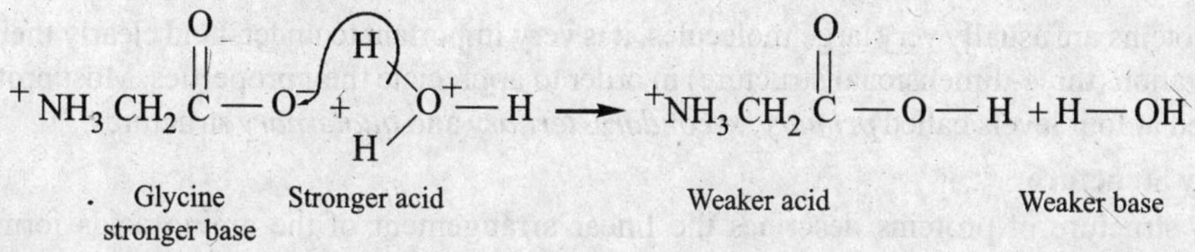

Amino acids, therefore, have the ability to neutralize both strong acids and strong bases and can, thus, serve as buffers. Some of the proteins in the bloodstream serve chiefly as buffers, constituting the third buffer system of blood.

STRUCTURE OF PROTEINS

Proteins are made up of a large number of L-amino acids linked with each other through peptide bond ($-\overset{O}{\overset{\|}{C}}-NH$). If the carboxyl group COOH) of one amino acid is acted upon by the α- amino group (NH_2) of another amino acid, an *amide linke* is formed between the two by replacement of one molecule of water, which is shown as under:

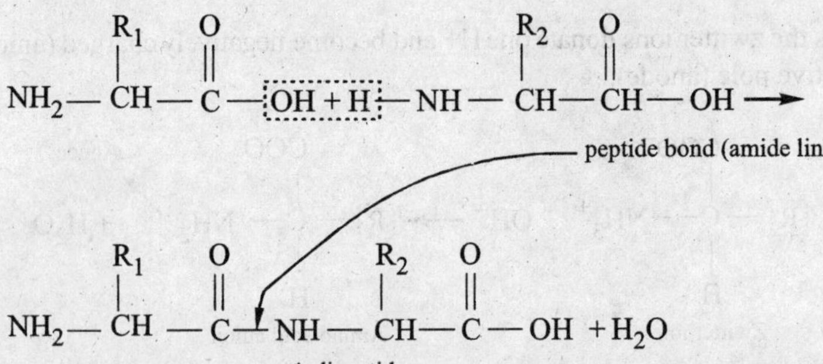

A dipeptide

In protein chemistry the amide link is called the *peptide bond.* When two amino acids are joined by a peptide bond it is known as a *dipeptide;* when three are joined it is called *tripeptide* and when more than three amino acids are joined it is known as a *polypeptide.* Then following figure shows a simple polypeptide chain (protein).

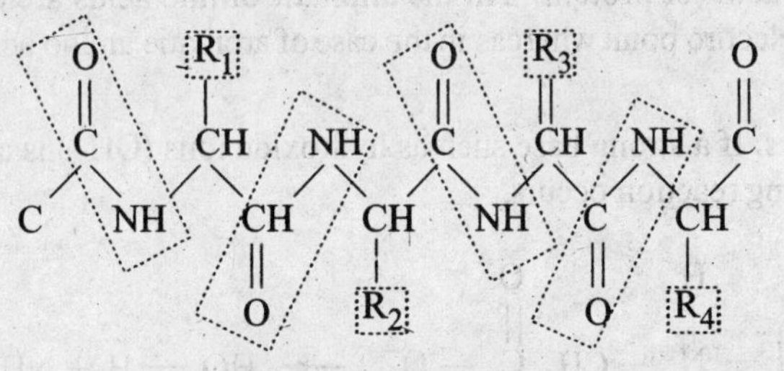

(A simple polypeptide chain)

Since millions (20^{20}) of possible polypeptide chains can be made with only 20 amino acids, both the sequence and number of amino acids are important in determining the structure and properties of a polypeptide.

The shape of proteins is very variable ranging from long ribbon like structure, *e.g., fibrinogen* to spherical structures such as *globulins.*

Since proteins are usually very large molecules, it is very important to understand clearly their spatial configuration (three-dimensional structure) in order to appreciate their properties. Most proteins are organised at four levels called *primary, secondary, tertiary* and *quaternary* structures.

Primary Structure

Primary structure of proteins describes the linear arrangement of the amino acids forming the polypeptide chain. No other forces and bonds, except the peptide bonds, the associated with the primary structure.

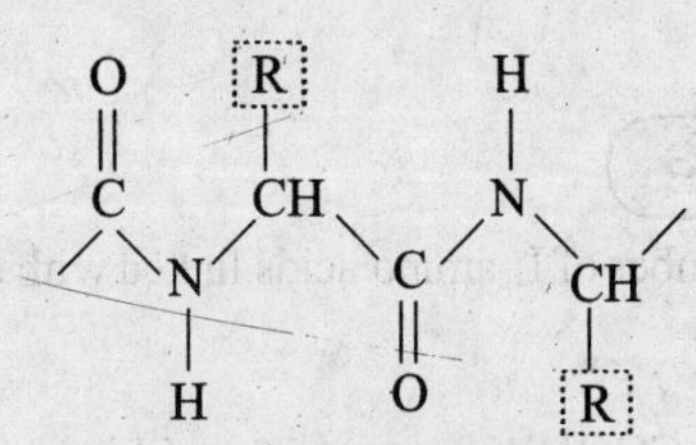

(A primary structure)

Secondary Structure

The polypeptide chain usually does not exist in a straight manner because there are various molecular forces in the chain acting on each other. One such force is exerted by the carbonyl group (C = O). The oxygen of the carbonyl group shares with the hydrogen of the *imido* (NH) grup of the third amino acid forming a hydrogen bond which partially twists the molecule. Consecutive hydrogen bonds coıl the entire molecule in a spring-shaped (helical) structure called the α-helix (Fig. 2.1).

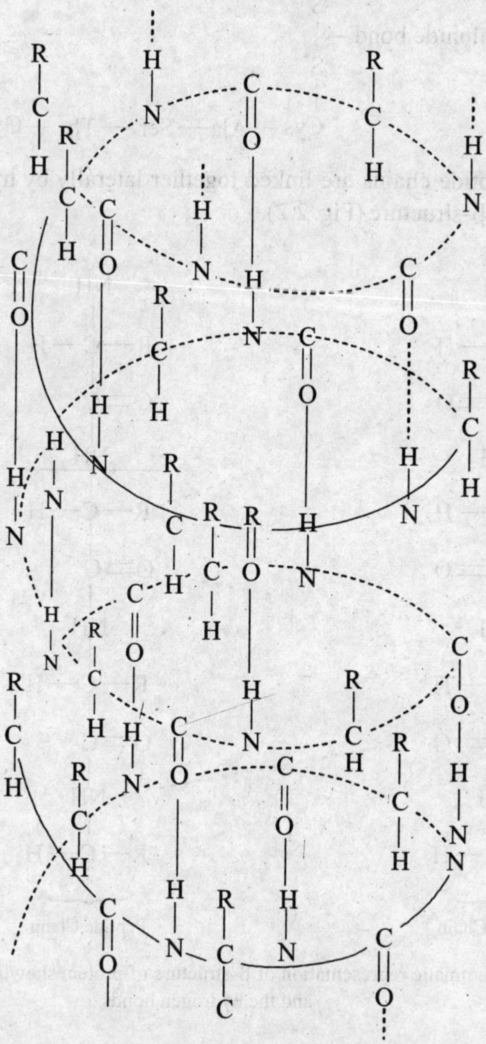

Fig. 2.1. Secondary structure of protein showing α-helix and hydrogen bonds.

From the above discussion it is obvious that the imido (NH) group essential for hydrogen bonding of the above type. The residue of the imino acids (proline and hydroxyproline) have only a nitrogen and so they cannot from a hydrogen bond with carbonyl grup. Peptide chains where proline and hydroxyproline predominate, therefore, do not have a typical α-helical structure.

Besides this, cystine (Dicysteine) is a molecule which is folded on itself at the disulphide bond. Peptide chains which have cystine molecule fold back on themselves. Cystine, thus, can also link two peptide chains at the disulphide bond.

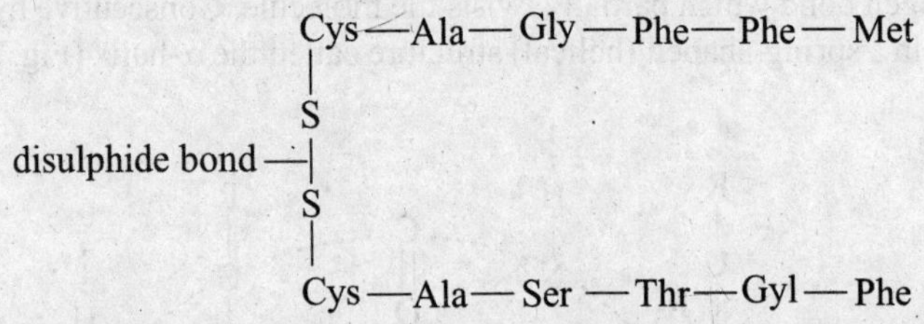

When two or more peptide chains are linked together laterally by hydrogen bonds, the resulting structure is known as a β-structure (Fig. 2.2).

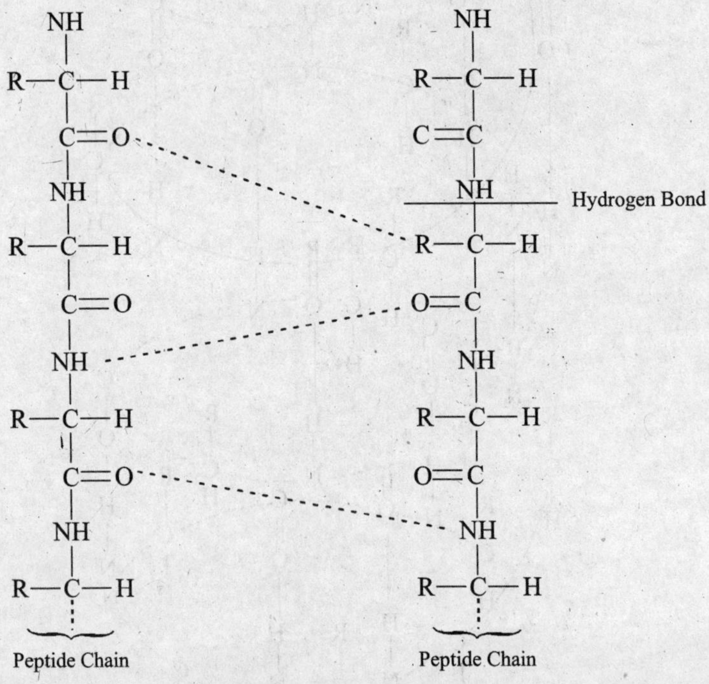

Fig. 2.2. Diagrammatic representation of β-structure of protein showining two peptide chains and the hydrogen bonds.

Tertiary Structure

Tertiary structure is found specially in the case of globular proteins. When a long peptide chain, with the without a helix, is coiled and variously folded in itself, the resulting structure of protein is termed as the tertiary structure (Fig. 2.3). Four types of bonds are responsible for maintaining the tertiary structure of protein molecule. They are the *hydrogen, ionic, hydrophobic* and *disulphide bonds.*

The hydrogen bonds are usually formed by OH groups of hydroxy amino acids, e.g.; tyrosine, threonine and serine, and the carboxyl (COOH) group of acidic amid acids. **Ionic** bonds are formed by the NH_2 group of acidic amino acid residues, but are not very important as they tend to lie on the extension of the molecule. The **hydrophobic bonds** are very important as the tertiary structure of protein molecules is mostly depends on these bonds. These bonds are formed when similar (R) groups lie in close proximity of each other. In spite of the fact that these bonds are hydrophobic, they are stable because they lie in the interior of protein molecule where there is less water.

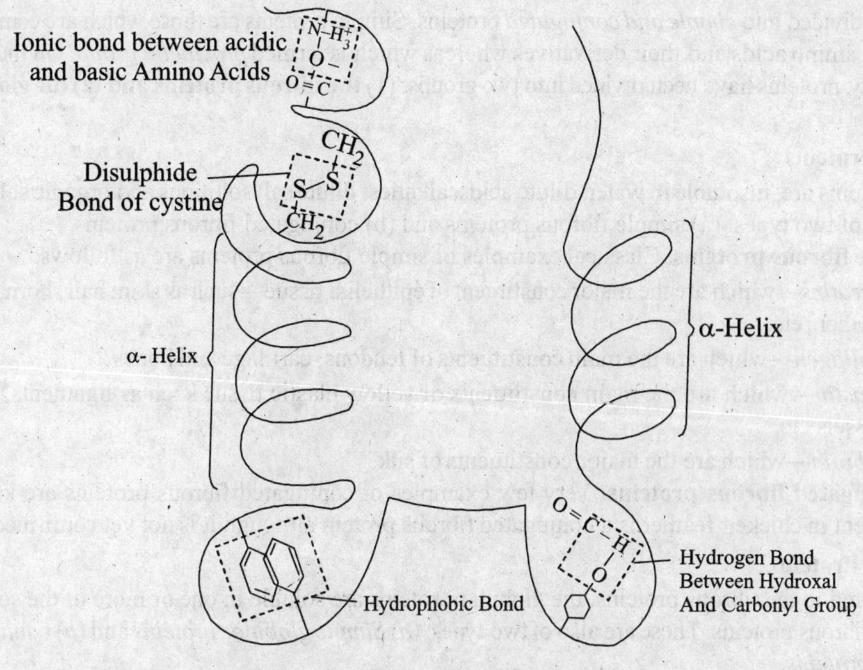

Fig. 2.3. Diagrammatic representation of tertiary structure of a protein showing ionic, disulphide, hydrogen and hydrophobic bondings.

Quaternary Structure

The quaternary structure is formed by polymerization of several tertiary structures. For example, the enzyme phosphorylase contains four sub-units which are identical to each other but separately inactive catalytically. However, when they are joined as tetramer, it becomes an active enzyme. This type of structure where all the units are similar is termed as *homogeneous quaternary structure* (Fig. 2.4). But, when the units are dissimilar, e.g., the tobacco mosaic virus, the structure is termed as *heterogeneous quaternary structure*.

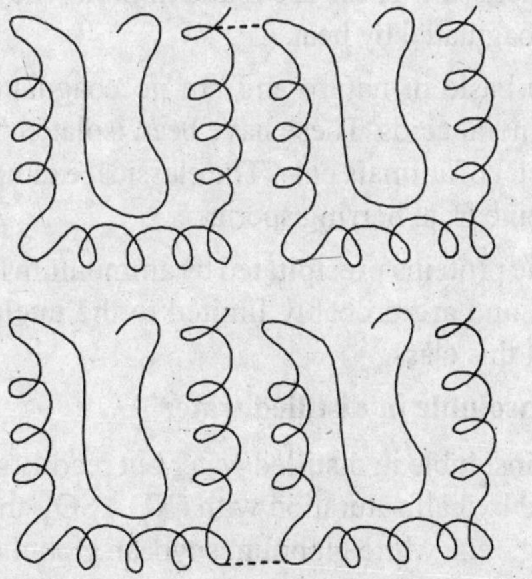

Fig. 2.4. Diagrammatic representation of the homogenous quaternary structure of protein.

CLASSIFICATION OF PROTEINS

The classification of proteins is based on their structure and solubility. On the basis of structure they have been divided into *simple and conjugated* proteins. Simple proteins are those which are composed only of α- amino acids and their derivatives whereas which is termed *prosthetic group.* On the basis of solubility, proteins have been divided into two groups. (1) **the fibrous proteins and (2) the globular proteins.**

Fibrous Proteins

These proteins are, insoluble in water, dilute acids, alkalies, dilute salt solutions and organic solvents. These are of two types. (a) simple fibrous proteins and (b) conjugated fibrous proteins.

(a) Simple fibrous proteins. Classical examples of simple fibrous proteins are as follows.

(i) *Keratins*—which are the major constituent of epithelial tissue—such as skin, hair, horn, nails, feathers etc.

(ii) *Collagen*—which are the main constituents of tendons, cartilage and bones.

(iii) *Elastin*—which are the main constituents of yellow elastic tissue such as ligaments, blood vessels etc.

(iv) *Fibroin*—which are the major constituents of silk.

(b) Conjugated fibrous proteins. Very few examples of conjugated fibrous proteins are known. The pigment in chicken feather is a conjugated fibrous protein although it is not yet confirmed.

Globular Proteins

As compared to the fibrous proteins, the globular proteins are soluble in one or more of the solvents listed for fibrous proteins. These are also of two types: (*a*) *Simple globular proteins* and (*b*) *conjugated globular proteins.*

(*a*) **Simple globular proteins**—On the basis of their solubility, the simple globular proteins can be divided into two major groups:

(1) Those soluble in distilled water and (2) Insoluble in distilled water.

(3) Simple globular proteins soluble in distilled water

(i) *Albumins*—These are highly soluble in water and can be precipitated by saturated solutions of acid salts like $(NH_4)_2\ SO_4$. These can be coagulated by heat. Albumins are found in white of egg, blood serum and soybean etc.

(*ii*) *Pseudoglobulines*—These can be precipitated from a water solution by one-fourth to three-fourth saturation by an acid salt. These are found in milk whey and are rather uncommon in nature. These are also coagulated by heat.

(*iii*) *Protamines*—These are basic in nature and are no coagulated by heat. Protamines form crystalline salts with mineral acids. These have been isolated from ripe sperm cells and may be limited to the cell nuclei of animals only. The classical examples are *salmin* separated from salmon sperm and clupeine from herring sperm.

(*iv*) *Histones*. These are basic proteins precipitated by ammonium hydroxide. Very few examples of this class are known and are probably limited to the nuclei of cells. Nucleohistones of thymus is an example of this class.

(4) Simple globular protein insoluble in distilled water

(*i*) *Euglobulins*. These are insoluble in distilled water but readily soluble in dilute salt solutions. They can be precipitated by half saturation with $(NH_4)\ SO_4$. this class of proteins are widely distributed in nature, *e.g.*, egg white globulin, soybean globulin, blood serum globulin etc.

(*ii*) *Protamines*. These are insoluble in distilled water but soluble in dilute alkali and 60-80% alcohol solutions. They have been isolated from plant seeds, *e.g.*, zein from corn and gliadin from wheat endosperm.

(*iii*) Gluteins. These are insoluble in distilled water and alcohols but are readily soluble in dilute solutions of acids and alkalies. They have also been isolated form plant speeds, *e.g.*, *glutelin* from corn, and glutenin from wheat endosperm.

(b) **Conjugated globular proteins.** When globular proteins are combined with certain non-protein (prosthetic group) groups, they form conjugated globular proteins. For example, the protein flobin is combined with an iron containing porphyrin compound *heme* to form *haemoglobin*. The conjugated proteins have been classified according to the nature of the prosthetic group.

(*i*) *Chromoproteins*. These include the proteins which are combined with pigments *e.g.* haemoglobin, cytochrome, flavoproteins etc.

(*ii*) *Glycoproteins*. These are proteins which are combined with carbohydrate e.g. mucin.

(*iii*) *Phosphoproteins.* These are proteins which are combined with a phosphorous containing radical (e.g., casein of milk) other than a phospholipids and nucleic acid.

(*iv*) *Lipoproteins*. These are proteins which are combined with lipids, *e.g.*, serum lipo-protein.

(*v*) *Nucleoproteins*. These are the proteins which are attached to nucleic acids (see Chapter 1).

PROTEIN DENATURATION AND REVERSAL

Denaturation

Any change in structure and/or composition of a natural protein is called *denaturation*. Agents which bring about these changes are termed *denaturing agents*. As described earlier, the protein molecule is held in shape due to four types of linkages—the *disulphide, hydrophobic, ionic* and *hydrogen* During denaturation some linkages responsible for maintaining the structure of proteins are broken, hence, the original structure of the protein is lost.

When the globular proteins, which have closely folded peptide chains, are denatured, they get unfolded and changed into insoluble protein of the fibrous nature. The commercial protein fibres are obtained in the manner.

There are a very large number of denaturing agents which act in various way. Table 2.1 given a representative list of denaturing agents and their probable made of action.

Table 2.1 Denaturing agents and their probable mode of action

Denaturing agents	*Probable made of action*
Heat	Terminal splitting of ionic linkage
Mineral acids and alkalies	Coverage of ionic linkage
Acetone	Cleavage of ionic linkage
Alcohol	"
Urea	"
Picric, tungstic, trichloroacetic acid	Cleavage of ionic-linkage
Vigorous grinding	Unknown
Shaking or stirring	Unfolding of peptide chain
Ultrasonic vibrations	Mechanical disruption, thermal effects etc.
Ultraviolet light, X-rays	Ionization, splitting of bonds.

Reversal

Denaturation takes place in several steps so that if the changes are not too extensive, some of the properties of original proteins can be restored. For example, when pepsin is moderately heated, it loses its proteolytic (catalytic) property; but when it is cooled, the catalytic activity is restored. This phenomenon is known as reversal. Similarly, the oxygen carrying capacity of haemoglobin can be adventured by treatment with salicylate, However, this is reversible. Although many of the properties of a reversed protein are similar to the original protein, it is still questionable whether the reversed protein is identical to the native protein or not. It seems that the characteristics of the reversed protein largely depends upon the time taken for denaturation and the denaturing agent employed.

Biological Importance of proteins

Proteins make up nearly 80% of the solid matter of living cell. Some proteins such as collagen are insoluble and serve as support material. Other, like globular proteins, being soluble serve in a variety of ways. Some have the ability to catalyze chemical reactions and, thus serve as *enzymes*. Globular proteins in the blood plasma of animals serve to regulate water movement between the capillary blood vessels and the tissue. This is possible because proteins cannot cross the walls of the capillaries, whereas water and ions can. The respiratory protein *haemoglobin* contain a pigmented prosthetic group belonging to a class of compounds known as *ferroporphyrins*. This group has the capacity to bind and release molecular oxygen without difficulty, but this property of ferroporphyrin is possible only when haemoglobin is present.

There are some other special type of proteins *e.g.*; the contractile elements of muscles, which have the ability to transform chemical energy into mechanical energy leading to muscle contraction. Other special proteins are *hormones e.g.*, *insulin* and a variety of other hormones produced by pituitary and other glands.

A foreign protein evokes the formation of antibodies if it is introduced into the organism. Such a protein is called antigenic. *Antigen—antibody* reactions are highly specific and immunological tests have become one of most valuable method of characterizing proteins.

3 Chapter

Carbohydrates

The term **carbohydrate** originally stood for compounds which could be described as 'hydrates of carbon' having a general formula Cx (H_2O)y. According to this definition only those compounds could qualify as carbohydrates which were composed of carbon, hydrogen and oxygen, and were the ratio of hydrogen to oxygen was the same as in water. This definition did not include compounds where the ratio of hydrogen to oxygen was not 2 : 1, but which showed the properties of a carbohydrate. Furthermore, with the knowledge of carbohydrates containing sulphur and nitrogen, the above definition of a carbohydrate become unsuitable.

As biochemistry stands today, a more meaningful definition of carbohydrates would be: compounds which have at least one hydroxy group and at least one ketone (C = O) or aldehyde (—C(=O)H) group. The simplest carbohydrates have at least two carbons. Carbohydrates on hydrolysis yield ketones and aldehydes. The names of carbohydrates end with the suffix *ose*, such as *glycerose*, glucose, sucrose, amylase (starch) etc. Those carbohydrates have an aldehyde group are called *aldoses* and those have a ketone group are called *ketoses*. The simpler carbohydrates are called sugars. Following are some of the simplest carbohydrates.

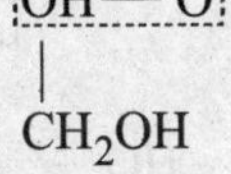

Hydroxy acetaldehyde (aldose)

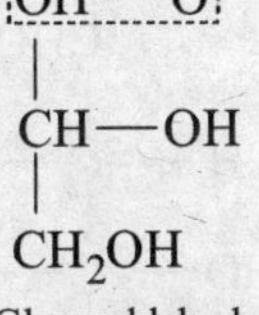

Glyceraldehyde (aldose)

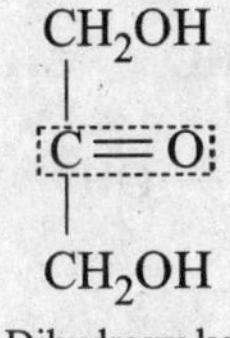

Dihydroxy ketone (ketose)

Classification

Carbohydrates can be classified into three major groups depending upon their complexity. The are *monosaccharides, oligosaccharides* and *polysaccharides.*

Monosaccharides are simple sugars which can be further hydrolysed into smaller units under mild conditions. The monosaccharides are further classified on the basis of number of carbon atoms in the molecule, *e.g., trioses* having three carbon atoms; *tetroses* having four carbon atoms; *pentoses* having five carbon atoms; *hexoses* containing six carbon atoms and so on, up to *octoses* or *nanoses.*

Oligosaccharides are compound sugars which on hydrolysis yield two six molecules of simple sugars. Thus a *disaccharide* is an oligosaccharide which yields two molecules of monosaccharide on hydrolysis. Similarly *trisaccharides* and *pentosaccharides* yield three and five molecules of monosaccharides, respectively. Monosaccharides and oligosaccharides are generally crystalline compounds, readily soluble in water and usually have a sweet taste.

Polysaccharides are polymers of monosaccharide which yield a large number of monosaccharides on hydrolysis. They are amorphous compounds, relatively insoluble and tasteless having high molecular weights.

Oligo-and polysaccharides are formed by the condensation of monosaccharides with the elimination of one molecule of water between adjacent monosaccharide units.

ISOMERISM

The phenomenon of isomerism due to asymmetrical carbon atom has already been explained with reference to proteins.

The compounds which possess the same molecular formula but differ from each other in physical and chemical properties, due to different arrangement of atoms within the molecule, are called *isomers* of each other and the phenomenon is known as *isomerism.* Isomerism is of two types: (1) Geometric (spatial) isomerism or stereoisomerism and (2) optical isomerism.

Geometric Isomerism

The compounds which are identical in their structure but differ from one another in spatial configuration are termed as geometric isomers or stereoisomers. Two such isomers of glucose are shown below which are mirror images of each other. Isomers which are mirror images of each other are called *enantiomeric pairs.*

As already mentioned (see page 16), the number of isomers of a compound depend on the number of asymmertial carbon atoms present in it. The number of possible isomers follow the Le Bel and Van't Hoff's rule (i.e., the number of isomers is 2^n, where n is the number of asymmetrical carbons). Hence, glucose, which has 4 asymmetrical carbon atoms 2, 3, 4, and 5) occurs in $(2)^4$,, i.e. 16 stereoisomeric forms.

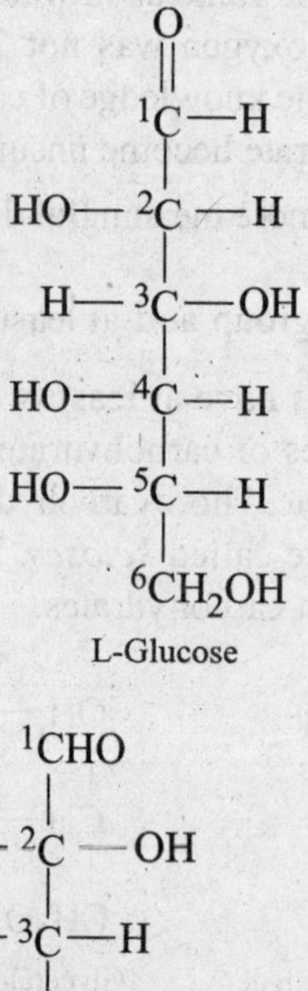

L-Glucose

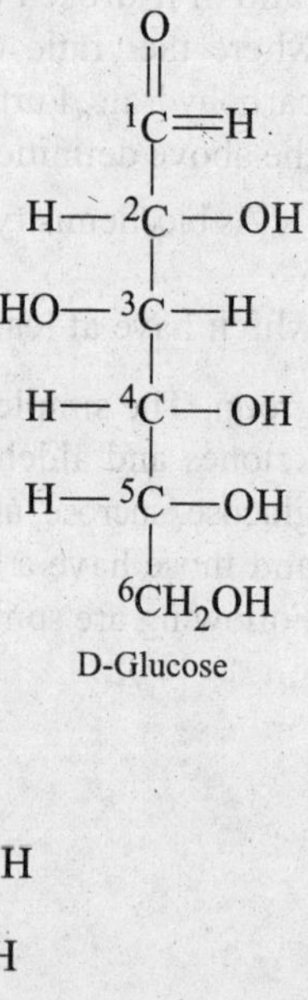

D-Glucose

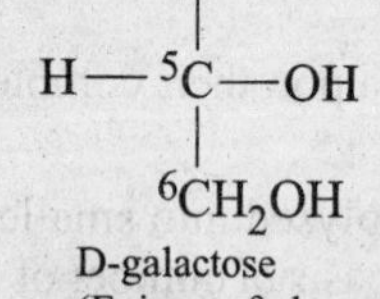

D-galactose
(Epimer of glucose differing on the 2nd carbon atom only)

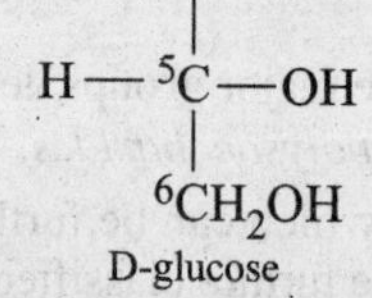

D-glucose

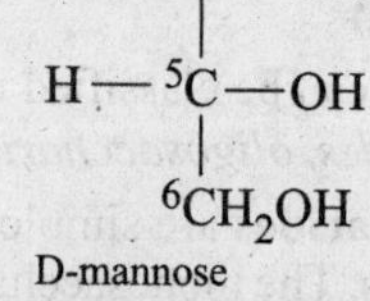

D-mannose
(Epimer of glucose differing on the 2nd carbon atom only)

The sugars which differ from one another only in the configuration of one carbon atom are termed as *epimers* of each other, Examples of epimers are glucose and galactose which differ in the configuration around the 4th carbon atom; and glucose and mannose which differ with respect to 2nd carbon atom

The designation of the isomers as *D*- and *L*-forms depends upon the orientation of *H* and *OH* groups around the carbon atoms just adjacent to the terminal primary alcohol group. These symbols (*D* and *L*) do not stand for any words and do not give any indication about the sign of rotation. They are merely small capital letters designating families of configuration. The L and D forms of glyceral dehyde are given below:

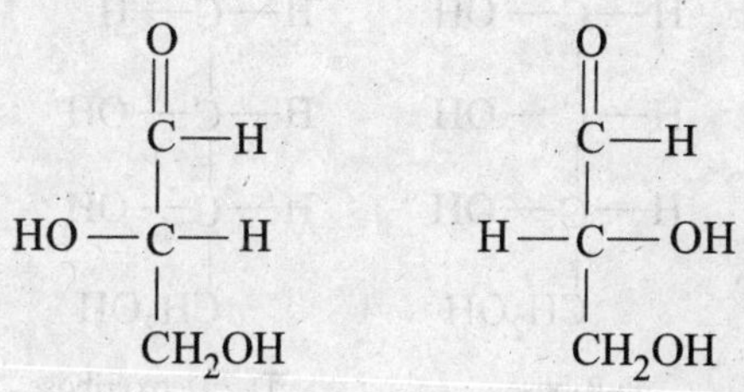

L-Glycerose (L-Glyceraldehyde) D-Glycerose (D-Glyceraldehyde)

When the OH group is present on the left side of the penultimate carbon atoms, the sugar belongs to L-series and when the OH group is on the right side of this carbon atom, the sugar is a D-sugar.

For instance, in the case of glucose, the configuration around the 5th carbon atom (penultimate) determines the L and D configuration. However, majority of sugar found in the body belong to D-series.

Optical Isomerism

As said before, geometric isomers exist in pairs which are mirror images of each other. Such isomers are called *enantiomers* of each other. Their properties like melting and boiling points, and solubility in different solvents are same. However, they differ in one important aspect, *i.e.*, they rotate the plane of polarized light in different direction *to the same extent*. Because of this difference in their optical property, they are termed as *optical isomers*. When a beam of polarized light is passed through an optically active compound, it rotates the light either to the left or the right. When the light is rotated to the left, the compound is said to be *levorotatory* and usually designated by the (–) sign. On the contrary, when the light beam is rotated to the right, the compound is said to be *dextrorotatory* and designated by the (+) sign.

When equal concentration of dextrorotatory and levorotatory isomers are present in a solution, it has not optical activity as they neutralize the effect of each other. Such mixture are called DL or *racemic* mixtures. Synthetically prepared compounds from a racemic (DL). The separation of optically active compounds from a racemic mixture is termed as *resolution*.

As said before, the symbols L and D represent only the molecular configuration and do not reflect the optical properties of a molecule. A dextrorotatory sugar should not be confused with the D-from. From this, it follows that a sugar of the D-series can be levorotatory and designated as D (–). Similarly L-sugar can be dextrorotatory and designated as L (+).

MONOSACCHARIDES

As mentioned earlier, the monosaccharides are those carbohydrates which are not easily hydrolysed into simpler carbohydrate units. They include trioses, tetroses, pentoses, hexoses, etc. having three, four, five and six carbon atoms, respectively. The important monosaccharides are trioses, pentoses and hexoses. The most important trioses are the aldo-and keto-derivatives of glycerol.

Pentoses

Pentoses are important constituents of coenzymes and nucleic acids, and are formed during the phosphogluconate pathway (HMP shunt) which is an important alternative for glucose oxidation (see Respiration). This pathway provides both oxidative and non-oxidative routes for the synthesis of pentoses from hexoses. The structure of 4 important pentoses is presented below.

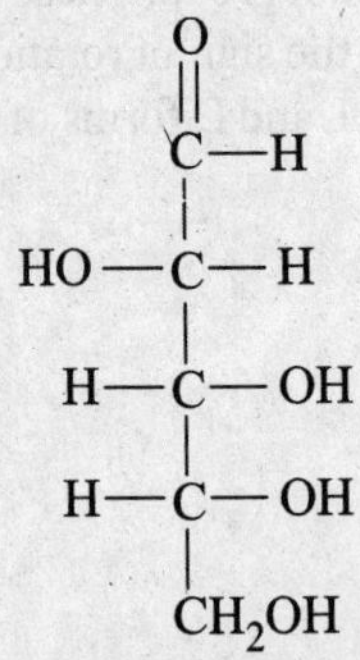
D-Arabinose

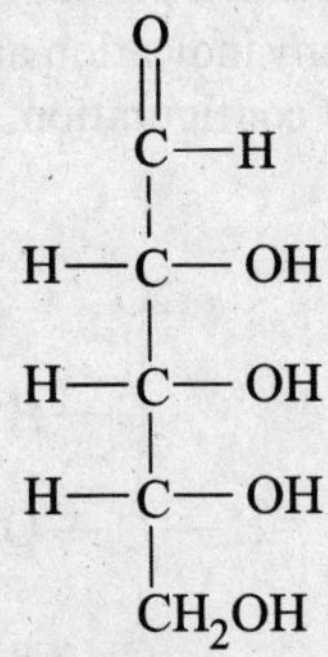
D-Ribose

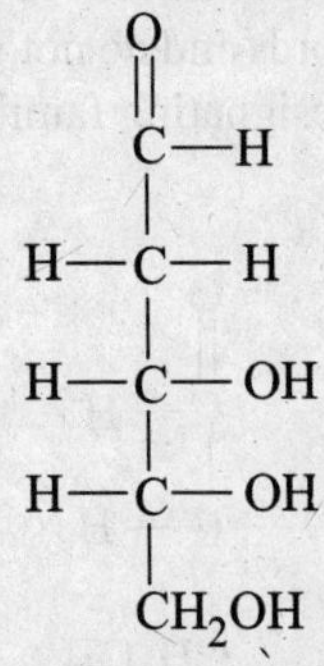
D-2 Deoxyribose

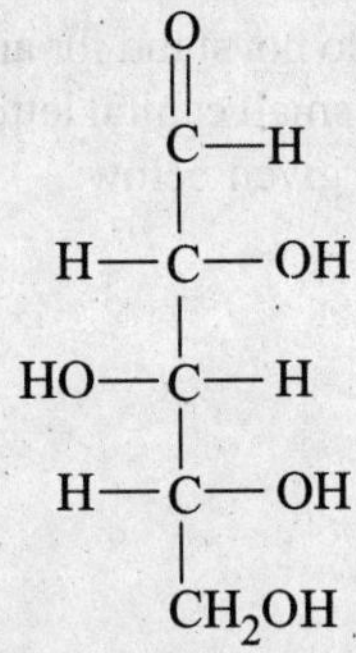
D-Xylose

The L-forms of these pentoses are also found.

Hexoses

Hexoses are the most common monosaccharides found in nature. Combinations of hexoses make up most of the biologically important di-and poly-saccharides. The structure of certain hexoses is given below:

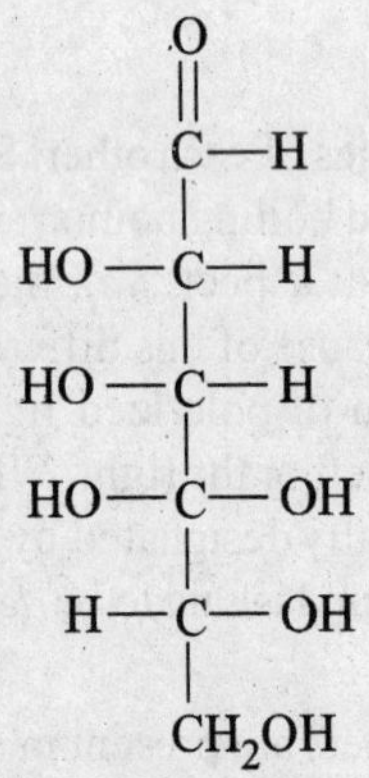
D-Talose

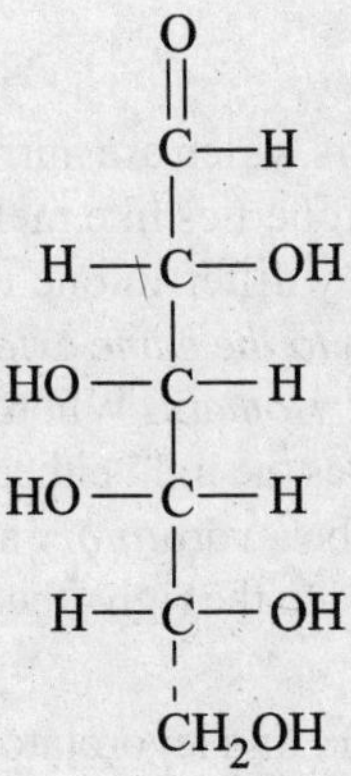
D-Galactose

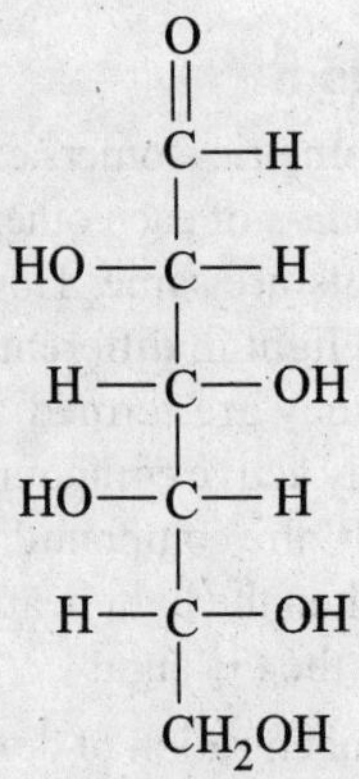
D-Idose

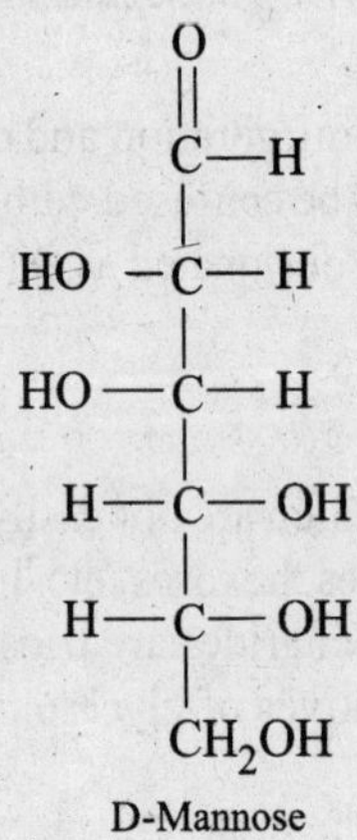
D-Mannose

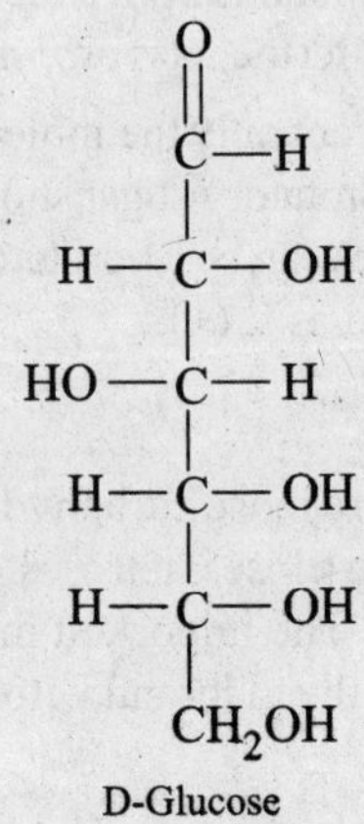
D-Glucose

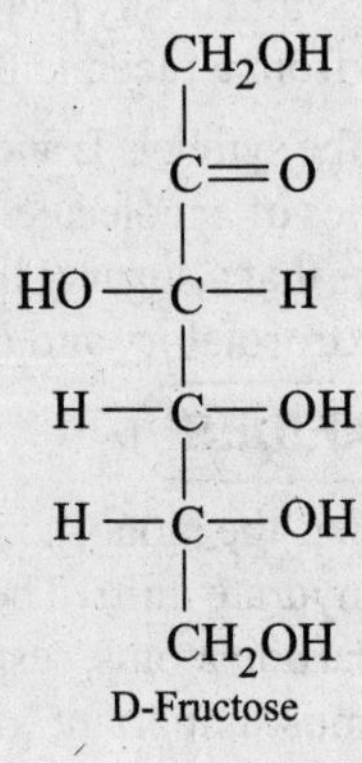
D-Fructose

STRUCTURE OF CARBOHYDRATES

Straight Chain Structure

So far the structure of carbohydrates has been depicted in the form of a straight chain. However, this straight chain structure is insufficient to explain certain properties of pentoses and hexoses, *e.g., mutarotation.* In order to understand this, it is essential to review the history of the structure of monosaccharides. In 1886, Kiliani was able to synthesize tetroses from glyceraldehydes by the addition of HCN in the following manner:

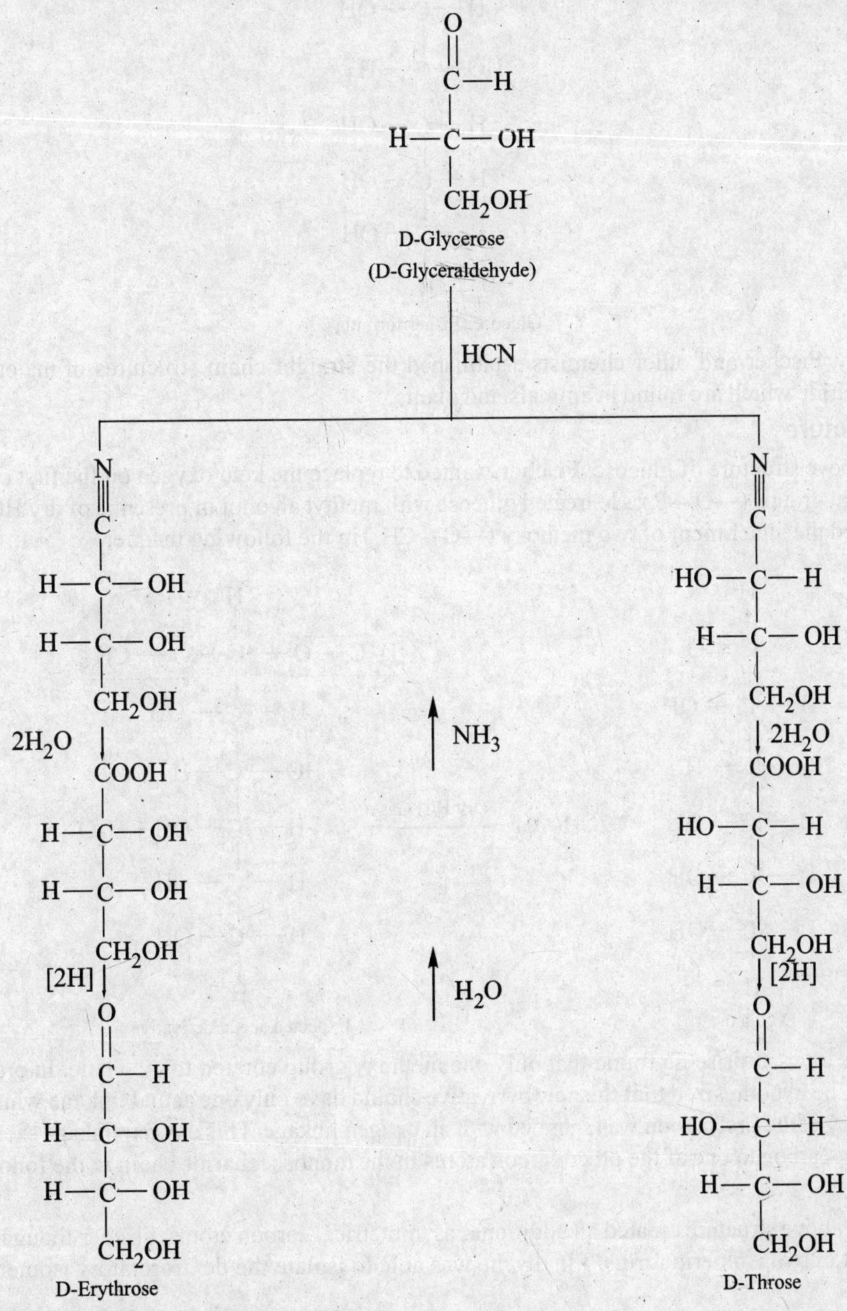

In a similar manner the synthesis of glucose could also be carried on. Kiliani, however, was unable to assign the correct position of H and OH to the four asymmetrical carbon atoms.

Ten years later **Emil Fischer** assigned the correct position of H and OH groups to the four asymmetrical carbon atoms of glucose in the following manner:

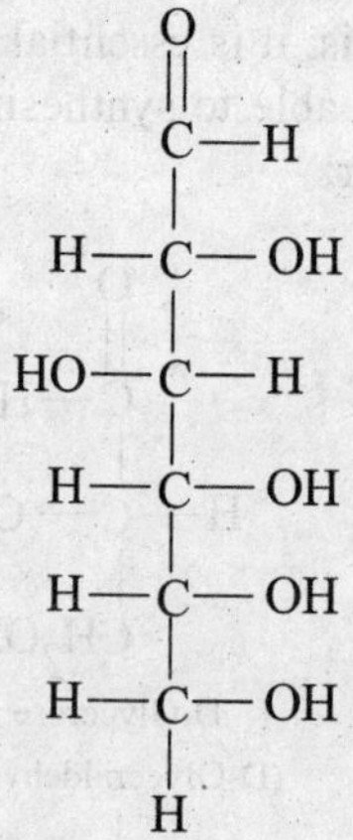

Glucose (Fisher formula)

Similarly, Fischer and other chemists established the straight chain structures of majority of monosaccharide which are found in animals and plants.

Ring Structure

From the above structure of glucose, **Fischer** wanted to replace the keto oxygen on the first carbon by two acetal groups (—O—R). He treated glucose with methyl alcohol in presence of dry HCl gas and expected the attachment of two methoxy (—O—CH_3) in the following manner:

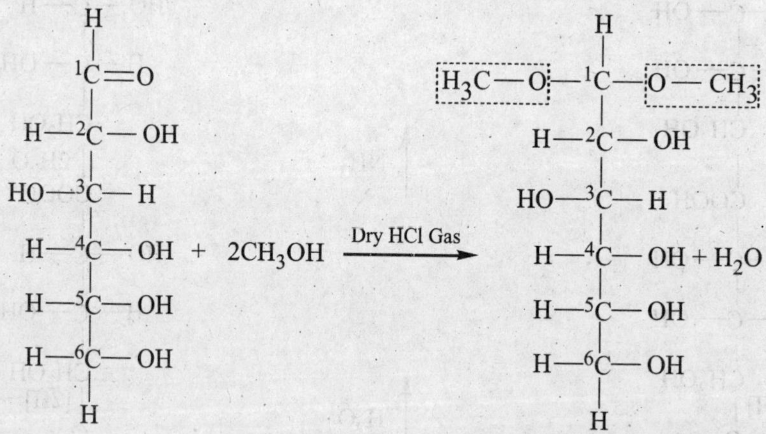

D-Glucose

Expected acetal derivative

But to his utter surprise he found that only one methoxy group entered to molecule. In order to explain this, he hypothesized that the new derivative should have only one actual linkage while the fourth valency of the first carbon was satisfied with an oxygen linkage. This oxygen linkage extended from the first carbon to one of the other carbon atoms in the monosaccharide chain in the following manner:

Since the above structure created an additional asymmetrical carbon atoms, Fischer thought that it must exist in two isomeric forms. Finally, he was able to isolate the dextrorotatory isomer (α -

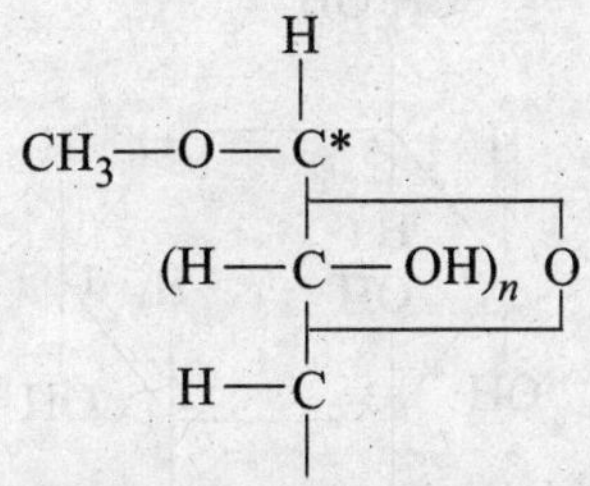

Possible ring structure by Fischer

methyl glucoside). About a year later the levorotatory isomer (β -methyl glucoside) was isolated by **Ekenstein.**

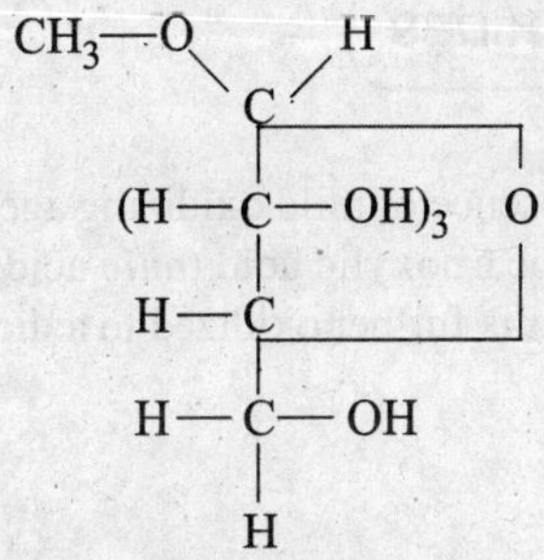

β-Methyl glucoside

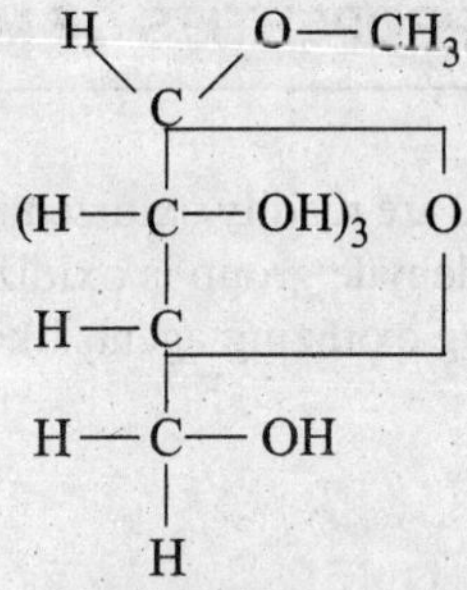

α-Methyl glucoside

The discovery of these two glucose derivatives solved the mystery of the phenomenon known as *mutarotation.* (A freshly prepared solution of D-glucose has a specific rotation of +113°. But when the solution is kept for some time at room temperature the rotation gradually drops to a value of + 52.5°. In contrast to the above, another form of glucose isolated in 1897 by **Tanret** had a very low rotation (+ 19°). However, on standing, this isalso charged to + 52.5°. Therefore, it was concluded that glucose exists in two forms. The form having a rotation of + 113° was named as α -D-glucose and that having a rotation of + 19° was named as β -D-glucose. Mutarotation, then, can be defined as the change in optical rotation of two isomers of a substance to common value).

The points of attachment of oxide ring to the Fischer's straight chain formula was first proposed by **Haworth** and **Hirst** (1926). The terminology of the sugars proposed by Haworth was base on the simple organic compounds exhibiting similar ring structure, *i.e. pyran and furan.*

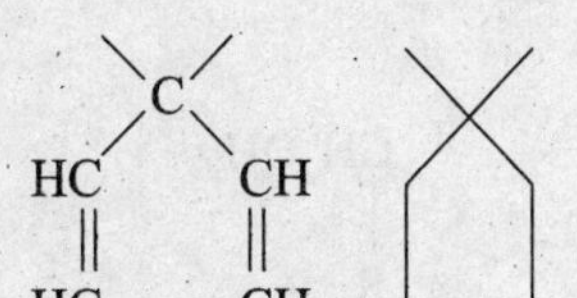

Pyran

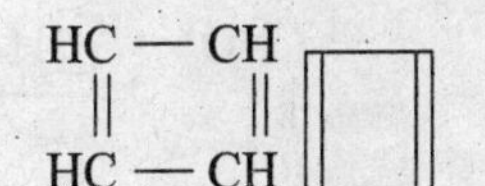

Furan

The spatial configuration of the groups around the asymmetrical carbon atoms of glucose, as proposed by Haworth, is given on next page

* Indicates asymmetrical carbon atom.

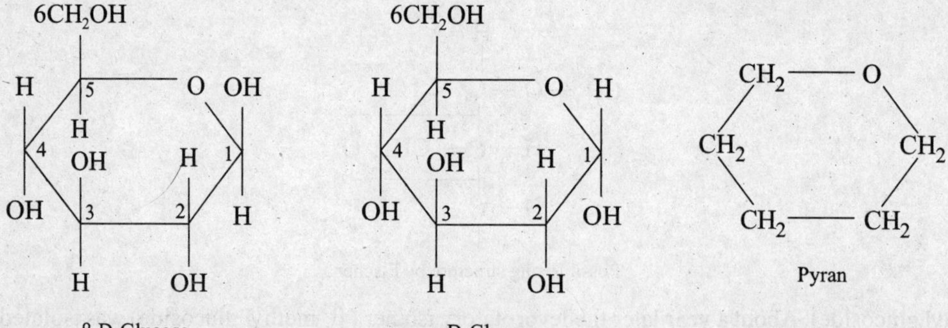

β-D-Glucose α-D-Glucose

Pyran

CHEMICAL PROPERTIES OF MONOSACCHARIDES

Oxidation

Monosaccharides are readily oxidized in acids in the presence of mild oxidizing agents. Under this condition the aldehyde group is oxidized to form monocarboxylic acid (*onic* acid). However, in presence of strong oxidizing agents like HNO_3, the sugar is further oxidized to a dicarboxylic acid (*aldaric* acid).

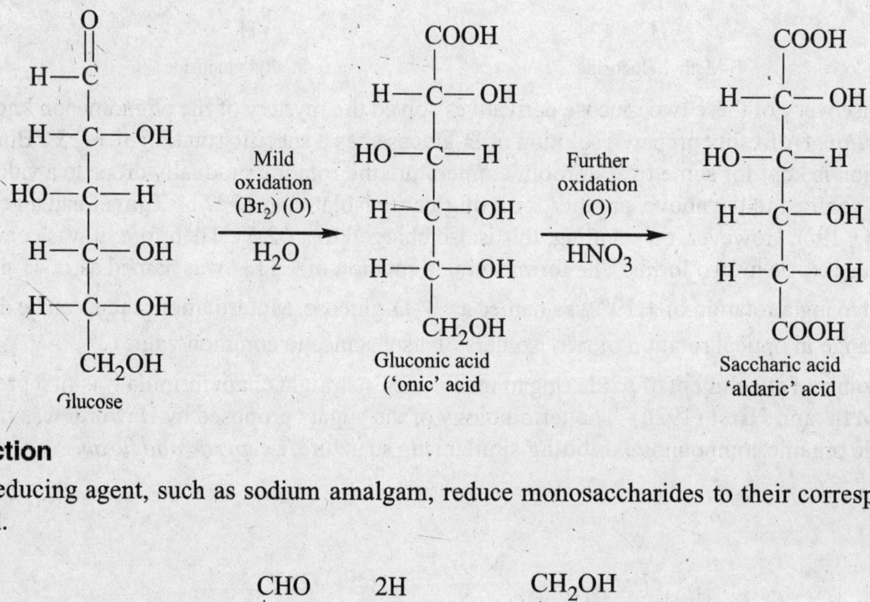

Glucose Gluconic acid ('onic' acid) Saccharic acid 'aldaric' acid

Reduction

Mild reducing agent, such as sodium amalgam, reduce monosaccharides to their corresponding alcohol.

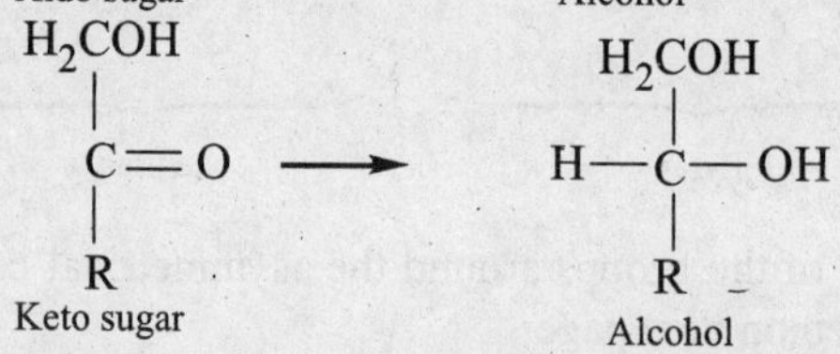

Aldo sugar Alcohol

Keto sugar Alcohol

Thus, glucose yields sorbitol; mannose yields mannitol and fructose yields a mixture of sorbitol and mannitol.

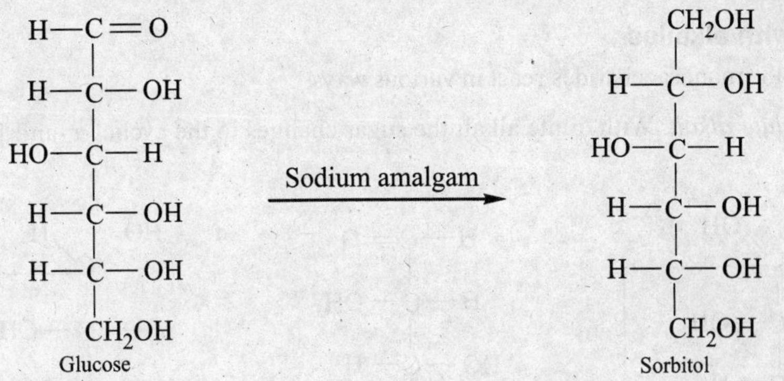

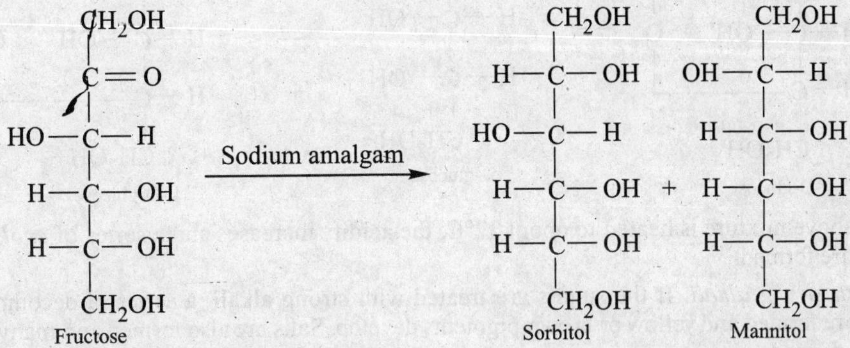

Reactions with strong mineral acids

When the aldoses are heated with strong mineral acids, a furfural derivative is formed.

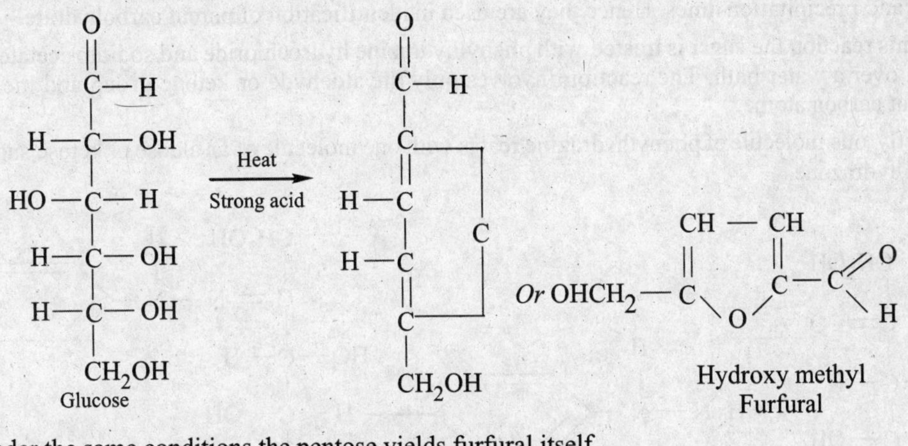

Under the same conditions the pentose yields furfural itself.

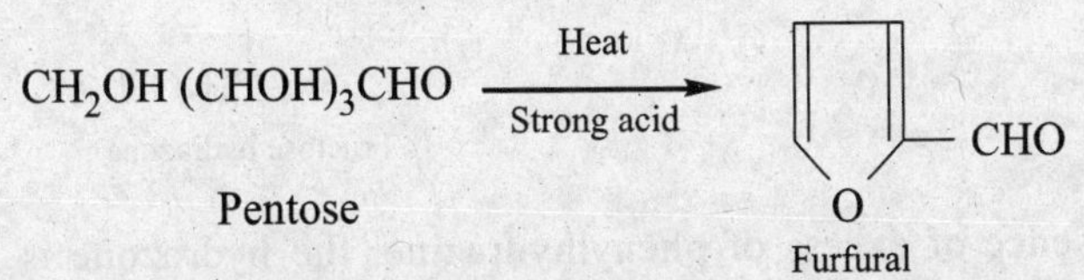

If the furfurals are treated with amines, coloured products are formed. These reactions are important for qualitative and quantitative tests of carbohydrates.

Reactions with alkalies

With alkalies the monosaccharides react in various ways.

(*a*) *With dilute alkali.* With dilute alkali the sugar changes to the cyclic α -and β -forms (i.e., mutarotation).

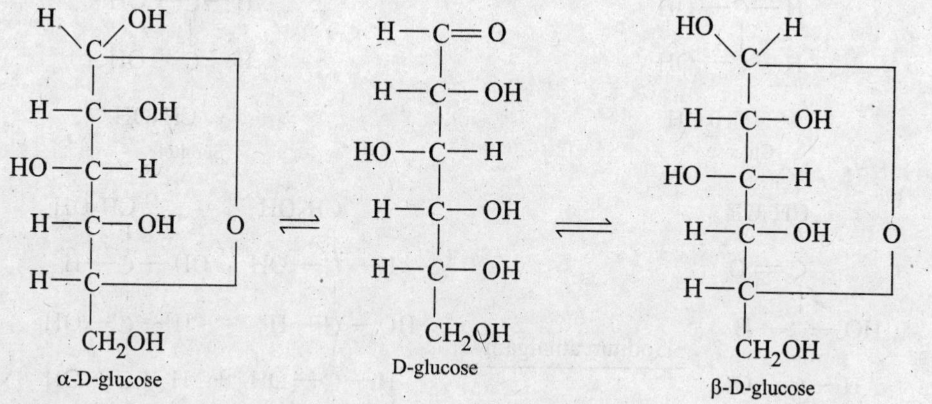

If the above mixture is heated to about 37°C, the acidity increases and a series of *enols* called '*endiols*' are formed.

(*b*) *With strong alkali.* If the sugars are treated with strong alkali, a series of decomposition products are formed and yellow or brown pigments develop. Salts are also formed and many double bonds are formed.

Osazone Formation

The osazone compound of sugars is coloured; possesses characteristic crystalline forms, melting points and precipitation times. Hence they are used in identification of parent carbohydrate.

In this reaction the sugar is treated with phenylhydrazine hydrochloride and sodium acetate, and heated over a water bath. The reaction involves only the aldehyde or ketone group and the next adjacent carbon atom.

Firstly, one molecule of phenylhydrazine reacts with one molecule of an aldose or ketose sugar to form a hydrazone.

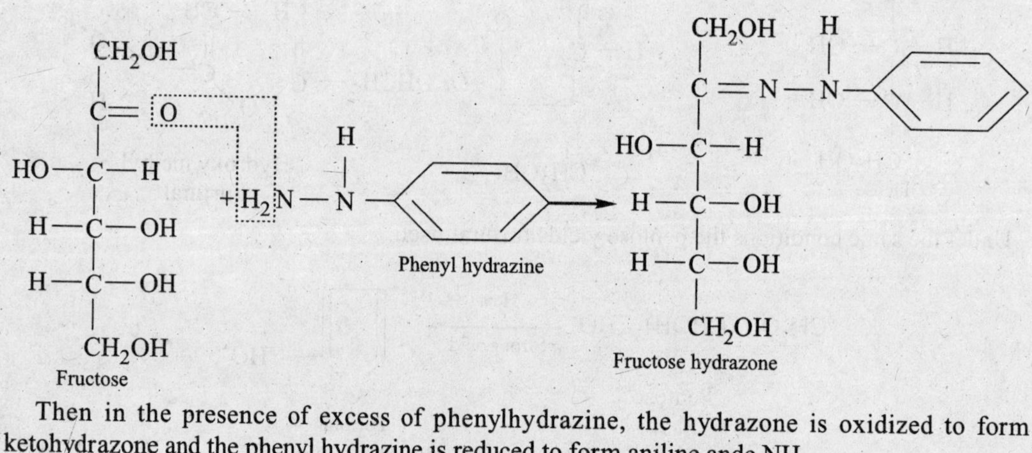

Then in the presence of excess of phenylhydrazine, the hydrazone is oxidized to form ketohydrazone and the phenyl hydrazine is reduced to form aniline ande NH_3.

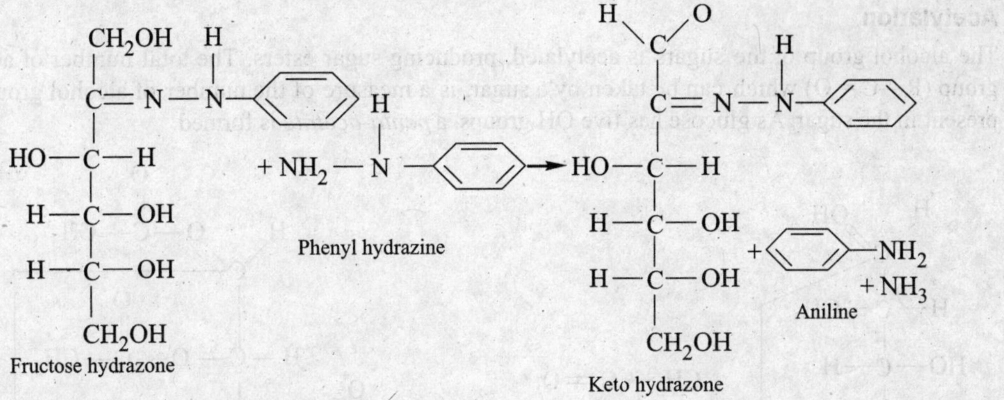

Finally, third molecule of phenylhydrazine reacts with ketohydrazone to yield the characteristic brilliantly coloured *osazone.*

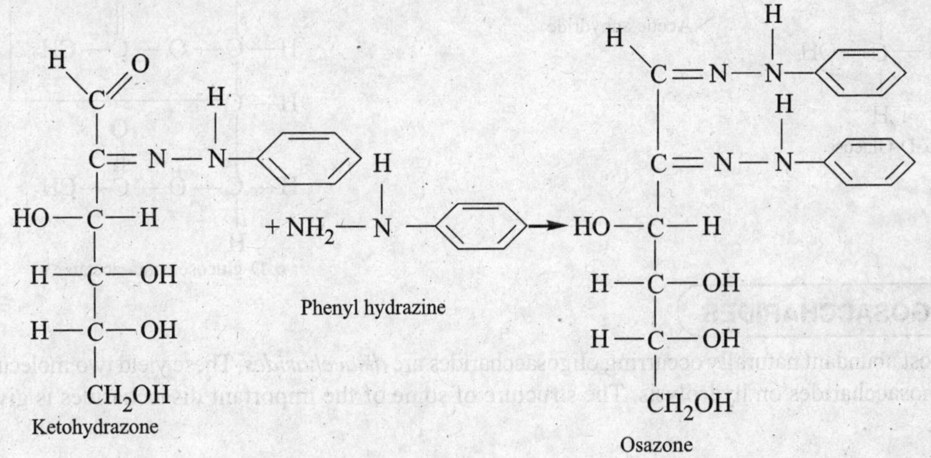

It may be mentioned here that the osazones of glucose, fructose and mannose are identical.

Interconversion

As stated earlier, weak alkali solutions convert sugars into endiols. The endiols of glucose, fructose and mannose are interconvertible as explained below:

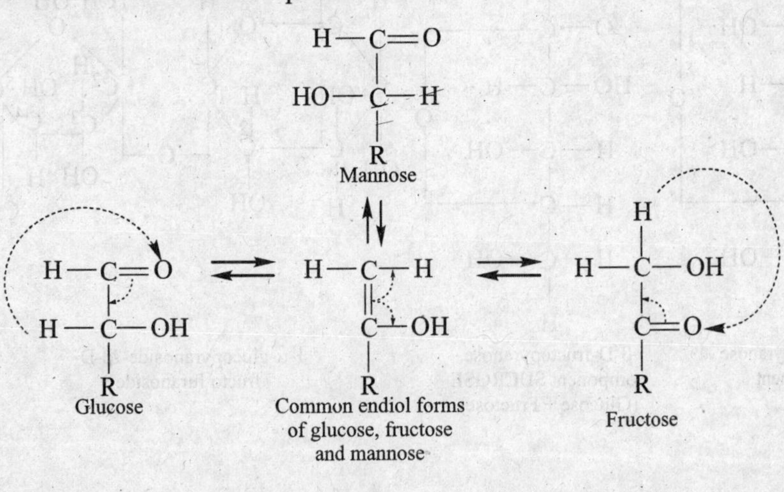

Acetylation

The alcohol group of the sugars is acetylated, producing sugar esters. The total number of acyl group (R—C = O) which can be taken by a sugar, is a measure of the number of alcohol groups present in the sugar. As glucose has five OH groups, a *penta-acetate* is formed.

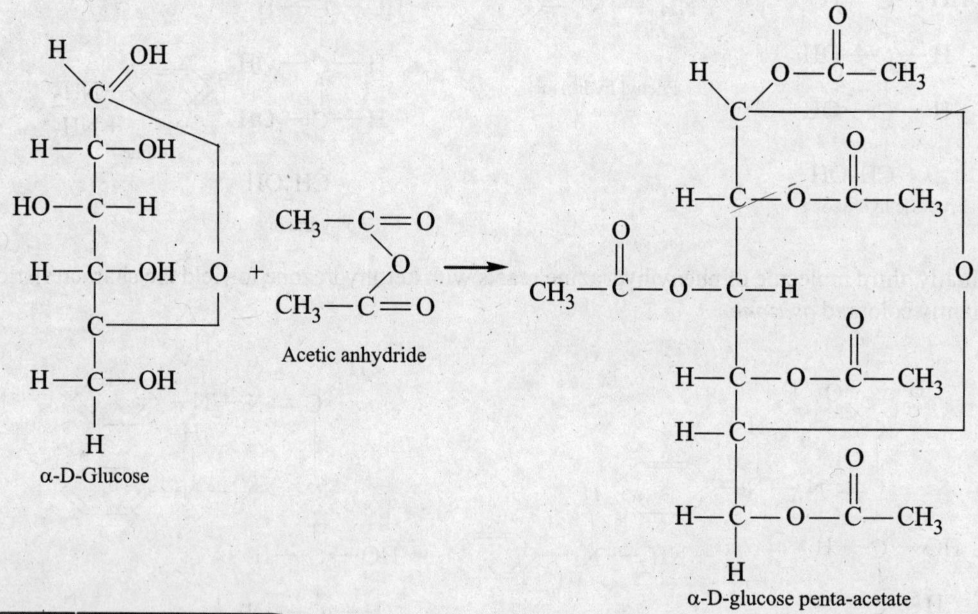

OLIGOSACCHARIDES

The most abundant naturally occurring oligosaccharides are *disaccharides*. These yield two molecules of monosaccharides on hydrolysis. The structure of some of the important disaccharides is given below:

FISCHER'S FORMULA

HAWORTH FORMULA

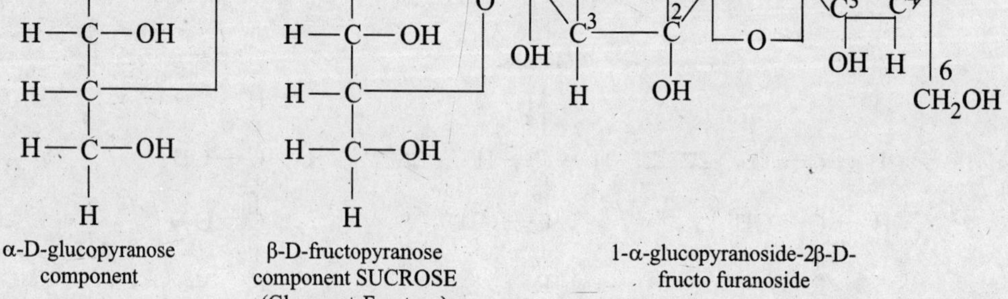

α-D-glucopyranose component

β-D-fructopyranose component SUCROSE (Glucose + Fructose)

1-α-glucopyranoside-2β-D-fructo furanoside

FISCHER'S FORMULA

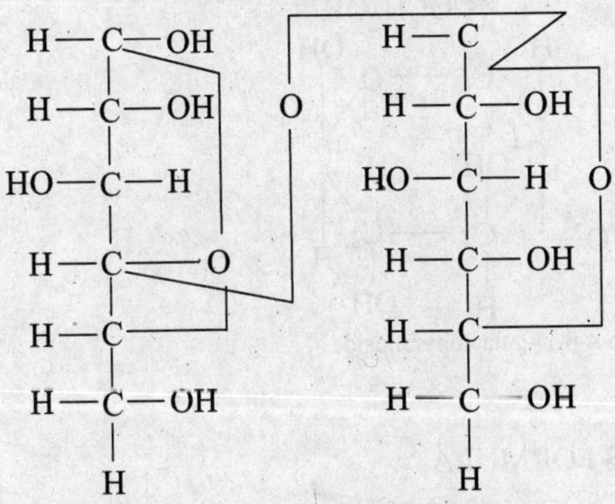

Two α-D-glucopyranose components

HAWORTH FORMULA

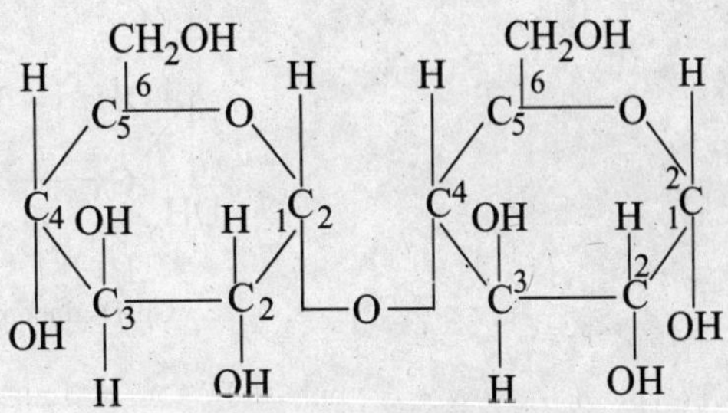

1-α-D-glucopyranoside-4-D-Glucopyranoside.
MALTOSE (Glucose + Glucose)

FISCHER'S FORMULA

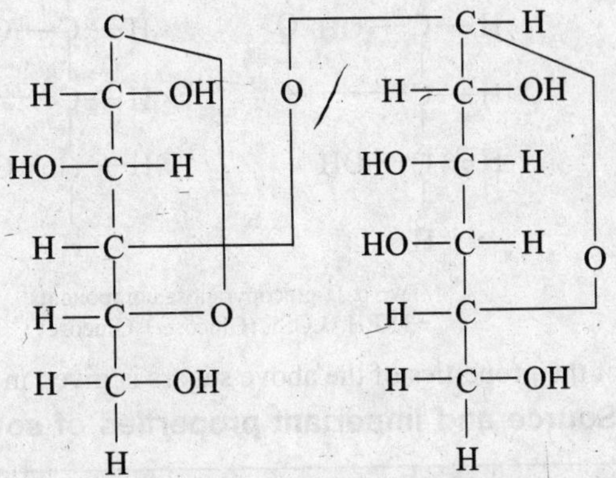

β-D-glucopyranose, components β-D-galactopyranose, components

HAWORTH FORMULA

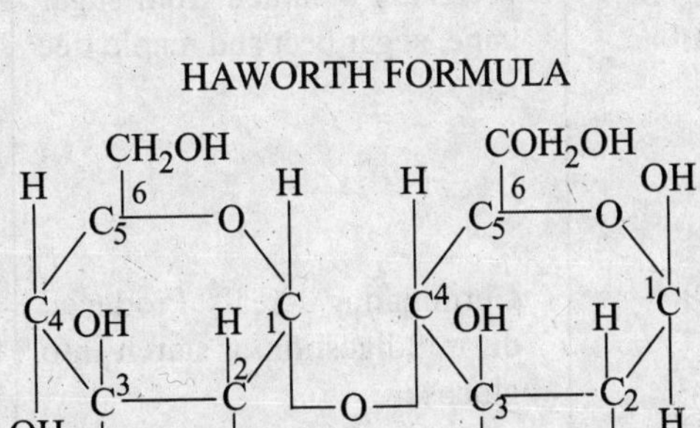

1-β-D-glucopyranosido-4-β-D-galacotopyranoside
Lactose (Glucose + Galaclose)

HAWORTH FORMULA

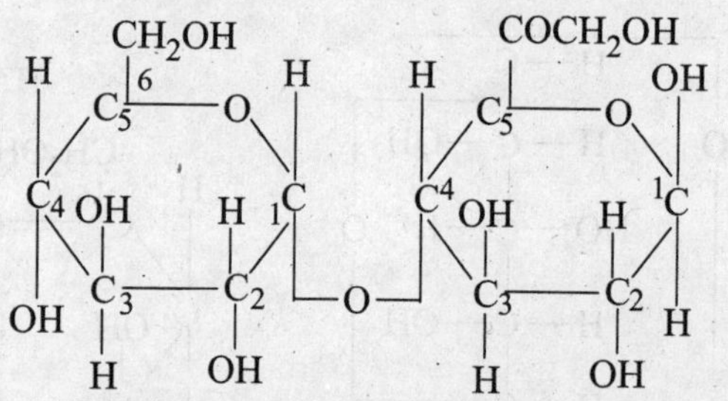

1-β-D-glucopyranosido-4-β-D-galactopyranoside

FISCHER'S FORMULA

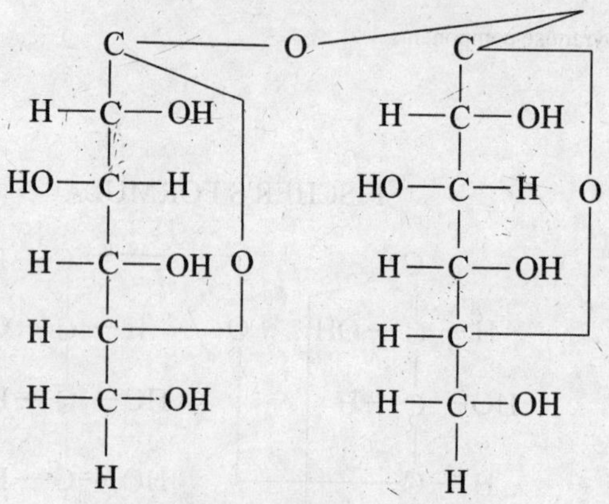

Two α-D-glucopyranose components
TREHALOSE (Glucose + Glucose)

A brief summary of the properties of the above sugars is given in Table 3.1.

Table 3.1 Source and important properties of some disaccharides

Name of sugar	*Occurrence*	*Properties*
SUCROSE (Cane sugar or beet sugar)	Widely found in plants. Commercially obtained from sugar cane, sugar beet and maple tree sap.	White sweet crystalline. Melting point 160-186°C- Nonreducing since both reducing groups involved in linkage. Does not form osazone. Yields a molecule each of glucose and fructose on hydrolysis.
Maltose (Malt sugar)	Germinating cereals. Produced during digestion of starch into glucose.	Hard, crystalline, less sweet than sucrose. Reducing sugar, because only one reducing group forms the linkage. Form osazones. Yields two molecules of glucose on hydrolysis.

Name of sugar	*Occurrence*	*Properties*
Lactose (Milk sugar)	Milk of mammals. Formed in the body from glucose. By-product in the manufacture of cheese.	Hard, gritty, crystalline; less sweet than sucrose. Melting point 203°C. Reducing sugar because the reducing group of galactose does not take part in linkage. Form osazones. Yields one molecule each of glucose and galactose on hydrolysis.
TREHALOSE	Major sugar on insect haemoloymph. Sect Young Mushrooms and yeasts.	White, crystalline. Melting point 210-215°C. Non-reducing sugar because both the reducing groups of glucose are involved in linkage. Does not form osazone. Yields two molecules of glucose on hydrolysis.

POLYSACCHARIDES

Polysaccharides are defined as the polymeric anhydrides of monosaccharides. Some of the important polysaccharides are mentioned below.

Starch

It is the chief carbohydrate reserve of plants and is stored in the form of granules. It is abundantly found in cereals, legumes, potatoes, etc. Natural starch is insoluble in water and gives blue colour when treated with iodine solution. Starches are of two types: (*a*) the straight chain polysaccharide known as *amylose* and (*b*) branched chain polysaccharide called *amylopectin*. Most natural starches are a mixture of these two. Potato and cereal starches are 20—30% amylose and 70—80% *amylopectin*.

Both amylose and amylopectin are composed of α-D-glucose units joined by 1—4 linkages. The molecular weight of amylose ranges between 10,000—58,000. A condensed structure of amylose is given aside:

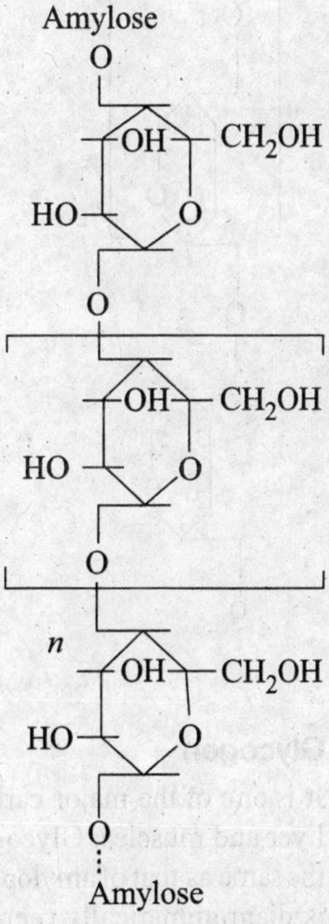

The amylopectin molecule possesses the same basic chain of α-1-4 glucosidic linkage but has in addition a number of side chains. The side chains are attached to the main chain by 1-6 glucosidic linkage as shown on the next page. The molecular weight of amylopectin may vary greatly and be up to 5,00,000 or more.

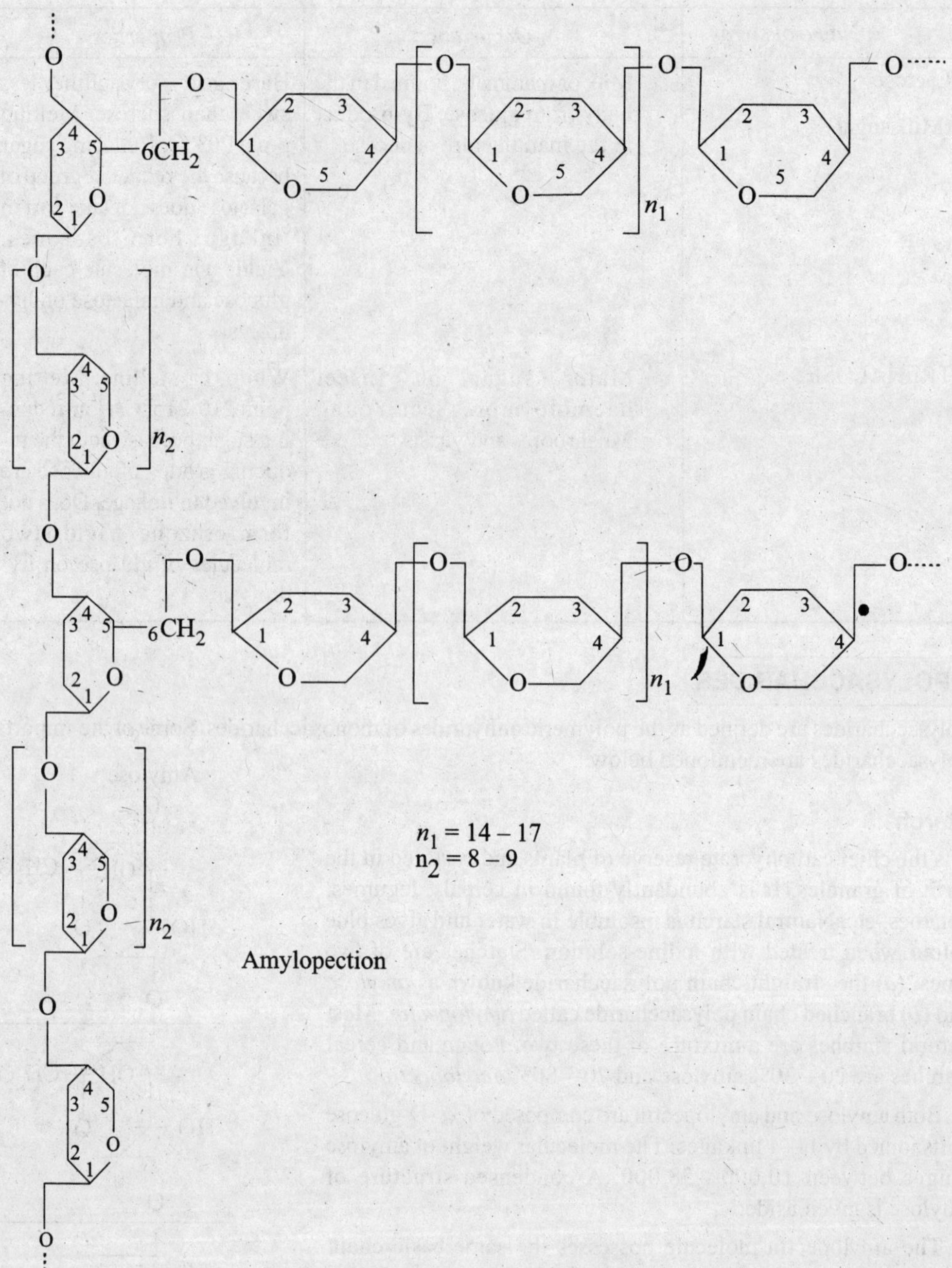

$n_1 = 14 - 17$
$n_2 = 8 - 9$

Amylopection

Glycogen

It is one of the major carbohydrate reserves in animals. In most vertebrates glycogen is stored in liver and muscles. Glycogen is also polymer of α -D-glucose units and its structure is essentially the same as that of amylopectin, but its chain length is usually shorter and more branched. Its structure is diagrammatically represented in Fig. 3.1, where each bead of the chain represents a glucose molecule.

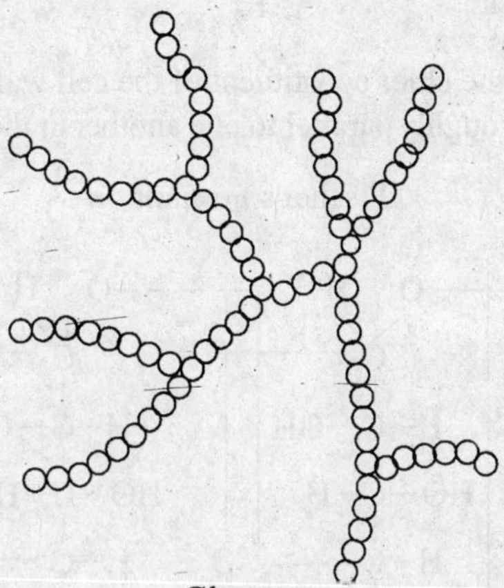

Glycogen

Fig. 3.1. Structure of a glycogen molecule. Each circle represents a glucose residue

It may be mentioned here that branching in polysaccharide molecule determines its solubility. In general, the more the branching; the more the solubility. That it why glycogen is more soluble than amylopectin and amylopectin is more soluble than amylose.

Inulin

This is a polymer of β -D-fructose, soluble in water. This occurs as stored food material in the tubers of may plants especially the members of Compositate family (dandelions, dahilias, artichokes, etc.). The molecular weight of inulin is about 5,000. The structure of inulin is shown on the next page.

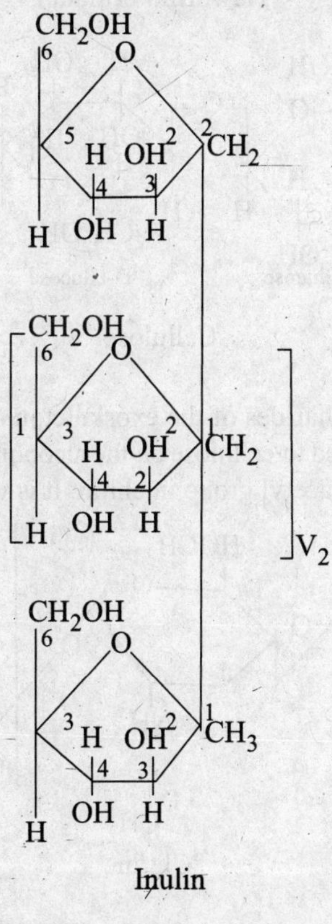

Inulin

Cellulose

Cellulose is found in nature as the chief constituent of the cell walls of plants. This is a polymer of the beta glucose units arranged roughly parallel to one another in the form of strong fibrous structure.

Fischer's Formula

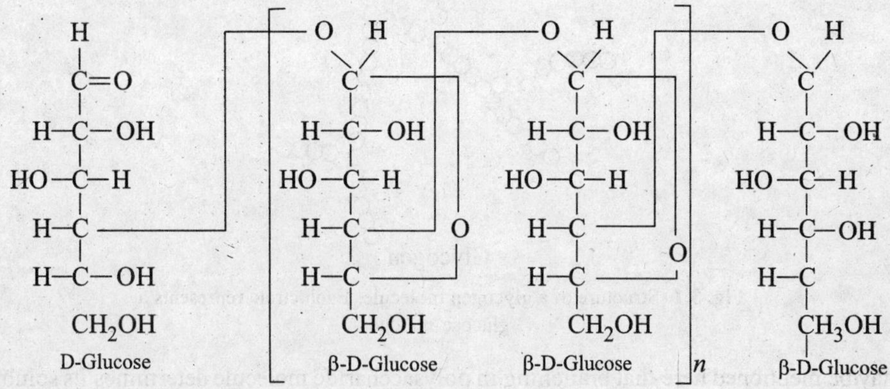

Cellulose

It gives no colour with iodine and is not soluble in water. It is not hydrolysed by the enzymes of man and thus constitutes the 'bulk' of human faeces. It is estimated that cellulose molecule contains between 1,600 and 2,700 β -D-glucose units joined by 1-4 glucosidic linkage. It is a long thread-like molecule which accounts for the fibrillar structure of its products, e.g., cotton, linen, etc.

Haworth Formula

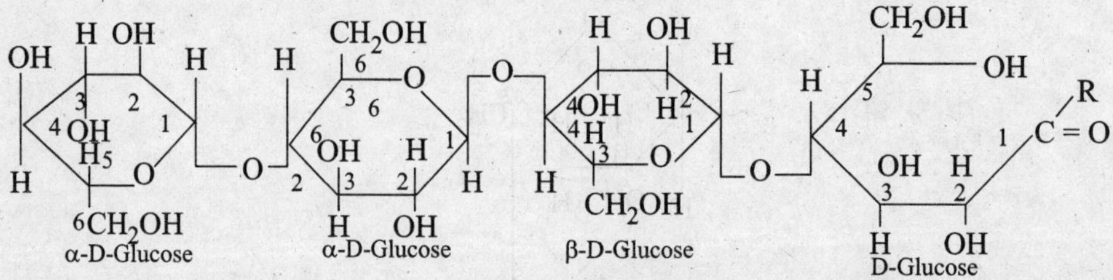

Cellulose

Chitin

It is one of the important polysaccharides of the exoskeleton of invertebrates (mostly insects and crustaceans). Chitin is closely related to cellulose bu the alcoholic hydroxyl group on the 2nd carbon atom of cellulose is replaced by N-acetyl group in chitin. It is very resistant to hydrolysis.

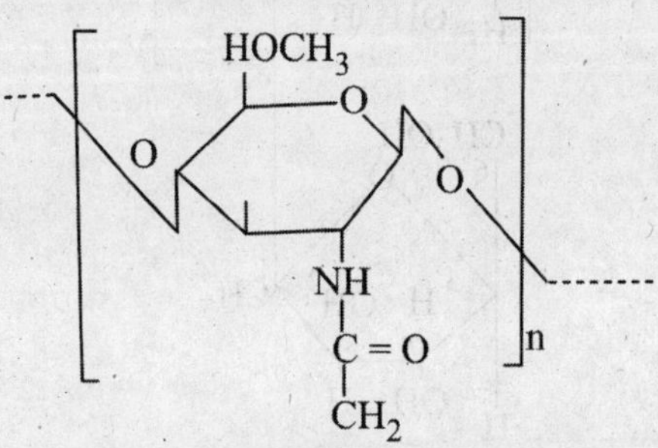

°Chitin

MUCOPOLYSACCHARIDES

These are polymers of amino sugars, uronic acids and monosaccharides. They can be divided into three groups.

(1) *Neutral Mucopolysacchrides.* These contain an N-acetyl hexosamine and a hexose. They are important constituents of the immunological factors of blood groups.

(2) *Hyaluronic acid.* This polysaccharide is associated with the structure of animal tissues and are anologous to the cellulose of plant cells. They are especially found in the tissues of eye and umbilical cord. Hyaluronic acid has a high molecular weight up to 5,000,000 and consists of alternating units of D-glucuronic acid and N-acetyl-D glucosamine (NAG).

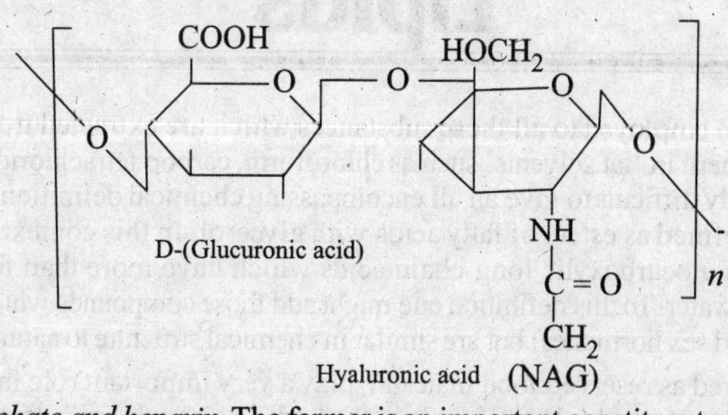

Hyaluronic acid

(3) *Chondrin sulphate and heparin.* The former is an important constituent of cartilage, skin, tendons, and cornea. It is made of N-acetyl-D-galactosamine, glucuronic acid and sulphuric acid. Heparin is a naturally occurring anticoagulant found in the liver, lung, spleen, kidney, etc. It is a polymer of glucosamine-N-sulphate, glucuronic acid and sulphuric acid.

4 Chapter

Lipids

Lipid is a general term employed to all those substances which are extracted from plant and animal tissues by dissolving them in 'fat solvents' such as chloroform, carbon tetrachloride, benzene, acetone, ether etc. It is extremely difficult to give an all encompassing chemical definition to lipids. Generally, however, lipids are defined as esters of fatty acids with glycerol. In this context, the fatty acids may be defined as those monocarboxylic long-chain acids which have more than four carbons and are relatively insoluble in water. To this definition one might add those compounds which are synthetic (*e.g.*, synthetic cortisones and sex hormones) but are similar in chemical structure to naturally occurring lipids.

Lipids, largely stored as reserved food material, play a very important role in metabolism. Along with carbohydrates, they act as biological fuels in the body. Although lipids form a large proportion of the mass of all cells, they are not always present in free state. In combination with proteins (lipoproteins), they are an integral part of the cell membrane of all animals. A group of lipids, called *steroids,* are important hormones.

Classification

Lipids have been classified into three classes, *viz., (i) simple lipids, (ii) compound lipids and (iii) derived lipids.*

SIMPLE LIPIDS

Simple lipids are esters of one or more fatty acids and an alcohol, mostly glycerol. They are made up of carbon, hydrogen and oxygen. This group includes *fats* and *oils* (triglycerides) and *waxes.*

Fats and oils (triglycerides)

These are mixtures of lipids obtained from biological sources. The difference between fats and oils is purely physical, fats are solids and oils are liquids at room temperature. The commercially important fats and oils include lard, olive oil, cotton seed oil, soybean oil, butter fat, corn oil, peanut oil, coconut oil. Fats and oils are esters of fatty acid and glycerol. Three molecules of fatty acid combine with one molecule of glycerol to form a triglyceride. In this process three molecules of water are lost.

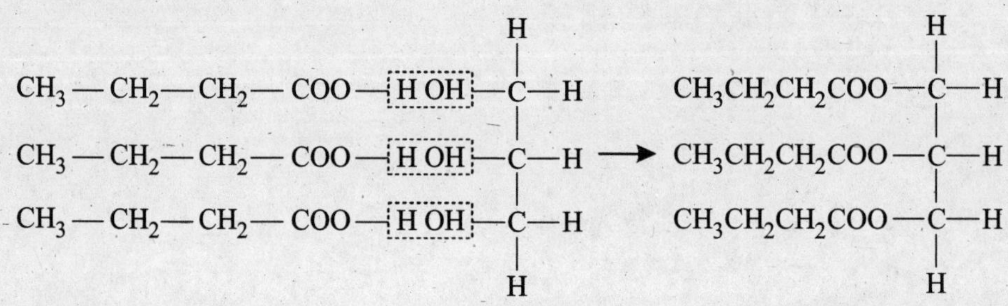

Butyric acid — Glycerol — Tributyrin (a simple lipid found in butter) + $3H_2O$ water

Most of the oils and fats are mixed triglycerides, made up of two or three different fatty acids instead of a single fatty acid.

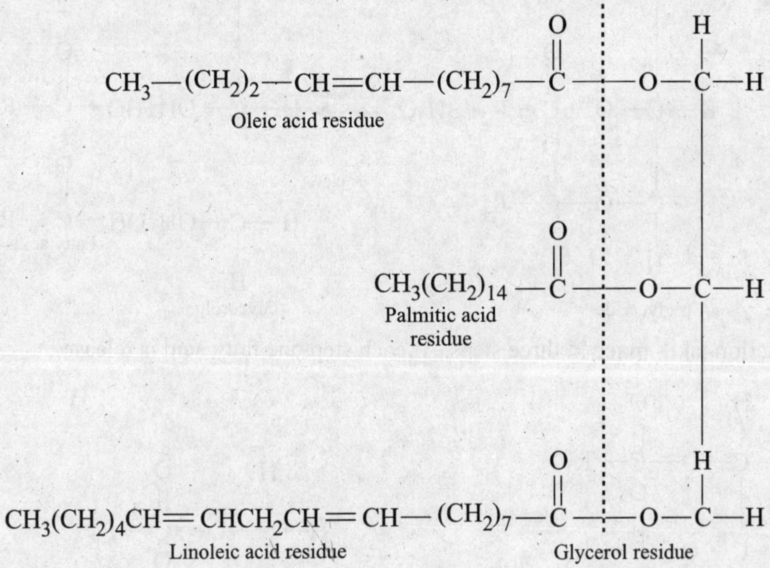

(A mixed triglyceride)

Glycerol is common to all fats and oils and their properties (fats and oils) are dependent only on the fatty acid component. Most of the fatty acids forming oils and fats have the following properties—

(i) They are monocarboxylic with a variable number of carbon atoms.
(ii) The number of carbon atoms is nearly always *even* ranging from 4 —30.
(iii) The carbon atoms from a side chain which is usually unbranched.
(iv) The side chains may be saturated or unsaturated. Unsaturated fatty acids have a variable number of double bonds depending on the degree of saturation.

PROPERTIES OF TRIGLYCERIDES

Fats and oils float on water because their specific gravity is always less than one. The melting point of lipids depends upon the degree of saturation of the fatty acids which form them. Lipids composed of saturated fatty acids are solids whereas those made up of unsaturated fatty acids are liquids at room temperature. Unsaturated fatty acids have a lower melting point as compared to saturated fatty acids. The melting point is dependent on the number of double bonds present in the fatty acids (Table 4.1).

Table 4.1 Showing relationship between degree of saturation and melting point of fatty acids

Fatty acids	*Number of double bond*	*Melting point (°C)*
Stearic acid	Zero (saturated)	70
Oleic acid	one	4
Linoleic acid	two	– 5
Linolenic acid	three	– 11

Chemical Reactions

Enzyme Hydrolysis. Enzymes (lipases) in the alimentary tracts of animals act as efficient catalyst for the hydrolysis of the ester bonds of triglycerides releasing fatty acids and glycerol.

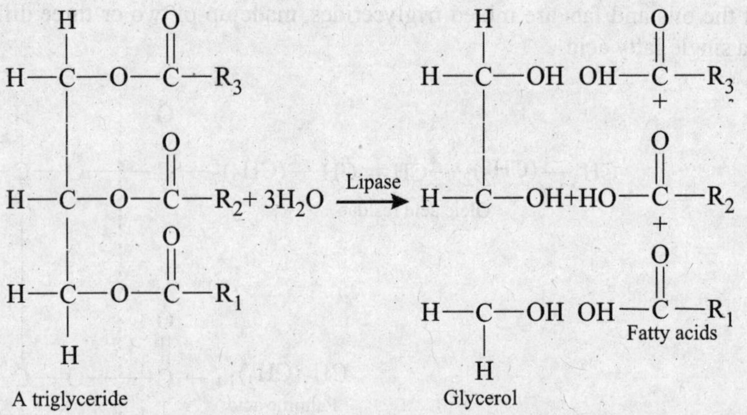

This reaction takes place in three steps. At each step one fatty acid is released.

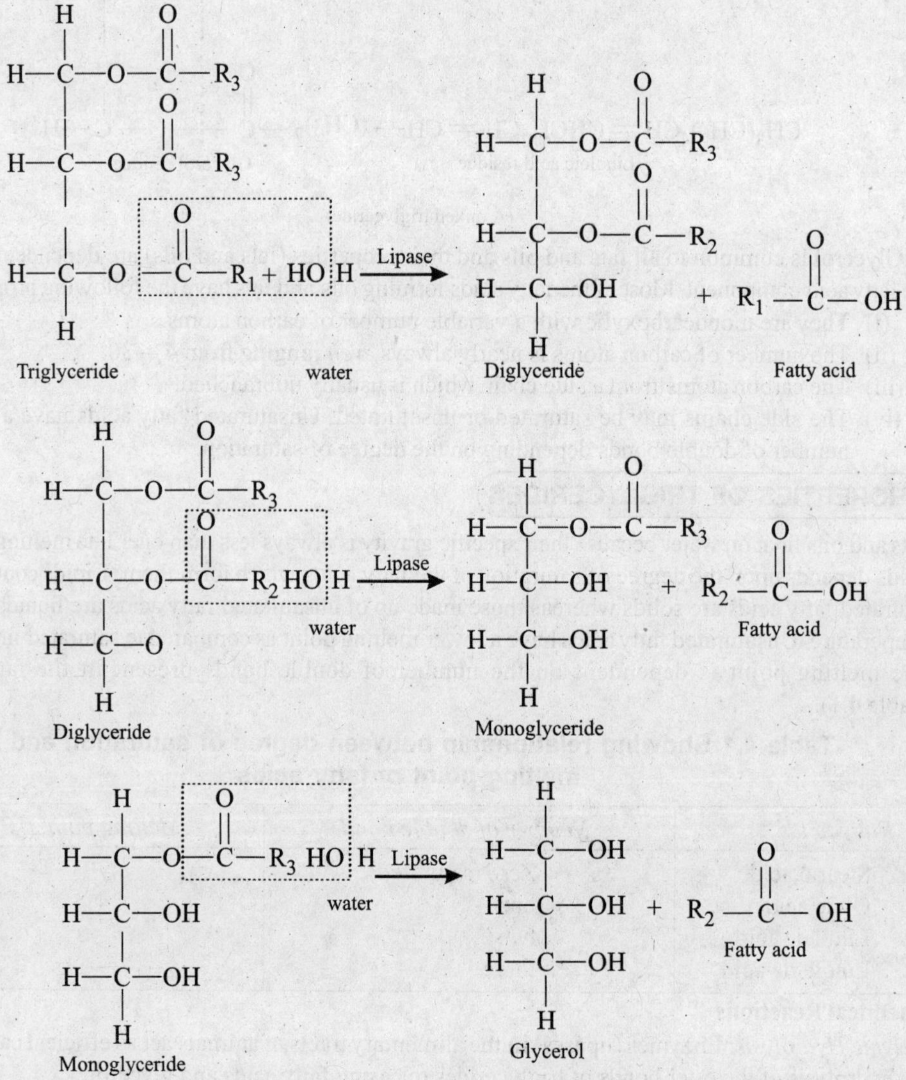

However, in animals, absorption of mono and diglyceride is very rapid and, therefore, very little glycerol is formed in the alimentary canal.

Hydrolysis with Alkalies (*Saponification*). The hydrolysis of triglycerides with strong bases (NaOH and KOH) is termed *saponification.* When the hydrolysis is complete, glycerol and the salts of fatty acids (soaps) are formed. This reaction is used for the commercial manufacture of soaps. Ordinary soaps are salts of sodium and potassium and are water soluble. The term 'soap' is also applied to salts of other metals, *e.g.,* calcium, barium, aluminium, magnesium. etc., but these are unlike the ordinary soaps.

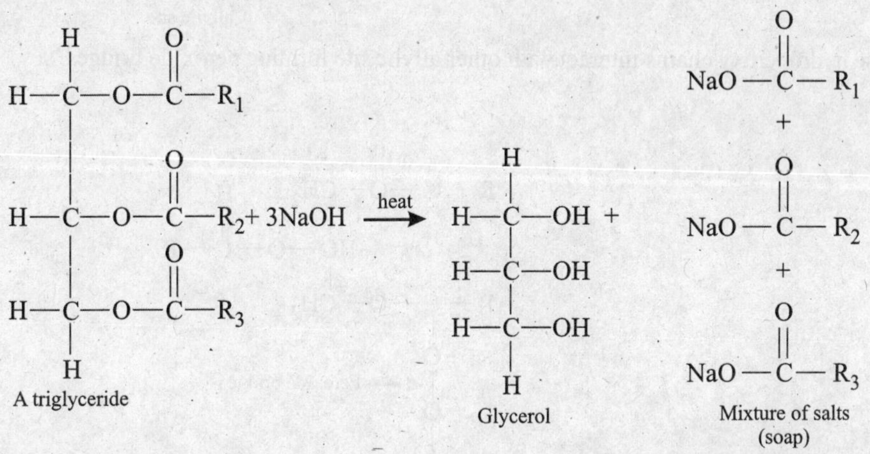

Rancidity. When oils and fats are stored in warm and moist air, they become malodorous. This phenomenon is termed *rancidity.* Two types of reactions are mainly responsible for making a lipid rancid. They are (i) hydrolysis of ester bonds due to micro-organisms and (ii) by oxidation of double bonds in the unsaturated fatty acids.

Hydrolytic rancidity is caused by lipases secreted by microorganism which contaminate the fat. This type of hydrolysis is possible only when there is water in the lipid. The lipase produced by the microbes, hydrolyses the triglyceride to di-and monoglyceride, glycerol and free fatty acids. If by any case the fatty acid produced has low molecular weight (4—10 carbon atoms), it imparts a rancid odour. Hydrolytic rancidity can be reduced by refrigeration; by exclusion of water or by destruction of microbes. All these three processes are used for the preservation of butter, lard etc.

During *oxidative rancidity*, the atmospheric oxygen attacks the unsaturated side chains of mixed triglycerides resulting in the formation of volatile carboxylic acids and aldehydes, which impart bad odour.

Hardening of Oils 'Drying': The air drying of oil paints, varnishes etc., is a phenomenon of common occurrence. When oxygen attacks the unsaturated oils at the carbons present at allylic[1] position, hydroperoxy groups are formed.

$$CH_3\ \overset{*}{CH_2}\ CH = CH\ \overset{*}{CH_2}\ CH = CH\ \overset{*}{CH_2}\ CH = CH\ \overset{*}{CH_2}\ (CH_2)_6\ CO_2H$$

Linolenic acid (*indicate allylic sites)

1. The position on the chain directly attached to the double bond is known as allylic position.

2. 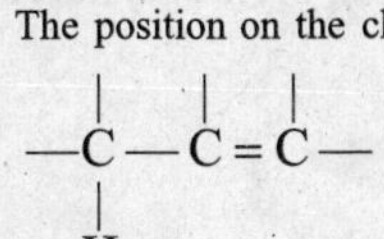

Allylic position ← Allylic hydrogen

Oxygen first attacks these allylic sites to form Hydroperoxides (—OOH).

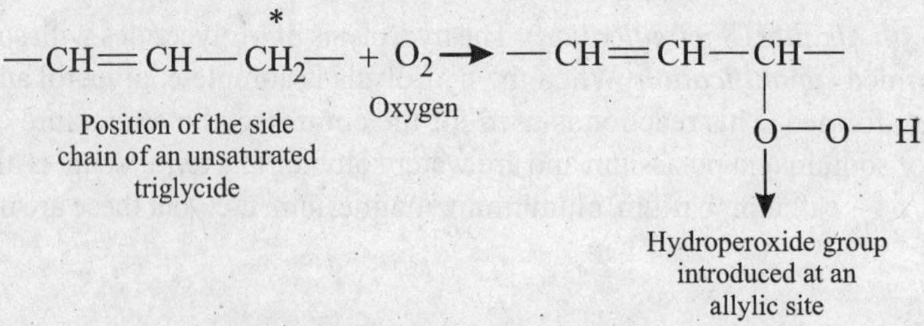

These hydroperoxy chains interact with other allylic site forming peroxide bridges.

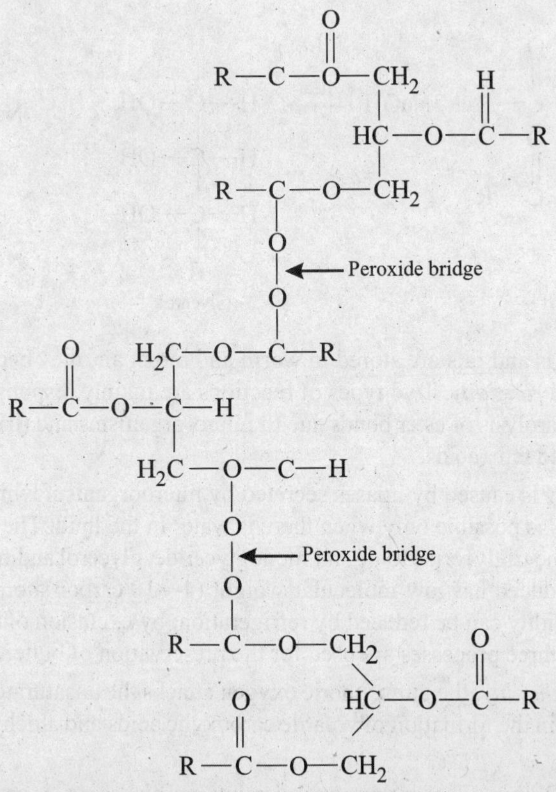

Such oxidized oils, when spread, dry out and form a hard, glossy and waterproof coating. Because of this property, unsaturated oils like, linseed oil, are widely used for manfacture of paints, varnishes, linoleum and oil cloth.

Waxes

Waxes are the esters of higher fatty acids (than those found in fats and oils) and high molecular weight monohybrid alcohols. The classical examples are beeswax and *spermaceti* (sperm whale wax). These are esters of palmitic acid and alcohols having 20—40 carbons, *e.g.*, cetyl alcohol (C_{16} $H_{32}OH$), hexacosanol ($C_{26}H_{53}OH$), myricyl alcohol ($C_{37}H_{61}OH$) etc.

* Indicates allylic site.

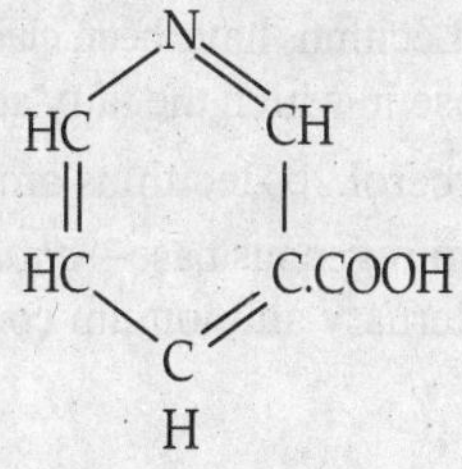

Nicotinic acid

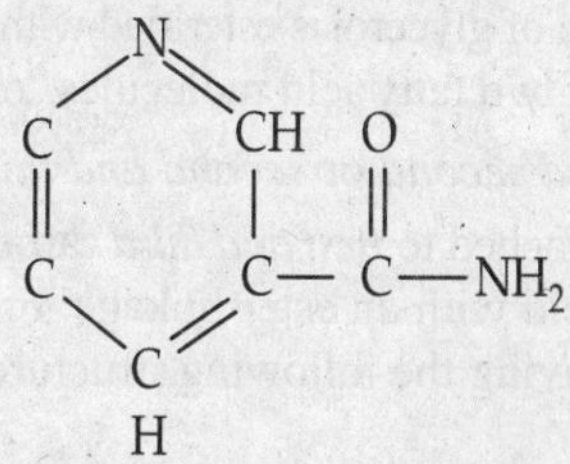

Nicotinic acid amide

(Cetyl palmitate—one of the chief constituents of spermaceti).

Waxes are found at several places in animals, plants and microbes. They may form a protective covering, over leaves and fruits or may be present in the form of waxy secretions in animals, microbes, etc.

Waxes are relatively inert, insoluble in water and resistant to atmospheric oxygen. Saponification of waxes is difficult. Due to these properties, they are used in furniture and automobile polish. Wax papers are commonly used for packing perishable materials.

COMPOUND LIPIDS

Like simple lipids, these are also esters of fatty acids and other hydroxy compounds. However, they are *united* with other *phosphoric acid, sugar, sulphuric acid and proteins.* Compound lipids have been classified on the basis of the nature of these additional moieties into four groups, *viz., phospholipids, glycolipids, sulpholipids and lipoproteins.*

Phospholipids

Lipids containing phosphorous are called phospholipids. The phospholipids are always present in the animal cell membrane where they are responsible for the selective permeability of the membrane.

The phospholipids can be further sub-divided into following groups:

Phosphatidic acids. Phosphates of glycerol are formed during the synthesis of simple lipids. These are called phosphatidic acids. Although not found in high concentrations in the body they are important during fat synthesis. The typical formula of phosphatidic acid is given below:

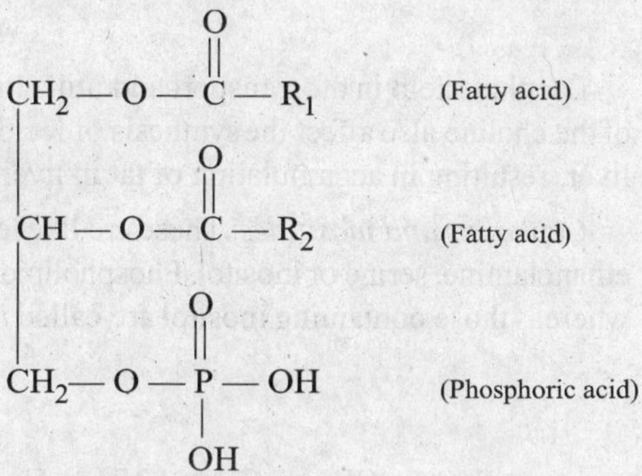

Phosphoric acid

Lecithins: This is the most common phospholipids present in animals and plants. These contain fatty acids, phosphoric acids, glycerol and a nitrogenous base *choline.* A typical structure of lecithin is given below:

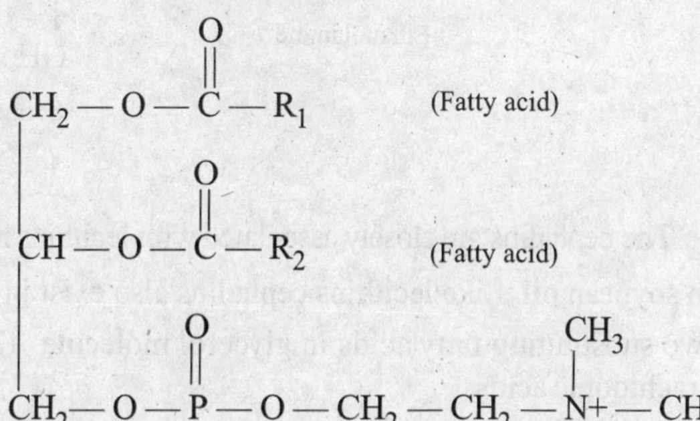

Phosphatidyl choline

The fatty acid commonly found in lecithin are palmitic, stearic, oleic, linolenic, arachidonic acids etc. As it is clear from the above structure, only two molecules of fatty acids are present in a lecithin molecule. The remaining carbon of glycerol is esterified with phosphoric acid. Lecithins have been classified according to the position of two fatty acid molecules. α -**lecithins** are those in which the fatty acids are attached either to *first and second* or *second and third* carbons of glycerol. β -lecithins are those when the fatty acids are attached to *first and third* carbons of glycerol. A nitrogenous base—*choline*—is attached to phosphoric acid with an ester linkage. Free choline is quaternary ammonium compound with an alcohol group having the following structure.

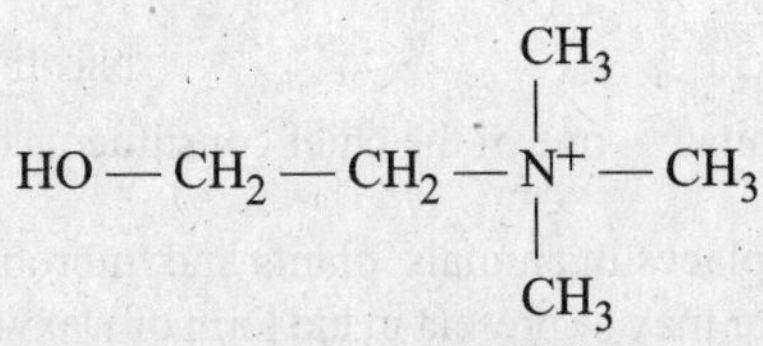

(Choline)

Choline can be synthesized in the body provided other compounds containing methyl groups, *e.g.*, methionine, are present in the diet. However, deficiency of choline is extremely rare as it is abundantly found in all food substances. The acetyl derivative of choline—*acetyl choline*—is undoubtedly the most important neurotransmitter in animals (see Chapter 19).

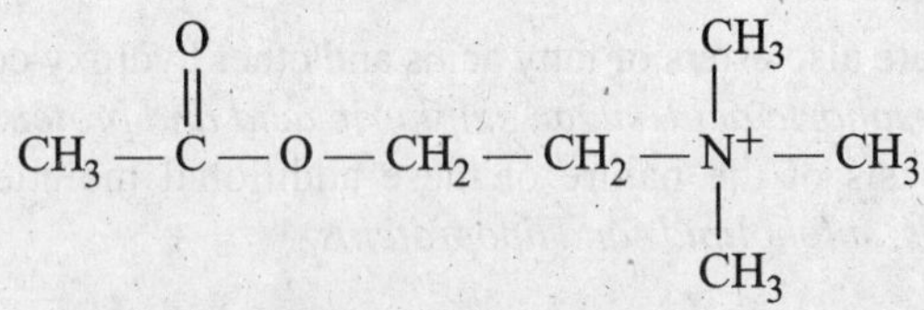

Acetylcholine

Lecithins help in the transport and utilization of other lipids. Factors which interrupt the synthesis of the choline also affect the synthesis of lecithins. This disrupts normal transport of lipid to and from liver, resulting in accumulation of fat in liver giving rise to a condition known as 'fatty liver'.

Cephalins and inositides: These are like lecithins in which the choline moity is replaced by either ethanolamine, serine or inositol. Phospholipids containing ethanolamine or serine are called *cephalins* whereas those containing inositol are called *inositides*.

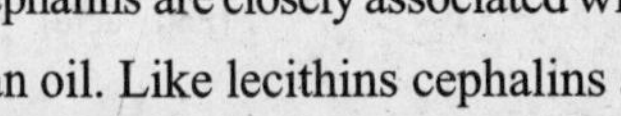

Ethanolamine

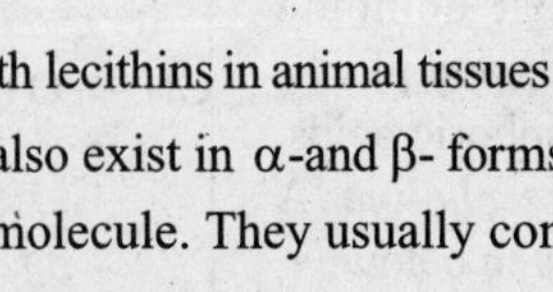

Serine

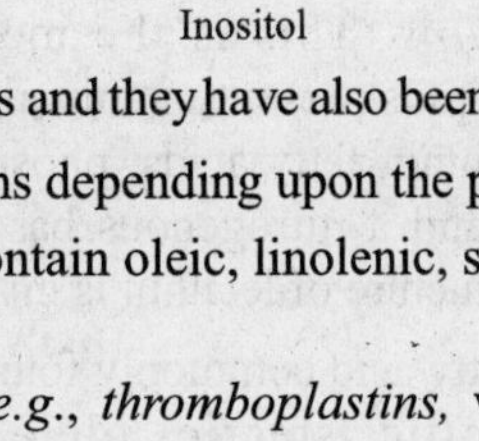

Inositol

The cephalins are closely associated with lecithins in animal tissues and they have also been identified in soybean oil. Like lecithins cephalins also exist in α-and β- forms depending upon the position of two substituting fatty acids in glycerol molecule. They usually contain oleic, linolenic, stearic and arachidonic acids.

Cephalins are components of some important lipoproteins, *e.g.*, *thromboplastins,* which are important in blood coagulation. Cephalins as well as inositids are closely associated with nervous tissue. However their exact functions are not known.

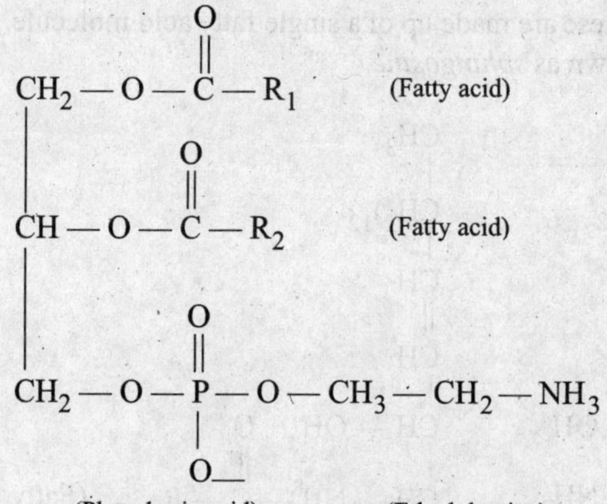

(Phosphoric acid) (Ethadolamine)

Phosphatidyl ethanolamine
(A cephalin)

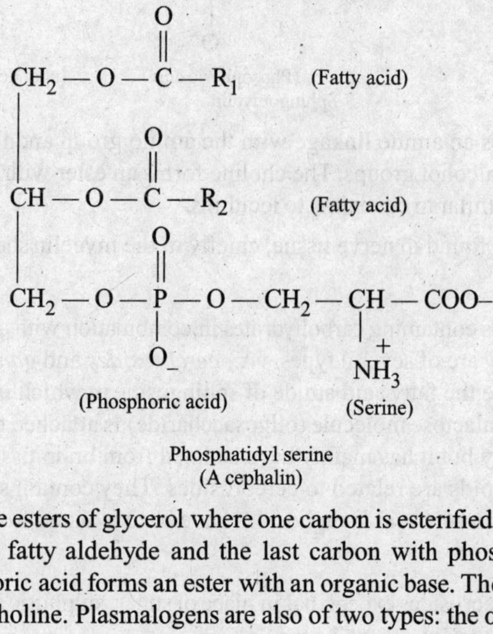

(Phosphoric acid) (Serine)

Phosphatidyl serine
(A cephalin)

Plasmalogens: These are esters of glycerol where one carbon is esterified with a fatty acid, another carbon with a long chain fatty aldehyde and the last carbon with phosphoric acid. Like other phospholipids, the phosphoric acid forms an ester with an organic base. The organic base in the case is either ethanolamine or choline. Plasmalogens are also of two types: the α and β-type as described for lecithins or cephalins.

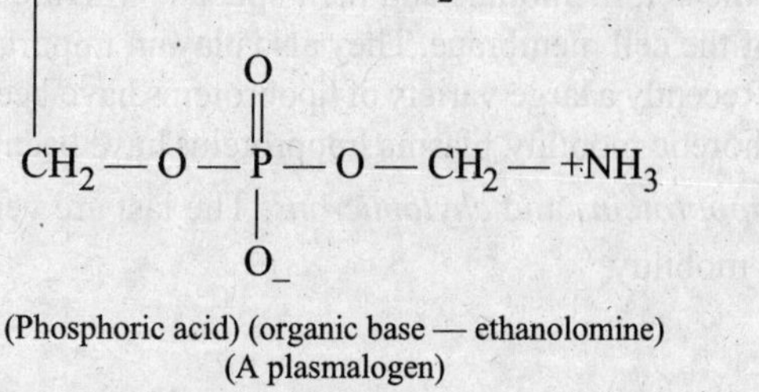

(Phosphoric acid) (organic base — ethanolomine)
(A plasmalogen)

Sphingomyelins: These are made up of a single fatty acid molecule, phosphoric acid, choline and a nitrogenous base known as *sphingosine*.

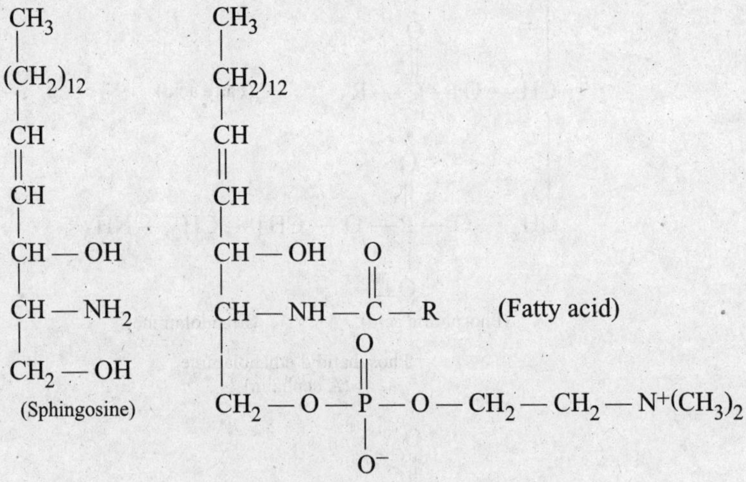

(Sphingosine) (Phosophoric acid) (choline)

Sphingomyelin

The fatty acid makes an amide linkage with the amino group and the phosphoric acid forms an ester with the primary alcohol groups. The choline forms an ester with phosphoric acid (phosphoryl choline) in a manner similar to cephalins to lecithins.

Sphingomyelins are found in nerve tissue, chiefly in the mycelin sheath of the nerve.

Glycolipids

These are complex lipids containing carbohydrates in combination with an ester of long-chain aliphatic acid and alcohols. They are of several types, *viz., cerebrosides* and *gangliosides*.

Cerebrosides: These are the fatty acid amide of sphingosine in which instead of phosphoryecholine, as in spingomyelin, a galactose molecule (oligosaccharide) is attached to the terminal alcohol group. This lipid occurs widely but it has mainly been isolated from brain tissue and hence the name.

Gangliosides: These lipids are related to cerebrosides. They contain sphingosine, long-chain fatty acids, hexoses and nuraminic acid Gangliosides have been isolated from nerve and spleen tissue.

Sulpholipids

These are similar to cerebrosides except that in place of sugar, sulphuric acid is present in the molecule in the form of cerebronic acid.

Lipoproteins

Liporoteins are composed of lipid materials bound with proteins. The lipid component is usually a phospholipids or a cholesterol ester. In the case of phospholipids derivatives, the protein molecule is bound with the highly charged oxygen in the phosphate by non-covalent forces. Lipoproteins are an important component of the cell membrane and their special structure is responsible for the active permeability properties of the cell membrane. They also play an important role in electron transport system of mitochondria. Recently a large variety of lipoproteins have been reported in the plasma. On the basis of their electrophoretic mobility, plasma lipoproteins have been divided into α *-lipoproteins*, β *-lipoproteins*, *pre-* β *-lipoproteins* and *chylomicrons.* The last are very large molecules which do not show electrophoretic mobility.

DERIVED LIPIDS

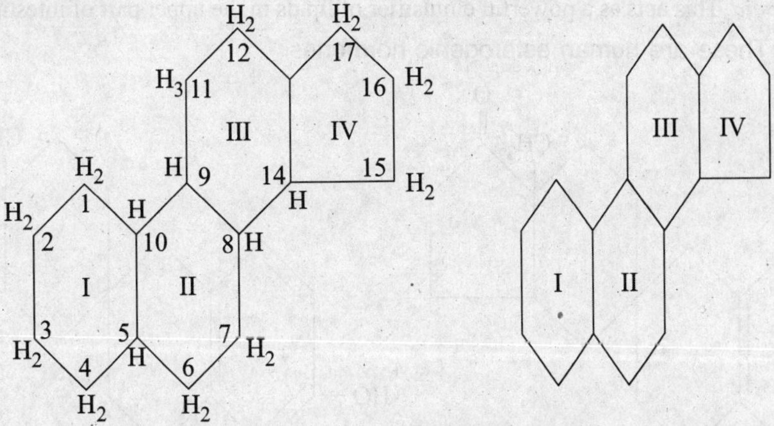

Cyclopentanoperhydrophenanthrene nucleus
with carbon atoms numbered.

The derived lipid group is generally termed as the "catch all" group. These are derivatives of simple or compound lipids but do not have any ester linkage. Most of these are complex compounds having an active hydroxyl group. Derived lipids include steroids, sterols, certain fat soluble vitamins (*e.g.* vitamins *A D*, *E* and *K*), some plant and animal pigments and prostaglandins. A brief description of steroids, sterols and prostaglandins is given below.

Sterols and Steroids

These are widely distributed in living organisms. They are wax-like compounds found both in free forms and also as fatty acid esters. All sterols and steroids have in common the base ring structure which is known as cyclopentanoperhydrophenanthrene nucleus.

In the structure shown on the prevous page, the 1st, IInd and IIIrd rings constitute a perhydrophenanthrene ring whereas the IVth is a cyclopentane ring. In most sterols and steroids a side chain is attached to the carbon atom.

Both sterols and steroids have the same cyclopentanoperhydrophenanthrene. However, sterols have a long side chain and a single functional hydroxyl group.

Steroids: Steroids have a wide range of biological activity. Some steroids are fat emulsifer fund in bile (cholic acid): others are important hormones such as sex hormones (estrogen, progesteron, testosteron, etc.) and hormones of adrenal cortex (glucocorticoids and mineralocorticoid) — (see Chapter 11). The structure, name and properties of some of the steroids are given below:

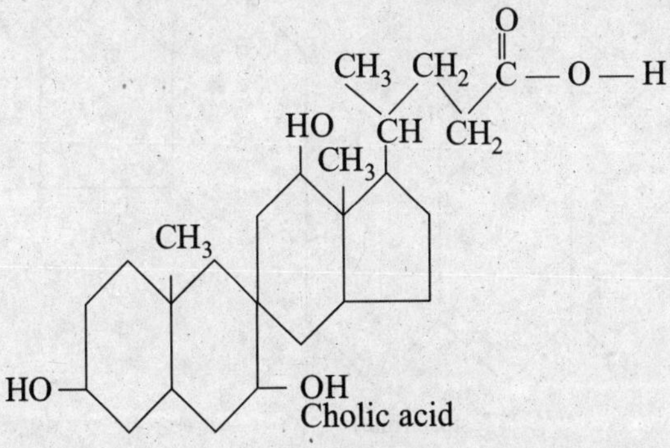

Cholic acid

Cholic acid

Cholic acid is found in bile in the form of sodium salts. As it is found in bile, it is sometimes also called as bile *acid*. This acts as a powerful emulsifier of lipids in the upper part of intestine.

Estrogens: These are human esterogenic hormones

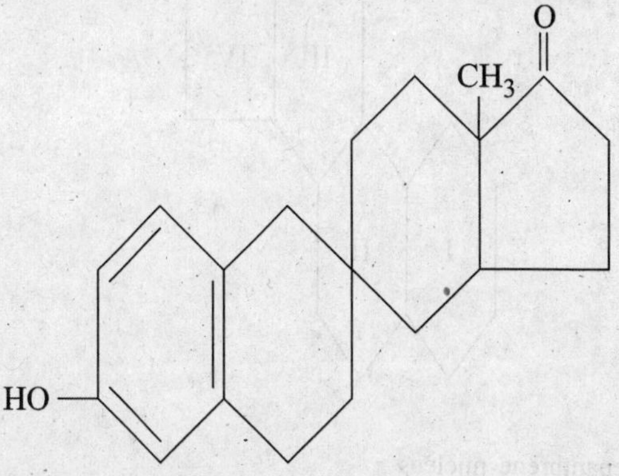

(Estrone)

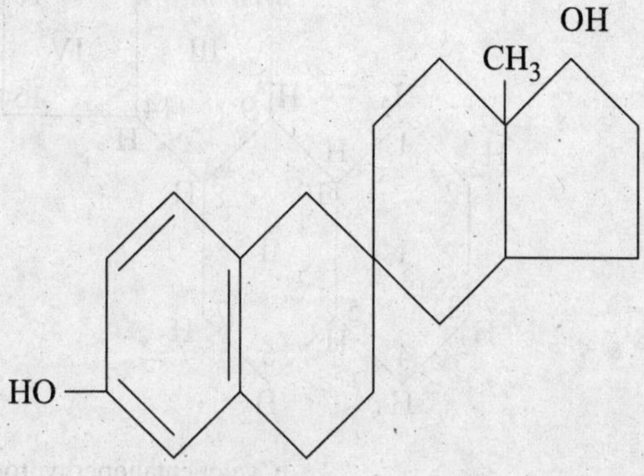

Esteradiol 17-β

Progesterone:

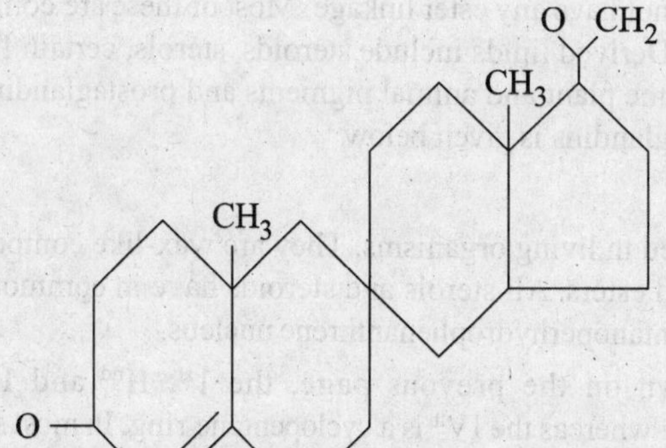

(Progesterone)

This is the human pregnancy hormone and is secreted by corpus luteum.

Testosterone:

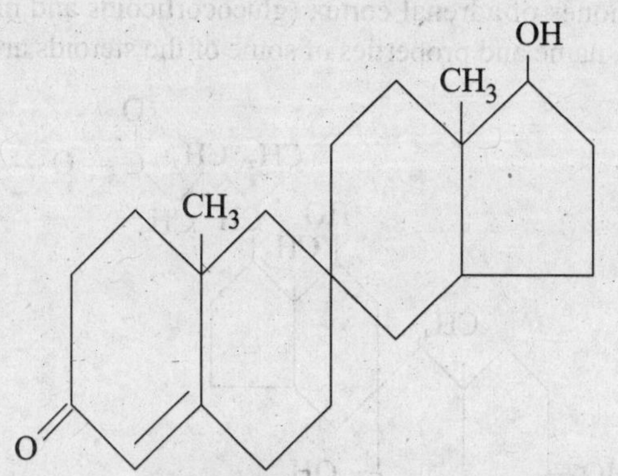

(Testosterone)

This is the male sex hormone which regulates the development of secondary sexual characters and gonads.

Cortisol (*Glucocorticoid*):

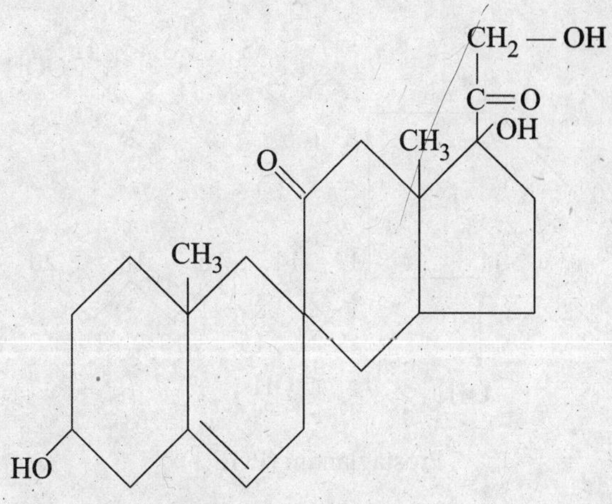

(Cortisol)

This is one of the important cortical hormones. It controls the carbohydrate metabolism but is also effective in relieving the symptoms of rheumatoid arthritis.

Aldosterone (*Mineralocorticoid*):

This hormone controls electrolite balance in body and is secreted by adrenal cortex.

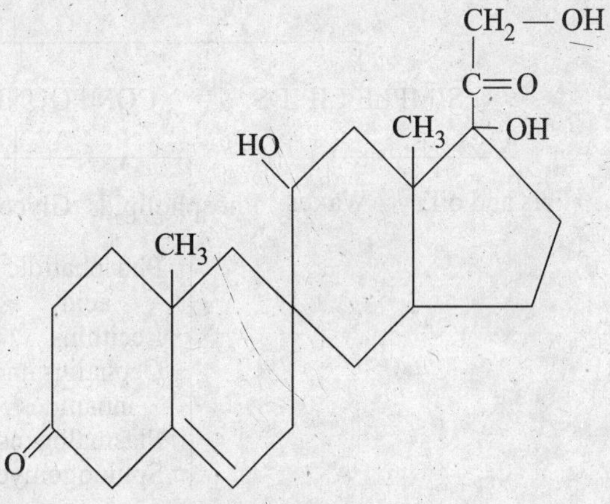

(Aldosterone)

Sterols

Sterols are alcohols of steroids. The common examples of sterols are *cholesterol found in animals* and *ergosterol* and *stigmasterol* found in plants.

Cholesterol

Cholesterol is found in nearly all vertebrates. It has been isolated in great quantities from the extracts of brain and spinal chord, and is one of the main constituents of gallstones (stones found in gall bladder). Cholesterol also causes hardening of arteries *arterosclerosis.* In tissue it is found both in free and combined state. Cholesterol is the precursor of all the steroids found in the body. 7-dehy-drocholesterol is found in the skin where it is responsible for the synthesis of vitamin D in the presence of sunlight.

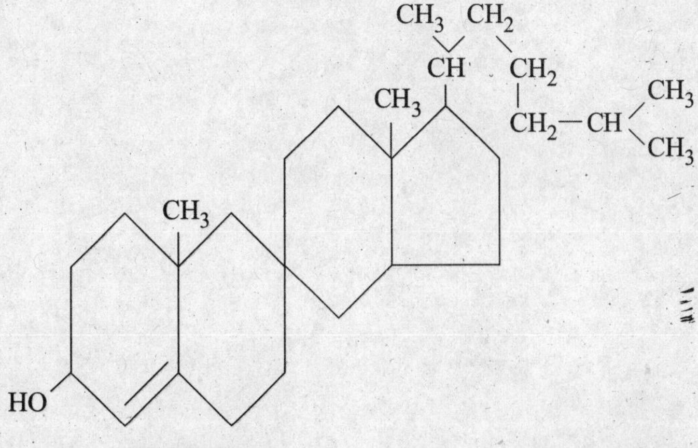

(Cholesterol)

Prostaglandins

These were first discovered in the human semen in 1933. However, now it has been reported in a variety of tissues. Many different varieties of prostaglandins are known. These are hydroxyl derivatives of polyunsaturated fatty acids.

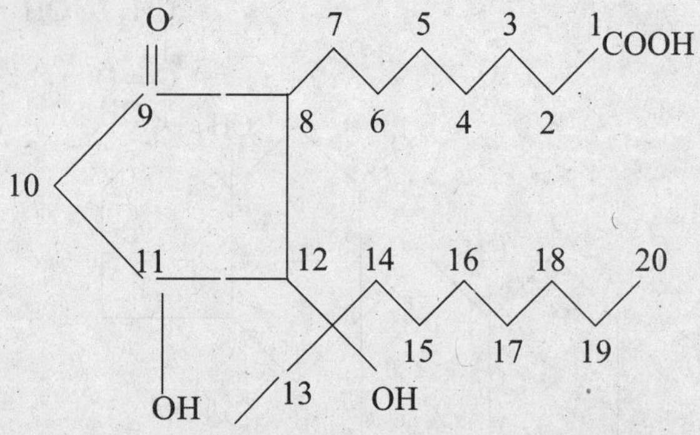

Prostaglandin (PGE 1)

CLASSIFICATION OF LIPIDS

LIPIDS

- SIMPLE LIPIDS
  - Fats and oil
  - Waxes
- COMPOUND LIPIDS
  - Phospholipids
    - Phosphatidic acid
    - Lecithins
    - Cephalins and inositides
    - Plasmalogens
    - Sphinogomyelins
  - Glycolipids
    - Cerebrosides
    - Gangliosides
  - Sulpholipids
- DERIVED LIPIDS
  - Steroids
  - Sterols
  - Prostaglandins

5 Chapter

Nucleic Acids and Protein Synthesis

Deoxyribonucleic (DNA) and ribonucleic (RNA) acids, referred to as the nucleic acids, are long-chain polymers built by a large number of units called nucleotides. The hereditary (genetic) units in all forms of life are called genes. These, in fact, are molecules of deoxyribonucleic acids. However, in most plant viruses, certain bacteriophages, viroids and virosoids (satellite RNAs) the hereditary material is ribonucleic acid termed as ***genetic RNA***. Biochemists have synthesized DNA molecules by putting together the nucleotides into the sequence in which they occur in natural genes. A gene faithfully copies itself (DNA replication) at the time of every cell division. Gene function is identified with the formation of proteins of definite structure. During this process RNA plays an important role.

GENERAL CONSIDERATIONS

In bacteria (*prokaryons = no nucleus exists*) and higher cells (*eukaryons* = with nucleated cells) both DNA and RNA are present. However, some viruses, *Tobacco Mosaic Virus* (TMV), poliomyeletis virus, and certain others contain only RNA; whereas bacteriophages, vaccinia and adenoviruses contain DNA alone. Deoxyribonucleic acid is present in the nucleus and forms part of the chromosomes. In nucleus of plants and animals DNA is present in combination with chromosomal proteins forming *nucleoproteins*. However, in mycoplasmas and bacteria the DNA does not appear to be associated with any other molecule. Deoxyribonucleic acid is also present in mitochondria, chloroplasts and probably in other self-replicating cytoplasmic organelles, which show extranuclear, extrachromosomal or cytoplasmic inheritance. Ribonucleic acid is found both in the nucleus and the cytoplasm. In the nucleus RNA is present in the nucleolus, nucleoplasm, and chromatin; in the cytoplasm it forms a large part of the ribosomes.

Both the nucleic acids are organic polymers composed of repeating units of *nucleotides*. Every nucleotide is made up of an organic nitrogenous base, a pentose sugar, and phosphate. Within the nucleotide the combination of a pentose with a nitrogen base constitutes a *nucleoside* (Fig. 5.1).

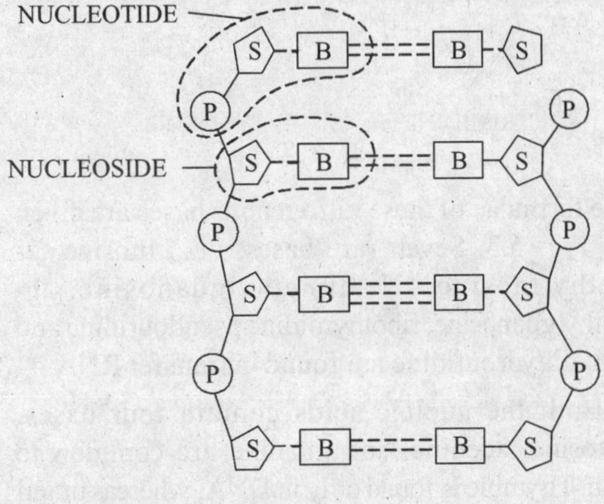

Fig. 5.1. Part of a polynucleotide chain of DNA showing the two strands connected by hydrogen bonds. The letters B, S, P represent the nitrogenous base, the pentose sugar, and the phosphate groups, respectively.

The two nucleic acids – DNA and RNA are distinguished on the basis of the pentose sugar component and by the differences in their nitrogenous bases. The pentose sugars concerned are D-ribose (or ribose) characteristic of RNA and deoxy-D-ribose (or deoxyribose) characteristic of DNA. The structural difference between the two sugars is in the number 2 carbon. In deoxyribose the oxygen on carbon 2 is lacking (Fig. 5.2).

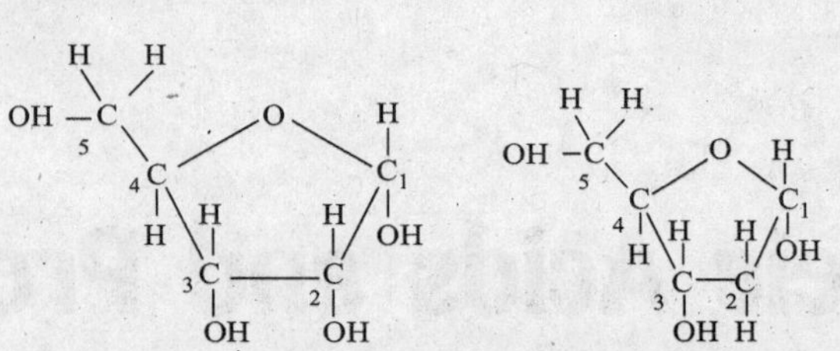

Fig. 5.2. The two pentose sugars present in the RNA and DNA

The nitrogenous bases are of two types, *viz.*, purines (double ring molecules) pyrimidines (single-ring molecules). The most commonly occurring purines are adenine and guanine; the pyrimidines are thymine, cytosine and uracil. These are conventionally abbreviated as A, G, T, C and U, respectively.

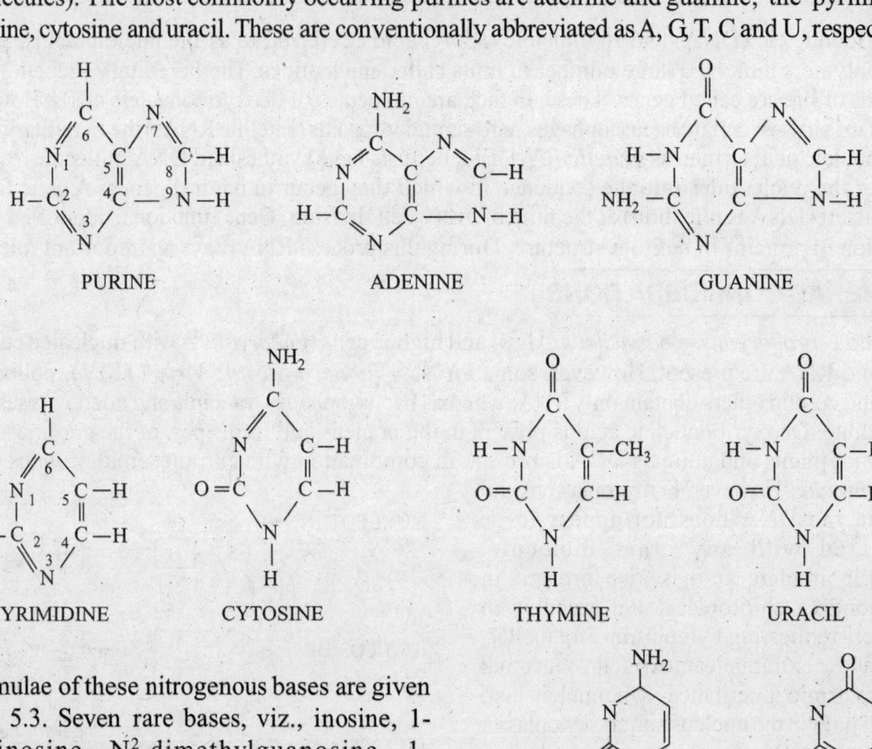

The formulae of these nitrogenous bases are given in Fig. 5.3. Seven rare bases, viz., inosine, 1-methylinosine, N^2-dimethylguanosine, 1-methylguanosine, ribothymidine pseudouridine and 5, 6, dihydrouridine are found in transfer RNA.

Both the nucleic acids contain four bases. Cytosine, adenine, and guanine are common to both. Thymine is found only in DNA, whereas uracil is exclusive to RNA. Thus, eight different nucleotides are commonly found in the nucleic acids (Table 5.1). The general structure of the nucleotides is shown in Fig. 5.4.

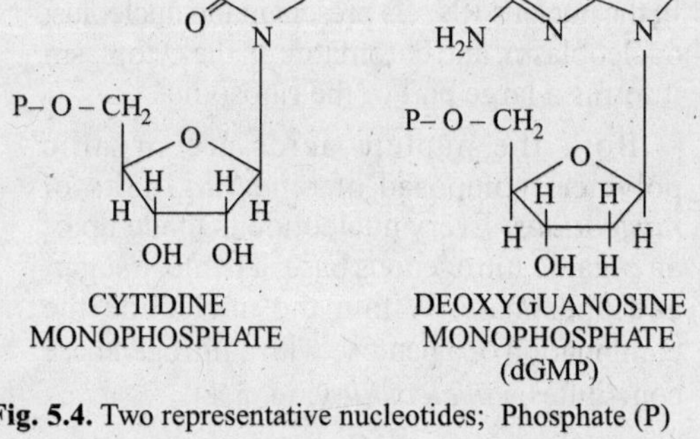

Fig. 5.4. Two representative nucleotides; Phosphate (P)

Table 5.1 Nucleotides of DNA and RNA

Base-sugar-phosphate	*Chemical Name*	*Nucleic Acid*
Adenine-ribose-phosphate	Adenosine monophosphate (AMP)	RNA
Adenine-deoxyribose-phosphate	Deoxyadenosine monophosphate (dAMP)	DNA
Guanine-ribose-phosphate	Guanosine monophosphate (GMP)	RNA
Guanine-deoxyribose-phosphate	Deoxyguanosine monophosphate (dGMP)	DNA
Cytosine-ribose-phosphate	Cytidine monophosphate (CMP)	RNA
Cytosine-deoxyribose-phosphate	Deoxycytidine monophosphate (dCMP)	DNA
Uracil-ribose-phosphate	Uridine monophosophate (UMP)	RNA
Thymine-deoxyribose phosphate	Thymidine monophosphate (dTMP)	DNA

STRUCTURE OF DNA

The molecular organization DNA was explained by Watson and Crick, a concept for which James D. Watson of the United States, Francis H.C. Crick and Maurice H.F. Wilkins of England shared the Nobel Prize of 1962. According to Watson – Crick model the DNA molecule consists of two parallel polynucleotide strands twisted around each other like a 2-ply wool to form a *double-stranded helix* or spiral (Fig. 5.5a). The DNA molecule exhibits *plectonemic coiling* of two hydrogen-bonded antiparallel strands wound into a right-handed configuration constituting a *double helix*. A plectonemic spiral is

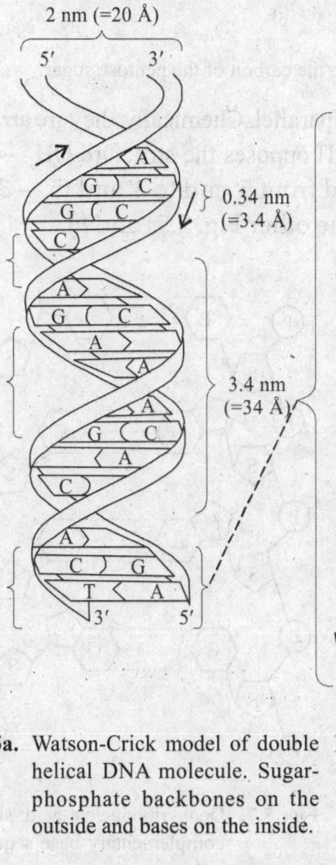

Fig. 5.5a. Watson-Crick model of double helical DNA molecule. Sugar-phosphate backbones on the outside and bases on the inside.

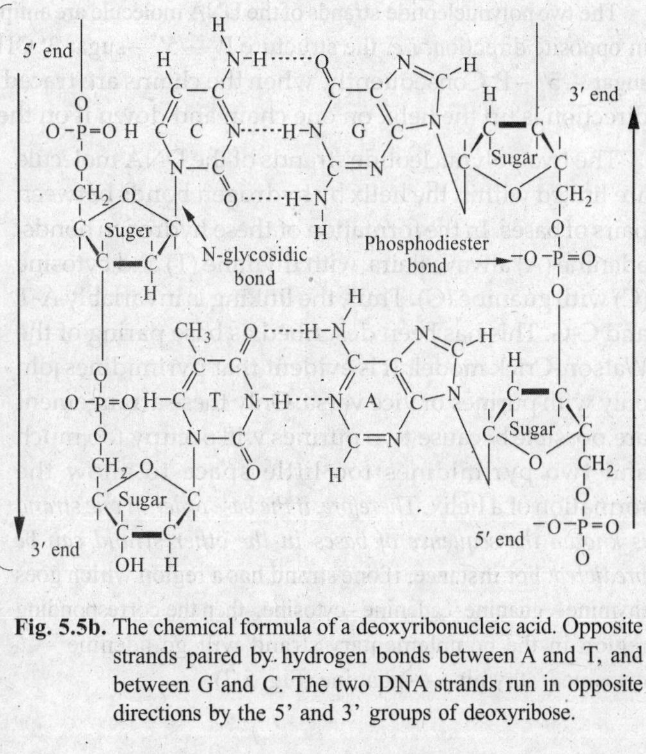

Fig. 5.5b. The chemical formula of a deoxyribonucleic acid. Opposite strands paired by hydrogen bonds between A and T, and between G and C. The two DNA strands run in opposite directions by the 5' and 3' groups of deoxyribose.

formed by two parallel threads which coil in the same direction about one another and cannot be separated unless uncoiled; antiparallel strands of DNA helix is that in which one strand is oriented left to right 5′ to 3′ and the complementary strand is oriented right to left 3′ to 5′; details are given in text and Fig. 5.5 a and b.

Both polynucleotide strands remain separated by a distance of 2 nm (= 20Å). A complete turn (a *pitch*) of the helix occurs every 3.4 nm (= 34 Å). As each nucleotide occupies 0.34 nm (= 3.4 Å) distance along the length of a polynucleotide strand, ten mononucleotides occur per complete turn. It follows, therefore, that the base pairs are rotated 36° with respect to each adjacent pair. The helix has two external grooves, namely, a *minor groove* and a *major groove* (Fig. 5.5a).

In each of the two strands the adjacent nucleotides are linked with each other by sugar-phosphate bonds. Each phosphate links the 5-carbon atom on one sugar to 3-carbon atom on the sugar of the next nucleotide forming 5′ – 3′ *phosphodiester bonds*, while the other strand has phosphodiester linkage in just reversed order or 3′ – 5′ direction. (Fig. 5.5b). The nitrogenous bases are attached through one of their nitrogens by an *N-glycosidic* bond to carbon 1′ of the pentose sugar (Fig. 5.5b.)

The schematic diagram of nucleotide chain as shown below (Fig. 5.6) is frequently useful; it can be drawn either in a horizontal or vertical direction.

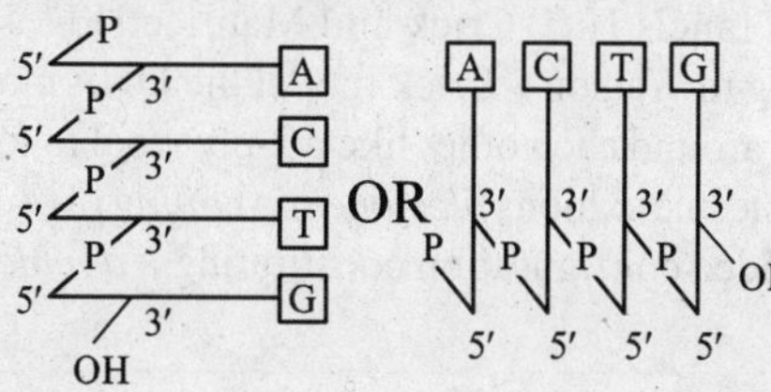

Fig. 5.6. Schematic diagrams of nucleotide chain. 3′ to 5′ are the carbon of the pentose sugar.

The two polynucleotide strands of the DNA molecule are antiparallel. Chemically, they are arranged in opposite direction, *i.e.* the structure P — 5′ —sugar 3′ OH opposes the structure OH — 3′ — sugar —5′ —P. Consequently, when the chains are traced from 5′ end to 3′ and (5′ —3′) the direction is up the helix on one chain and down it on the other Fig. 5.5*a* and *b*).

The two polynucleotide strands of the DNA molecule are linked within the helix by hydrogen bonds between pairs of bases. In the formation of these hydrogen bonds, adenine (A) always pairs with thymine (T) and cytosine (C) with guanine (G). Thus, the linking is invariably A-T and C-G. This has been described as base paring of the Watson-Crick model. It is evident that pyrimidines join only with purines or vice versa. Only these arrangement are possible because two purines will occupy too much and two pyrimidines too little space to allow the formation of a helix. *Therefore, if the base order in one strand is known the sequence of bases in the other strand can be predicted*. For instance, if one strand has a region which goes thymine – guanine – adenine –cytosine, then the corresponding region in the complementary strand will go adenine — cytosine — thymine — guanine (Fig. 5.7).

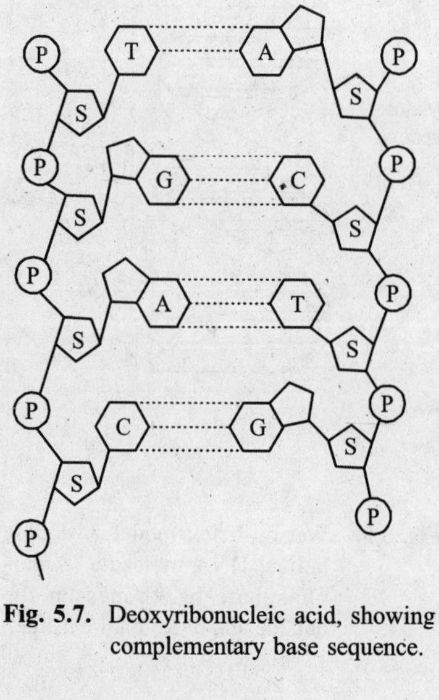

Fig. 5.7. Deoxyribonucleic acid, showing complementary base sequence.

The hydrogen bonds, described above, are formed by the sharing of hydrogen atoms of the members of a base. In case of A-T linkage there are two hydrogen bonds, whereas in the C-G linkage there are three such bonds (Fig. 5.5b and 5.7). The hydrogen bonds holding the purines and pyrimidines are weaker than the bonds holding these to the sugar-phosphate groups. Therefore, the DNA molecule easily splits into its two strands.

In short, the full structure of DNA resembles a winding stair-case, with the sugar and phosphate molecules of the nucleotides forming the railings and the linked nitrogen bases (either adenine-thymine or guanine-cytosine) forming the steps (Figs. 5.5a).

Deoxyribonucleic acid molecules may be very large. For example, in the bacterium *Escherichia coli,* DNA is about 1000μm (1mm) in length, and contains about 40,00,000 base pairs. Plant and animal cells may possess DNA molecules which are even longer.

The types of DNA described above is the B-from (B-DNA) that occurs under hydrated conditions and is thought to be the principal biological conformation. A modification of B-form, known as A-form (A-DNA) occurs under less hydrated conditions. Like the B-form, it too is a right-handed double helix, however, it is more compact, with 11 base pairs per turn of the helix. The bases of A-DNA are tilted 20 Å away from perpendicular and displaced leterally in relation to the diad axis. A third from of DNA, called Z-form (Z-DNA) also exists. It is a left-handed helix. It has 12 base pairs per turn of the helix and presents a zigzag conformation (hence the symbolic name). Unlike the B-DNA, Z-DNA is antigenic. V. Sarisekharan (1976-1980) proposed *RL model* of double heiical B-DNA. In this model the B-DNA structure is visualized as alternating right-hand (R) and left-hand (L) segments of five base pairs in a repeat of ten base pairs.

The different forms of double helical DNA represent *linear* DNA molecule. DNA may have a closed structure without any free ends both in prokaryotes and eukaryotes forming circular or large-looped conformations. The DNA in such cases becomes *supercoiled.* The torsion produced by supercoiling can be eased by a break in one of the two strands; such DNA molecule is said to be *relaxed.*

From the base-pairing rule it follows that the A : T (A/T) and G : C (G/C) ratio should always be 1 : 1 on any DNA molecule. But the (A + T) : (G + C) (or A + T)/(G + C) ratio is not always one. Higher plants and animals have an excess of A + T over (G + C) in their DNA content. For example, the (A + T) : (G + C) in wheat is 1.22, in man it is 1.40. But among the viruses, bacteria, and lower plants there is much variation; both (A + T) – rich and (G + C) – rich DNAs have been reported. For instance, in *Pseudomonas aeruginosa* and *Bacillus megaterium* the (A + T) : (G + C) are 0.51 and 1.66, respectively.

DNA REPLICATION

The sequence of nucleotides on the DNA strand determines the nature of the offspring. Therefore, the exact sequence of nucleotide bases in DNA of an organism is transmitted unchanged from one generation to the next. To ensure this, DNA must be copied very exactly during the life cycle of a cell before division. This process is known as *replication.*

When DNA is replicated the nucleotide strands separate at one end by the breaking of the hydrogen bonds between the bases. This separation signals the synthesis of two daughter strands complementary to the two parent strands. The synthesis of the new chains goes on hand in hand with unwinding of the old chain. The model was termed *semiconservative replication, i.e.,* one strand of a parent DNA molecule is conserved in each daughter molecule (Fig. 5.8 *a-e*). Each new molecule of DNA is known as hybrid duplex.

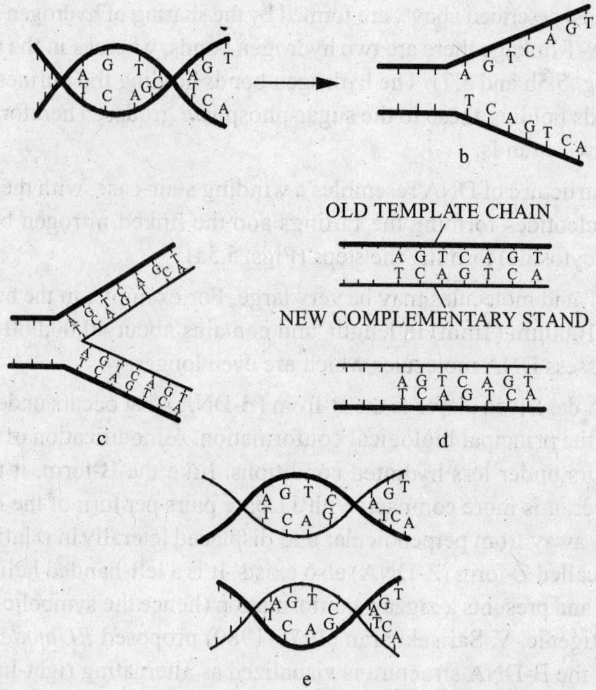

Fig. 5.8. Semiconservative replication of DNA molecule, *a*, A portion of the DNA molecule; *b*, separation of two strands; *c* the formation of a new strand with deoxynucleotides; *d*, formation of two daughter strands; *e*. each of the two new DNA strands twist around the parent molecule to form the double helix.

Another model of DNA replication, known as *conservative replication*, was proposed in which both old parental strands are retained in one sibling (daughter) cell, while the other gets the two newly synthesized strands. A third type of replication, called *dispersive replication*, was also proposed in which parental and newly synthesized daughter molecule are interspersed in an essentially random fashion. However, there is no evidence that conservative and dispersive kind of replications actually take place.

M. Meselson and F.W. Stahl in 1958 demonstrated semiconservative replication of DNA in prokaryotes (*Escherichia coli*) by using heavy nitrogen (N^{15}) labelled DNA and, also, by another principle utilizing caesium chloride density gradient. Semiconservative replication was also demonstrated by J. Cairns (1963) using the technique of autoradiography in bacterial chromosome labelled with tritiated thymidine (H^3 – TdR). Semiconservative replication in eukaryotes (*Vicia faba*) was also demonstrated by J.H. Taylor by autoradiography experiment in chromosomes labelled with H^3 – TdR. Several other experiments too proved the validity of semiconservative mode of replication of the DNA molecule. One such technique was the use of *harlequin chromosomes* in Chinese hamster ovary (CHO) cell. The method involves two rounds of replication in a 5-bromodeoxyuridine (BUDR) solution and thereafter, staining the replicated chromosomes with a fluorescent dye and Giemsa stain. This produces harlequin chromosomes in which the two sister chromatids stain differentially, demonstrating semiconservative replication.

The molecular mechanisms accounting for semiconservative replication are complex. Nevertheless, the first major breakthrough occurred in 1957 when A. Kornberg discovered an enzyme called DNA *polymerase* that was shown to have the capacity to polymerize the nucleotides of dATP (*deoxyadenosine triphosphate)*, dTTP (deoxythymidine triphosphate), dCTP (*deoxycytidine triphosphate*) and dGTP (*deoxyguanosine triphosphate*), collectively designated as dNTP (deoxynucleoside 5′ triphosphates), into chains of DNA in the presence of Mg^{++}. This was accomplished *in vitro* (*i.e.*, outside the living cell) as shown in Fig. 5.9.

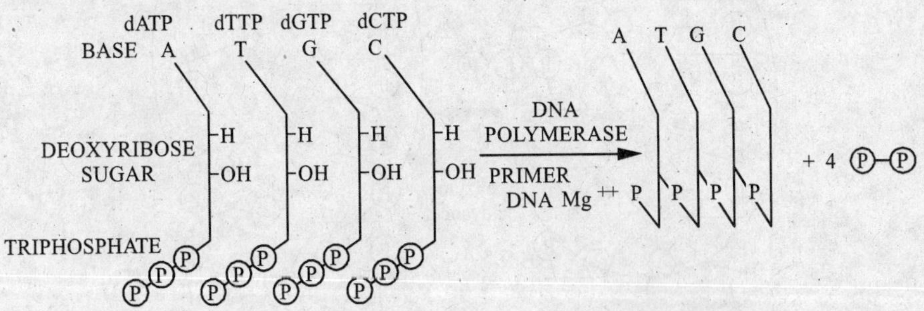

Fig. 5.9. Kornberg's *in vitro* system of nucleotide chain formation

Kornberg's enzyme, now called DNA polymerase I (pol I), is no longer considered to be the main synthesizing enzyme because bacterial mutants lacking this enzyme synthesize DNA. Polymerase I (pol I) now considered to be concerned mainly with DNA repair. DNA polymerase called pol III is essentially meant for DNA synthesis.

A model of prokaryotic DNA polymerase I enzyme was given by Kornberg (1974). It has five different sites (Fig. 5.10a): (i) a site for a portion of the DNA template strand, (ii) a site for the growing chain, the *primer*, (iii) a site with special recognition for the 3′ –OH of terminal nucleotide of the primer, the *primer terminus,* (iv) a site for a deoxynucleo-side triphosphate, and (v) a site for the 5′ – 3′ exonucleolytic cleavage or removal of a mispaired nucleotide. A schematic representation of polymerization of nucleotides on the primer site of DNA polymerase I is shown in (Fig. 5.10b and c).

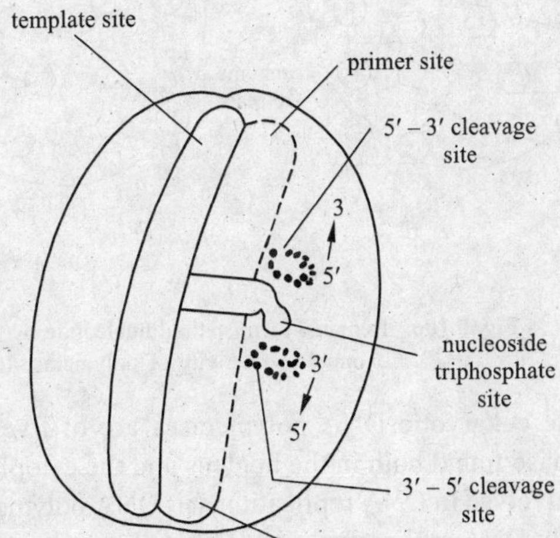

Fig. 5.10a. A model of prokaryotic polymerase-I enzyme, showing its five different sites.

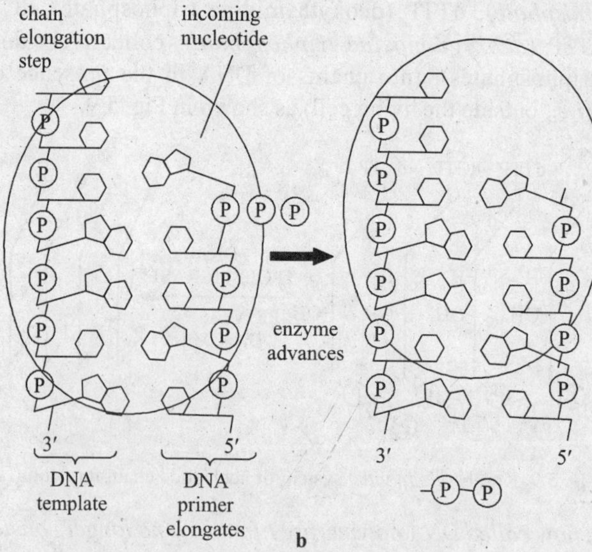

Fig. 5.10b. Mode of polymerization of nucleotides on the primer site of polymerase-I

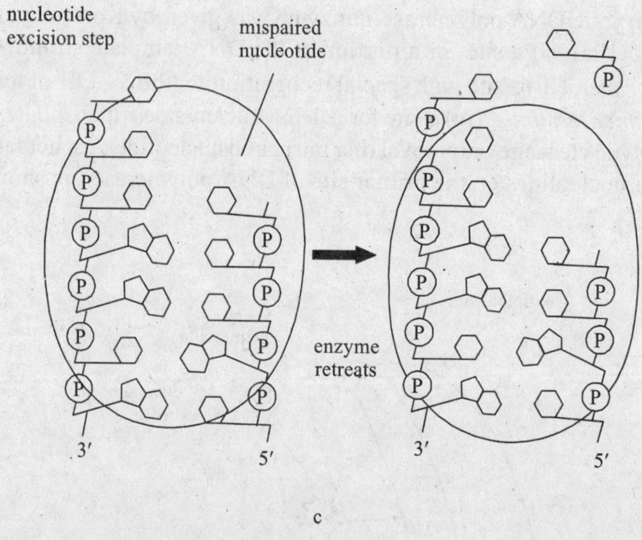

Fig. 5.10c. Removal of mispaired nucleotide by exonuclease activity of polymerase-II

Unlike prokaryotes, the eukaryotic DNA polymerases are of five different types: (i) DNA polymerase α (alpha) primase found both in the nucleus and the cytoplasm; it is the predominant DNA polymerase mainly involved in DNA replication, (ii) DNA polymerase β (beta) found only in vertebrates and takes part in DNA replication, (iii) DNA polymerase γ (gamma), a mitochondrial polymerase, (iv) DNA polymerase δ (delta), found in mammalian cells and (*v*) DNA polymerase ε (epsilon); it has been reported from mammalian **HeLa** cells (cells originating from a cervix carcinoma patient named **He**nrietta **La**cks).

Two fundamental properties are associated with all known polymerases. These are: (i) they all synthesize DNA only in the 5′ to 3′ direction, adding a dNTP to the 3′ hydroxyl group of a growing chain and (ii) all polymerases can add a new deoxyribonucleotide only to a preformed primer strand that is hydrogen-bonded to the template (Fig. 5.11). It may be emphasized here that polymerases are unable to initiated DNA synthesis *de novo* by catalyzing the polymerization of free dNTPs. The two properties of the polymerases are of great significance for DNA synthesis.

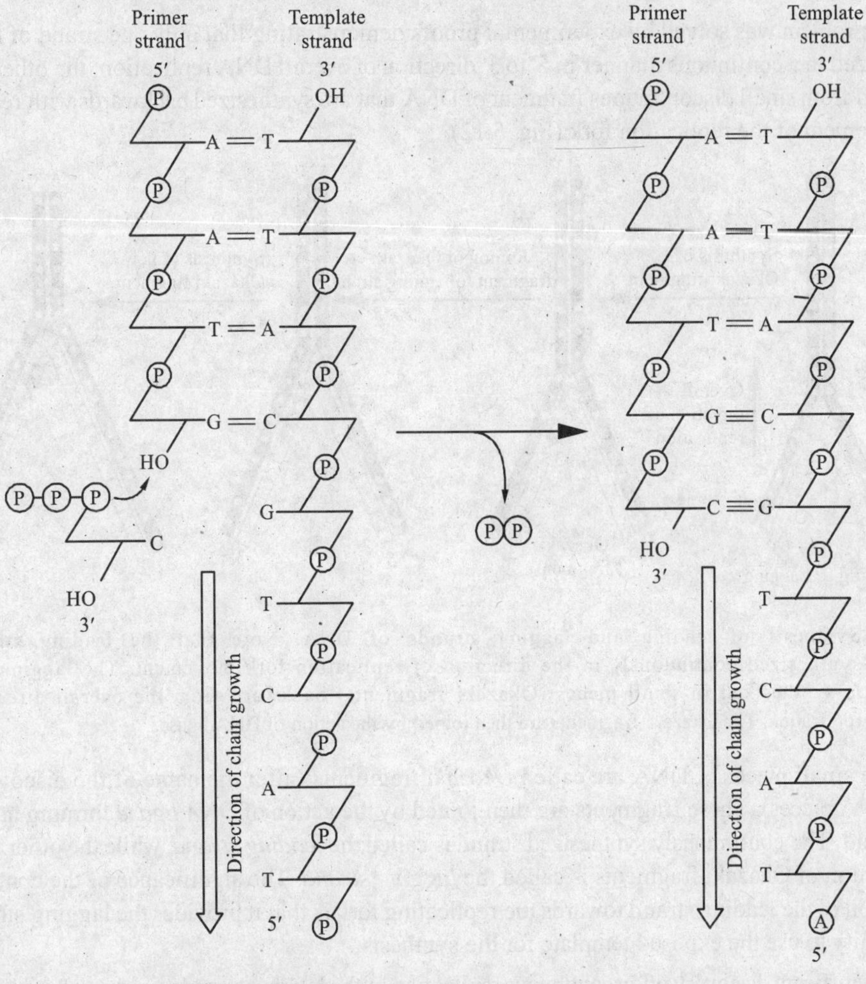

Fig. 5.11. All known DNA polymerase add a deoxyribonucleoside 5′- triphosphate to the 3′ hydroxyl group of a growing DNA chain (the primer strand).

At this Juncture, the reader may be reminded to the autoradiographic studies in ^3H – thymidine treated *E. coli*, confirming semiconservative replication. Cairns observed complete circular DNA molecule in the process of replication. These DNA molecules contained *two replicating forks* representing the sites of active DNA synthesis. At each fork the parental strands of DNA separated and two new daughter strands were synthesized.

The replication of both prokaryotic and eukaryotic DNAs starts at a unique sequence of base pairs of DNA called the *origin of replication* (*Ori*), which serves as a specific binding site for proteins that initiate the replication process. Only in vitro study of *Ori in E. coli* is briefly discussed below.

The synthesis of new DNA strands complementary to both strands of DNA template, as observed by Cairns, raised problems to explain DNA replication. It is known that the two strands of helical DNA are antiparallel and, therefore, synthesis of new strands at the replication fork will require that one strand is synthesized in 5′ to 3′ direction and the other in 3′ to 5′ direction. But as described above, DNA polymerase always catalyses the polymerization of dNTPs only in the 5′ to 3′ direction. The question arises, how can the other new strand of DNA be synthesized?

This problem was solved by experimental proofs demonstrating that only one strand of DNA is synthesized in a continuous manner in 5′ to 3′ direction of overall DNA replication; the other strand is formed from small discontinuous fragment of DNA that are synthesized backwards with respect to the movement of the replication fork (Fig. 5.12).

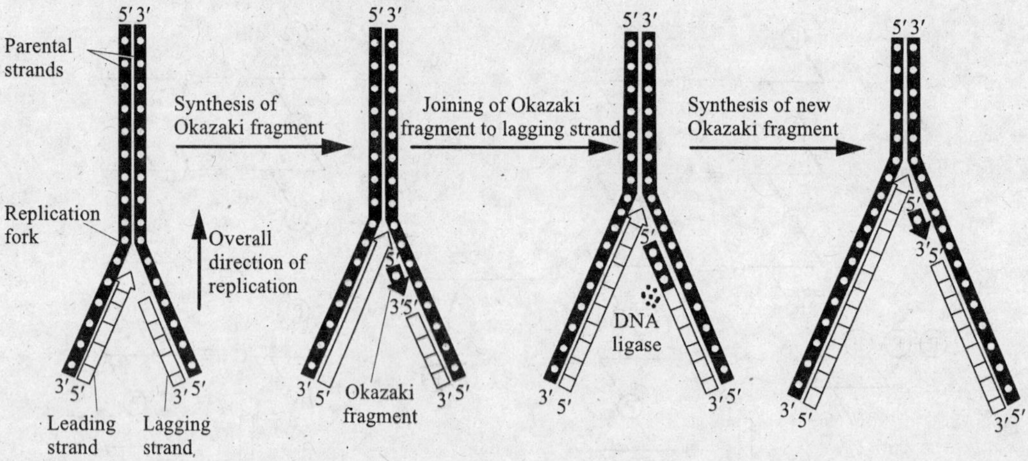

Fig. 5.12. Synthesis of leading and lagging strands of DNA. Note that the leading strand is synthesized continuously in the direction of replication fork movement. The lagging strand is synthesized in small pieces (Okazaki fragments) backward from the overall direction of replication. The Okazaki fragments are then joined by the action of DNA ligase.

These small pieces of DNA are called *Okazaki* fragments (after the name of the discoverer of these DNA pieces). These fragments are then joined by the action of *DNA ligase* forming an intact new strand. The continuously synthesized strand is called the *leading strand*, while the other formed by the union of Okazaki fragments is called the *lagging strand.* The significance of the continuous elongation of the leading strand towards the replicating fork is that it provides the lagging strand an opportunity to use the exposed template for the synthesis.

The significant discovery of discontinuous replication of the lagging strand raised another important question. As already discussed, the DNA polymerase requires a primer and cannot initiate synthesis *de novo*, how then is the synthesis of Okazaki fragments initiated? This answer is that short pieces of DNA (about 3 – 10 nucleotide long) serve as primers for DNA replication (Fig. 5.13). In contrast to DNA synthesis, the RNA synthesis can initiate *de novo*. This provides an opportunity for an enzyme called *primase* to synthesize short fragments of RNA to the lagging strand template at the replication fork. The Okazaki fragments are then synthesized via extension of these RNA primers by DNA polymerase. This would predict that the newly synthesized Okazaki fragments contain an RNA – DNA hybrid. The existence of such RNA – DNA hybrids has been experimentally demonstrated, providing evidence for the role of DNA primers in DNA replication (Fig. 5.13).

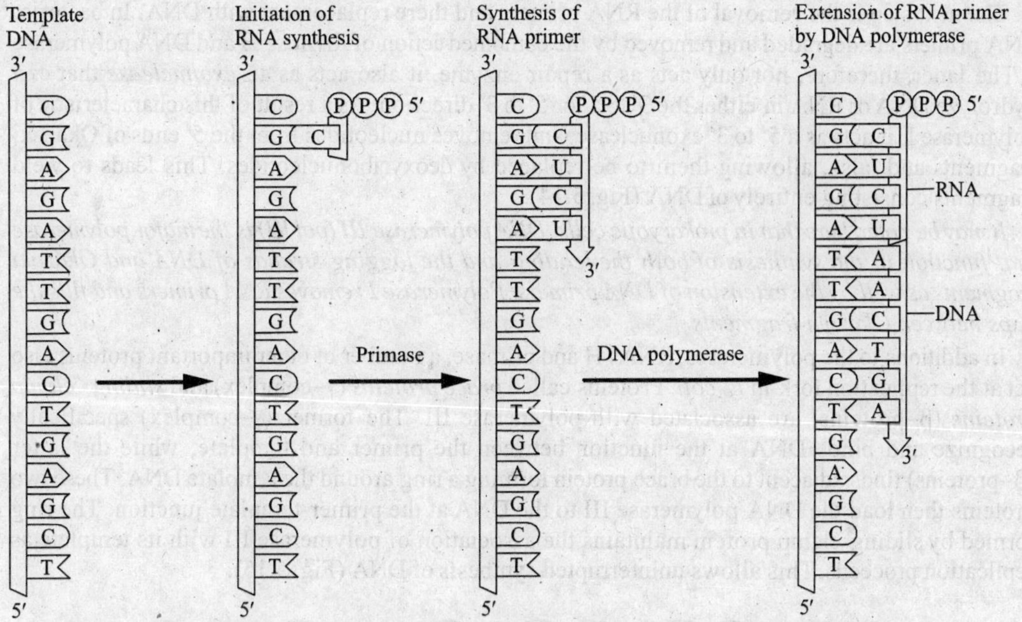

Fig. 5.13. Initiation of Okazaki fragments with RNA primers. Short fragments of RNA serve as primers that can be extended by DNA polymerase

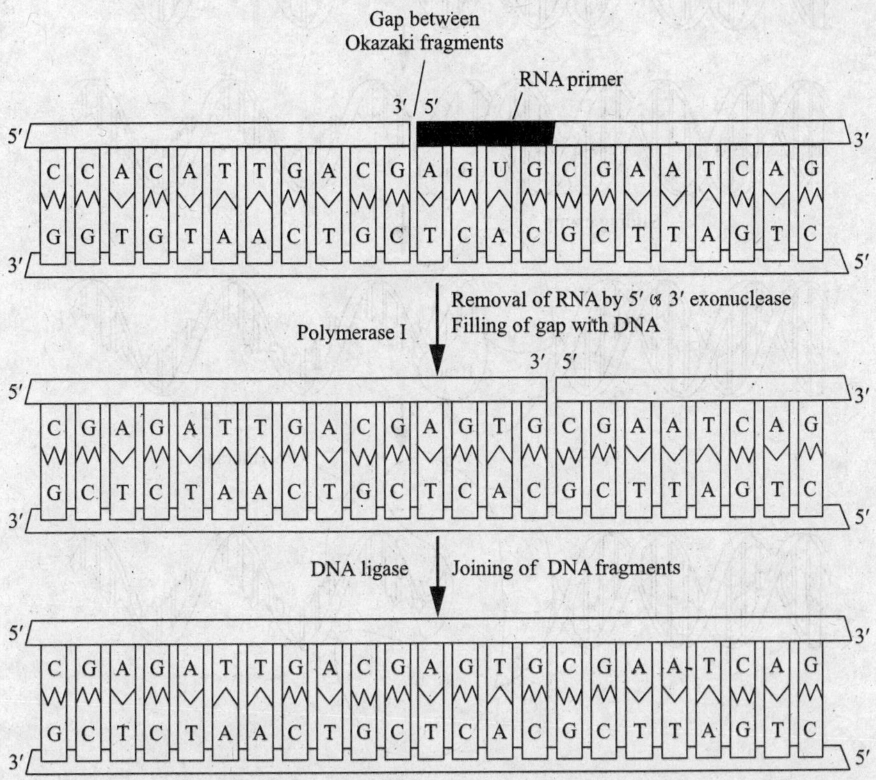

Fig. 5.14. Removal of RNA primers and joining of Okazaki fragments. The 5′ — 3′ exonuclease activity of DNA polymerase I removes RNA primers and fills the gap between Okazaki fragments with DNA. The DNA fragments are than joined by DNA ligase.

The next step is the removal of the RNA primers and there replacement with DNA. In bacteria, RNA primers are degraded and removed by the combined action of *RNAase H* and DNA polymerase I. The latter, therefore, not only acts as a repair enzyme, it also acts as an *exonuclease* that can hydrolyze DNA or RNA in either the 3′ to 5′ or 5′ to 3′ direction. As a result of this characteristic of polymerase I, it acts as a 5′ to 3′ exonuclease and removes nucleotides from the 5′ ends of Okazaki fragments and, also, allowing them to be replaced by deoxyribonucleotides. This leads to yield fragments consisting entirely of DNA (Fig. 5.14).

It may be noted here that in prokaryotic cells, DNA polymerase III (pol III) is the major polymerase that function in the synthesis of both the leading and the lagging strands of DNA and Okazaki fragments as well by the extension of DNA primers. Polymerase I removes RNA primers and fills the gaps between Okazaki fragments.

In additions to the polymerases, RNAs H and primase, a number of other important proteins also act at the replication fork in *E. coli*. Proteins called *brace proteins* (γ–complex) and *sliding –clamp proteins* (β–proteins) are associated with polymerase III. The former (γ–complex) specifically recognize and binds DNA at the junction between the primer and template, while the latter (β–proteins) finds adjacent to the brace protein forming a ring around the template DNA. These two proteins then load the DNA polymerase III to the DNA at the primer-template junction. The ring formed by sliding-clamp protein maintains the association of polymerase III with its template as replication proceeds. This allows uninterrupted synthesis of DNA (Fig. 5.15).

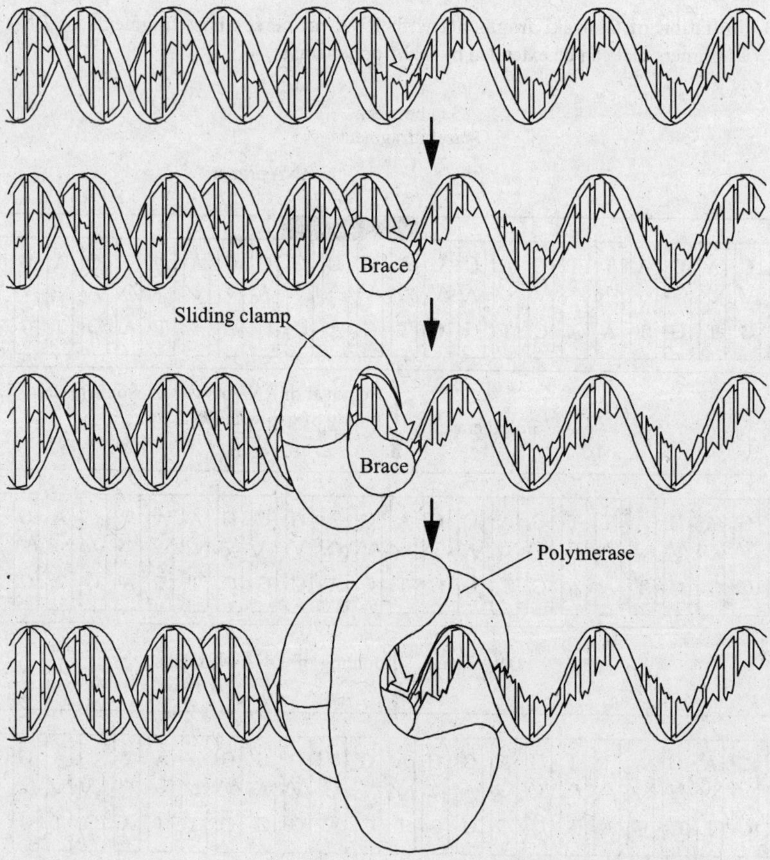

Fig. 5.15. Polymerase acessory proteins. The brace protein finds DNA at the junction between primer and template. The sliding-clamp protein finds adjacent to the brace protein. DNA polymerase then binds to the brace-clamp complex.

Another group of proteins unwind the template DNA and stabilize the resulting single-stranded regions. Enzymes, known as *halicases* catalyze the unwinding of parental helical DNA, coupled to the hydrolysis of ATP, ahead of the replication fork. *Single-stranded DNA binding proteins* then stabilize the unzipped template DNA and, also, keeping in an extended single-stranded state so that it can be copied by polymerase III. (Fig. 5.16).

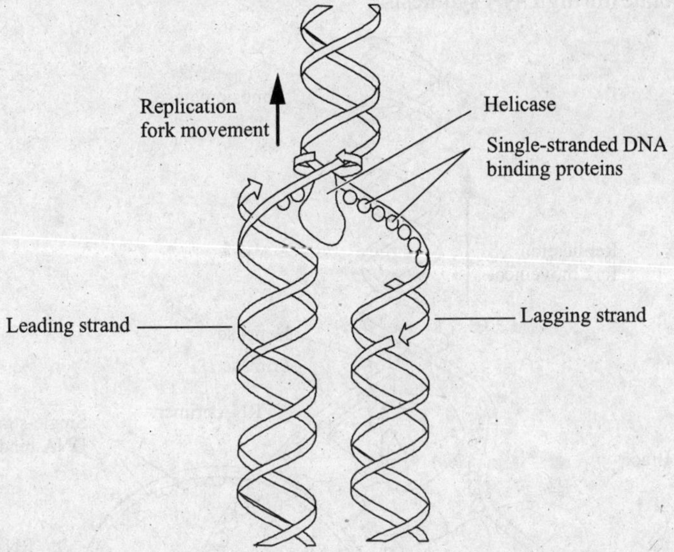

Fig. 5.16. Action of helicases and single-stranded DNA-binding proteins. Helicases unwind the two strands of parental DNA ahead of the replication fork. The unwound DNA strands are then stablized by single-stranded DNA-binding proteins so that they can serve as templates for new DNA synthesis.

When the parental DNA unwinds, the DNA helix ahead of the replicating fork is forced to rotate. If this is not checked, the circular DNA molecule of DNA in *E. coli* becomes twisted and replication would be blocked (Fig. 5.17). This problem is resolved by topoisomerases. These are enzymes which catalyse reversible breakage and rejoining of DNA strands. Topoisomerases and divided into

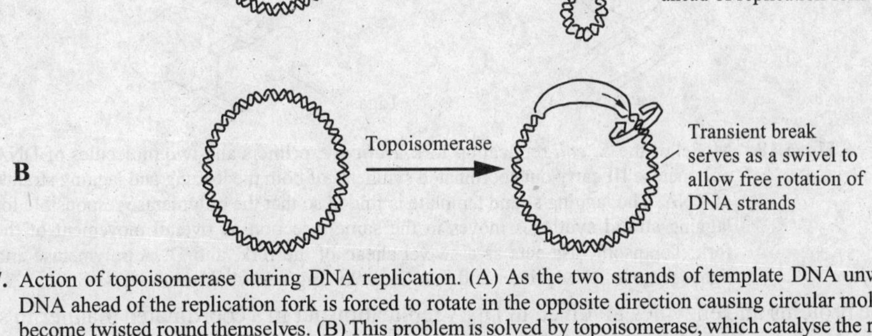

Fig. 5.17. Action of topoisomerase during DNA replication. (A) As the two strands of template DNA unwind, the DNA ahead of the replication fork is forced to rotate in the opposite direction causing circular molecules to become twisted round themselves. (B) This problem is solved by topoisomerase, which catalyse the reversible breakage and joining of DNA strands. The transient breaks introduced by these enzymes as swivels that allow the two strands of DNA to rotate freely around each other.

two classes.; type I topoisomerases make a break in one strand of the helix whereas type II topoisomerases introduce breaks in both the strands. Such breaks by type I and type II enzymes serve as "swivels" that allow the two strands of template DNA to rotate freely around each other allowing replication to proceed ahead of the replication fork (Fig. 5.18). Topoisomerases are also required for the replication of linear DNA (in eukaryotes), otherwise the whole chromosome would continuously rotate during DNA synthesis.

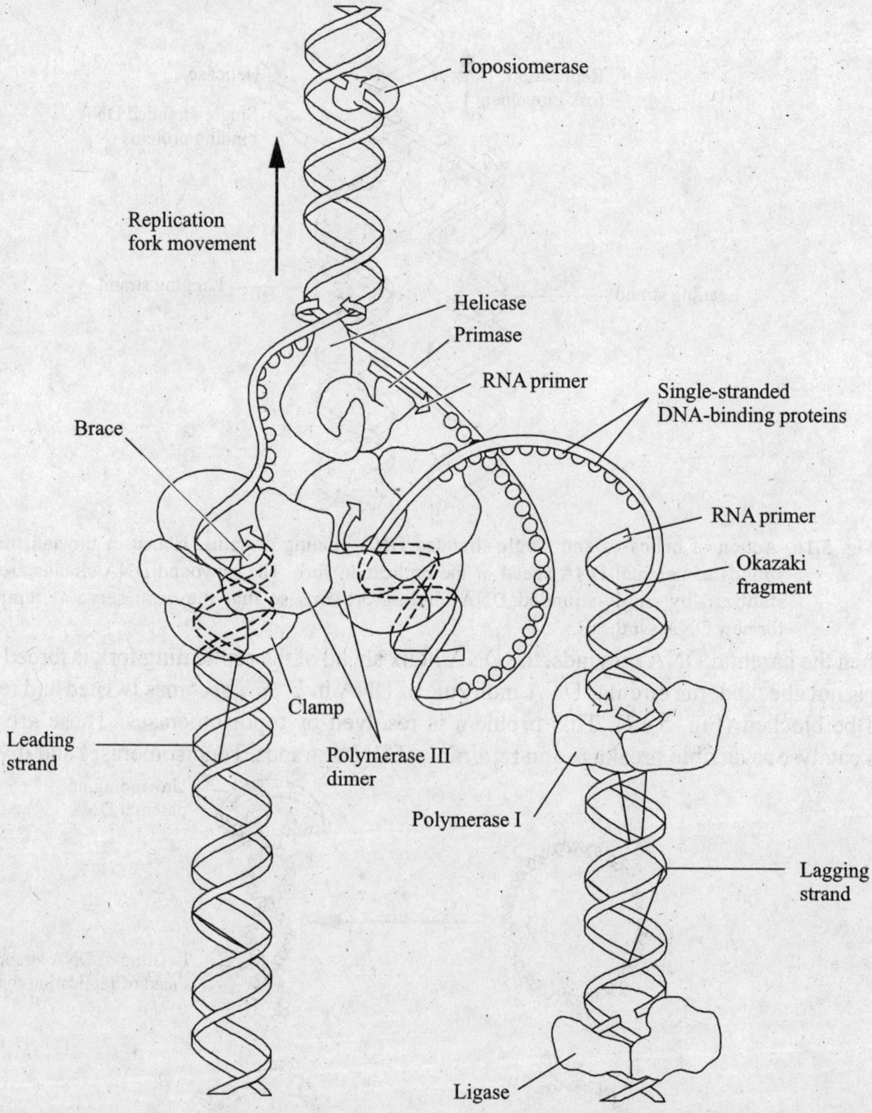

Fig. 5.18. Model of the *E. coli* replication fork. *Helicase*, primers and two molecules of DNA polymerase III carry out coordinated synthesis of both the leading and lagging strands of DNA. The lagging strand template is folded so that the polymerase responsible for lagging strand synthesis moves in the same direction as overall movement of the fork. Topoisomerase acts as a swivel ahead of the fork, and DNA polymerase and ligase remove RNA primer and join Okazaki fragments behind the fork.

The proteins and enzymes involved in DNA replication act in a coordinated manner to synthesize both the leading and lagging strands simultaneously at the replication fork. *A simplified and unified scheme of events taking place at the replication fork is given in* Fig. 5.18. It may he noted that replication is accomplished by the formation of *dimers* of DNA polymerase III, each with its accessory proteins. Each of the two molecules of the polymerase dimer synthesizes its corresponding leading and lagging strands. It may further be noted that the lagging strand template forms a *loop* at the replication fork so that its polymerase molecule and that of the leading strand move in unison in the same overall direction of replication fork movement.

STRUCTURE AND SYNTHESIS OF RNA

As already mentioned, two aspects in the building blocks of RNA distinguish it from DNA; the sugar component of RNA is ribose (Fig. 5.2) instead of deoxyribose, the bases adenine, guanine, and cytosine are common to both RNA and DNA, but the fourth base in RNA is *uracil* (Fig. 5.3) instead of thymine (Fig. 5.19). Moreover, in contrast to DNA, RNA exists as a single-stranded molecule. In a few instances, the single RNA strand is apparently bent back upon itself, and weak hydrogen bonds serve to hold it in a loose double helix.

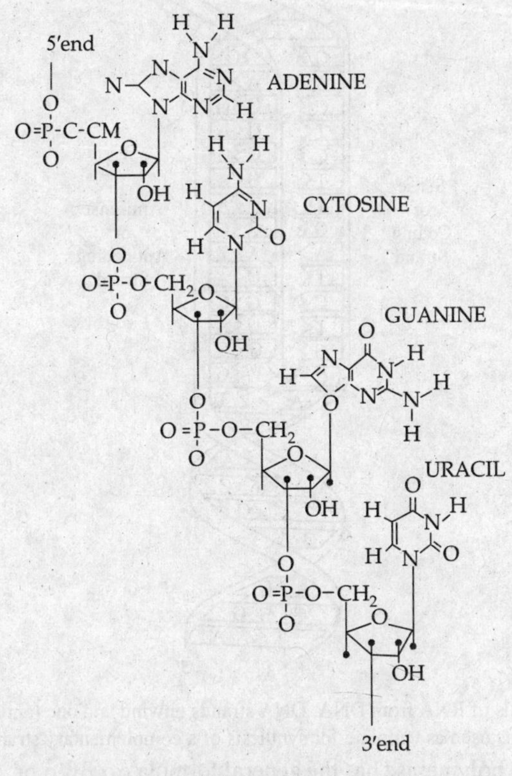

Fig. 5.19. The chemical formula of polyribonucleotide.

The fundamental mechanism for the synthesis of non-genetic RNA is very similar to that of DNA replication. A section of DNA strands separate and one of the strands functions as a template on which ribonucleotides are attracted by DNA; base pairing (adenine with uracil, and guanine with cytosine) then takes place (Fig. 5.20) Ribonucleic acid synthesis is, therefore, a process of copying (*transcription*) of one of the strands of DNA. The first step in expression of a gene (DNA) is its transcription into RNA.

The principal enzyme responsible for RNA synthesis is *RNA polymerase* which catalyzes the polynucleoside 5′-triphosphates (NTPs) as directed by DNA template. RNA polymerase catalyzes the growth of RNA chain always in 5′ to 3′ direction. Unlike DNA polymerase (as already discussed), transcription initiates *de novo* at specific sites at the beginning of genes.

RNA polymerase transcribes an RNA molecule from the strand of *antisense* or *anticoding strand* of a DNA molecule (Fig. 5.20). It may be pointed out here that the strand of the duplex DNA that has the same nucleotide sequence as RNA except that thymine (T) substitutes in DNA for uracil (U) in RNA is called the *sense* or *coding strand*. The other *strand*, which is the actual template for RNA synthesis, is the *antisense* or *anticoding strand*.

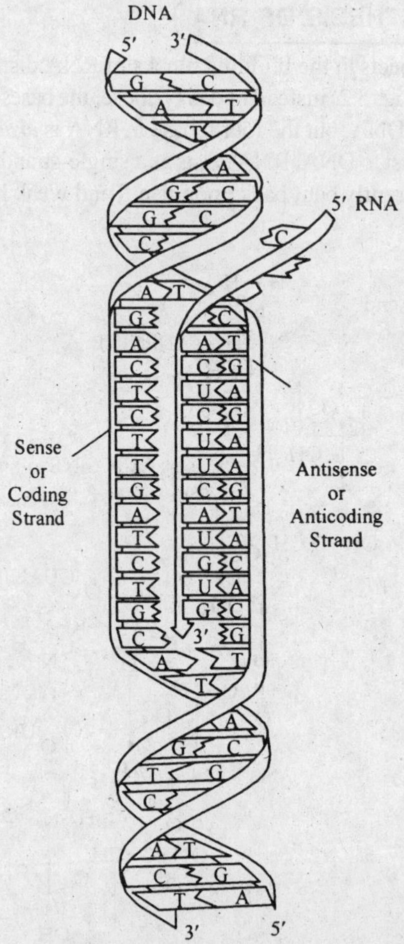

Fig. 5.20. Synthesis of RNA from DNA. DNA strands unwind and one (anticoding or antisense strand) is used as template for synthesis of a complementary strand of RNA.

The structure of RNA polymerase has the general formula αα ββ′σ or α_2ββ′σ, that is it contains five polypeptide chains (Fig. 5.21). The σ subunit is weakly bound and can be easily separated from the other subunits. Therefore, the two α, β and β′ form the *core enzyme.*

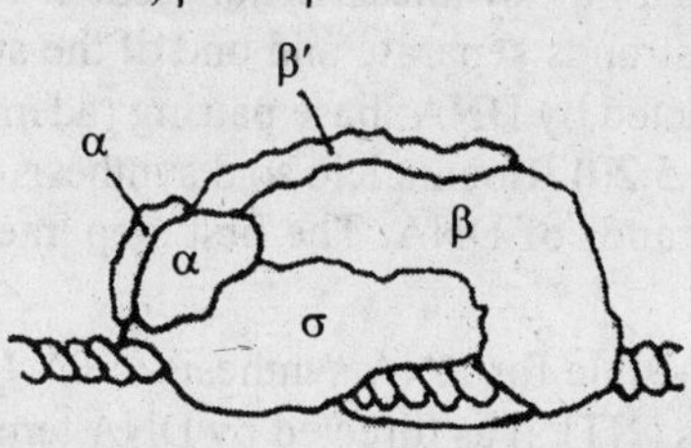

Fig. 5.21. *E. coli* RNA polymerase. The complete enzyme consists of five subunits two α, one β, one β′ and one σ. The σ subunit is relatively weakly bound and can be dissociated from the other four subunits, which constitute the core polymerase.

The core polymerase is fully competent to catalyze the polymerization of NTPs into RNA. Thus, σ is not needed for the basic catalytic activity of the enzyme. The polymerase does not specifically bind to the DNA sequence that signal initiation of transcription. Instead, the σ subunit is required to identify the correct sites for initiation of transcription so that functional RNA be formed at the beginning of a gene.

The DNA sequence to which RNA polymerase binds to initiate transcription of a gene is called a *promoter*. Two promoter segments have been identified and characterized by two sets of six nucleotides each, located 10 and 35 base pair upstream before the initiation point. They are called – **10** and – **35** elements denoting their position with respect to the transcription initiation site which is called as **+ 1** position (Fig. 5.22). The bases most commonly found at – 10 and – 35 position are TATAAT (*Pribnow box*) and TTGACA. These are therefore, known as *consensus sequences*. The promoters are *cis-dominant* (not involved in protein synthesis).

– 35	– 10	+ 1
TTGACA	TATAAT	
AACTGT	ATATTA	

+ 1 → Transcription start site

Fig. 5.22. Sequences of *E. coli* promoters. *E. coli* promoters are characterized by two sets of sequences located 10 and 35 base pairs upstream of the transcription start site (+1). The consensus sequences shown correspond to the base most frequently found in different promoters.

Initiation and elongation of RNA synthesis is *E. coli* (Fg. 5.23) involves five phases: (i) RNA polymerase (holoenzyme) binds nonspecifically to DNA template and migrates along the molecule

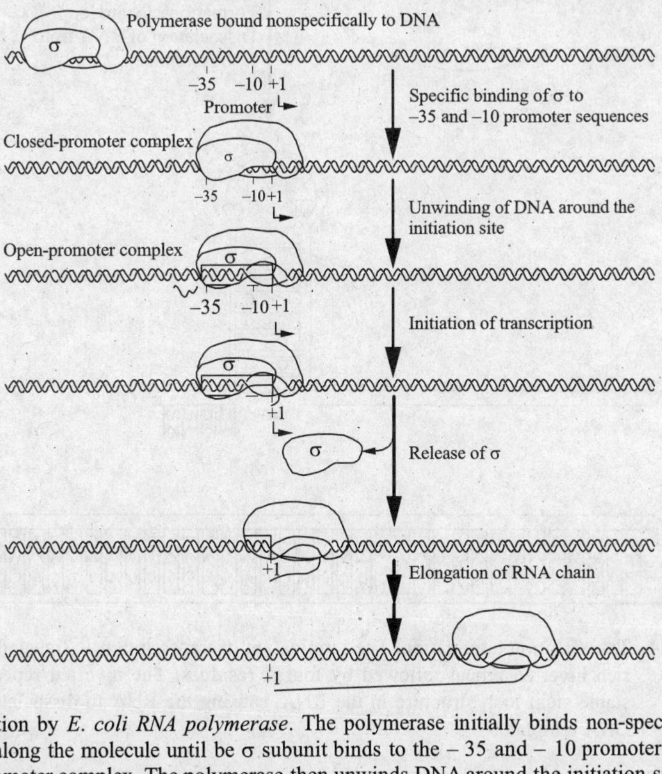

Fig. 5.23. Transcription by *E. coli RNA polymerase*. The polymerase initially binds non-specifically to DNA and migrates along the molecule until be σ subunit binds to the – 35 and – 10 promoter elements, forming a closed-promoter complex. The polymerase then unwinds DNA around the initiation site, and transcription is initiated by the polymerization of free NTPs. The σ subunit then dissociates from the core polymerase, which migrates along the DNA and elongates the growing RNA chain.

until the σ subunit binds the promoter segments – 35 and – 10; (ii) the RNA polymerase binds at the promoter side forming a *closed-promoter complex* in which the DNA remains helical; (iii) the polymerase unwinds DNA strands to form *open-promoter complex* in which about 15 bases are available on the antisense or anticoding single-stranded DNA molecule; (iv) transcription by joining of two free NTPs (v) after the addition of about 8 more nucleotides, the σ is released from the polymerase; and the core enzyme moves along the DNA to add more nucleotides to the nascent RNA.

As the core enzyme moves, it unwinds the DNA template ahead it and winds the DNA behind it. RNA synthesis continues until the polymerase encounter a *termination signal* when transcription stops. The synthesized RNA is rebased from the polymerase and the enzyme dissociates from the DNA template. The most common terminator signal in *E. coli* consists of a GC-rich inverted repeat followed by four or more A (adenine) residues on the antisense strand of DNA template (Fig. 5.24). The inverted repeat form a *stable stem loop* in the RNA by complementary base pairing. This disurbs its association with the DNA template and terminates transcription.

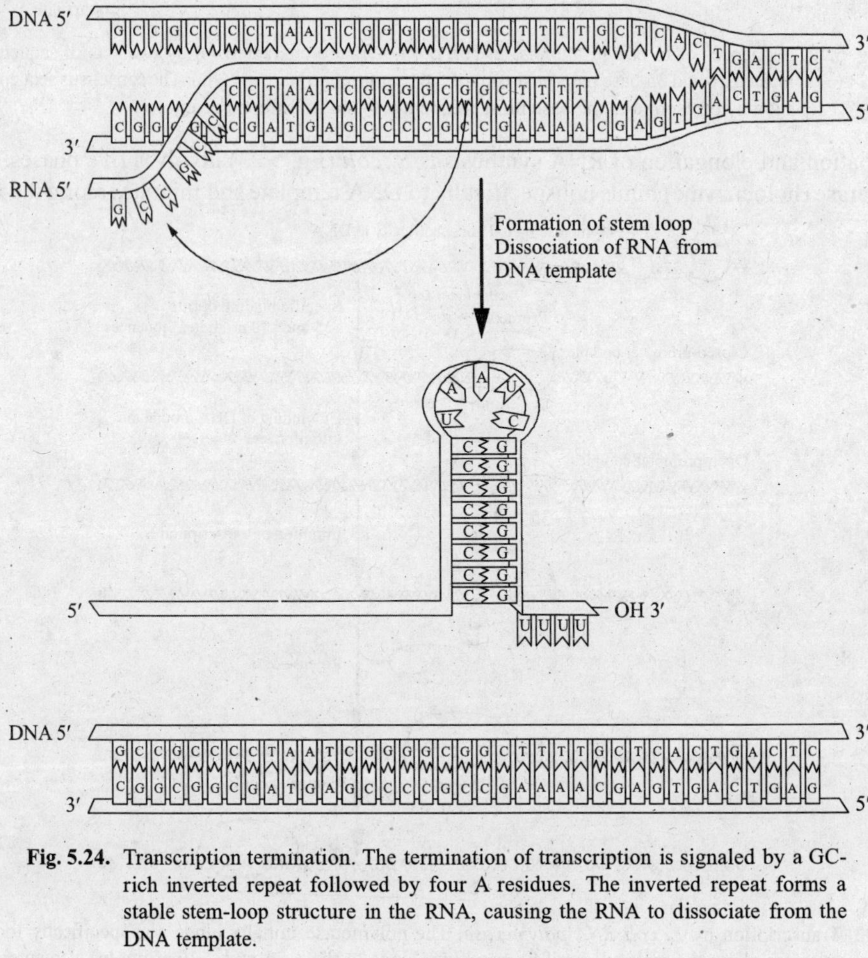

Fig. 5.24. Transcription termination. The termination of transcription is signaled by a GC-rich inverted repeat followed by four A residues. The inverted repeat forms a stable stem-loop structure in the RNA, causing the RNA to dissociate from the DNA template.

In those organisms in which the genetic material is RNA, the synthesis of RNA takes place on RNA template itself.

Three major types of ribonucleic acids are involved in protein synthesis in the cytoplam; these are *messenger, transfer*, and *ribosomal RNAs*; function of these three ribonucleic acids will be described in the following pages on genetic code and protein synthesis.

It may be noted here that in addition to the three major types of RNA, several other non-genetic RNAs are present in cells for specific functions. There are: (i) *Antisense RNA* (*asRNA*) having a nucleotide sequence complementary to a specified mRNA. Some bacteria generate *asRNA* for gene regulation. (ii) *Hetrogeneous nuclear RNA* (*hnRNA*) consisting of primary transcripts or to their modified products alone in the nucleus in higher eukaryotes. (iii) *Small nuclear RNA* (*snRNA*), pronounced "snurps" and *small cytoplasmic RNA* are found in nucleus and cytoplasm, respectively and bind to nuclear or cytoplasmic small *ribonucleoproteins* (*RNPs*). The *U* class of *snRNA* (*Usn RNA*) associated with their proteins (*snRNPs*); the latter make up one-third mass of *spliceosomes* (organelles in which excision and splicing reactions to remove *introns* from premessenger RNA occur). (iv) *guide RNA* (*gRNA*) synthesized *from mitochondrial DNA* (*mtDNA*) modify RNA transcribed from DNA.

THE GENETIC CODE AND PROTEIN SYNTHESIS

Different Species of RNA

Three classes of ribonucleic acids are found in living cells. These are messenger RNA (mRNA), *transfer* RNA (tRNA), and *ribosomal* RNA (rRNA). This nomenclature is based on functional and structural characteristics of the different RNAs. All the three types of RNAs are synthesized on the DNA template in the nucleus. Messenger RNA is found either free in the cytoplasm or associated with ribosomes. Transfer RNA is also found free in the cytoplasm of all cells. Approximately 20 different tRNA molecules, corresponding to the 20 amino acids, have been isolated. Ribosomal RNA is a structural component of the ribosomes.

The Genetic Code

The participation of all three species of RNA is essential for protein synthesis. The detailed process of protein synthesis will be described later in this chapter. Briefly, however, the mechanism is as follows: The mRNA is synthesized on a DNA template. The DNA being the carrier of genetic information transcribes this on the mRNA. This information is, in fact, in the form of a code for protein synthesis. Cells of the same organism, therefore, differ from each other on the basis of the segment of DNA that is active in making mRNA. The relationship of mRNA with protein synthesis is usually summarized by the following flow diagram:

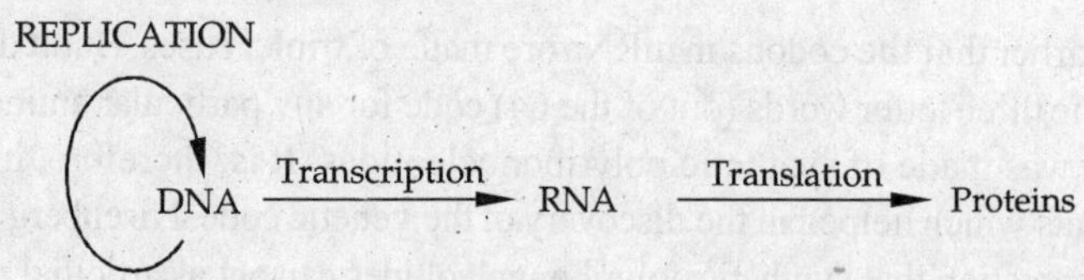

This concept is considered as the *central dogma* (one way flow of information) in molecular biology. According to the flow diagram, genetic code is transferred in three steps: (1) the replication of the central informational molecule — DNA; (2) the language of the DNA nucleotide sequence is faithfully *transcribed* into the language of mRNA nucleotide sequence (except that where DNA has thymine, RNA has uracil), and the mRNA carries the information of the transcribed DNA segment to the protein-synthesis apparatus in the cytoplasm; and (3) the mRNA nucleotide language is *translated*

to the language of amino acid sequence by the protein-synthesis apparatus. The latter consists of ribosomes and tRNA.

Evidence is available where RNA can control synthesis of DNA (reverse flow of transcriptional information) by an enzyme *reverse transcriptase* (RNA-dependent DNA polymerase now called RNA directed DNA polymerase). Such enzymes occur in oncogenic viruses (retroviruses, tumour or cancerous viruses). The functioning of such an enzyme contradicts the central dogma in the sense that the direction of information exchange between DNA and RNA is reversed as shown in the following flow diagram:

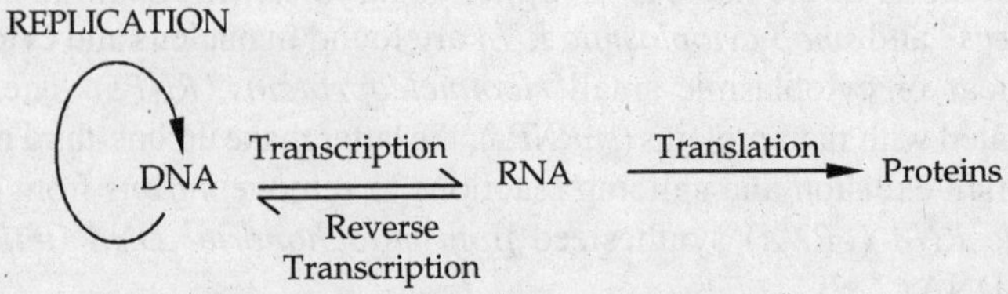

This discovery was demonstrated by H. Temin and D Baltimore in 1970 who shared the Nobel Prize in medicine with R. Dulbecco in 1975. Their work is significant for the understanding of cancer. The actual steps in protein synthesis are very complex. Energy and specific enzymes are needed for the various specific reactions involved.

Triplet Code

It has already been seen that the sugar-phosphate backbone is the same in all DNA molecules. Therefore, it cannot carry any genetic information. The information (*code*), therefore, has to be carried in the sequence of the four bases of DNA which is transcribed on the mRNA. However, the translation of the genetic code involves a change from the nucleotide language of mRNA to the amino acid language of proteins. Therefore, the four bases of the mRNA— adenine (A), cytosine (C) guanine (G), and uracil (U), must code in some way for the 20 amino acids. The question arises how is this achieved? If one nucleotide base (*singlet code*) coded for one amino acid, only four amino acids could be specified. If a combination of two bases (*doublet code*) were used, there would be only $4 \times 4 = 16$ different ways of arranging the four bases. This number (16) again is not sufficient to code for 20 amino acids. But a combination of three different bases (*triplet code*), $4 \times 4 \times 4 = 64$, is more than enough to designate the 20 amino acids. Elegant genetic experiments of Crick and his coworker in 1961 confirmed this triplet-code hypothesis. Each triplet (combination of three bases) which codes for a particular amino acid is called a *codon.*

It was noted earlier that the codons in mRNa are made of triplet bases. But it does not tell us which combination of the three letter words (out of the 64) code for any particular amino acid. To obtain this information, use was made of synthetic polyribonucleotides. It is, therefore, important to describe certain experiments which helped in the discovery of the genetic code. Nirenberg and Matthaei (1961) made the first observation that synthetic polyribonucleoltides can act as articifial mRNA and stimulate the synthesis of peptides from amino acids. The use of synthetic mRNAs of known composition was made possible by a previous discovery by Ochoa (1961) that the enzyme *polynucleotide phosphorylase* can link nucleotides added to the medium. Nirenberg and Matthaei used the homo-polymer of uridine monophosphate (uridylic acid) termed *poly-U* as mRNA, and synthesized poly-phenylalanine (a peptide chain made of phenylalanine alone). This experiment proved that the codon for phenylalanine was UUU. Similarly, experiment with *pol-C* and *poly-A* assigned codons for proline (CCC) and lysine (AAA).

By 1963, the experiments with synthetic mRNAs done in the laboratories of Nirenberg and Ochoa established most of the codon sequences. A cell-free extract consisting mainly of ribosomes was used for these experiments. The supernatant contained the activating enzymes, tRNA, energy source, radioactively labelled amino acids and the synthetic mRNA. The amount of synthesis was measured by the radioactivity found in the protein precipitated. Another method for the recognition of codons was designed by Nirenberg and Leder (1965). They prepared an artificial mRNA from the base uracil and guanine in the proportion of two of uracil to one of guanine. From this they obtained three combination CUU, UGU and UUG. Again using a system containing tRNA, radioactively labelled amino acids, and enzymes, they were able to show that only GUU codes for the amino acid valine.

In the laboratory of Khorana (1966), polydeoxynucleotides and polyribonucleotides with alternating *doublets* (e.g.-UU or UA) or triplets of known sequences of bases (copolymers) were synthesized and used in cell-free protein synthesis. With the help of these experiments polypeptides having predetermined amino acid sequences were fabricated with the help of codons contained in synthetic DNAs. Consequently, Khorana's work opened up the possibility of synthesizing DNAs having gene function. For this remarkable work Hargovind Khorana, American of Indian origin, shared the Nobel Prize for medicine with M.W. Nirenberg and R.W. Holley in 1968.

The results from the work of Nirenberg, Ochoa, and Khorana led to the unravelling of the *genetic dictionary* or *genetic code* for the 20 amino acids (Table 5.2).

Table 5.2 ***The Genetic Code Table.*** **The first base of a codon is given in the vertical column on the left of the table, the second base in the horizontal row at the top, and the third base in the vertical column on the right. Ter or termination codons (UAA, UAG and UGA); met (methionine) and f met (*N-formyl methionine*) are coded by initiation codon IC (AUG). See text for details.**

		Second Base			*Third*
First Base	*U*	*C*	*A*	*G*	*Base*
U	Phe	ser	tyr	cys	U
	Phe	ser	tyr	cys	C
	leu	ser	Ter	Ter	A
	leu	ser	Ter	trp	G
C	leu	pro	his	arg	U
	leu	pro	his	arg	C
	leu	pro	gln	arg	A
	leu	pro	gln	arg	G
A	ile	thr	asn	ser	U
	ile	thr	asn	ser	C
	ile	thr	lys	arg	A
	met and f met (IC)	thr	lys	arg	G
G	val	ala	asp	gly	U
	val	ala	asp	gly	C
	val	ala	glu	gly	A
	val	ala	glu	gly	G

It is obvious from Table 5.2 that several RNA codons code for a single amino acid in some cases. Leucine, for example, may be coded by UUA, UUG, CUU, CUC, CUA and CUG. This fact is referred to as *degeneracy of the genetic code*. In most cases the synonymous codons differ only in the base occupying the third position of the triplet; hence called *third base degeneracy*. Since the same amino acid is coded by synonymous codons, it is logical to assume that mutations due to replacement of the third-base will go unnoticed in the phenotype.

In general, the following properties are associated with the genetic code: (i) the code is triplet, i.e., a unit of three successive bases in DNA or RNA that codes for a single amino acid; (ii) the code is *degenerate* because more than one nucleotide triplet codes for the same amino acid (for example, arginine is coded by CGU, CGC, CGA, CGG, AGA, AGG); (iii) the code is *non-overlapping* in the sense that a base in an mRNA is not shared by two adjacent codons; had there been such sharing (overlapping) two successive codons would code for four amino acids (for instance, a sequence GUU AAC will be read as GUU UUA UAA and AAC); this does not occur. However, there are examples where nucleotide sequences overlap to some degree. The overlap may involve *control genes*, such *as tryptophan operater* and promoter regions on *E. coli* but they are translated in different reading frames which means that the same base cannot be used for two different codons during the synthesis of the same protein; (iv) the code is *commaless* that is successive codons that are contiguous (next in sequence) not separated by noncoding bases or groups of bases. However in most eukaryotes the genes contain coding regions called *exons* (specifying amino acids) interrupted by noncoding regions called *introns*. In such situations the code is said to contain commas. Bacterial genes do not have introns and, therefore, the sequence of amino acids in polypeptides and codons in the genes and arranged in the same linear order (colinear); (v) the code is *non-ambiguous* which simply specifies that a particular codon will always code for the same amino acid; however, it has been shown that same ambiguity exists as that both AUG and GUG may code for methionine as initiating codon although GUG is exclusively meant for valine; (vi) the code is *universal* because the same genetic code is used for all forms of life *with few exception*. Thus, for example in yeast mitochondria CUA codes for threonine instead of leucine; mammalian mitochondria AUA codes for methnione instead of isoleucine and codon AGA and AGG signal termination instead of coding for argenine; UGA codes for tryptophan instead of signaling termination in mitochondria from both yeast and mammals.

Each mRNA chain, like a DNA chain, has direction specified by the orientation of the sugar-phosphate backbone. The chain end terminated by the 5-carbon of ribose is called the 5′ end, while the end containing the 3-carbon atom is called the 3′ end (Fig. 5.26). The mRNA is read in the 5′ to 3′ direction (Fig. 5.12). The end having carbon 5′ always codes for the — NH_2 terminal of the polypeptide chain, whereas the carbon 3′ end of the mRNA codes for the —COOH terminal of the chain. (*It is well known that every polypeptide chain has a free — NH_2 end and a free—COOH terminal*). Many long chain mRNA are known to code for more than one polypeptide chain. This demands that the mRNA chain should have points along its length which can initiate and terminate the synthesis of polypeptide chain. The codon AUG for methionine (met) in eukaryote (N-Formyl methionine in bacteria) is known to be the *initiation codon* (IC) *or start codon* as the polypeptide synthesis begins at this codon. UAA, UAG and UGA (also referred to as 'ochre', 'amber' and 'opal') which do not code (*stop codons* or *nonsense codons*) for amino acid are known as *termination codons* (Ter) as one of these is always present at the end of the code for polypeptide chain. It follows from this that these always precede the initiation codon shown as under:

DNA
5′– A T G G T T A A C   C T C T A A A T G A C A A A A   T A T – 3′
3′– T A C C A A T T G   G A G A T T T A C T G T T T T   A T A – 5′

TRANSCRIPTION

mRNA 5′– A U G G U U A A C   C U C U A A A U G A C A A A A   U A U – 3′
IC   TER

TRANSLATION

(NH_2 - end) f met val asn..... leu   fmet thr lys....... tyr (COOH-end)

POLYPEPTIDE 'a'   POLYPEPTIDE 'b'

It is obvious that mRNA nucleotide sequences are written 5′ to 3′ left to right, respectively, because that is the direction in which translation occurs. Thus, as shown above, the mRNA segment is 5′ — 3′ whereas its complementary antiparallel DNA template strand is 3′ — 5′. The end result is the synthesis of polypeptides 'a' and 'b' due to the presence of the termination codon UAA on the mRNA.

PROTEIN SYNTHESIS APPARATUS

Ribosomes

Ribosomes comprise about a quarter of the total cell mass. This fact reflects the importance of protein synthesis in cellular activity. As already said the ribosomes are composed of roughly equal parts of RNA (ribosomal RNA or rRNA) and protein. Electron microscopy indicates that ribosomes are composed of two unequal subunits bound together by Mg^{++}, measuring about 200 Å in diameter in prokaryotes (Fig. 5.25).

As shown in Table 5.3 at least four classes of ribosomes exist that can be characterized by the *sedimentation coefficients* of their constituent rRNAs. The *sedimentation coefficient* (S, usually given in terms of Svedberg units, depends upon the size and density of particle; it is constant for a given particle in a given medium in the ultracentrifuge) of the intact prokaryotic ribosome is 70S, while the subunits have values of 50S (large subunit) and 30S (small subunit); see Fig. 5.25. The aggregation of the two subunits depends upon the concentraton of Mg^{++} ion; at low concentration of Mg^{++} the two subunits dissociate. Two sizes of RNA are found in the ribosomes. The smaller ribosomal subunit (30S) contains, besides protein, rRNA with a sedimentation value of 16S; the larger one contains rRNA with a 23S value. The larger subunit contains an additional 5S RNA in both prokaryotes and eukaryotes. In eukaryotes ribosomes are larger (80S), with 40S and 60S subunits. Animal ribosomes also contain a 5.8S rRNA and is hydrogen bonded to the 28S rRNA. The role of rRNA in protein synthesis will be discussed later.

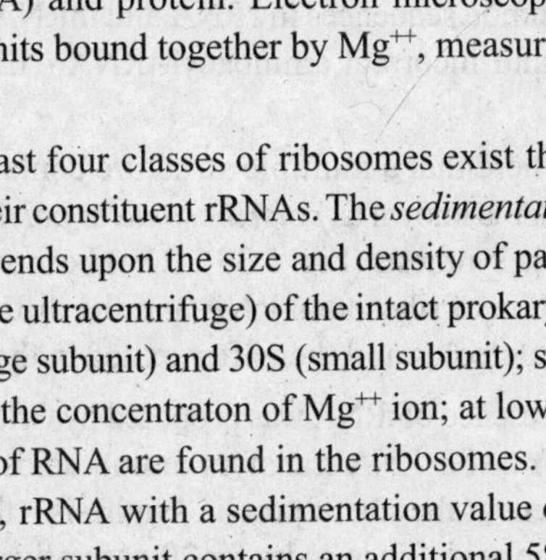

Fig. 5.25. Prokayotic ribosome, showing influence of Mg^{++}.

Table 5.3 Sedimentation coefficients (S) and characteristics of four classes of ribosomes.

Ribosomal Source	*Ribosome*	*Large subunit*	*Small subunit*	*Large rRNA*	*Small rRNA*
Bacteria, chloroplasts	70	50	30	23 + 5	16
Plant cytoplasm	80	60	40	25	16
Animal cytoplasm	80	60	40	28 + 5+ 5.8	18
Mitochondria	81	61	47	25	19

Several models, based on electron microscopy, have been proposed for ribosome. The most accepted model has been given by **J.A. Lake** (1981) in E. coli. In this model the smaller subunit comprises a *head,* a *base* and a *platform*. The larger subunit consists of a *central protuberance* planked by a ridge and a stalk.

As stated earlier, the ribosomal components are RNA and proteins. The smaller (30S) and larger (50S) subunits of prokaryotic ribosomes when dissociate give rise to *inactive core particles* with sedimentation coefficient of 23S and 40S, respectively. The former contains 16S and the later 23S RNAs and some *core proteins*. In addition to the core proteins, *split proteins* (SP) are also present in both the ribosomal subunits. In total the smaller and larger subunits contain 21 and 31 different proteins, respectively. Their corresponding designations are S1 to S21 and L1 to L34 (and not L1 to L31 due to overlap in their identification). *However in eukaryotes there are approximately 33 and 45 proteins in the smaller and larger subunits, respectively*. Proteins in the ribosomal subunits play both structural and functional roles. The latter are designated as *active centres*. Thus, for mRNA-ribosome binding there are common sequences in rRNA and mRNA. Ribosomes can also distinguish between the entry of correct and incorrect aminoacyl-tRNA - EF – Tu – GTP complexes (to be discussed later).

It may also be pointed out here that a multiple structure containing a group of ribosomes held together by a molecule of mRNA is known as *polyribosome* or *polysome*. These associations actively engaged in translation or protein synthesis (Fig. 5.33). It has been suggested that more than 50 genes may be involved in the formation of ribosomal RNA and the different proteins. In eukaryotes, the subunits are processed in the nucleolus and thereafter exited in the cytoplasm.

Transfer RNA (tRNA): The presence of tRNA was proposed by Crick and demonstrated by Hoagland and his coworkers in 1955. Transfer RNA transfers amino acids present in the cytoplasm to the mRNA which are already present at the site of protein synthesis – the ribosomes. There is one tRNA molecule specific for each amino acid. Since codes for 20 naturally occurring amino acids are present there should be at least 20 different tRNAs. The number of tRNAs may be more than 20 because there may be several tRNA for a single amino acid. However, the number of tRNAs is not necessarily as high as the number of codons with the result that one tRNA species can recognize several synonymous codons meant for the some amino acid. The basic structure of all tRNAs is similar; 5′ end chain of every tRNA molecule has guanine (–G), and 3′ end has a sequence of three aired base, cytosine, cytosine and adenine (-C-C-A). The amino acids always attach at the 3′ end. The specific site on the tRNA molecule which attaches to mRNA is in the form of three bases. The sequence of these bases is complementary to the codon on mRNA, and is known as the *anticodon* (Fig. 5.26); for example, phenylalanine whose codon is –UUU has an anticodon AAA on the tRNA.

Before amino acids can be transferred to the protein building sites, they are attached to the tRNA. A specific adapter enzyme, *amino-acyl-tRNA synthetase,* is required to couple the amino acid to its specific tRNA at the —C—C—A end.

The first step in translation (protein synthesis) is the *activation* of the amino acids. This involves coupling of a specific amino acid with ATP by specific aminoacyl.– tRNA synthetase to form an aminocyl AMP synthetase intermediate. The second step is the attachment of the activated amino acid to its own species of tRNA forming an aminoacyl tRNA. This reaction is also mediated by aminoacyl – tRNA synthetase. It follows that the activating enzyme aminoacyl tRNA synthetase must have two sites : one, to recognize the specific amino acid and, second, to recognize the specific tRNA. The link formed between the amino acid and its specific tRNA is a covalent bond between the carboxyl group of the amino acid and the ribose of the 3′ terminus of adenine nucleotide of the tRNA. The tRNA with its amino acid is referred to as charged tRNA. Four components are, therefore, necessary for this reaction : a specific amino acid (AA_1), ATP, specific aminoacyl tRNA synthetase (Enz_1) and the specific tRNA ($tRNA_1$), as shown below:

$$AA_1 + ATP \xrightarrow{Enz_1} \underset{\text{(Aminoacyl AMP synthetase)}}{(AA_1 - AMP)\ Enz_1} + 2Pi$$

$$(AA_1 + AMP)\ Enz_1 + tRNA_1 \longrightarrow \underset{\text{(Charged tRNA)}}{(AA_1 - tRNA_1)} + AMP + Enz_1$$

Each tRNA contains approximately 77 - 80 nucleotides. The first nucleotide sequence of tRNA was established by Holley in 1964 for a tRNA from yeast. However, the sequence of several tRNAs is now known. Holley suggested a "clover leaf "– model of tRNA with four regions of hydrogen bonding (Fig. 5.26). The tRNA, in addition, is characterized by the presence of several unusual bases other than A, U, G, and C. These unusual bases include pseudouridine (ψ), inosine (I), methylguanosine (G^m) methylcytosine (C^m), methylthymine (T^m) and others. Many of these base differ from normal bases by the presence of one or more methyl (CH_3) groups. Inosine has the same base pairing properties as guanine (Fig. 5.26), and ψ as the base uracil.

As described earlier, ribosomal RNA forms an integral part of the ribosomes. When rRNA is removed, the ribosomes collapse. The two different sizes of rRNA (16S and 23S – Table 5.3) have molecular weights of about 16×10^5 and 1×10^6, respectively. Both contain regions where the rRNA molecule appears double stranded due to hydrogen bonding as in tRNA. The rRNA is synthesized in the nucleolus.

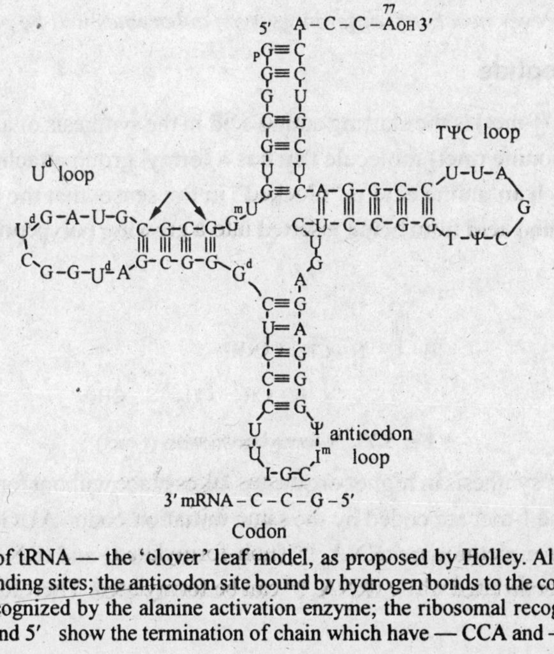

Fig. 5.26. The structure of tRNA — the 'clover' leaf model, as proposed by Holley. Alanine tRNA showing : the aminoacyl binding sites; the anticodon site bound by hydrogen bonds to the codons of alanine; the region of tRNAs recognized by the alanine activation enzyme; the ribosomal recognition site common to all tRNAs ; 3′ and 5′ show the termination of chain which have — CCA and —G, respectively.

TRANSLATION IN PROKARYOTES

The preceding discussion has provided the understanding for the mRNA code for polypeptides. The next step in protein synthesis involves a change from the nucleotide language of mRNA to the amino acid language of proteins. This process is called *translation*. In order to fully understand protein synthesis, the role of mRNA, tRNA, rRNA, and the various other factors (energy and specific enzymes) must be clearly defined; all these together contribute to protein synthesis.

The two Nobel laureates – F. Jacob and J. Monod (1961) proposed the name "messenger" to the RNA which carries the message for protein synthesis. Depending on the size of the polypeptide chain for which they code, mRNA molecules are of varying lengths. Most of the polypeptide chains contain at least 100 amino acids. Therefore, very few mRNA molecules contain less than 3 × 100 nucleotides in it. Many mRNA molecules are considerably longer. The length of mRNA (or the segment of DNA) which carries the information necessary to determine the complete polypeptide chain is called a *cistron*. This definition was subsequently expanded to include the start and stop signals. In cases where a mRNA encodes two or more proteins, it is referred to as polycistronic. The proteins specified by polycistron are often enzymes that function in the some metabolic pathway. Some mRNA molecules carry the information for synthesis of more than one polypeptide produced by adjacent cistrons is the same operon. Such mRNAs are called *polycistronic*.

Studies using cells incubated with radioactive building units of RNA (C^{14}—uridine, etc.) show that labelled material is first incorporated into the nucleus. Later, it appears in the cytoplasm (ribosomal). This indicates that the mRNA is first transcribed from DNA in the nucleus; it then leaves the nucleus through the nuclear membrane-pores into the cytoplasm. In the cytoplasm the mRNA moves to the site of protein synthesis—the ribosomes. Magnesium (Mg^{++}) is involved in the formation of the mRNA-ribosome complex.

The *translation* in both prokaryotes and eukaryotes is a complex process which involves (i) initiation, (ii) *elongation* and (iii) *termination* of polypeptide chains, these steps will be described briefly in the prokaryotes (*and wherever necessary differences from eukaryotes will be pointed out*).

Initiation of Polypeptide

N-formyl-methionine (f-met) is the starting amino acid in the synthesis of all bacterial polypeptides. It is a modified methionine (met) molecule that has a formyl group attached to its terminal amino group (Fig. 5.27). Such an amino acid is "blocked" in the sense that the absence of a free amino group prevents the amino acid from being inserted into a growing polypeptide chain.

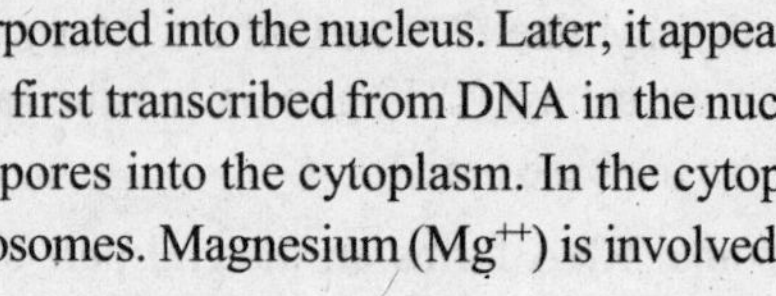

Fig 5.27. N-formylmethionine (f-met)

Initiation of protein synthesis in higher organisms takes place without formylation of methionine. However, both met and f-met are coded by the same initiation codon AUG in an mRNA. In *E. coli* there are two tRNAs for methionine $tRNA_m^{met}$ (non-formylated) and $tRNA_m^{met}$ (formylated). The $tRNA_m^{met}$ cannot be formylated while $tRNA_f^{met}$ can be formylated. Therefore, it is $tRNA_f^{met}$ which deposits f-met as the starting amino acid whereas $tRNA_m^{met}$ deposits met at the internal (intercalary) positions only. *Eukaryotes have* $tRNA_i^{met}$ only.

The initiation codon AUG is preceded by a specific sequence **AGGAGGU** in the mRNA. The sequence is complementary to a sequence at the 3′ end of 16S rRNA and thereby serves as binding site for ribosome called as *ribosome binding site* or **RBS**, the sequence AGGAGGU is called *Shine Delgarno* (**SD**) sequence (Fig. 5.28).

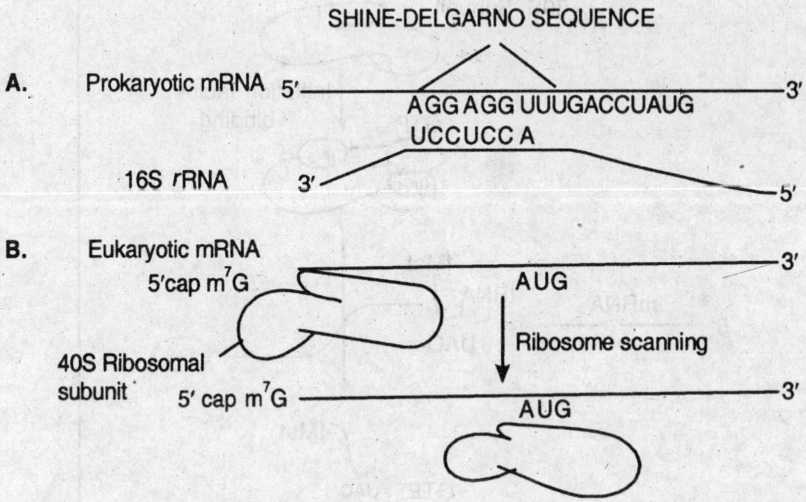

Fig. 5.28. Signals for translation initiation sites. A. In prokaryote mRNA, these sites are characterized by a Shine-Delgarno sequence that precedes the AUG initiation codon. Base-pairing between the Shine-Delgarno sequence and a complementary sequence near the 3' terminus of 16S rRNA aligns the mRNA on the ribosome. B. In contrast the eukaryotic mRNAs are bound to the 40S ribosomal subunit by their 5' 7-mithyl guanosine caps. The ribosome then scans along the mRNA until it encounters an AUG initiation codon.

Following the formation of met-$tRNA_f^{met}$, it undergoes formylation with the help of the enzyme *transformylase* (*which is absent is animals*) in the presence of a formate source.

$$\text{met} - \text{tRNA}_f^{met} + \text{formate} \xrightarrow{\text{transformylase}} \text{f-met-tRNA}_f^{met}$$

Several proteins are required for the initiation of protein synthesis. These are IF1, IF2 and IF3 in prokaryotes (and ten eIF in eukaryotes). The protein IF3 is required first to dissociate the 70S ribosome into 50S and 30S subunits

$$70S \xrightarrow{IF3} 50S + 30S.$$

IF1, IF2 and IF3 then bind to the 30S ribosomal unit. The IF2 is bound to GTP.

$$\text{GTP} - \text{IF2} + \text{IF1} + \text{IF3} + 30S \longrightarrow \text{GTP} - \text{IF2} - \text{IF1} - \text{IF3} - 30S$$

This step is followed by binding of mRNA and the initiation of N-formylmethionine – $tRNA_f^{met}$ (fmet-$tRNA_f^{2met}$) which is recognized by IF2 bound to GTP, IF3 is then released.

$$\text{GTP} - \text{IF2} - \text{IF1} - \text{OF3} - 30S + \text{mRNA} + \text{f-met} - \text{tRNA}_f^{met}$$

$$\rightarrow \text{GTP} - \text{IF2} - \text{IF1} - 30S - \text{mRNA} - \text{f-met}_f^{met} + \text{IF3}$$

Therefore, the initiation complex GTP – IF2 – IF1 – 30S – mRNA - f-met-tRNA$_f^{met}$ associates with 50S subunit to from a complete ribosome (70S), utilizing energy by the hydrolysis of the GTP molecule. This is followed by the release of IF1, IF2 and GDP (Fig. 5.29).

$$\text{GTP} - \text{IF2} - \text{IF1} - \text{30S} - \text{mRNA} - \text{f-met tRNA}_f^{met} + \text{50S.}$$
$$\rightarrow \text{70S} - \text{mRNA} - \text{f-met} - \text{tRNA}_f^{met} + \text{IF1} + \text{GDP} - \text{IF2} + \text{Pi}$$

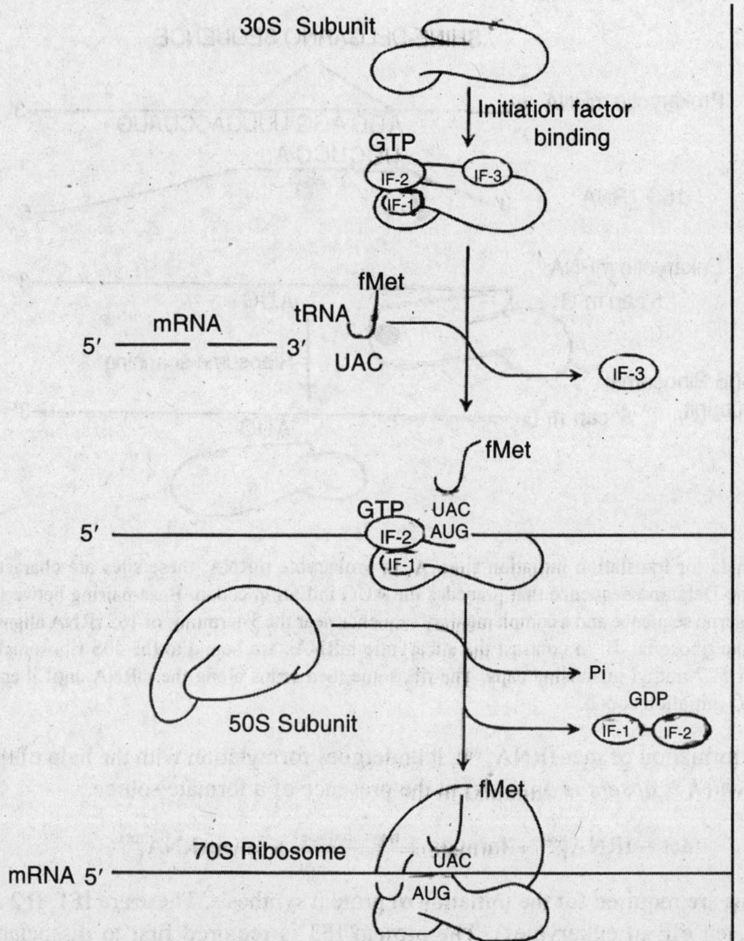

Fig 5.29. Initiation of translation in bacteria. For details, see text. f^{met} = N-formylmethionyl.

Elongation of Polypeptide

With the formation of 70S – mRNA – f-met – tRNA$_f^{met}$ initiation complex, translation proceeds by elongation of the polypeptide chain. The ribosome has three binding sites designated **P** (peptidyl), **A** aminoacyl) and **E** (exit) sites. The initiator f-met-tRNA$_f^{met}$ is positioned in the P site, leaving an empty *A* site the second aminoacyl-tRNA (for example, glycine-tRNA, having anticodon CCC)is then brought to the A site and pairs with its complementary codon (GGG in this case) on the mRNA. The second aminoacyl-tRNA is escorted by an elongation factor *EF-Tu* which is complexed with GTP. The GTP is hydrolysed to ***GDP*** as the second aminoacyl-tRNA (Glycine-tRNA) is inserted into the A site of the ribosome; the elongation factor EF-Tu complexed with GDP is released. A peptide bond is then formed between the initiator f-met-$tRNA_f^{met}$ at the P site and the second aminoacyl-tRNA at the A site. The formation of the peptide bond is catalyzed by the enzyme *peptidyl transferase* present in the 50S subunit of the ribosome. This results in the transfer of f-met to the second aminoacyl-tRNA at the A site of the ribosome, forming a peptidyl-tRNA at this position (A site) and leaving the uncharged initiator $tRNA_f^{met}$ (without its amino acid f-met) at the P site.

The next step in elongation is translocation which requires elongation factor ***EF-G*** (old name translocase) coupled to GTP hydrolysis. During translocation the ribosome moves three nucleotides along the m-RNA this bring the next codon (third) in the empty A site. Moreover, this step also moves the peptidyl $tRNA_f^{met}$ from the P site to the E site. Consequently, the ribosome is left with a peptidyl tRNA (glycine-tRNA) bound at the P site and an empty A site. The binding of a new (third) aminoacyl-tRNA induces the release (or exit) of the uncharged $tRNA_f^{met}$ from the E site. This leaves the ribosome ready for the insertion of the next (third) aminoacyl-tRNA in the growing polypeptide chains. Fig. 5.30 shows the elongation stages of translation.

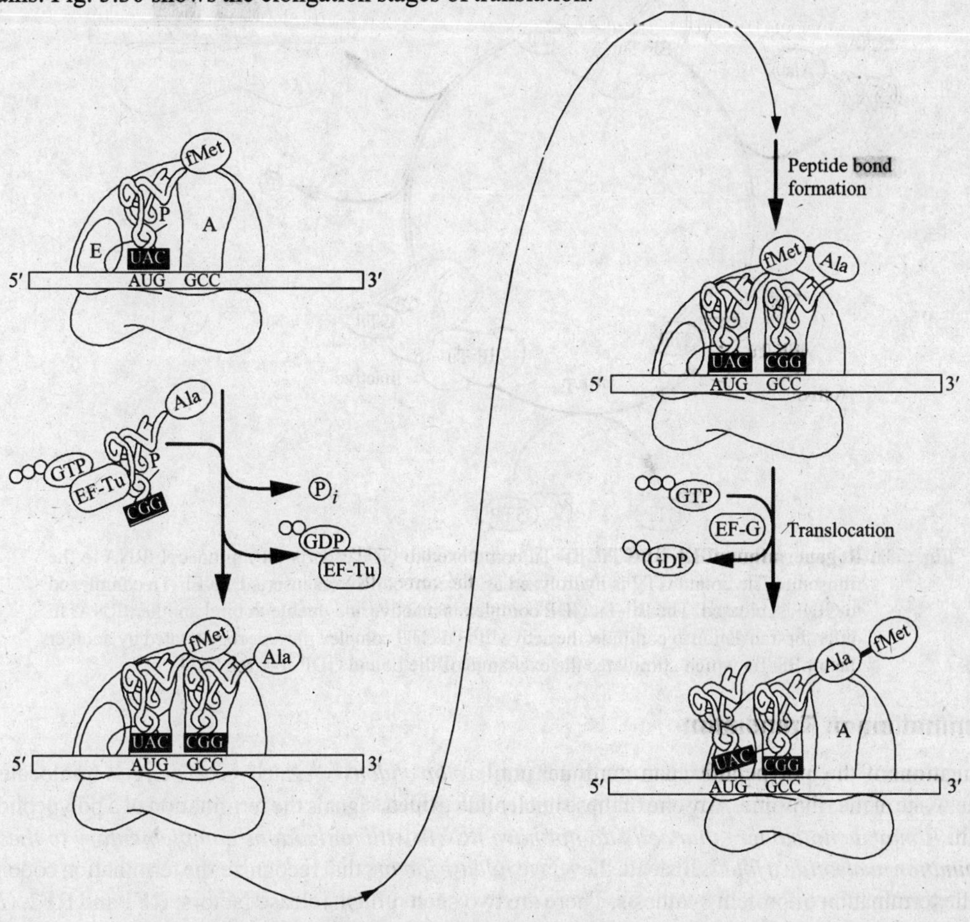

Fig. 5.30. Elongation stage of translation. The ribosome has three tRNA-binding sites, designated P (peptidyl), A (aminoacyl), and E (exit). The initiating N-formylmethionyl tRNA is positioned in the P site, leaving an empty A site. The second amino-acyl tRNA (e.g., alanyl tRNA) is then brought to the A site by EF-Tu (complexed with GTP). Following GTP hydrolysis, EF-Tu (complexed with GDP) leaves the ribosome, with alanyl tRNA inserted into the A site. A peptide bond is then formed, resulting in the transfer of methionine to the aminoacyl tRNA at the A site. The ribosome then moves three nucleotides along the mRNA. This movement translocates the peptidyl (Met-Ala) tRNA to the P site and the uncharged tRNA to the E site, leaving an empty A site ready for addition of the next amino acid. Translocation is mediated by EF-G, compled to GTP hydrolysis. The process, illustrated here for prokaryotic cells, is very similar in eukaryotes.

One significant step of the elongation process is the *regeneration* of EF-Tu/GTP. The has already been explained that as elongation continues the EF-Tu released from the ribsome is bound to GDP (EF-Tu/GDP). This form in *inactive* and *unable* to bind another tRNA. Therefore, EF-TU/GDP must be converted to the **active** from EF-TU/GTP so that translation continues. The active EF-Tu/GTP is regenerated by another factor **EF-Ts** which stimulates the exchange of the bound GDP for free GTP. This exchange results in the regeneration of EF-Tu/GTP which can escort a new aminoacyl-tRNA to the A site of the ribosome, beginning a new cycle of elongation of the polypeptide chain (Fig. 5.31).

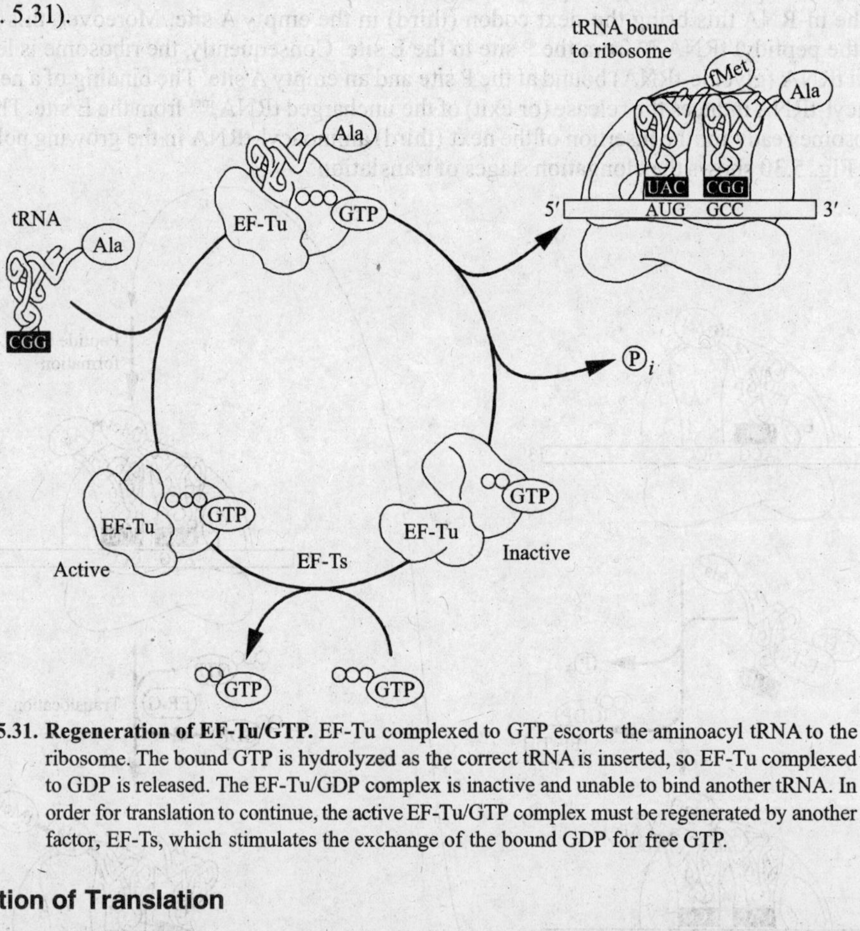

Fig. 5.31. Regeneration of EF-Tu/GTP. EF-Tu complexed to GTP escorts the aminoacyl tRNA to the ribosome. The bound GTP is hydrolyzed as the correct tRNA is inserted, so EF-Tu complexed to GDP is released. The EF-Tu/GDP complex is inactive and unable to bind another tRNA. In order for translation to continue, the active EF-Tu/GTP complex must be regenerated by another factor, EF-Ts, which stimulates the exchange of the bound GDP for free GTP.

Termination of Translation

Elongations of the polypeptide chain continues until a *stop codon* (UAA, UAG or UGA) is tranlocated in the A site of the ribosome. Any one of these nucleotide triplets signals the termination of a polypeptide chain. *It may be noted here that cells do not have tRNAs with anticodons complementary to these termination nucleotide triplets.* Instead, they have *release factors* that recognize the termination codons for the termination of protein synthesis. There are two such protein release factors: **RF1** and **RF2**. (*In eukaryotes only one release factor eRF has been identified*). RF1 recognizes UAA and UAG and RF2 recognizes UAA or UGA. The corresponding release factors bind to a termination (or stop) codon at the A site and stimulate hydrolysis of the bond between the tRNA and the polypeptide chain at the P site. This results in the release of the completed polypeptide chain from the ribosome. Thereafter the ribosomes dissociate with the help of the initiation factor EF3 (discussed already). The various steps of termination of translation are shown in Fig. 5.32.).

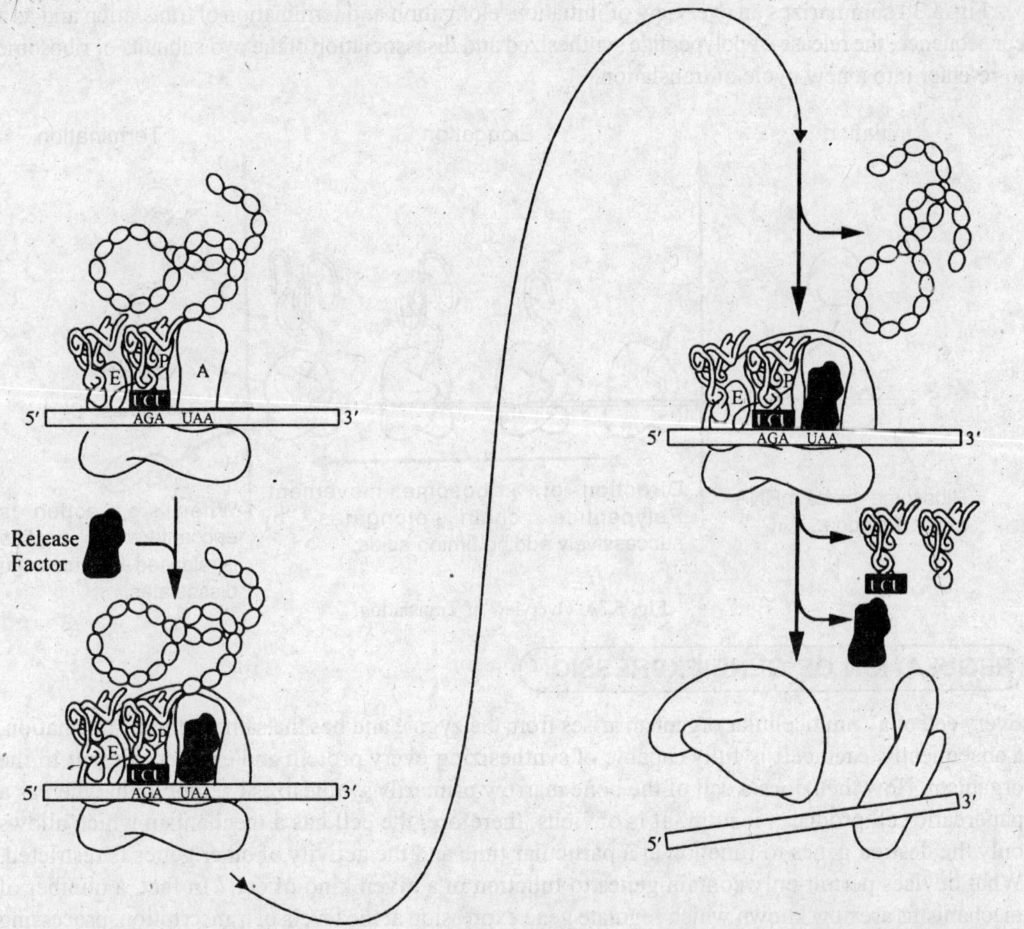

Fig. 5.32. Termination of translation. A termination codon (e.g. , UAA) at the A site is recognized by a release factor rather than by a tRNA. The result is the release of the completed polypeptide chain, followed by the dissociation of tRNA and mRNA from the ribosome.

It may be noted here that mRNA can be translated simultaneously by several ribosomes. Once one ribosome has moved away from the initiation site another can bind to the mRNA and begin synthesis of a new polypeptide chain. Usually, mRNAs are translated by a series of ribosomes, Such ribosomes are called **polysomes or polyribosomes.** Each ribosome within the group functions independently to synthesize a separate polypeptide chain. The direction of movement of the ribosome is from 5′ to 3′ and of the mRNA. It is for this reason that ribosomes closer to the 3′ end of mRNA have longer polypeptide chains (Fig. 5.33).

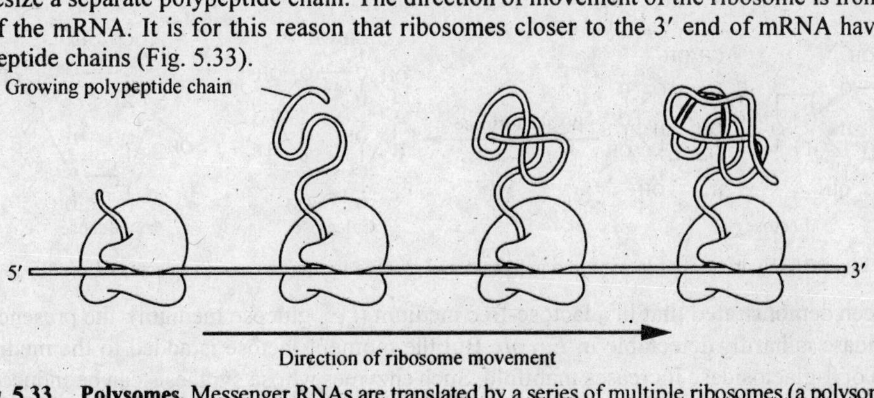

Fig. 5.33. **Polysomes.** Messenger RNAs are translated by a series of multiple ribosomes (a polysome). Note that the ribosomes closer to the 3’ end of the mRNA have longer polypeptide chains.

Fig. 5.34 summarizes an overview of initiation, elongation and termination of translation and as a consequence, the release of polypeptide synthesized and disassociation of the two subunits of ribosome to re-enter into a new cycle of translation.

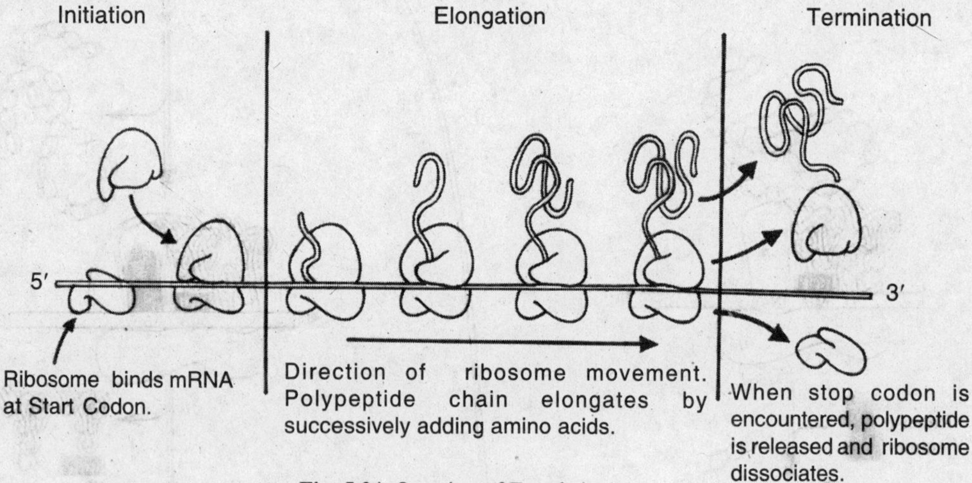

Fig. 5.34. Overview of Translation.

REGULATION OF GENE EXPRESSION

Every cell of a multicellular organism arises from the zygote and has the same genetic information. Consequently, each cell is fully capable of synthesizing every protein and enzyme present in the organism. How then does a cell of the bone marrow primarily synthesizes haemoglobin whereas a pancreatic cell produces insulin? It is obvious, therefore, the cell has a mechanism which allows only the desired genes to function at a particular time and the activity of other genes is restricted. What devices permit only contain genes to function in a given kind of cell? In fact, a number of mechanisms are now known which regulate gene expression at the levels of transcription, processing of mRNA and translation. Abundant information is available on the regulation of gene expression in prokaryotes by pioneering researches done by **F. Jacob** and **J. Monod** in 1961 on *E. coli*.

The Development of Operon Concept

Induction and Repression: In E. coli, growing on a lactose medium the enzyme β-galactosidase is responsible for the hydrolysis of the disaccharide lactose (a β-galactoside) into two monosaccharides **galactose** and **glucosec** (Fig. 5.35).

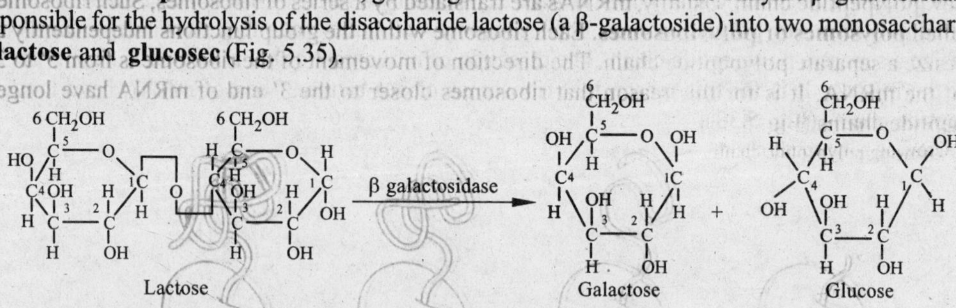

Fig. 5.35. Hydrolysis of lactose into galactose and glucose by the enzyme β-galactosidase.

It has been demonstrated that in a lactose-free medium (*i.e.*, glucose medium), the presence of β-galactosidase is hardly detectable in *E. coli*. But the moment lactose is added to the medium, production of β-glactosidase increases manifold. Such enzymes whose synthesis can be induced by adding the substrate are known as **inducible enzymes.** The genetic system responsible for the synthesis of such enzymes is known as **inducible system.** The substrate (lactose in *E. coli*) whose, addition induces the synthesis of the *enzyme* (β-galactosidase) is called an *inducer*. (*There are constitutive enzymes as well which are synthesized at a constant rate irrespective of the cell's environment. This will be discussed later*).

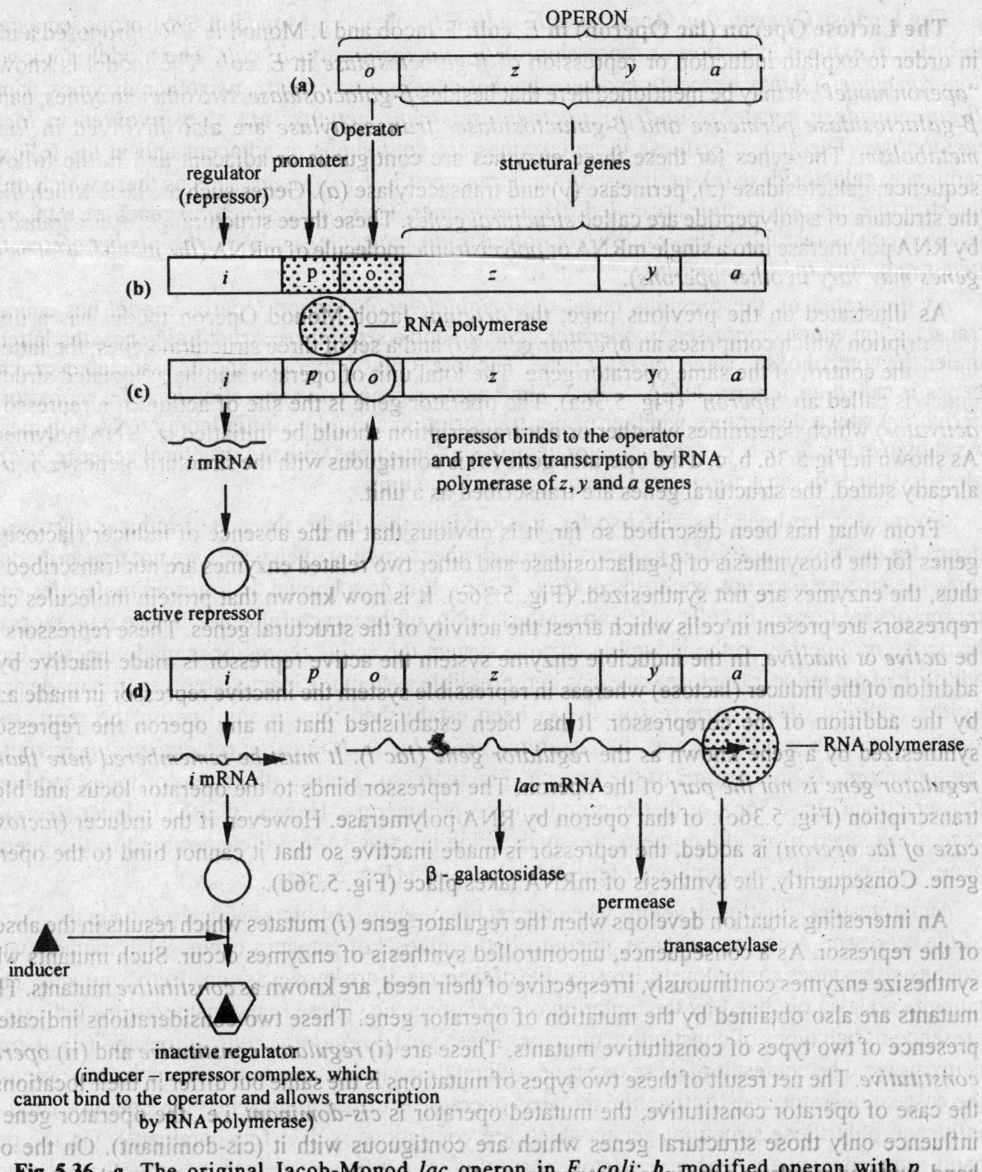

Fig. 5.36. ***a,*** The original Jacob-Monod *lac* operon in *E. coli*; ***b,*** modified operon with *p* (promoter) and *i* (regulator) genes; ***c,*** lac operon in repressed state; ***d,*** lac operon in induced state. The inducer (lactose) changes the shape of repressor, so repressor can no longer bind to the operator and allows transcription by RNA polymerase.

In another situation, when no amino acids are supplied from outside, *E.coli* cells synthesize all the enzymes required for the synthesis of different amino acids. However, if a particular amino acid like *histidine* is added to the medium, the synthesis of histidine synthesizing enzymes decrease markedly. Thus, the addition of the end product of a biosynthetic pathway blocks the synthesis of the enzymes needed for its biosynthesis. Such enzymes whose, synthesis is stopped by adding the end product in the medium are known as **repressible enzymes** and the specific genetic septem responsible for blocking the synthesis of enzymes by adding the end product is known as **repressible system.** In this case where the end product which blocks the synthesis of its own biosynthetic enzymes is known as a **corepressor.**

The Lactose Operon (lac Operon) in *E.* **coli:** F Jacob and J. Monod in 1961 proposed a model in order to explain induction or repression of *β-galactosidase* in *E. coli.* The model is known as "*operon model*". It may be mentioned here that besides *β-galactosidase*, two other enzymes, namely, *β-galactosidase permease and β-galactosidase transacetylase* are also involved in *lactose metabolism*. The genes for these three enzymes are contiguous or adjacent and in the following sequence: galactosidase (*z*), permease (*y*) and transacetylase (*a*). Genes such as these which dictate the structure of a polypeptide are called *structural genes*. These three structural genes are transcribed by RNA polymerase into a single mRNA or *polycistronic* molecule of mRNA (*the number of structural genes may vary in other operons*).

As illustrated on the previous page, the *original* Jacob-Monod Operon model has a unit of transcription which comprises an *operator gene* (*o*) and a set of three structural genes; the latter are under the control of the same operator gene. The total unit of operator and its associated structural genes is called an "*operon*" (Fig. 5.36a). The operator gene is the site of action of a repressor (*or activator*) which determines whether or nor transcription should be initiatled by RNA polymerase. As shown in Fig 5.36. b, c, d the operator gene (*o*) is contiguous with the structural genes (*z, y, a*). As already stated, the structural genes are transcribed as a unit.

From what has been described so far, it is obvious that in the absence of inducer (lactose) the genes for the biosynthesis of β-galactosidase and other two related enzymes are not transcribed and, thus, the enzymes are not synthesized. (Fig. 5.36c). It is now known that protein molecules called repressors are present in cells which arrest the activity of the structural genes. These repressors may be *active* or *inactive*. In the inducible enzyme system the active repressor is made inactive by the addition of the inducer (lactose) whereas in repressible system the inactive repressor in made active by the addition of the corepressor. It has been established that in any operon the repressor is synthesized by a gene known as the *regulator gene* (*lac i*). *It must be remembered here that the regulator gene is not the part* of the operon. The repressor binds to the operator locus and blocks transcription (Fig. 5.36c). of that operon by RNA polymerase. However, if the inducer (*lactose in case of lac operon*) is added, the repressor is made inactive so that it cannot bind to the operator gene. Consequently, the synthesis of mRNA takes place (Fig. 5.36d).

An interesting situation develops when the regulator gene (*i*) mutates which results in the absence of the repressor. As a consequence, uncontrolled synthesis of enzymes occur. Such mutants which synthesize enzymes continuously, irrespective of their need, are known as *constitutive* mutants. These mutants are also obtained by the mutation of operator gene. These two considerations indicate the presence of two types of constitutive mutants. These are (i) *regulator constitutive* and (ii) *operator constitutive.* The net result of these two types of mutations is the same but differ in their locations. In the case of operator constitutive, the mutated operator is *cis-dominant*, i.e., the operator gene can influence only those structural genes which are contiguous with it (cis-dominant). On the other hand, mutated regulator gene in regulator constitutives can influence even those genes that are present on a different chromosome (*trans position*).

The Jacob-Monod operon model has been altered little which is a remarkable tribute to their research and creative data. However, the one major addition to the original Jacob-Monod operon model is the assignment of a *promoter gene* (*p*) which is located at the operator end, adjacent but external (*to the left*) of the operator gene. (Fig. 5.36 b, c, d). The promoter is a region on a DNA molecule to which an RNA polymerase binds and moves from the promoter and initiates transcription.

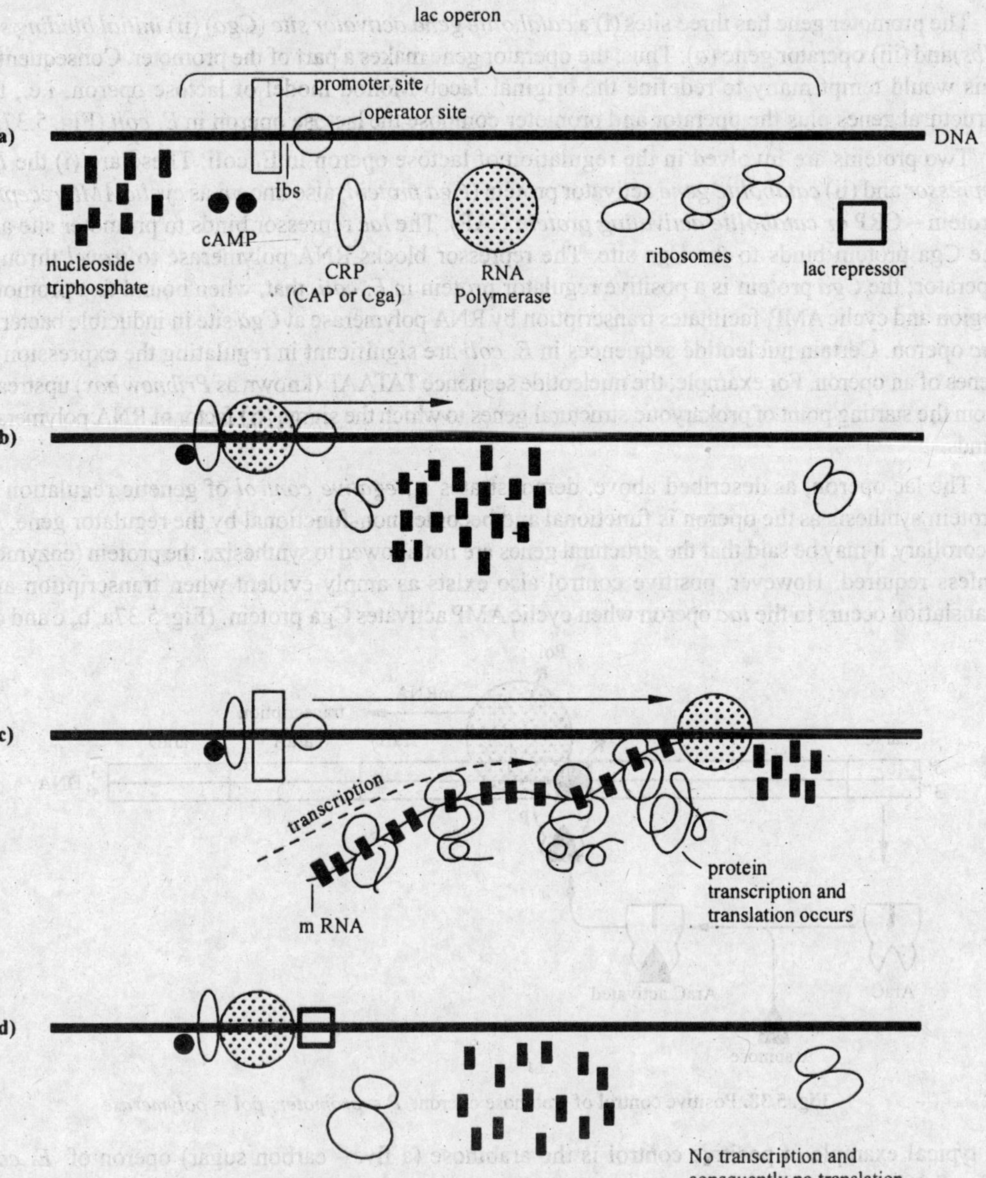

Fig. 5.37. Positive control of genetic regulation of transcription and translation in *E. coli lac* operon: ***a,*** various components for its regulation; ***b,*** and ***ĉ,*** only in the absence of the repressor, RNA polymerase enzyme can travel and transcribe *lac* operon end protein synthesis takes place; ***d,*** presence of the repressor on the operator site in the path of RNA polymerase does not allow transcription and, as a result, no protein synthesis takes place

The nucleotide sequence of the promoter determines both the nature of the enzyme that attaches to it and the rate of RNA synthesis. It has already been made clear that the operator is site of action of the repressor. The repressor when attached to the operator prevents the movement of RNA polymerase (Fig. 5.36c).

The promoter gene has three sites (i) a *catabolite gene activator site* (*Cga*) (ii) *initial binding site* (*Ibs*)and (iii) operator gene (*o*). Thus, the operator gene makes a part of the promoter. Consequently, this would tempt many to redefine the original Jacob-Monod model of lactose operon, i.e., the structural genes plus the operator and promoter compose the lactose operon in *E. coli* (Fig. 5.37a).

Two proteins are involved in the regulation of lactose operon in E. coli. These are (i) the *lac repressor* and (ii) *catabolite gene* activator protein (*Cga protein*; also known as *cyclic AMP receptor* protein – CRP *or catabolite activating protein* CAP). The *lac* repressor binds to promoter site and the Cga protein binds to the Cga site. The repressor blocks RNA polymerase to *travel* through operator; the *Cga* protein is a positive regulator protein in *E. coli*, that, when bound to a promoter region and cyclic AMP, facilitates transcription by RNA polymerase at *Cga* site in inducible bacterial *lac* operon. Certain nucleotide sequences in *E. coli* are significant in regulating the expression of genes of an operon. For example, the nucleotide sequence TATAAT (known as *Pribnow box*) upstream from the starting point of prokaryotic structural genes to which the sigma (σ) factor of RNA polymerase binds.

The lac operon, as described above, demonstrates a *negative control* of genetic regulation of protein synthesis as the operon is functional and becomes non-functional by the regulator gene. As a corollary, it may be said that the structural genes are not allowed to synthesize the protein (enzymes) unless required. However, positive control also exists as amply evident when transcription and translation occurs in the *lac* operon when cyclic AMP activates Cga protein. (Fig. 5.37a, b, c and d).

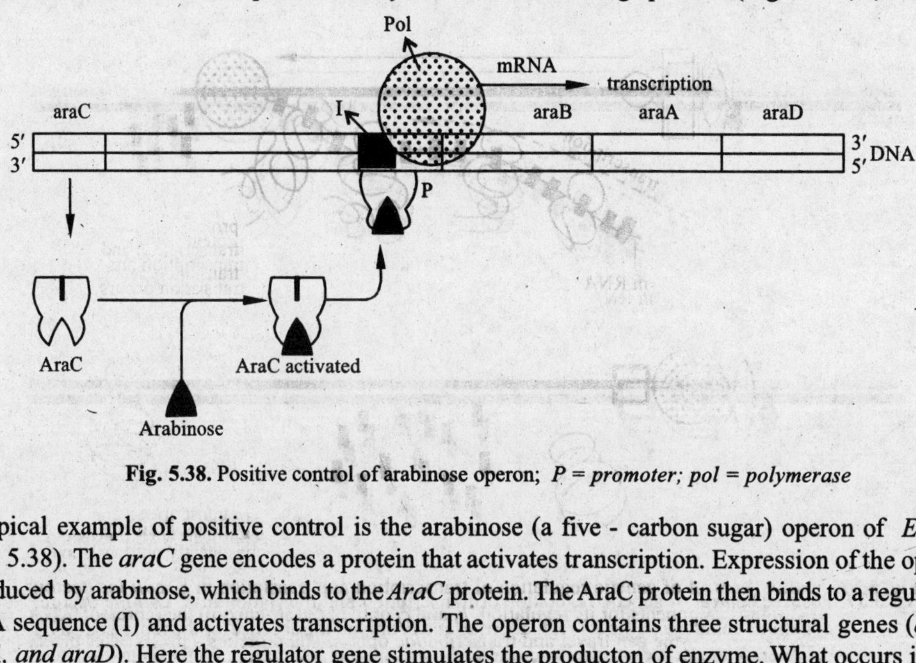

Fig. 5.38. Positive control of arabinose operon; *P = promoter; pol = polymerase*

A typical example of positive control is the arabinose (a five - carbon sugar) operon of *E. coli* (Fig. 5.38). The *araC* gene encodes a protein that activates transcription. Expression of the operon is induced by arabinose, which binds to the *AraC* protein. The AraC protein then binds to a regulatory DNA sequence (I) and activates transcription. The operon contains three structural genes (*araB, araA, and araD*). Here the regulator gene stimulates the producton of enzyme. What occurs in this case is that the gene without the activator will be inactive. *It may be emphasized here that the genetic regulation of gene expression is mostly operative by negative control.*

FEEDBACK INHIBITION OF ENZYME ACTIVITY

Besides genetic regulation of gene expression, a very simple form of regulation is the control of enzyme catalysis by *feedback inhibition*. Feedback inhibition is a typical example of a simple feedback loop of *allosteric regulation*, in which a regulatory molecule binds to a site on an enzyme that is distinct from the catalytic site (*alla* = "*other*"; *steric* = "*site*"). The binding of such regulatory

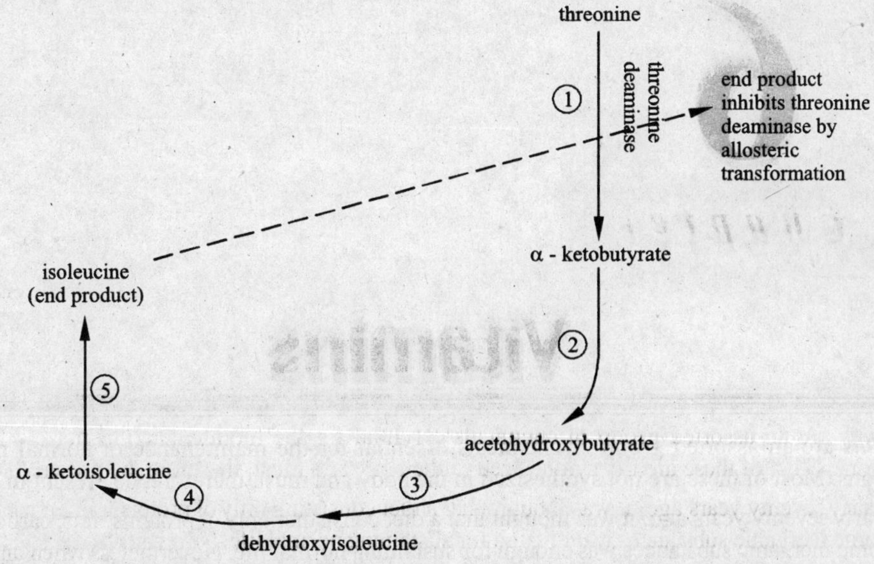

Fig. 5.39. Feedback inhibition. The end *product (isoleucine)* of a biochemical pathway acts as an allosteric inhibitor of the enzyme (threonine deaminase) that catalyzes the first step in its synthesis.

molecule alters the conformation of the protein, thereby changing the shape of the catalytic site and affecting catalytic activity. Thus, in *E.coli*, the pathway of *isoleucine* biosynthesis is a five-step enzymatic process utilizing *threonine* as the substrate and form intermediates It has been demonstrated that *E. coli* cells growing in isoleucine blocks the activity of *threonine deaminase* which is involved in the first step of the biosynthetic pathway of isoleucine (Fig. 5.39). In other words, it means that accumulation of isoleucine stops its own synthesis. Isoleucine does this by inhibiting threonine deaminase by *allosteric transformation* so that the enzyme is unable to perform its catalytic function. This is also referred to as *end product inhibition*. In summary, the end product of a reaction sequence inhibits the catalytic activity of the enzyme that catalyzes the first reaction of the sequence, i.e., the signal to stop making the end product itself. However, when the end product has beeh used (as a source of energy or as building block), the regulatory enzyme (threonine deaminase in this example) is freed to initiate synthesis (of isoleucine) again. Thus, in feedback inhibition the enzyme does not function, while in repression no mRNA synthesis occurs and hence no enzyme synthesis takes place. Therefore, this is not a case of control of gene activity, but gene expression is controlled only at post-translational level.

6

Chapter

Vitamins

Vitamins are an assorted group of chemicals essential for the maintenance of normal metabolic functions. Most of these are not synthesized in the body and must, therefore, be present in the diet.

Nearly seventy years ago, it was thought that a diet consisting only of proteins, fats, carbohydrates and some inorganic substances was enough for sustaining normal life. Nevertheless when an artificial diet consisting of these materials was given to rats they died after sometime. However, when small amount of milk was added to the above diet the rats die not die. The unspecified substances provided by milk were called *accessory food factors.* About the same time it was also found that exclusive eating of polished rice (rice in which the pericarp or bran was removed) caused a disease known as *beri-beri.* Addition of polishings to the diet cured this disease. This discovery led to the concept of *deficiency diseases.* Funk (1911) isolated a concentrate from rice polishing which was found to contain nitrogenous bases (amines). As these substances were vital for sustaining life, they were called *vitamines.* Later, when it was discovered that all these essential substances, were not exclusively amines the letter 'e' was dropped and the name *vitamin* was given to all the *essential substances which must be present in the diet in small quantities and whose deficiency causes disease.*

The vitamins are divided into two major categories—(a) fat soluble, and (b) water soluble vitamins. (Table 6.1)

Table 6.1 List of Vitamins

Fat soluble vitamins	*Water soluble vitamins*
Vitamin A	Thiamine—Vitamin B_1
Vitamin D	Riboflavin— Vitamin B_2
Vitamin E	Nicotinic acid—Niacin
Vitamin K	Pyridoxine— Vitamin B_6
Vitamin C	Pteroyl glutamic acid—Folic acid
Lipoic acid	Pantothenic acid
	Cyanocobalamin— Vitamin B_{12}
	Biotin
	Choline
	Inositol
	Para amino benzoic acid—PABA (All these along with Lipoic acid from Vitamin B complex)

The knowledge of the chemical structure, physiological role, deficiency diseases and available sources of vitamin is essential in nutritional studies. A detailed description of the different vitamins follows:

FAT SOLUBLE VITAMINS

Vitamin A (Antixeropthalmic vitamin)

Vitamin A is a primary unsaturated alcohol which exists in nature in two forms, vitamin A_1 and vitamin A_2 called *retinol* and *dehydroretinol* respectively.

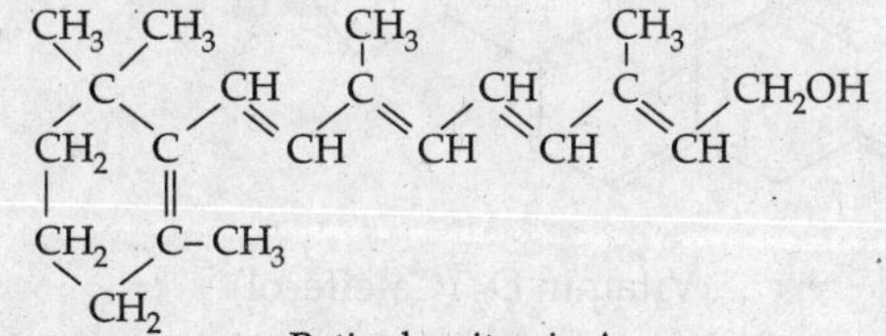

Retinol – vitamin A_1

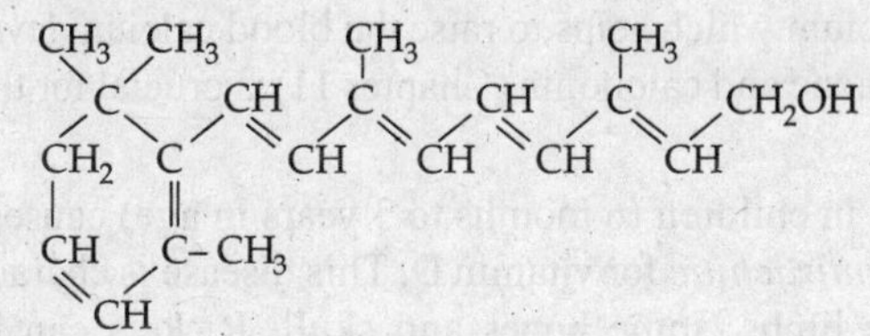

(3, Dehydroretinol)

Vitamin A exists in several *cis-trans* isomers which are easily interconvertible in the body. This vitamin is essential for vision and for the maintenance of the various epithelial cells. The rods of the retina (Chapter 18) contain a pigement *rhodopsin* or visual purple. Rhodopsin is a combination of the protein *opsin* and a compound called *retinene.* Retinene in fact is the aldehyde of *cis* vitamin A. Similarly the cones of the retina contain another pigement called *iodopsin.* This pigment is responsible for colour vision. It is made of *photopsin* and retinene. When human beings are fed on diets deficient in vitamin A their ability for dark adaptation is lost which results in *night blindness*. Severe deficiency of vitamin A causes keratinization of corneal cells resulting in a disease called *rexopthalmia.*

Absence of vitamin A also causes atrophy of epithelial cells. In the alimentary canal its deficiency results in the damage of intestinal mucosa resulting in diarrhoea. Defects in the dentine of the teeth or even its absence is caused by acute deficiency of vitamin A.

The daily human requirement of vitamin A is 6,000 international units (I.U. = 0.30 micrograms). Very high consumption of vitamin A, on the other hand, causes a disease *hypervitaminosis A*. This is marked by deep swellings in the forearms, shanks and feet which are very painful.

Most vitamin A is the stored in the liver. Cod liver oil, halibut liver oil etc. are commercially sold as sources of this vitamin. In addition to the above, edible fats like milk, butter, ghee etc. also contain vitamin A. This vitamin is slowly destroyed in the fats and oils even at room temperature. Plants too, contain vitamin A in the form of pigements. Two such sources of the vitamin are **β-carotene** and **cryptoxanthin** from the yellow-green parts of plants, which are converted into vitamin A by liver and alimentary canal.

Vitamin D (Calciferol)

There are at least ten different substances possessing properties of vitamin D. All these have a general structure similar to the sex hormones and cortisones, and are derivatives of **cyclopentaphenanthrene**

(Chapter 11). With the help of ultraviolet radiation present in the sun rays vitamin D is synthesized in the skin by molecular rearrangement in already existing sterols. The two most important sterols which are converted into the vitamin are *ergosterol* and *7-dehydroxycholesterol.* These two on exposure to ultraviolet rays, change into vitamin D_2 (calciferol) and vitamin D_3 respectively.

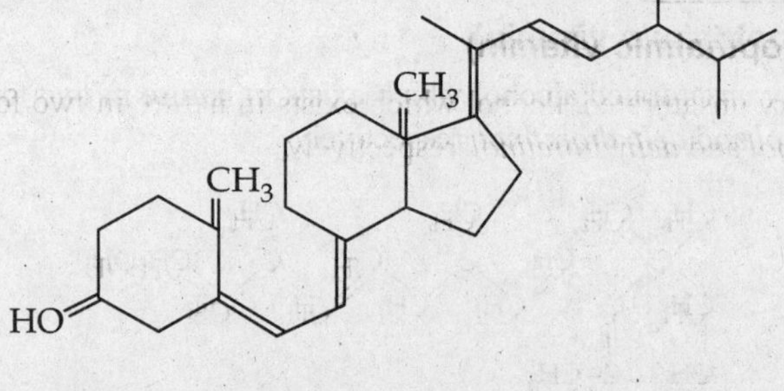

Vitamin D_2 (Caleiferol)

Vitamin D_2 helps in maintaining the levels of calcium and phosphorous in the body and is involved in resorption of calcium from calcified bones by *osteolytic cells.* It also increases the permeability of the intestinal mucosa to calcium which helps to raise the blood calcium level of the body. Vitamin D along with parathyroid hormone and calcitonin (Chapter 11) is crucial for the metabolism of calcium and phosphorous.

Deficiency of vitamin D in children (6 months to 3 years in age) causes a disease called *rickets* (*rachitis*), hence the name *antirachitic* for vitamin D. This disease is characterised by deformities of the skeleton especially the limbs, spine bones and skull. Rickets can be cured rapidly by the administration of vitamin D. Since ultraviolet radiation helps in the synthesis of vitamin D in the body, exposure of rickety children to sunlight may help in the cure of the disease. Another ricket-like disease *osteomalcia* is found in adults in India and China.

The daily human requirement of vitamin D is 500 I.U. (One I.U. is equivalent to 0.025 microgram of vitamin D_3 contained in U.S.P. vitamin D reference standard), Excess vitamin D (150,000 I.U./day or more) causes a disease called *hypervitaminosis D.* In this, the bones become weak (*osteoporosis*) due to calcium resorption and there is calcification of soft tissue especially the kidney, heart, lungs etc.

The important sources of vitamin D are the liver oils of tunny, halibut, cod and other fishes. Of these, tunny liver oil is the richest source of this vitamin. Besides these egg yolk and milk also contain vitamin D.

Vitamin E (Tocopherols)

There are three types of vitamin E (tocopherols). These are α, β and γ tocopherols. The structure of α tocopherol is given here.

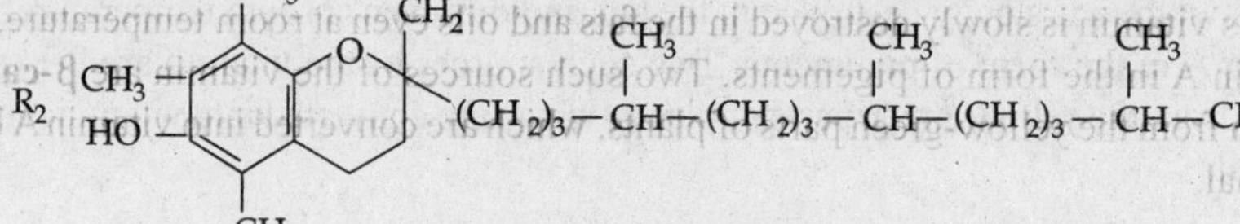

α-tocopherol

The β and γ tocopherols differ from the α compound in having two instead of three methyl groups.

The tocopherol has only R_2, R_3, R_1 and H. The γ tocopherol has R_1, R_2, R_3 and H of these the α tocopherol isbiologically active. Its deficiency in experimental animal causes degeneration of the germinal epithelium of both sexes. In males it results in immobile sperms. Although vitamin E deficient females can have fertilization, the embroyo gets aborted after some time. Vitamin E, therefore, is also referred to as *antisterility* vitamin. Its utility in human beings has not been conclusively proved.

Vitamin E is found in most vegetables, milk and meat. It is found abundantly in vegetable oils. Of these wheat germ oil is the richest source of vitamin E.

Vitamin K

Vitamin K occurs in two forms, vitamin K_1 (farnoquinone) and vitamin K_2 (phylloquinone). Both of these contain a 2 methyl 1, 4 naphthoquinone. They differ from each other in their side chain R.

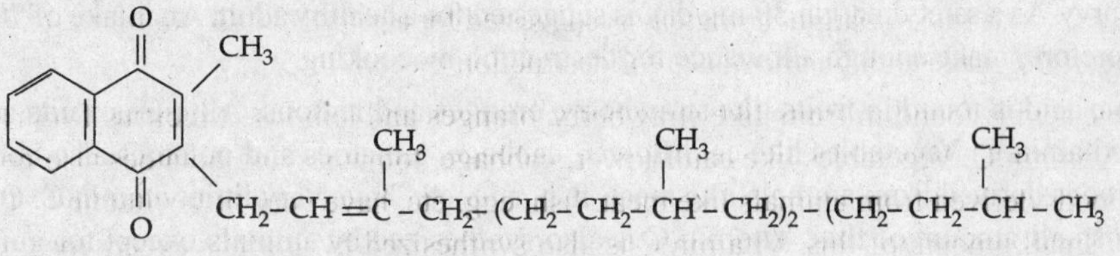

2-methyl-1, 4 naphthoquinone

Vitamin K plays an important role in blood coagulation. Deficiency of vitamin K reduces the concentration of prothrombin and factors VII, IX and X in the liver (Chapter 14). As these factors are essential for blood clotting, deficiency of vitamin K increases the clotting time. New born infants deficient in vitamin K are in danger of bleeding to death. Administration of the vitamin in such cases may prevent the tragedy.

The amount of vitamin K required by an individual is not known. However, a normal diet provides enough of the vitamin. Vitamin K is abundantly found in green vegetables like spinach, cauliflower, cabbage and alphalpha. Besides this, the intestinal bacteria can also synthesize this vitamin.

Vitamin C (Ascorbic Acid)

Vitamin C is both water and fat soluble. It has the following structural formula:

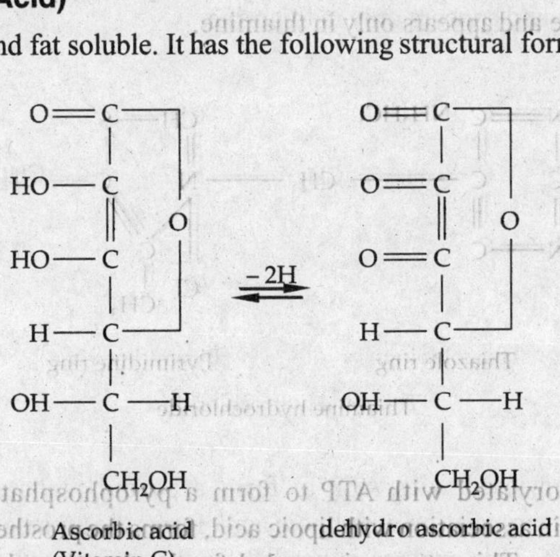

Ascorbic acid (Vitamin C) dehydro ascorbic acid

Ascorbic acid has an asymmetrical carbon and is, therefore, found both in L and D forms. However, only the L-isomer is found in nature and has biological activity. Ascorbic acid is a powerful reducing agent as it gives a pair of hydrogen atoms and gets oxidized to *dehydroascorbic acid.* Because of this property ascorbic acid is unstable and gets easily oxidised during cooking. Freshly chopped vegetables and fruits contain an enzyme *ascorbic acid oxidase* which is capable of oxidising this vitamin.

The exact mode of action of ascorbic acid is not known. However, it is throught that vitamin C may have an important role in maintaining the SH-containing enzymes in the reduced form. In high concentration in the adrenals, it prevents epinephrine from being oxidized. Besides this, vitamin C may also have a role in the synthesis of *adreno-corticotrophic hormone* (ACTH). It is essential for the maintenance of collagen fibres, epithelial tissue, bone matrix and capillary lining.

In case of vitamin C deficiency the capillary linings develop leaks causing haemorrhage. The deficiency of this vitamin causes a disease called *scurvy.* This is characterized by spongy and bleeding gums, bleeding under the skin and extreme weakness. Since vitamin C is continuously oxidised in the body, the daily requirement of the vitamin is rather high. 10 mg/day of the vitamin is enough to prevent scurvy. As a safety margin 30 mg/day is suggested for a healthy adult. An intake of 70 mg/day should, therefore, make enough allowance for destruction by cooking.

Ascorbic acid is found in fruits like strawberry, oranges and lemons. All citrus fruits are a rich source of vitamin C. Vegetables like cauliflower, cabbage, tomatoes and potatoes also contain the vitamin. Foods derived from animals like meat, fish, egg, etc. have very little vitamin C. Only milk contains a small amount of this. Vitamin C is also synthesized by animals except for guineapigs, frogs, apes and man.

WATER SOLUBLE VITAMINS

The Vitamin B Complex

The vitamin B complex comprises of many vitamins, all grouped together because they are water soluble and can be obtained from the same source *i.e.* liver and yeast. However, these two are not the only sources of these vitamins. Several of these vitamins are essential parts of important enzymes. The vitamin Bs known so far are given in Table 6.1. A detailed discussion of these is given below.

Thiamine: Vitamin B_1 or Aneurin

Thiamine was the first vitamin B to be identified, hence the name vitamin B_1. Thiamine is a complex molecule having a pyrimidine and a thiazole nucleus joined by a methyl bridge. The thiazole nucleus as far as is known is unique and appears only in thiamine.

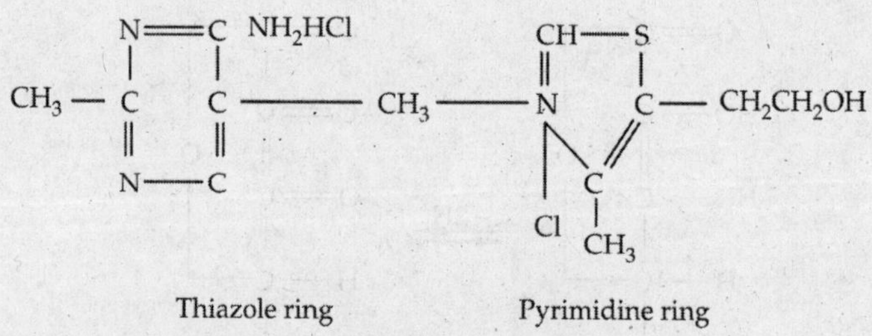

Thiamine hydrochloride

Thaiamine is phosphorylated with ATP to form a pyrophosphate ester called *thiamine pyrophosphate.* The latter in association with lipoic acid, forms the prosthetic group *co-carboxylase* for the enzyme *carboxylase.* This enzyme is needed for the decarboxylation of α keto acids like pyruvic acid (Chapter 13). Thiamine is, therefore, very important in carbohydrate metabolism.

Thiamine deficiency leads to a disease known as *beri-beri* (Singhalese-beri=weakness). *Beri-beri* is a disease of the nervous system which is characterised by partial paralysis of the extremities, emaciation and anemia. This is caused due to the degeneration of peripheral nerves (*polyneurities*). Even before the concept of vitamins was developed *beri-beri* was described as a deficiency disease. It was found that it occurred in individuals eating polished rice. Administration of rice polishing cured the disease.

The normal person needs about 3331.U. (1 mg thiamine hydrochloride) of the vitamin every day. Thiamine has the highest concentration in the heart followed by liver, kidney and brain. It is also present in nuts, cereals, pulses and yeast. The bran of pulses and grains is a very rich source of this vitamin; it is because of this that it is advisable to eat whole flour and unpolished rice.

Raw flesh of fish and molluscs contains an enzyme *thiaminase* which destroys the vitamin in the body. People eating raw fish and molluscus, therefore, are liable to suffer thiamine deficiency.

Lipoic acid (Thioctic acid)

This is an exception of the vitamin B series as it is fat soluble rather than water soulble.

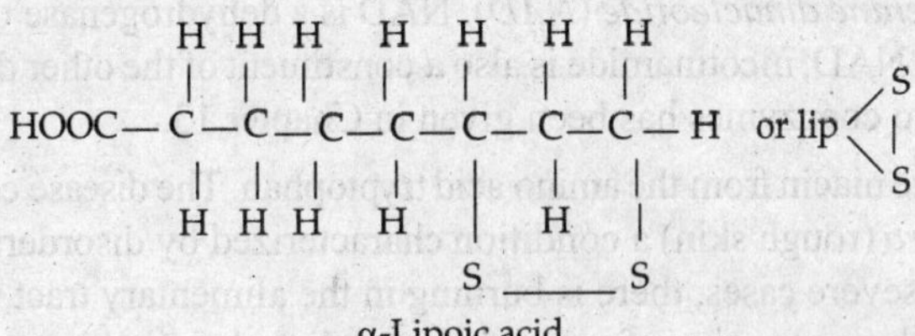

α-Lipoic acid

As said before, lipoic acid in association with thiamine forms the prosthetic group of caboxylase. Lipioc acid is synthesized by most animals, hence, it is not an essential factor of the diet. It is found in the liver of most edible animals.

Riboflavin (Vitamin B_2)

Riboflavin is a yellow coloured substance having the following structure.

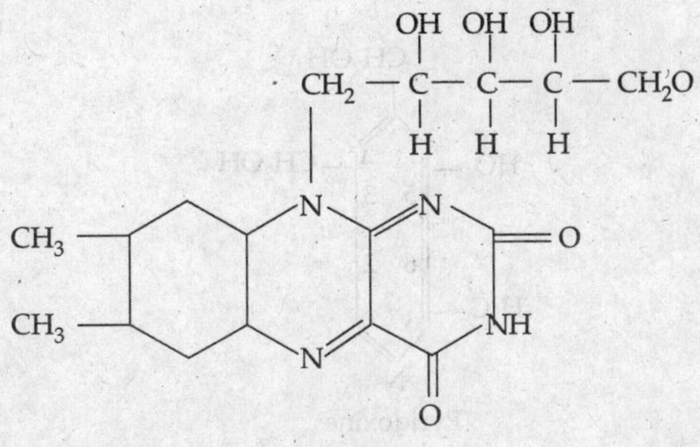

Riboflavin

Riboflavin is a component of all oxidative enzymes called *flavoproteins* which consist of a coenzyme *flavin-adenine dinucleotide* (FAD). The flavoproteins play a very important role in the oxidation of hydrogen (electron trnasfer). Since all energy eventually comes from electron transfer, FAD occupies a key place in energy metabolism. Details of this enzyme have been given in Chapter 13.

Riboflavin deficiency results in the growth of fine blood vessels in the cornea which give it a blood shot appearance and the eyes become photophobic. In severe deficiency the skin becomes scaly, the lips become red and cracked, and the tongue enlarged and magenta coloured.

The daily human requirement is about 2 mg. Riboflavin is found in germinating seeds and growing plants. The most important source of this vitamin are peas, beans, grain and yeast. Milk, meat and egg yolk are also rich in this vitamin.

Nicotinic acid (Niacin)

Nicotinic acid is a pyridine derivative.

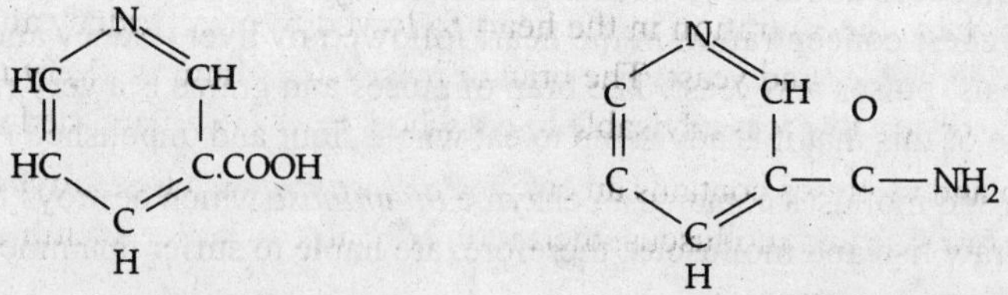

Nicotinic acid Nicotinic acid amide

The amide of nicotinic acid combines with one molecule each of adenine, a pentose and phosphate to form *nicotinamide adenine dinucleotide* (*NAD*). NAD is a dehydrogenase of extreme importance in metabolism. Similar to NAD, nicotinamide is also a constituent of the other dehydrogenase, NADP. A description of these two coenzymes has been given in Chapter 13.

The liver can synthesize niacin from the amino acid tryptophan. The disease caused by the deficiency of niacin is called *pellagra* (rough skin) a condition characterized by disorders of the skin, digestive and nervous systems. In severe cases, there is burning in the alimentary tract, nausea and vomiting; skin becomes red and itchy and various forms of mental disturbances appear. The average human requirement of niacin is 10-15 mg/day. The richest sources for this are liver, kidney, yeast, wheat and green vegetables; milk and meat also contain this vitamin.

Pyridoxine (Vitamin B_6)

Pyrodoxine like niacin is a pyridine derivative. This vitamin exists in three forms, *pyridoxine, pyridoxal* and *pyridoxamine.* These differ from each other in the nature of the substituent on the fourth carbon.

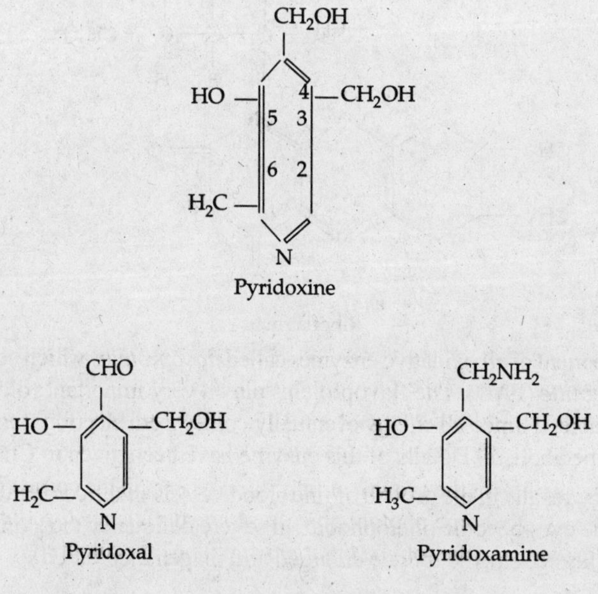

Pyridoxine

Pyridoxal Pyridoxamine

Phosphorylated pyridoxine and its derivatives are coenzymes for *transaminases* (Chapter 8), As the intestinal bacteria are capable of synthesizing vitamin B_6, its deficiency is not normally found in human beings. In cases where deficiency does occur, the skin near the eyes, nose etc. develops *seborrhea*—like (only crusts or scales on the skin) symptoms. Nervous convulsions have also been reported in experimental animals having vitamin B_6 deficiency.

The daily human requirement ranges from 0.2 to 1 mg. It is abundantly found in foods of both plant and animal origin.

Pteroylglutaminc acid (Folic acid)

Folic acid contains a pteridine nucleus linked to *p-aminobenzoic acid* to form pteroic acid. The latter is attached to glutamic acid.

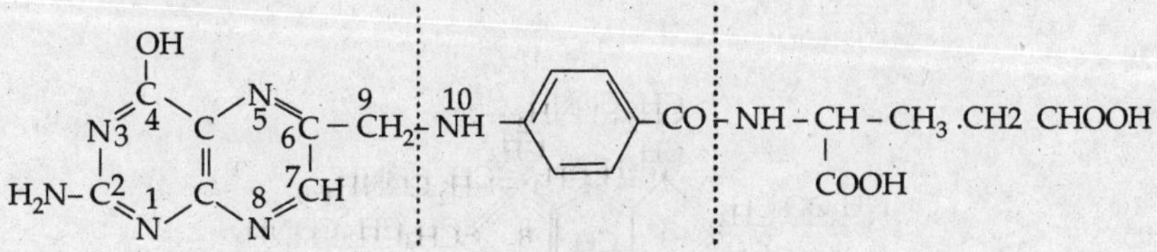

Pteridyl Para-aminobenzoic acid Glutamic acid

Folic acid itself is not an active vitamin in mammals. However, it takes up hydrogen to get converted to *folinic acid* or *citrovorum factor* (C.F.) which along with vitamin B_{12} acts as coenzyme F. The latter is essential for the methylation of pyrimidince to form thymine and the formation of purines by introducing to 2 and 8 carbon atoms in the molecule. It is also important in the synthesis of serine and choline. It is believed that its ability to synthesize the above depends on its property of lifting one-carbon units from the so called formate pool. Pyridoxine is thus extremely important in the synthesis of pyrimidines and purines which form the nucleic acids. Deficiency of folic acid leads to disturbances arising out of incomplete nucleic acid synthesis, such as failure of mitosis and meiosis. This deficiency is most marked in tissues which have a rapid rate of cell division. The first evidence of folic acid deficiency is seen in impaired erythrompoiesis (production of R.B.C). The latter causes *anemia*. Other complications of folic acid deficiency are diarrhoea and gingivitis.

The daily human requirement is 0.1 mg to 0.2 mg. Folic acid is synthesized by the microorganisms of the human colon. Important sources of folic acid are liver and green vegetables, especially spinach.

Pantothenic acid

This is a complex organic acid having the following structure.

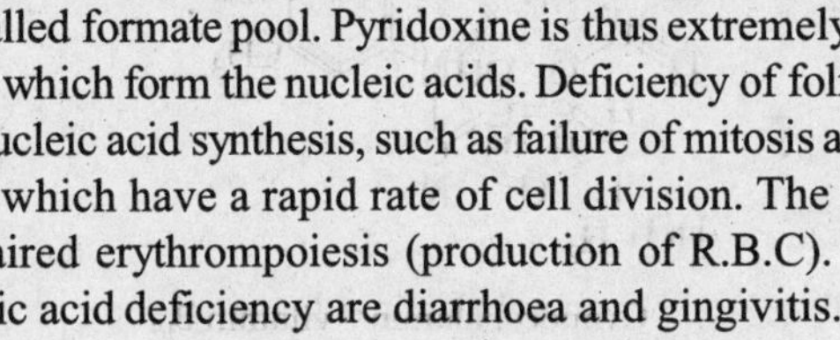

Pantothenic acid

This vitamin is bound with adenylic acid, 2-mercapto ethylamine (HS—CH_2CH_2—NH_2) and three phosphates to form coenzyme A, which is important for fat and carbohydrate metabolism (Cha pater 13). Thus five members of vitamin B Complex *i.e.*, thiamine, lipoic acid, riboflavin, niacin and pantothenic acid are essential for cellular metabolism.

Pantothenic acid deficiency is not known in man. However, in experimental animals it causes degenerative changes in the skin, heart, kidneys, lungs and the nervous system.

The daily human requirement of pantothenic acid is 5 mg. It is synthesized in the colon of man by symbiotic bacteria. It is found in yeast, most vegetables and foods of plant sources.

Cyanocobalamin (Vitamin B_{12})

A number of compounds called *corrinoids* come under the common name of vitamin B_{12}. All corrinoids essential have a Corrine nucleus. This has four pyrrole rings linked in a manner similar to the porphyrin nucleus of haemoglobin. In this, however, porphyrin is formed around cobalt instead of iron.

The various members of the vitamin B_{12} group are formed by the replacement of CN by other groups. The molecule which contains CN is called cyanocobalamin. Vitamin B_{12b} is hydroxycobalamin since the CN in this is replaced by a hydroxyl group. Similarly vitamin B_{12c} is nitrocobalamine since. CN in this is replaced by NO_2.

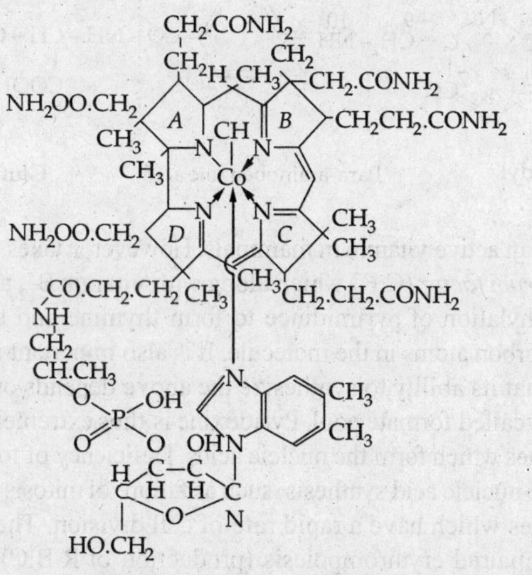

Cyanocobalamin – vitamin B_{12}

Vitamin B_{12} is converted into a coenzyme called coenzyme B_{12}. This coenzyme converts methyl malonyl coenzyme A to succinyl coenzyme A which is essential for the maintenance of the SH groups of enzymes in the *reduced form.* This form is necessary for the functioning of many enzymes. Vitamin B_{12} is essential in nutrition and growth. It is essential in highly proliferating tissues like bone marrow which produces about 200,000,000 R.B.C. per minute. Deficiency of vitamin B_{12} therefore, causes *pernicious anemia.* This disease is characterized by changes in the bone marrow and reduction in the number of circulating R.B.C. Vitamin B_{12} deficiency also causes destruction of the lining of the alimentary canal. In the stomach, secretions of pepsin and HCl are reduced. In server cases of vitamin B_{12} deficiency myelinated nerve fibres are destroyed, resulting in various kinds of nervous disorders.

The average human requirement per day in only 1 μg of vitamin B_{12}. This vitamin is generally absent in most plant products. Kidney, liver and milk contain more than the body requirements of the vitamin. A normal meat diet, therefore, is sufficient for supplying the requisite amount of vitamin B_{12}.

Biotin

Biotin is a complex organic acid which is optically active.

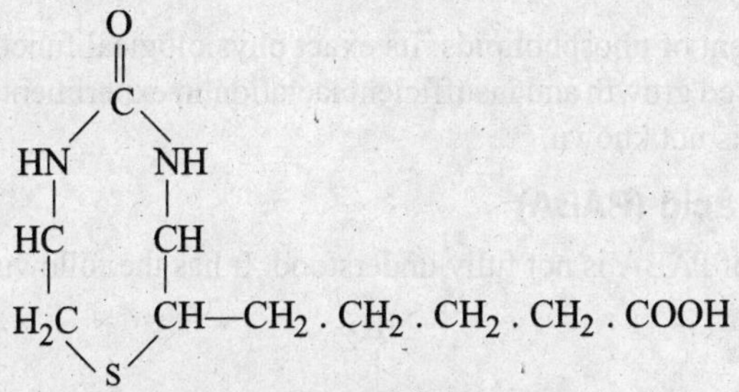

Biotin

Biotin acts as a coenzyme in the carboxylation of pyruvate to oxaloacetic acid. Biotin deficiency is normally not found in human beings. In experimental animals its deficiency produces greasiness of the skin, drowsiness, muscular pain and heart distress.

The intestinal bacteria synthesize biotin in such high quantities that the amount excreted in urine and feces exceeds the intake. This is the reason why the daily human requirement is not known. Biotin is found in yeast, liver, kidney and egg yolk.

Choline

Choline is a quaternary ammonium compound.

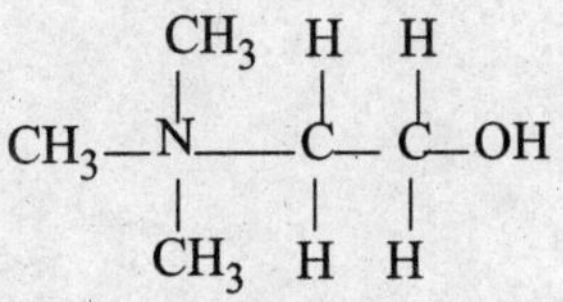

Choline

The chemistry of this compound has been dealt in Chapter 4. It is an important lipotropic agent, and participates in the mobilization of fat from the liver; its absene, therefore, causes accumulation of fat in the hepatic tissues. Whenever methylation is required, choline acts as an important methyl donor in intermediary metabolism. The ester of choline with acetic acid *i.e.*, acetylcholine is an important neurotransmitter (Chapter 9). Choline can be synthesized in the body by the methylation of ethanolamine and therefore, strictly speaking, it is not a vitamin. Its requirement ranges from 1.5 to 3 g/day. The best source for this are vegetable foods.

Inositol

Inositol is a hexahydroxycyclohexane. There are eight possible isomers of this. Of these only *i*-inositol or *meso* inositol is biologically active.

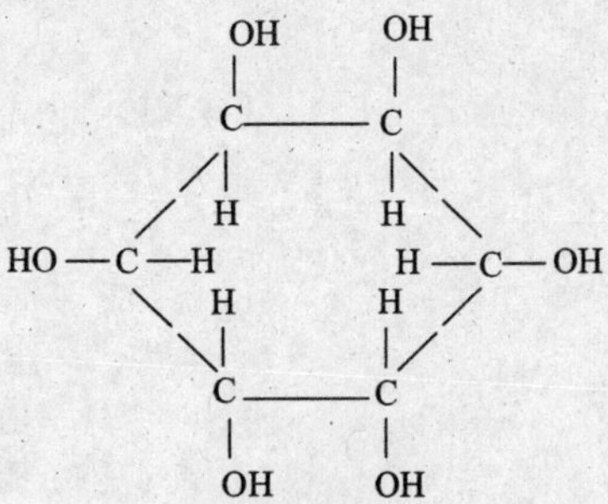

Meso-inositol
(i-inositol)

Inositol is a component of phospholipids. Its exact physiological function is not known. Inositol deficiency causes impaired growth and insufficient lactation in experimental animals. The amount of inositol needed by man is not known.

Para aminobenzoic acid (PABA)

The physiological role of PABA is not fully understood. It has the following structure:

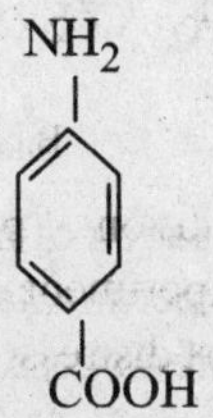

Para aminobenzoic acid

PABA is essential in microorganisms for the synthesis of folic acid. In case of human beings its presence helps in the synthesis of folic acid by intestinal bacteria. There is no clearly defined deficiency disease of PABA and there is no evidence that it is a dietary essential for human beings.

7
Chapter

Hydrogen Ion Concentration And Its Regulation

All living organisms are subjected to changes in temperature, osmotic pressure, oxygen and carbon dioxide concentrations in their *external environment.* Such changes are usually beyond the control of the individual organisms. In order to avoid the stresses produced by these changes, living organisms have developed means for keeping the *internal environment* of their body at a relatively constant level. Two factors can thus be distinguished—internal environment, by which is meant the composition of the cytoplasm, blood, lymph etc. and external environment which includes all factors which are present in the world arount the organsim. The study of the interaction of the latter with the organism comes under the realm of ecology and is beyond the scope of this book.

About a century ago, the great French physiologist, Claude Bernard, showed that the stability of the internal environment was crucial for the existance of every living organism. According to him, "All the vital mechanisms, however varied they may be, have only one object, that of preserving constant conditions of life in the internal environment (*milleu interieur*)". The mechanism by which a constant internal environment is maintained in the body is called *homeostasis.*

The factors which make up the internal environment are cytoplasm, intercellular fluids like lymph and extracellular fluids like blood and plasma. All these fluids are aqueous solutions which contain many salts, bases, organic compounds and colloidal particles (mostly proteins). The concentration of these substances determines the internal environment of the body.

Importance of Hydrogen Ion Concentration

All metabolic activities—digestion, excretion, synthesis of proteins and hormones, nerve conduction, growth and energy production involve enzymatic reactions. The structure and activity of enzymes, which are proteins, to a great extent depends on the ions present in solution. Amongst the various ions the hydrogen ion is very important as it changes the configuration of protein molecules. According to Bull (1964) and Pearson (1966) the importance of the hydrogen ion is much more than is commonly realised as its is a very *important catalyst in biological reactions.* It is, therefore, imperative that a proper concentration of H ions is maintained in the body fluids. The exact mechanism of regulation of hydrogen ion concentration, or for that matter any other equilibrium process, cannot be clearly followed unless the fundamental law i.e., 'law of mass action,' which governs this, is clearly, understood.

LAW OF MASS ACTION

A chemically reversible reaction reaches a *stable state* (equilibrium) which has characteristic chemical, kinetic, and thermodynamic properties. The law governing these properties was enunciated by Goldberg and Waage and it states that "*The rate at which a substance reacts is proportional to its active mass and the rate of a chemical reaction is proportional to the product of the reacting substances".* (By active mass we mean molecular concentration (gram moles/litre, and it is given by enclosing the reacting substances in square brackets [ ]. (Thus the active masses of A and B would be written as [A] and [B] respectively).

According to the law of mass action, in the reaction,

$$A + B \underset{V_2}{\overset{V_1}{\rightleftharpoons}} C + D$$

The velocity (V_1) of the reaction from left to right would be—

$$V_1 \propto [A] \times [B] \quad \dots(1)$$

Similarly the velocity (V_2) of the reaction from right to left would be—

$$V_2 \propto [C] \times [D] \quad \dots(2)$$

$$V_1 = K_1 \times [A] \times [B] \quad \dots(3)$$

$$V_2 = K_2 \times [C] \times [D] \quad \dots(4)$$

whereK_1 and K_2 are constants called *velocity constants.*

At equilibrium the rate of conversion of A + B to C + D would be the same as the rate of conversion of C + D to A + B. In other words equilibrium will be reached when $V_1 = V_2$.

or

$$K_1 \times [A] \times [B] = K_2 \times [C] \times [D] \quad \dots(5)$$

therefore,

$$\frac{[C] \times [D]}{[A] \times [B]} = \frac{K_1}{K_2} \quad \dots(6)$$

or

$$\frac{[C] \times [D]}{[A] \times [B]} = K_c \quad \dots(7)$$

K_c is called the *equilibrium constant* of a reaction. It is the ratio of the velocity constants of the opposing reactions. *K_c however, is constant only under isothermal conditions.* Equation (7) is the mathematical expression of the 'law of mass action'; by convention, the product of the products is written as numerator and the product of the reactants is written as denominator in this equation. The law of mass action is also applicable to the dissociation of dissociable compounds.

Dissociation of weak acids

When weak acids dissociate in an aqueous solution they give off hydrogen ions or protons.

$$HA \rightleftharpoons H^+ + A^- \quad \text{(where } A^- \text{ is anoin)}$$

According to the law of mass action,

$$\frac{[A^+] \times [A^{-1}]}{[HA]} = K_a \quad \dots(8)$$

K_a is called the dissociation constant of a weak acid. A strong acid gives off more H^+ and consequently the value of K_a is large. It follows from this that the larger the dissociation constant the stronger is the acid. (Table 7.1).

Table 7.1 Dissociation constants of some weak acids

Acid	*Dissociation constant* $\frac{[H^+]\times[A^{-1}]}{[HA]}=K_a$
Boric acid	6.4×10^{-10}
Carbonic acid	4.52×10^{-7}
Acetic acid	1.85×10^{-5}
Formic acid	2.4×10^{-4}
Chloracetic acid	1.38×10^{-3}

It should be noted that strong acids like HCl, H_2SO_5 etc. always dissociate completely and therefore there is no possibility of ionic equilibrium.

Dissociation of Weak Bases

Weak bases also dissociate in a manner similar to weak acids. In an aqueous medium these give off OH^- to water,

$$BOH \rightleftharpoons B^+ + OH^{-1} \quad \text{(where } B^+ \text{ is a cation)}$$

therefore,

$$\frac{[B^+]\times[OH^-]}{[BOH]}=K_b \qquad ...(9)$$

K_b is the dissociation constant of a weak base.

Dissociation of Water

Water dissociates to a very small extent but is unique as it gives off both H^+ and OH^-.

$$H_2O \rightleftharpoons H^+ + OH^-$$

Applying the law of mass action,

$$\frac{[H^+]\times[OH^-]}{[H_2O]}=K \qquad ...(10)$$

or

$$[H^+]\times[OH^-]=K\times[H_2O] \qquad ...(11)$$

Where K is the dissociation constant of water. At room temperature its value is 1.8×10^{16} moles/litre.

Since the dissociation constant of water is extremely low therefore most of the water is present in an undissociated state. If we take one litre of water then for practical purposes it can be said that the volume of undissociated water is also one litre.

In the equation,

$$[H^+]\times[OH^-]=K\times[H_2O] \qquad ...(12)$$

the value of H_2O is known *i.e.*, 1 litre. Since the weight of 1 litre of water is 1000 g and its molecular weight is 18, therefore the molar concentration of 1 litre of water is $1000 \div 18 = 55.5$. If these values are substituted in equation (11) then,

$$[H^+]\times[OH^-]=1.8\times10^{-16}\times55.5$$

or $$[H^+]\times[OH^-]=10^{-14} \quad ...(13)$$

or $$K.H_2O=10^{-14}$$

$K.H_2O$ is also written as K_w and is called *ionic product of water*. Therefore,

$$K_w=10^{-14} \quad ...(14)$$

Experimental data show that the concentration of H^+ and OH^- in pure water is equal and gives water its characteristic neutral property. The concentration of both H^+ and OH^-, therefore, will be 10^{-7} g/mole. Like other electrolytes the dissociation of water also increases with rise in temperature so that the H^- concentration rises with temperature. Hydrogen is a very important catalyst in biologieal reactions and it is quite possible that increase in biological reaction rates due to rise in temperature is partially due to increase in H^+ also.

The Common Ion Effect

If a salt of a weak acid is added to a solution of the same acid the dissociation of the acid is suppressed. This can be explained by taking the example of acetic acid, and sodium acetate which dissociates as follows:

$$CH_3\ COOH \rightleftharpoons CH_3\ COO^- + H^+$$

Sodium acetate on the other hand dissociates as follows:

$$CH_3\ COONa \rightleftharpoons CH_3\ COO^- + Na^+$$

In case of acetic acid the product of CH_3COO^- and H^+ is constant. Increase in the number of either of the two ions *i.e.* CH_3COO^- or H^+ reduces the number of the corresponding ions since their product is a constant (dissociation constant). Now if sodium acetate is added to a solution of acetic acid then the total number of acetate ions increases; because of increase in acetate ions the number of hydrogen ions is reduced and the dissociation of acetic acid is suppressed. Similarly, addition of a strong acid like HCl increases the number of H^+ which also suppresses the dissociation of acetic acid.

A similar parallel can be drawn to explain the dissociation of a weak base and its salt. The suppression of the dissociation of a weak acid or a weak base in the presence of its own salt is called the *common ion effect*.

Equations 12 and 13 state that the product of H^+ and OH^- is constant ($K_w = 10^{-14}$ at room temperature). This means that if the number of hydrogen ions is increased the number of hydroxyl ions will decrease and *vice versa*. Thus if in a solution the concentration of H_7 is 10^{-5} gram moles/litre then the concentration of OH^- will be 10^{-9} gram moles/litre (Table 7.2). Since it is inconvenient to write the concentration of hydrogen ions in negative exponential terms, therefore Sϕ rensen suggested that the hydrogen ion concentration should be expressed in a more convenient way called the *pH scale.*

THE pH SCALE

pH is the *negative logarithm of hydrogen ions to the base ten expressed as gram moles/litre.*

$$pH=-\log_{10}[H^+]$$

e.g. if the concentration of H^+ in a solution is 10^{-5} gram moles/litre then its logarithm will be – 5 and its negative logarithm will be 5. The pH according to the definition will, therefore, be 5.

The only reason, why a pH scale is used instead of a direct expression of H^+ concentration, is because it is more comprehensible. The main draw back of this system, however, is that while the $[H^+]$ may change ten fold it produces a change of only one unit on the pH scale. This makes the change apparently less impressive.

Since hydrogen ions impart acidic properties and hydroxyl ions alkaline properties to solutions the pH value is an expression of *acidity* and *alkalinity* of solutions. The pH of pure water, at room temperature, is 7 and at this pH the concentration of H^+ and OH^- is equal, hence pure water is netural. When pH is below 7 the solution is acidic and when it is above 7 it is alkaline as can be seen in Table 7.2.

Table 7.2 Relationship between hydrogen ion concentration, hydroxyl ion concentration and pH in aqueous solutions. If the concentration of H^+ and OH^- is equal the pH is 7 and the solution is neutral. Below pH 7 the concentration of OH^+ is more and the solution is acidic. Above pH 7 the concentration of OH is more and the solution is alkaline.

Hydrogen ion concentration	*Hydroxyl ion concentration*	*pH*
10^{-0}	10^{-14}	0 (Acidic)
10^{-1}	10^{-13}	1 (Acidic)
10^{-3}	10^{-11}	3 (Acidic)
10^{-5}	10^{-9}	5 (Acidic)
10^{-7}	10^{-7}	7 Neutral
10^{-9}	10^{-5} (Alkaline)	9 (Alkaline)
10^{-11}	10^{-3} (Alkaline)	11 (Alkaline)
10^{-13}	10^{1} (Alkaline)	13 (Alkaline)
10^{-14}	10^{-0} (Alkaline)	15 (Alkaline)

Just as pH *is the negative logarithm of hydrogen ion concentration, pOH is the negative logarithm of hydroxyl ions, pKa the negative logarithm of the dissociation constant of acids and pKb the negative logarithm of the dissociation constant of bases.*

Mixtures of Weak Acids and their Salts

An acid HA, ionizes as follows:

$$HA \rightleftharpoons H^+ + A^{-7}$$

According to the law of mass action,

$$\frac{[H^+]\times[A^{-7}]}{[HA]} = K_a \qquad ...(15)$$

or

$$\frac{[HA]}{[H^+]\times[A^{-7}]} = \frac{1}{K_a}$$

$$[H^+]\times[A^-] = K_a + [HA]$$

or

$$[H^+] = K_a \frac{[HA]}{[A^-]} \qquad ...(16)$$

Taking the negative logarithm of each side,

$$-\log [H^+] = -\log K_a - \log \frac{[HA]}{[A^{-7}]} \qquad \ldots(17)$$

By definition,

$$-\log [H^+] = pH$$

$$-\log [K_a] = pK_a \text{ therefore,}$$

$$pH = pK_a - \log \frac{[HA]}{[A^-]} \qquad \ldots(18)$$

since $$-\log \frac{[HA]}{[A^-]} = +\log \frac{[A^-]}{[HA]}$$

therefore,

$$pH = pK_a + \log \frac{[A^-]}{[HA]} \qquad \ldots(19)$$

If the salt of a weak acid is added to a solution of the same acid the dissociation of the acid is suppressed because of the common ion effect. If the salt is highly dissociable then most of it will be dissociated. Due to this, the anions will mostly come from the salt while the acid mostly remain undissociated.

So that $$[\text{Anion}] = [\text{Salt}]$$

[Undissociated acid = Total acid]

Now, $$pH = pK_a + \log \frac{[A]}{[HA]} \qquad \ldots(20)$$

Substituting salt for $[A^-]$ and Acid for [HA]

$$pH = pK_a + \log \frac{\text{Salt}}{[\text{acid}]} \qquad \ldots(21)$$

Equation (19) is known as the *Henderson-Hasselbalch equation for buffer action.* (It must be understood that this equation is only an approximation as it assumes that the salt dissociates completely while the acid does not dissociate at all).

When the molar concentration of salt and acid is equal then

$$\frac{\text{Salt}}{[\text{Acid}]} = \log 1 = 0$$

Equation (18) then becomes

$$pH = pK_a + \log 1$$

or $$pH = pK_a \text{ when the concentration of}$$

salt and acid is equal. In other words, pH is equal to pK_a when an acid is half titrated with a strong alkali.

pK_a can be easily measured by measuring pH after mixing equal concentrations of an acid and its salt. It can also be determined by half titrating an acid with an alkali and the measuring the pH.

Once the pK_a is known the exact quantities of salt and acid required to get a particular pH can be determined. However, since the Henderson Hasselbelch equation is an approximation it is most correct when the term $\log \frac{[\text{Salt}]}{[\text{Acid}]}$ is near zero, *i.e.*, when salt and acid are present in nearly equal concentrations.

BUFFERS

In many biochemical reactions it is essential to keep the pH of a solution constant even when small quantities of acids and alkalies are added to it. *Solutions which can resist changes in pH by dilution or by addition of small quantities of acids and alkalis are called buffers.* The most commonly used buffers are mixtures of weak acids and their salts. The mechanism of buffer function can be understood by taking the example of a buffer made up of acetic acid and sodium acetate. These two dissociate in the following manner.

$$CH_3COOH \rightleftharpoons CH_3COO^{-7} + H^+$$

$$CH_3COONa \rightleftharpoons CH_3COO^- + Na^+$$

If an acid *e.g.,* HCl is added to this mixture, the hydorgen ions from the acid react with sodium acetate to form acetic acid.

$$CH_3COONa + HCl \rightleftharpoons CH_3COOH + NaCl$$

If a base *e.g.*, NaOH is added then it reacts with acetic acid to form sodium acetate.

$$CH_3\,COOH + NaOH \rightleftharpoons CH_3COONa + H_2O$$

Thus the addition of acids or bases changes the salt-acid ratio of a buffer. However, as is obvious from the Henderson-Hasselbatch equation small additions of acids or bases bering about a very meagre change log pH.

Suppose in a buffer of acetic acid and sodium acetate the molar concentration of salt and acid is equal, then

PH = pK_a (from Henderson-Hasselbalch equation), pK_a for acetic acid is 4.73, therefore pH = 4.73.

Now, if an equal amount of acetic acid is added to this buffer then the salt-acid ratio changes to 1 : 2.

or $$pH = pK_a + \log \frac{1}{2}$$

$$pH = pK_a + \log 0.5$$

or $$pH = 4.73 + (-0.3)$$

therefore, $$pH = 4.73$$

Thus a change of only 0.3 pH units is brought about when the quantity of acid is doubled.

It should be noted that a buffer can work efficiently for one pH unit on either side of the pK_a *e.g.*, a buffer made up, of acetic acid will be efficient from pH 3.73 to 5.73. This is why different acids have to be employed for different buffer ranges.

Buffering Capacity

Although the pH of a buffer depends on the salt-acid ratio, (Henderson-Hasselbalch equation) its capacity to resist change in pH depends on the absolute amount of salt and acid. According to Van Slyke the *buffereing capacity* of a buffer is 1, if it requires 1 gram equivalent (1000 ml of 1N) acid or alkali to bring about a change of 1 pH unit.

Example 1. If 100 ml of buffer requires 5 ml of 1N NaOH for change of 1 pH unit, then its buffering capacity is:

100 ml of buffer requires 5 ml
1000 ml of buffer will require 50 ml

The buffering capacity will therefore be $\frac{50}{1000}$ or 0.05 to strong alkali.

The following example shows how the absolute concentration of salt and acid effect the buffering capacity.

Example 2. A buffer contains 20 ml of 0.1 N CH_3COOH and of 20 ml of 0.1 N CH_2COONa, then,

$$pH = pK_a + \log\frac{20}{20}$$

or

$$pH = 4.73 \ (pK_a \text{ for } CH_2COOH \text{ is } 4.73)$$

Now, if 5 ml of 0.1 N HCl is added then it will changes 5 ml of salt into acid and the new ratio of salt and acid will be $\frac{15}{25}$

$$pH = 4.73 + \log\frac{15}{25}$$

or

$$= 4.73 + \log 15 - \log 25$$
$$= 4.73 + 1.761 - 1.3979$$
$$= 4.5083 \text{ a change of approximately 0.22 p cal units.}$$

Now, suppose there is a buffer with 20 ml of 0.5 N CH_3COOH and 20 ml of 0.5 N CH_3COONa. The addition of 5 ml of 0.1 N HCl in this case will change only 1 ml of salt into acid. The new salt acid ratio will be $\frac{19}{21}$

$$pH = pK_a + \log\frac{19}{21}$$

or

$$pH = 4.73 + \log 19 - \log 21$$
$$= 4.73 + 1.2788 - 1.3222$$
$$= 4.6866$$

a change of 0.04 pH units only as compared to 0.22 pH units in the previous case.

Buffers of Polybasic Weak Acids

A weak acid with several dissociable hydrogens has a different pK_a for each hydrogen. Phosphoric acid, for example, dissociates in the following manner (Fig. 7.1):

(1)

$$H_3PO_4 \rightleftharpoons H^+ + H_2PO_4^-$$

$$\frac{[H^+] \times [H_2PO_4^-]}{[H_3PO_4]} = pK_{a1} = 2.12$$

(2)

$$H_2PO_4 \rightleftharpoons H^+ + HPO_4$$

$$\frac{[H^+] \times [HPO_4^-]}{[H_3\,HPO_4^-]} = pK_{a2} = 7.20$$

(3) $$HPO_4^- \rightleftharpoons H^+PO_4^-$$

$$\frac{[H^+] \times [PO_4^-]}{[HPO_4^-]} = pK_{a3} = 12.66$$

Buffers made of such acids are effective for one pH unit on either side of the pK_a values of an acid are close together then buffers made of such acids have a wide buffering range. In case of citric acid *e.g.*, the different pK_a values are close to each other so that buffers of citric acid and its salts have a wide buffering range.

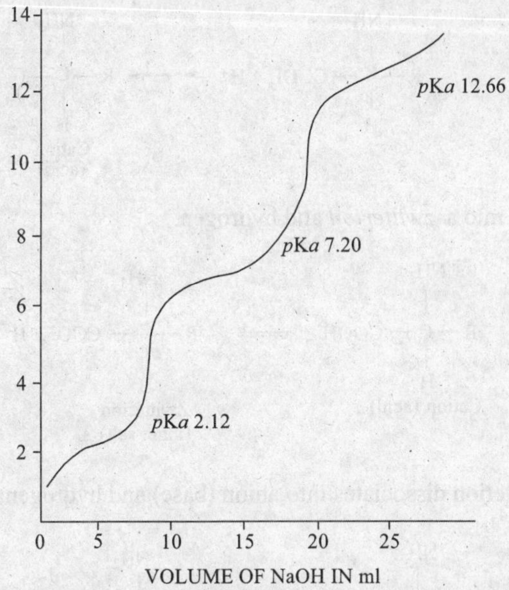

Fig. 7.1. It shows that phosphoric acid has three pK_a values. Note that when phosphoric acid is titrated with NaOH change in pH becomes slow around the pK_a values indicating that buffers of this value are most effective.

PHYSIOLOGICAL IMPORTANCE OF BUFFERS

The pH of most parts of the body, except for the digestive system, is maintained close to neutrality (Table 7.3). This is because the concentration of the hydrogen ions is very important. Firstly, because it is an important biological catalyst, and secondly, because it affects the structure of many biological molecules. It is not possible to give a complete list of all biological compounds whose structure is pH sensitive or to enlist all the reactions in which the hydrogen ion acts as a catalyst. However, the pH of the internal environment affects the body function in an all pervading manner as protein structure can change with change in pH value. This function of pH is very significant because (1) protein are the largest single group of compounds (besides water) found in the body; they form the bulk of the body, and as such any major alteration in their structure would have far reaching consequences. (2) All enzymes are proteins (or at least have a protein component). As the activity of enzymes is structure-dependent, it is very important that their correct structural entity is maintained.

In the light of the above statement it is crucial to understand the manner in which protein structure is dependent on pH.

pH Dependence of Protein Structure

As mentioned in Chapter 2, proteins are made up of amino acids which are *amphoteric* compounds *i.e.*, they can act both as acids and bases. The acid group in an amino acid is COOH and the basic group is NH_2. In an acidic pH amino acids accept a proton,

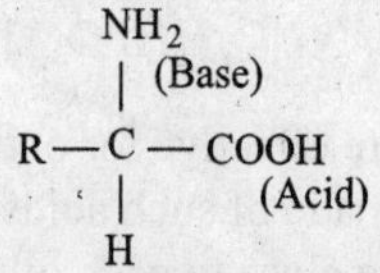

become positively charged (cation) and behave as acids:

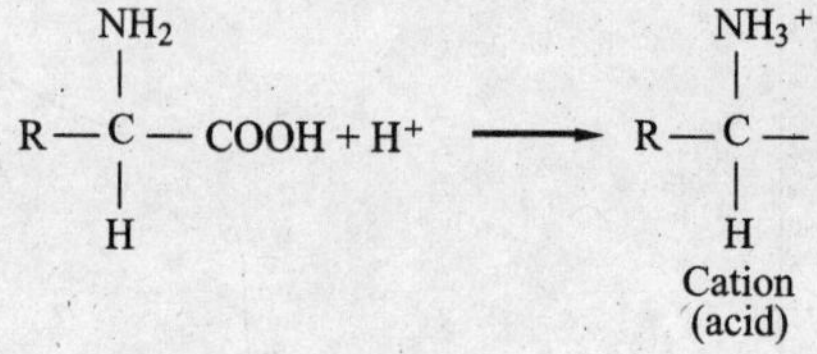

The cation dissociates into a *zwitterion* and hydrogen.

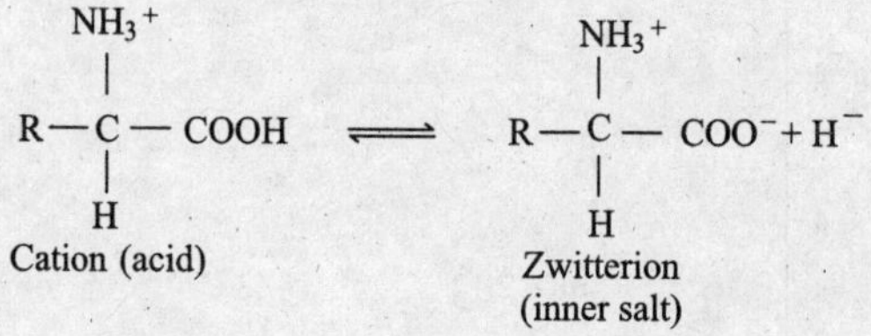

In a basic pH the zwitterion dissociates into anion (base) and hydrogen ion.

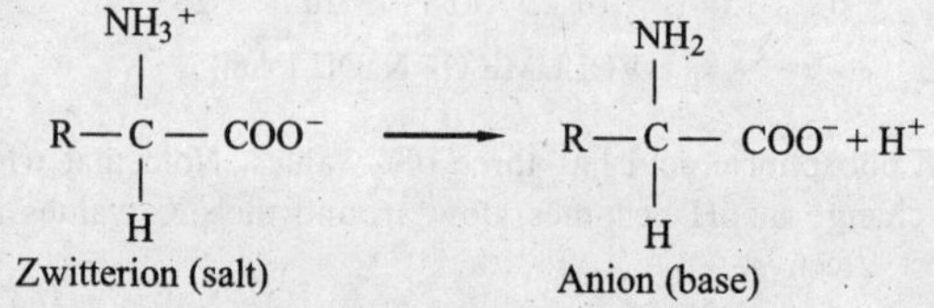

Because of a double dissociation amino acids have two dissociation constants:

$$\frac{[\text{Zwitterion}]\times[\text{H}^+]}{[\text{Cation}^+]}=\text{K}_1$$

$$\frac{[\text{Anion}^-]\times[\text{H}^+]}{[\text{Zwitterion}]}=\text{K}_2$$

The cations give H^+ to water and act as acids while the anions take up H^+ from water and act as base. Generally, the strength of acid and base in an aqueous solution of amino acids is not equal. When the cations are in excess the amino acids behave as acids and when anions are in excess they behave as bases. If the concentration of cations and anions is equal the amino acids behave as salts. Presence of cations, therefore, lowers pH and presence of anions raises pH. It is obvious that at some pH the number of cations and anions is equal. At this pH the net charge of the solution is zero. If the amino acid is placed in an electric field at this pH there is no accumulation of the amino acid either at the anode or the cathode. This pH is called the *isoelectric* point-*p*I. The *p*I for different amino acids is different; in case of mono-amino-monocarboxylic amino acids the isoelectric point lies midway between the two dissociation constants.

$$\text{Isoelectric point } (p\text{I}) = \frac{K_1 + K_2}{2}$$

All amino acids in a protein molecule do not dissociate as most of these are bound to each other by peptide linkage (Chpater 2). Only the terminal amino acids of peptide chains and the extra amino and carboxyl groups on polyamino and polycarboxylic acids are free to dissociate. These groups, therefore, are very important for maintaining the structure and function of the protein molecule.

Buffering Mechanisms of the Body

Buffering in the body is carried out at two levels: (1) Buffering within the cells (2) Buffering outside the cells.

There are several buffering systems in the cell which are made up of organic acids, fatty acids, amino acids and the salts of all of these. Phosphoric acid and its salts are important buffers of the cell. It is not very easy to determine the exact buffering mechanisms involved in the cell because the cell is a heterogenous assembly of organelles, having their own enzyme and buffer systems. Very little work has been done on the buffering mechanism of the intracellular fluids. It however, appears that phosphates form the most important buffering system within the cells.

Buffering Mechanisms Outside the Cell

The individual cells of the body are surrounded by, extra cellular fluid called *lymph,* which flows through lymphatic vessels. Lymph is an ultrafiltrate of blood and its composition is very similar to the former. The pH of lymph is also very close to that of blood. In order to maintain a constant pH of the intracellular fluid and thereby the entire body the pH of blood (plasma) is maintained very accurately. In case of man, for example, the pH of plasma is maintained accurately around 7.4 by powerful mechanisms. Even small changes in plasma pH are very serious and may cause death. The susceptibility of the body to small changes in plasma pH is so great that a pH of 7 causes acidity of plasma (*acidosis*) and a pH of 7.6 causes alkalinity of plasma (*alkalemia*) both of which can be fatal. It is apparent, however, that even in acidosis the blood may not *actually become acidic.*

Buffers of the blood. A large number of acids are produced during intermediary metabolism (Chapter 13). These tend to lower the pH of blood. Of all the acids which are produced, H_2CO_2 (carbonic acid) is most important as it is the end product of energy producing reactions. It is produced in large quantities and has to be continuously removed or neutralized in order to maintain the pH of blood.

The haemoglobin (HHb) present in R.B.C. is weakly acidic. It reacts, with bases, mostly potassium bases; present in the R.B.C. to form salts. These salts are called potassium haemoglobinates (KHb).

Thus a buffer system of $\frac{\text{KHb}}{\text{HHb}}$ is formed.

The erythrocyte (R.B.C) membrane is semipermeable. It is permeable to CO_2. H^+ and anions like HCO_3^-, Cl^- etc. but is impermeable to cations like Na^+, K^+, and Ca^{++}. The CO_2 produced during metabolism enters the erythrocytes. In the presence of an enzyme *carbonic anhydrase,* present in R.B.C., CO_2 combines with water to form carbonic acid.

$$CO_2 + H_2O \xrightarrow{\text{carbonic anhydrase}} H_2CO_3$$

H_2CO_3 is a stronger acid than HHb. It therefore reacts with KHb to form $KHCO_3$ and HHb

$$H_2CO_3 + KHb \rightleftharpoons KHCO_3 + HHb$$

The $KHCO_3$ thus formed, dissociates in the following manner:

$$KHCO_3 \rightleftharpoons R^+ + HCO_3^-$$

The HCO_3^- enters the plasma leaving charged K^+ behind. In order to maintain electrical neutrality Cl^- ions which are formed due to dissociation of NaCl enter the erythrocytes leaving Na^+ behind (because the erythrocyte membrane is impermeable to Na^+). There are thus 1.5% more chloride ions in the erythrocyte than in the plasma.

In the plasma the HCO3$^-$ combines with Na^+ left behind by Cl^-.

$$Na^+ + HCO_3^- \longrightarrow NaHCO_3$$

The level of $NaHCO_3$ rises in blood and it raises the blood pH. This entire mechanism is called *chloride shift* and was discovered by Hamberger in 1918. It is thus apparent that in response to the production of CO_2 (or for that matter H2CO_3) there is a continuous production of $NaHCO_3$ so that a buffer of $\frac{NaHCO_3}{H_2CO_3}$ is maintained.

The birccarbonates in blood are called the *alkali reserves* of the body.

Explusion of CO_2 in the lungs. The bicarbonates formed in the blood act with the acids which are present in it.

$$HA + NaHCO_3 \rightleftharpoons NaA + H_2CO_3$$

In the lung the carbonic anhydrase acts in a reverse manner,

$$H_2CO_3 \xrightarrow[\text{anhydrase}]{\text{carbonic}} CO_2 + H_2O$$

The CO_2 thus formed is expelled out through the lungs by exhalation.

Due to the expulstion of CO_2 the HCO_3^- level in the erythrocytes falls and the HCO_3^- passes from the plasma into the cell. Again in order to maintain electrical neutrality Cl^- which entered the cell goes back into the plasma. The NAHCO_3 of plasma is reduced and NaCl increases (Figure 7.2).

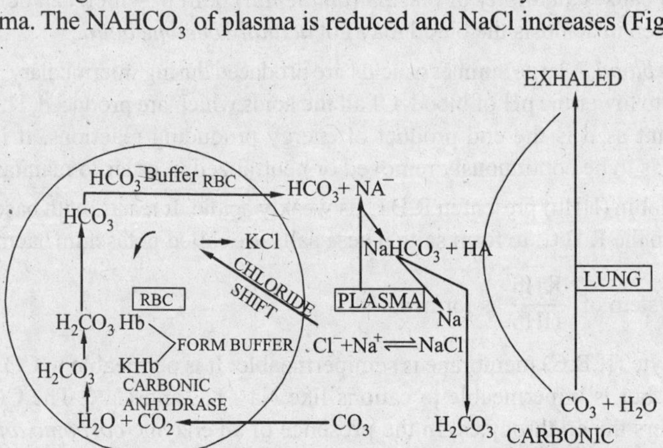

Fig. 7.2. Figure showing the buffering mechanisms found in the blood. (For details see text).

Excretion of Acids through Kidneys

The blood is always alkaline (pH 7.4) and the urine always acidic (pH 5 to 6). It appears that by extracting the acids from the blood the urine maintains its acidity while the blood maintains its alkalinity. Hydrogen ions are taken up in the proximal tubule of the renal tubule (Chapter 15). This upsets the electrolyte balance and in exchange Na^+ goes out from the tubules and enters the blood through the cells of the proximal tubule. In the blood Na combines with HCO_3^- to form $NaHCO_3$. The $NaHCO_3$ thus formed adds to the alkali reserves of the body.

The hydrogen ions which enter the tubule combine with bicarbonate ions to form carbonic acid. This has one of the following fates:

(1) It combines with phosphates,

$$HPO_4^- + H_2CO_3 \rightleftharpoons H_2PO_4^- + HCO_3^-$$

The phosphates are excreted with the urine and the bicarbonates go back to the blood. Only a very small amount of bicarbonate is excreted as such; during acidosis even this is reduced.

(2) Part of the hydrogen ions combine with ammonia which is continuously secreted by the tubules and forms the following buffer system.

$$pH = pKa + \log \frac{NH_3}{NH_4^+}$$

It has been observed by many workers that the secretion of NH_3 increases with increased acidity of the blood (Fig. 7.3).

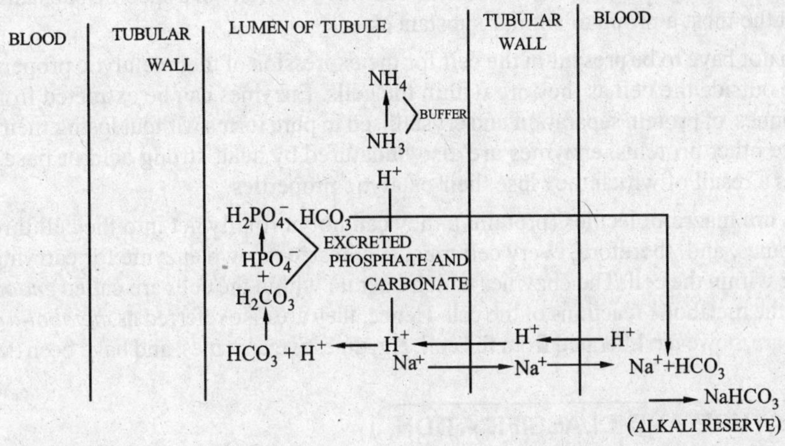

Fig. 7.3. Figure showing buffering mechanism in the kidney (For details see text).

8

Chapter

Enzymes

A large number of chemical reactions proceed at a very slow rate at room temperature and normal pressure. The velocity of such reactions can be raised by resorting to extreme conditions of temperature, pressure and pH. In living systems such conditions are not possible for obvious reasons. However, in the body, all the biochemical reactions proceed at a high rate with the help of organic catalytic agents called *enzymes.* The rate of these enzyme-catalysed reactions is sometime higher than those an organic chemist could obtain at very high pressure and temperature. Every living cell has its requisite enzymes which are proteins having catalytic properties. Like other catalysts, enzymes are required in extremely small concentrations and do not alter the *free energy change* (ΔG) of a reaction. These are highly specific as they catalyse only the reactivity of a specific substance (called *substrate*) or at the most a group of similar substances.

Enzymes do not have to be present in the cell for the expression of their catalytic properties; they are as effective outside the cell as they are within the cells. Enzymes can be extracted from cell by standard techniques of protein separation and crystallised in pure forms without losing their catalytic properties. Like other proteins, enzymes are also denatured by heat, strong acid or base, organic solvents etc., as a result of which they lose their catalytic properties.

As enzymes are macromolecules (proteins), they can not normally get into the cell through the plasma membranes, and, therefore, every cell must synthesize its own enzyme for carrying out the reactions going within the cell. The enzymes which operate within the cells are called *endoenzymes.* These catalyze the metabolic reactions of the cell. Hence, they are also referred as *metabolic enzymes.* Some enzymes are, however, liberated from the cell, e.g, digestive enzymes, and have been designated *exoenzymes.*

NOMENCLATURE AND CLASSIFICATION

For many years most enzymes were named after the substrate upon which the enzyme was acting by addition of the suffix '*ase*'. Thus, the enzyme which splits sucrose into glucose and fructose was termed as *sucrase.* Similarly, the enzyme hydrolysing maltose was called *maltase.* This nomenclature did not give adequate information regarding the chemical nature of reaction which was being catalysed.

However, the present system of nomenclature and classification of enzymes is based on the recommendations of the International Union of Biochemists (1961), according to which all enzymes are divided into *six* major classes namely,

(1) Transferases
(2) Hydrolases
(3) Oxidoreductases
(4) Lyases

(5) Isomerases and
(6) Ligases.

(1) Transferases

The enzyme catalyses the transfer of a specific group from one substrate to another.

$$A\ \boxed{Grp} + B \rightarrow B\ \boxed{Grp} + A$$

(where A and B are the two substrates and Grp is the group which is being transferred).

According to the previous nomenclature, the various transferases were called *transaminases, kinases, transacetylases* etc. This nomenclature did not give adequate information. Thus, the enzyme which catalyses the following reaction was called *acetyl kinase.* This has now been termed *ATP—acetate phosphotransferase,* which is more informative.

$$ATP + CH_3\,COOH \rightleftharpoons ADP + CH_3 - \underset{\underset{O}{\|}}{C} - OPO_3\,H_2$$

(2) Hydrolases

The hydrolases catalyse the hydrolysis (chemical degradation of a substance by water) of variety of compound like esters, peptides, glycosyls, acid-anhydrides, etc. The older names of the well known enzymes of this group are lipases, phosphatses, peptidases, nucleases, glycosidases, amidases etc. According to the recent nomenclature, *lipase* would be termed as *glycerol ester hydrolase.*

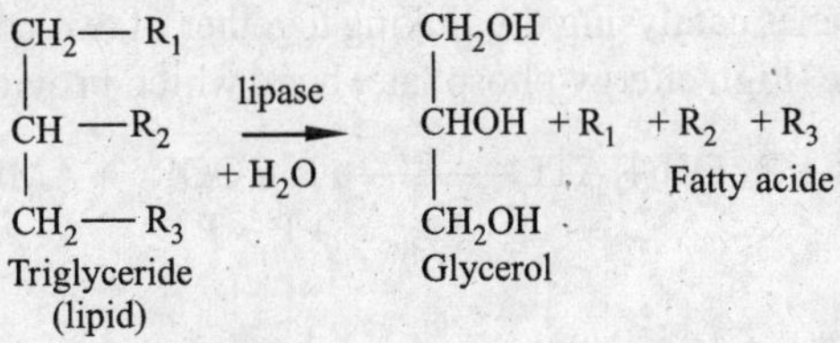

(3) Oxidoreductases

These are enzymes which catalyse oxidation-reducation reactions between a pair of substrates A and B.

$$\underset{\text{reduced}}{A} + \underset{\text{oxidized}}{B} \rightarrow \underset{\text{oxidized}}{A} + \underset{\text{reduced}}{B}$$

This includes a large number of enzymes formerly known as dehydrogenases, oxidases, oxygenases, peroxidases etc.

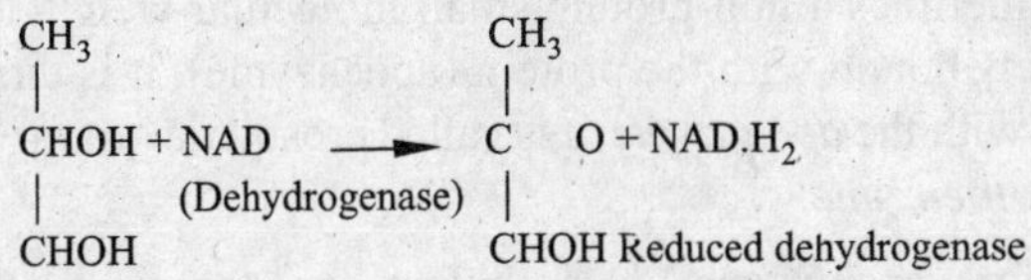

(4) Lyases

These are the enzymes which catalyse a reaction in which two parts of a molecule are separated with the formation of a double bond in one of them. Lyases catalysed reactions are commonly reversible.

$$CX - CY \rightleftharpoons XY + C = C$$

Thus, fumarase (L—malate hydrolyase) catalyse the removal of water from malate and yield fumarate.

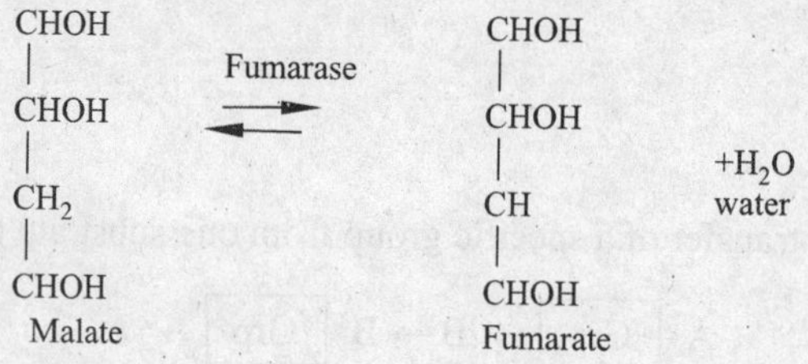

(5) Isomerases

These include a rather large variety of enzyme, catalysing the interconversion of optical, geometric or positional isomere.

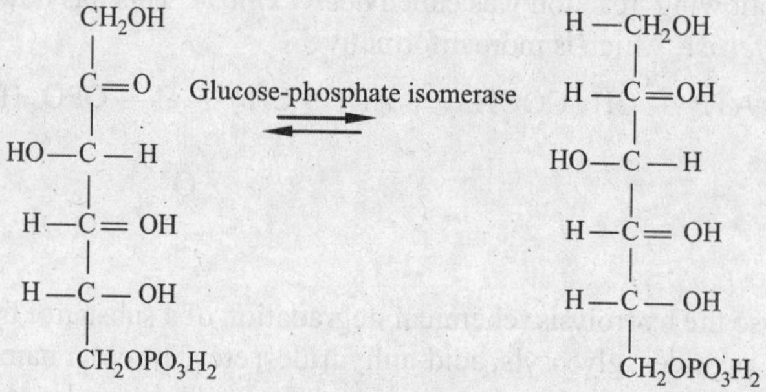

(6) Ligases

This class includes the enzymes catalysing the linking together of two compounds coupled with the breaking of a pyrophosphate (high energy phosphate) bond which provides energy for the reaction.

$$R.COOH + R_1\,OH + ATP \xrightarrow{\text{Ligases}} R\,COOR_1 + AMP + H_2O + P \sim P$$

↑ high energy bond

CHEMICAL NATURE OF ENZYMES

Repeated analyses have shown that enzymes are either simple or conjugated proteins. The physical and chemical properties of the enzymes are same as those of other proteins.

Many enzymes consist of large polypeptide or protein units (the protein portion is designated as *apoenzyme*) plus some other type of substance called as *cofactor.* Without its cofactor, these enzymes are functionally inactive. The cofactor may be some simple divalent metallic ion (e.g. Mg^{++}, Ca^{++}, Zn^{++}, Co^{++}, Mn^{++}) or sometimes a non-protein small molecular weight organic compound. If the organic cofactor is loosely bound with the protein (apoenzyme), it is termed *coenzyme* and if the cofactor is firmly bound with the *apoenzyme* it is called prosthetic group. The fully intact enzyme is sometimes referred as *holoenzyme*.

$$\text{cofactor} + \text{apoenzyme} \longrightarrow \text{holoenzyme (or enzyme)}$$

MECHANISM OF ENZYME ACTION

Energy of Activation

All the catalysts increase the rate of reaction by lowering down the *energy of activation* of the reaction. Energy of activation is the minimum amount of energy which must be present in two colliding molecules before they can proceed in a particular reaction. At any given moment some molecules have an energy level above the activation energy and some below this level. An enzyme (catalyst) lowers the energy of activation so that at any given temperature, a large number of molecules possess this energy and they therefore, enable the reaction to proceed at a greater velocity. The advantage of an enzyme, therefore, is that enzyme-catalysed reactions take place at lower temperature. For example, hydrogen peroxide requires an energy level equivalent to 18,000 calories/mol for its decomposition into oxygen and water. But, in the presence of an enzyme, (catalase), decomposition takes place only at an energy level equivalent to 5,500 calories/mol. Thus, in the presence of an enzyme, the activation energy is reduced to less than one-third of its original value. Similarly, casein is hydrolysed by hydrogen ions at the energy level equivalent to 20,600 calories/mol, whereas in the presence of typsin it is reduced to only 12,000 calories/mol.

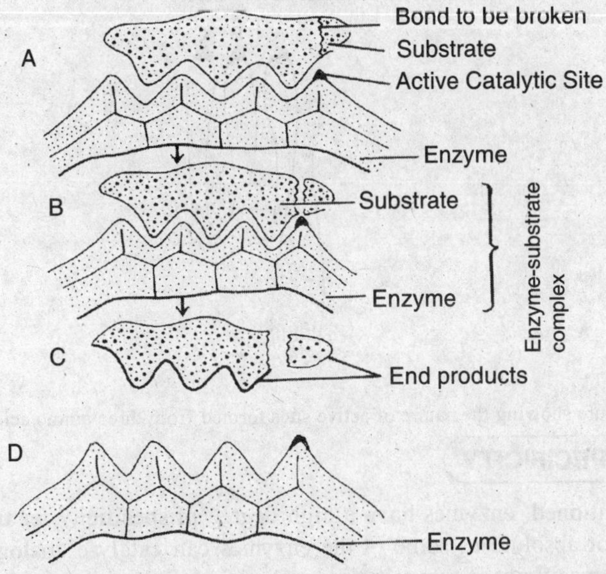

Fig. 8.1. Showing the mechanism of action of enzyme. The enzyme combines with substrate which has a complementary structure (A), and produces a strain on the bond to be broken. The bond finally breaks (B). The products then leave the enzyme (C) and the enzyme is then ready to get attached to another molecule of substrate (D). This is called the 'Lock and Key' theory of enzyme action.

It has been suggested that the enzymes temporarily combine with substrate. During this process it produces a strain on the bonds which is to be broken so that lesser amount of energy is required to break it. An enzyme will act on a substrate only if the substrate can *fit* on the surface of the enzyme molecule at *specific sites* (active sites) which have a structure complementary to the structure of substrate molecule (Fig. 8.1). This has been termed '*lock and key*' theory of enzyme action. In this process the enzyme works as a '*lock*' in which unless the substrate is able to fit itself as '*key*' the reaction will not take place.

The theory of an actual combination between the enzyme and substrate was proposed by Michaelis and Menten (1913) which can be explained by following equation.

$$\underset{\text{Enzyme}}{E} + \underset{\text{Substrate}}{S} \rightleftharpoons \underset{\text{Enzyme-substrate complex}}{ES} \rightleftharpoons \underset{\text{Products}}{P} + \underset{\text{Enzyme}}{E}$$

Nature of Active Sites (Centres)

The active site may be defined as a small part of the enzyme molecule where the substrate can attach itself by electrostatic hydrogen and/or hydrophobic bonds. This is followed by breaking of the bonds of the substrate. The exact chemical nature of active sites of most enzymes is not known. It has been proposed that the active site is formed by amino acid chains. Sometimes, more than one side chain may form the active site. These side chains are not necessarily adjacent to each other; they may be brought together by the tertiary folding of the enzyme protein (Fig. 8.2). The active site, in many cases, has the SH group as their component. Substances which bond the SH group inhibit the activity of many enzymes. Besides this, the OH group of hydroxy amino acids, NH_2 group of basic amino acids and carboxyl group of acidic amino acids have been shown to be important components of the active sites. Those enzymes which are conjugate proteins have their cofactor (co-enzyme or prosthetic group) as very important component as their active sites.

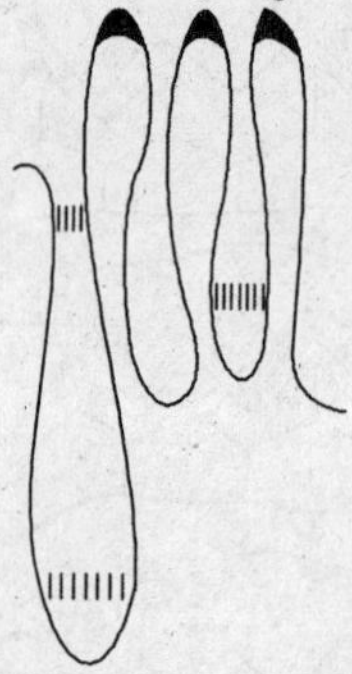

Fig. 8.2. Figure showing the nature of active sites formed from three amino acid residues (A, B and C).

ENZYME SPECIFICITY

As already mentioned, enzymes have a high degree of specificity for their substances. But this specificity is not absolute as some of the enzymes can catalyse analogous reactions of similar substrates. The specificity of enzymes can be classified in the following manner.

(A) Stereospecificity

Some enzymes exhibit optical specificity for a D or L optical isomer. Thus, the same amino acid has two oxidases; one for its L-isomer and the other for its D-isomer.

$$\text{L-amino acids} \xrightarrow[\text{L-amino acid oxidase}]{O_2} \text{D-Keto acids} + NH_3 + H_2\,O_2$$

$$\text{D-amino acids} \xrightarrow[\text{D-amino acid oxidase}]{O_2} \text{D-Keto acids} + NH_3 + H_2O_2$$

Although enzymes exihibit optical specificity, some enzymes such as amino acid racemases, interconvert optical isomers between the L and D forms. Thus, alanine racemase catalyses the following reaction.

$$\text{L-alanine} \rightleftharpoons \text{D-alanine}$$

Similarly some enzymes show specificities towards the geometric or cis-trans forms. For example, fumarase adds water to *trans* isomer (fumaric acid) but does not react with the *cis* isomer (maleic acid).

The above discussion deals with only those compounds where the carbon was asymmetrical.

However, enzyme specificity is sometimes found even in those substrates where the carbon is symmetrical, and, therefore, the substrate does not exist in isomeric form; for example, glycerol and citric acid, where group a_1 is identical with group a_2. But when enzymic reaction takes place the a_1 group of glycerol and citric acid is preferentially attacked whereas the group a_2 is not attacked.

General Case

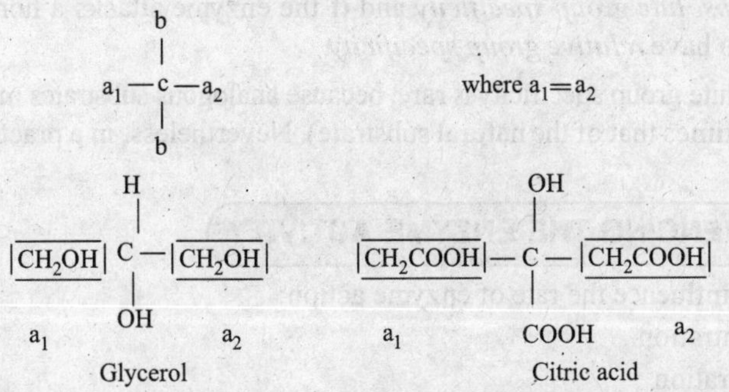

This curious phenomenon was explained by Ogston (1948) who stated that although a substrate may be symmetrical, but the enzyme substrate relationship is asymmetrical. According to him, the substrate is attached to the enzyme at three specific spatially interrelated sites. This is called the *three point attachment*. In the case of glycerol, for example, the three sites attached were the groups a_1, a_2 and H (Fig. 8.3). However, if the glycerol molecule is rotated, then a_1 comes to the site of a_3 ; a_2 comes to the site of a_1 ; but OH comes to the site of H and hence no attachment is possible and, consequently, there can be no enzymic activity.

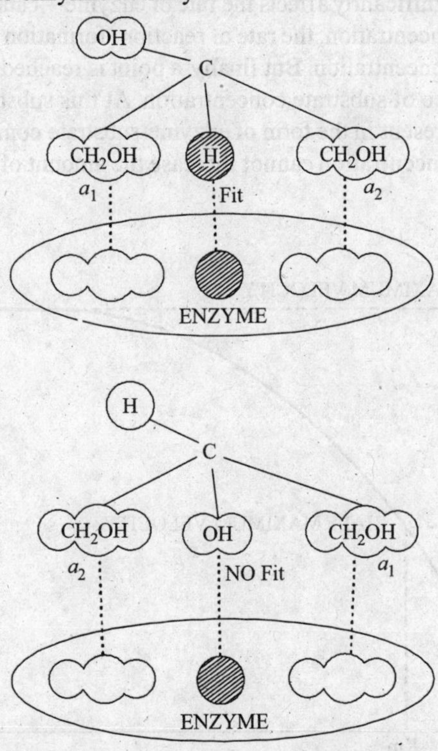

Fig. 8.3. Representation of three point attachment of a substrate with active site of an enzyme.

(B) Group Specificity

Certain enzymes exhibit group specificity, *i.e.*, they can act only on a certain chemical grouping. Thus, pepsin will attack the peptide bonds and estereases on ester linkages. Similarly, alcohol dehydrogenases will act on primary alcohols and glycosidases on glycosides.

If the enzyme attacks only one substrate for example, glucose and not any other related compound, it is said to have an *absolute group specificity* and if the enzyme attacks a homologous series of aldhexoses, it is said to have *relative group specificity.*

In strict sense, absolute group specificity is rare, because analogous substrates may also be attacked at slower rate (1/1000 times that of the natural substrate). Nevertheless, in a practical sense they can be said as *specific*.

FACTORS INFLUENCING THE ENZYME ACTIVITY

The following factors influence the rate of enzyme action:

(1) Substrate concentration.
(2) Enzyme concentration
(3) Tempeature,
(4) *p*H,
(5) Activators, and
(6) Inhibitors.

(1) Substrate Concentration

The substrate concentration significantly affects the rate of enzyme—catalysed reaction. As evident from Fig. 8.4 up to a certain concentration, the rate of reaction (formation of end products) increases with the increase of substrate concentration. But finally, a point is reached where there is no increase in reaction with further increase of substrate concentration. At this substrate concentration, all the enzyme gets saturated and is present in the form of enzyme-substrate complex (ES). Obviously, any further increase in substrate concentration cannot increase the amount of ES.

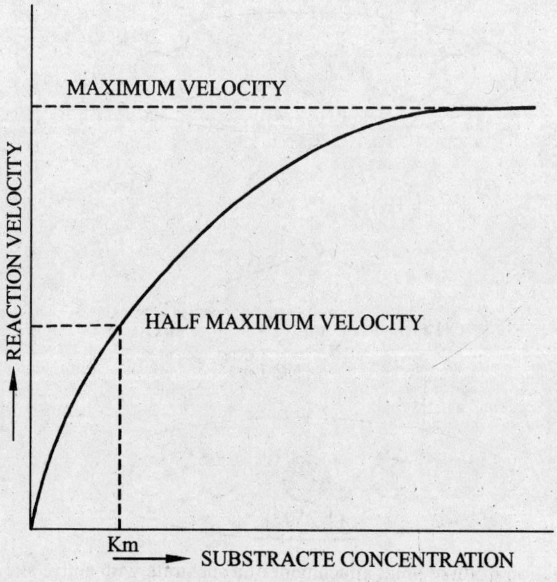

Fig. 8.4. Effect of substrate concentration on rate of enzyme catalysed reaction.

The Michaelis Equation: The first satisfactory explanation of the relationship between substrate concentration and velocity of enzyme reactions was provided by Michaelis and Menten equation.

The equation can be derived with the aid of certain assumptions. The first assumption is that the entire reaction takes place in two steps:

$$E + S \rightleftharpoons ES \quad \ldots(1)$$

$$ES \rightleftharpoons E + P \quad \ldots(2)$$

(where E is the enzyme; S is the substrate, ES is the enzyme-substrate complex and P is the end product.)

Experimentally it has been shown that the time taken for the formation of enzyme-substrate complex is negligible as compared to the time taken for the breaking up of enzyme-substrate complex into free enzyme and end products.

Hence, the velocity (V) of the total reaction is proportional to the concentration of enzyme-substrate complex (Law of mass action).

$$V \propto [ES] \quad \ldots(3)$$

or $$V = K \times [ES] \quad \ldots(4)$$

where K is constant.

The maximum reaction rate (V_{max}) will occur at a substance concentration sufficiently high to utilize all the available enzyme, so that all the available enzyme is present in the form of [ES]. Under the conditions, the concentration of [ES] will be equal to the total enzyme concentration [E_{total}].

$$[ES] = [E_{total}] \times K \quad \ldots(5)$$

Therefore, $$V_{max} = K \times [E_{total}] \quad \ldots(6)$$

Dividing equation (4) by (6) we get

$$\frac{V}{V_{max}} = \frac{K \times [ES]}{K \times [E_{total}]} \quad \ldots(7)$$

Now turning to the first step of enzyme-catalysed reaction, we can write the equilibrium constant for the dissociation of [ES] complex as

$$Km = \frac{[E] \times [S]}{[ES]} \quad \ldots(8)$$

In this equation (8) Km is a constant which is known as Michaelis constant. In this equation we have E rather than E_{total} because E is the concentration of *free enzyme*. But E_{total}, ES and E are related in following way.

$$[E] = [E_{total}] - [ES] \quad \ldots(9)$$

Now substituting the value of E in equation (8) we get

$$Km = \frac{[E_{total}] - [ES] \times [S]}{[ES]} \quad \ldots(10)$$

or $$Km = \frac{[S][E_{total}] - [S] \times [ES]}{[ES]} \quad \ldots(11)$$

or $$Km = \frac{[S][E_{total}]}{[ES]} - \frac{[S][ES]}{[ES]} \quad ...(12)$$

or $$Km = \frac{[S][E_{total}]}{[ES]} - [S] \quad ...(13)$$

or $$Km + [S] = \frac{[S][E_{total}]}{[ES]} \quad ...(14)$$

or $$\frac{1}{Km + [S]} = \frac{[ES]}{[S][E_{total}]} \quad ...(15)$$

or $$\frac{[S]}{Km + [S]} = \frac{[ES]}{[E_{total}]} \quad ...(16)$$

By equation (7) we know that the value $\frac{[ES]}{[E_{total}]} = \frac{V}{V_{max}}$

The substituting this value in equation (16) we get

$$\frac{V}{V_{max}} = \frac{[S]}{Km + [S]} \quad ...(17)$$

This is one form of *Michaelis-Menten Equation.*

According to another definition Km is the substrate concentration at which the reaction develops at half its maximum velocity (V_{max}).

$$\frac{V}{V_{max}} = \frac{[S]}{Km + [S]} \quad ...(18)$$

or $$V = \frac{V_{max} \times [S]}{Km + [S]} \quad ...(19)$$

Substituting $\frac{V_{max}}{2}$ for V in equation (19) we get

$$\frac{V_{max}}{2} = \frac{V_{max} \times [S]}{Km + [S]} \quad ...(20)$$

or $$\frac{1}{2} = \frac{[S]}{Km + [S]} \quad ...(21)$$

or $$2[S] = Km + [S] \quad ...(22)$$

or $$Km = [S] \quad ...(23)$$

Km is equal to [S] when the reaction velocity is half of the maximum velocity (Fig. 8.4). This equation tells us that the substrate concentration necessary to develop half maximum velocity is equal to the Michaelis constant (Km). The Michaelis constant is also measure of the affinity of the enzyme for its substrate.

(2) Enzyme Concentration

Increase in enzyme concentration accelerates the velocity of enzyme catalysed reaction. However, under practical condition it is not always possible to achieve very high concentration of enzyme, because these being proteins, their molecular weight is usually very high of the order of 1,000, 000 or more. Therefore, the preparation of 1 litre of 1 molar of solution of a typical enzyme would require 160 g of the enzyme, which is impracticable. Under normal condition the enzyme concentration rarely exceeds 0.001 M. Hence, we are restricted to work only with very low concentration of enzyme. Working with these conditions the rate of reaction is directly proportional to the concentration of enzyme (Fig. 8.5).

In Fig. 8.5 at zero enzyme concentration, the rate of activity has been indicated to be zero. This is always not true because many reactions, which are normally enzyme-catalysed take place even in the absence of the enzyme though at a very slow rate.

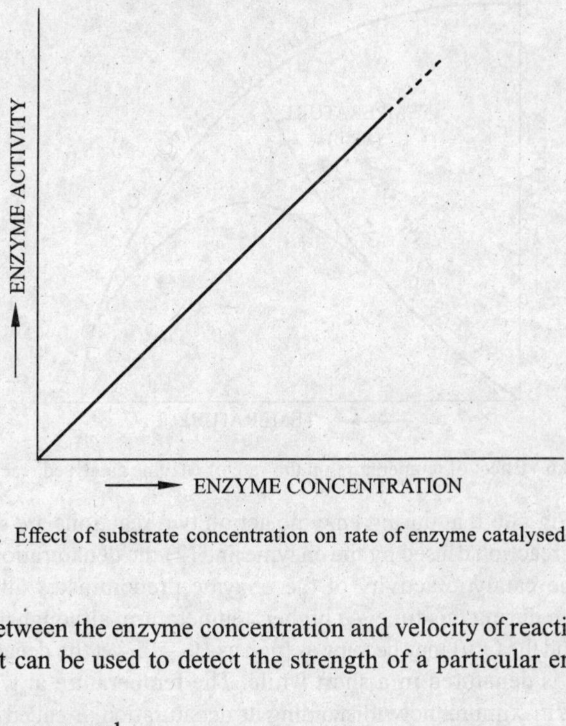

Fig. 8.5. Effect of substrate concentration on rate of enzyme catalysed reaction.

The relationship between the enzyme concentration and velocity of reaction is of great practical importance because it can be used to detect the strength of a particular enzyme in a mixture of proteins.

This relation can be expressed as

$$\underset{\text{Velocity}}{(V)} \propto \underset{\text{Enzyme concentration}}{E} \quad \ldots(1)$$

or

$$\underset{\text{Velocity}}{(V)} = \underset{\text{Constant}}{K} + \underset{\text{Enzyme concentration}}{[E]} \quad \ldots(2)$$

Therefore,

$$[E] = \frac{(V)}{K} \quad \ldots(3)$$

(V) can be expressed as

$$\left(V = \frac{\text{Amount of product}}{\text{time of reaction}} \right) \quad \ldots(4)$$

Substituting the value of (V) in equation (3) we get

$$[E] = \frac{\text{Amount of product}}{\text{Time of reaction} \times K} \quad \ldots(5)$$

(3) Temperature

Temperature is one of the important factors which always affects enzyme-catalysed reactions. As the temperature is raised, the rate of reaction increases until of maximum is reached. The rise in reaction rate with increase of every 10°C is expressed as Q 10. When the rate of reaction becomes double by a rise of 10°C, then the Q 10 is equal to 2. For most enzymes the Q 10 is approximately 2. But at higher temperature, the rate of reaction rapidly decreases and at still higher temperature it comes to an end. The temperature at which the rate of reaction is maximum is termed as the *optimum temperature* (Fig. 8.6).

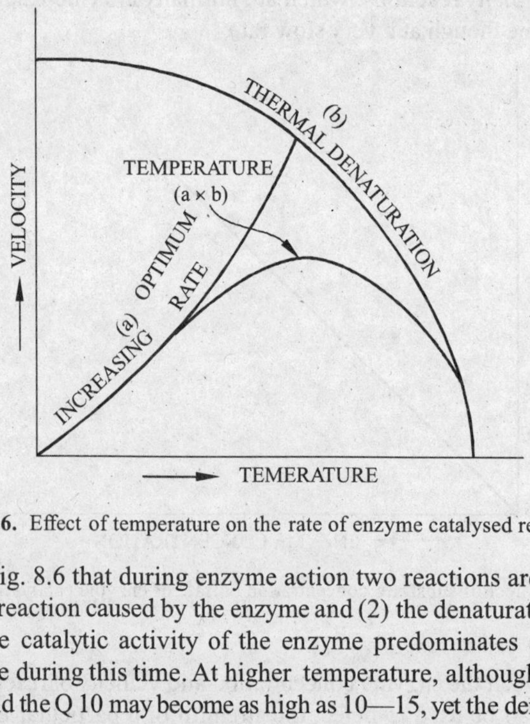

Fig. 8.6. Effect of temperature on the rate of enzyme catalysed reaction.

As it is clear from Fig. 8.6 that during enzyme action two reactions are simultaneously taking place. (1) The catalytic reaction caused by the enzyme and (2) the denaturation of the enzyme itself. At low temperature the catalytic activity of the enzyme predominates although some thermal denaturation takes place during this time. At higher temperature, although the catalytic activity of the enzyme increases and the Q 10 may become as high as 10—15, yet the denaturation predominates. Therefore, all enzyme is denatured in a short while. The temperature at which the net catalytic activity of the enzyme is maximum, notwithstanding its denaturation, is called *optimum temperature.*

The reason for the increased enzymatic activity with rise of temperature is as follows. The energy of molecule becomes greater which, in turn, enhances both the inherent reactivity of the molecules and the frequency of their collisions.

Because of the high Q 10 for enzyme denaturation, as mentioned earlier, an enzyme is stable for weeks at 0°C, for days at 10°C, for hours at 30°C but for fraction of second at 70°C. Usually, the reaction rate rises up to 40°C, starts declining from 40°C. The optimum temperature of most enzymes falls between 37—40°C.

(4) pH

Enzymes are *p*H specific as these function optimally within a narrow *p*H range (Table 8.1).

Table 8.1 Name of some of the enzymes and their optimum pH

Enzyme	*Optimum pH*
Ribonuclease	7.75
Lipase	7.5 – 8.0
Cholinesterase	7.0
α -amylase	6.9
Trypsin	7 – 9
Pepsin	1.5

Fig. 8.7. Effect of pH on the rate of enzyme catalysed reaction.

The hydrogen-ion concentration of an enzyme solution has a profound influence on enzymic activity. It has been seen in case of most of the enzymes that slow reaction occurs at strongly acidic and basic solutions. This is due to denaturation of enzyme protein in strongly acidic or basic media. But denaturation is not the only explanation for the striking change in reactions with only minor changes in hydrogn-ion concentration (Fig. 8.7). Usually maximum enzymic activity is obtained at or near the *isoelectric points* (Chapter 2) of the enzymes. Thus, trypsin, whose isoelectric point is 10.1, shows maximum activity at pH range 7 – 9.

The change of pH involves a change in the degree of ionisation of functional group of enzymes. This causes structural change in the protein molecule. For instance, a positively charged group and a negatively charged group are necessary at the active site of the enzyme, then in any enzyme if any one of these groups will be uncharged, the enzymes will be catalytically inactive.

(5) Activators

Certain substances either initiate or enhance the rate of enzyme-catalysed reactions. Such substances are called *activators.* The activators can be classified into two groups, (a) metallic inorganic activators, and (b) proenzyme activators.

(a) Metallic-ion activators. Several enzymes require specified metallic ions; for better activity. Thus, arginase requires cobalt, manganese or iron; leucyl peptidase requires manganese or magnesium ions; α -amylase requires an anion, the chloride ion etc.

It is generally believed that the activators combine with the enzyme to furnish an altered protein which is a better catalyst than the original. In some cases the metallic ion acts as a bridge between the substrate and the enzyme. The resulting complex is fragile and the substrate is easily broken. This is shown below by using leucylglycine as substrate, leucyl peptidase as enzyme and manganese as activator.

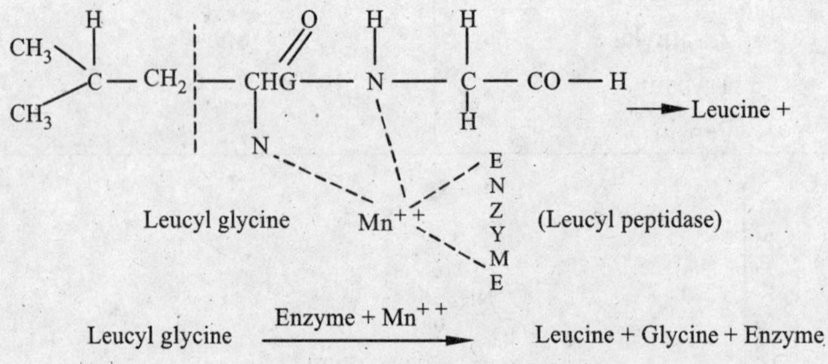

(b) Proenzyme activators. A variety of enzymes, particularly the digestive enzyme and the enzymes responsible for coagulation of blood are produced in inactive form. The inactive forms are known as *zymogens* or *proenzymes* examples of zymogens are tripsinogen, pepsinogen, prothrombin and chymotripsinogen. This type of enzyme activation involves the breaking of some peptide bond and the rearranged remaining molecule is the catalytically active enzyme.

The production of zymogens by the animals is a biological safeguard against autodigestion (or autolysis) of the cell producing the proenzyme.

(5) Inhibitors

Substances which inhibit the rate of enzyme reaction are called *inhibitors,* and theprocess is termed *enzyme-inhibition.* These are of two types: (a) competitive, and (b) non-competitive.

(a) Competitive inhibitors. Inhibitors (*e.g.* biological antagonastis and antimetabolites) which compete with the substrate for the same active site on enzyme molecule are called *competitive inhibitors*. Consequently, they leave very small number of active sites for the reaction with substrate. For example, D-histidine is a competitive inhibitors for the enzyme liver histidase.

General, competitive inhibitors are structural analogues of the substrate. Their inbibition caused by such inhibitors is reversible in the presence of large quantities of substrate.

(b) Non-competivie inhibitors. Inhibitors which can not be displaced from the enzyme molecules by the substrates are called *non-competitive inhibitors*. The non-competitive inhibitor often acts at a site other than the active site of the enzyme molecule. Although the active site is available to the substrate, no reaction occurs because the inhibitor alters the chemical nature of the enzyme. An interesting example of non-competivie inhibitors are two polypeptidases secreted by *ascaris;* one inhibits pepsin and other trypsin. This prevents the worms from being digested themselves by the action by these enzymes in the gut.

COENZYMES

Coenzymes are special activators in the sense that they can be reversibly removed from the protein molecule and are essential for the activity of enzymes. Coenzymes are non-proteins (prothatic) groups which usually have vitamin derivatives as one of their "active groups. The structure of some of the coenzymes is on next page.

(1) Coenzymes Acting as Hydrogen Acceptors

These enzymes are responsible for electron transfer during tissue respiration.

(a) Pyridine coenzyme. These include nicotinamide adenine dinucleotide (NAD) or conenzyme I and nicotinamide adenine dinucleotide phosphate (NADP) or coenzyme II. These act as hydrogen acceptors. For example, *ethyl alcohol* is oxidized to *acetaldehyde* by the enzyme *alcohol dehydrogenase* which has NAD as coenzyme.

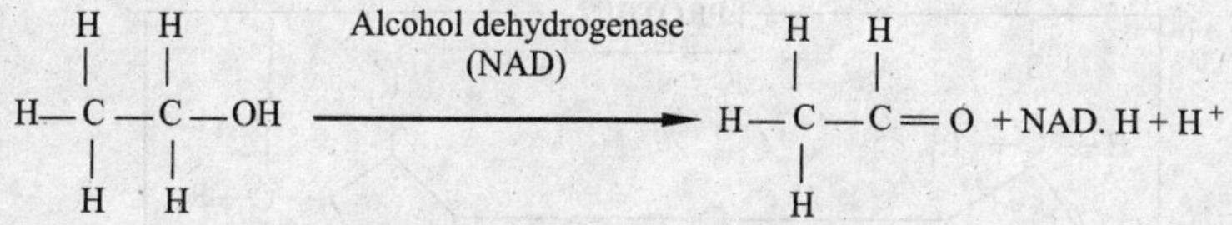

This part of NAD molecule which is affected by this reaction is nicotinamide, which is reduced as under.

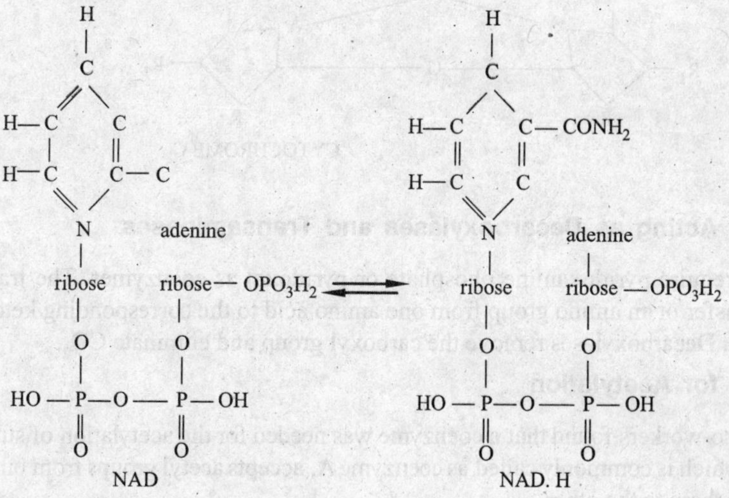

The reaction of NADP is similar to that of NAD.

(b) Riboflavius coenzymes. These include riboflavin mononucleotide (FMN) and riboflavin adenine nucleotide (FAD). They also act as hydrogen acceptor and isoalloxazine group is affected.

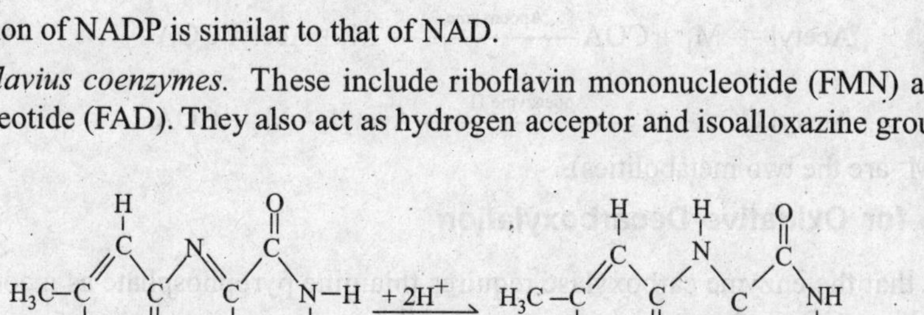

(c) Iron Porphyrin Compounds. In strict sense they can not be classified as coenzymes as they are so tightly bound to the apoenzyme that it is not possible to remove them by dialysis. The iron porphyrin part is called prosthetic group. Typical examples of enzymes requiring this prosthetic group are catalase, proxidase, and cytochromes. The cytochromes (a, a_3, b, c and c_1) are respiratory enzymes of the terminal respiratory chain (Chapter 13). The abbreviated structural formula of cytochrome c is given below.

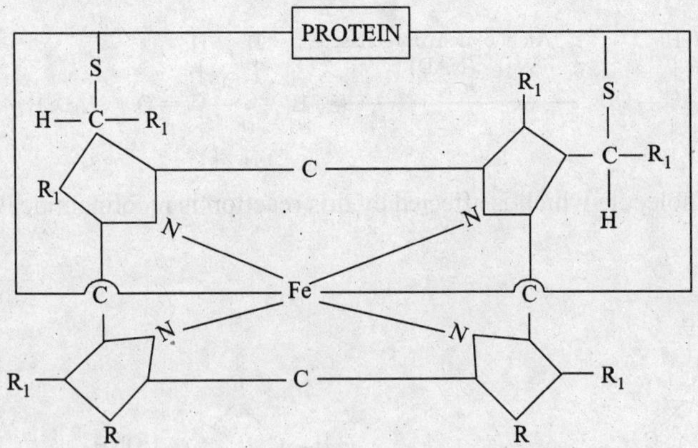

CYTOCHROME C

(2) Coenzyme Acting as Decarboxylases and Transaminases

These enzymes require pyridoxamine phosphate or pyridoxal as coenzymes. The transaminases catalyses the transfer of an amino group from one amino acid to the corresponding keto acid of an other amino acid. Decarboxylases remove the carboxyl group and eliminate CO_2.

(3) Coenzyme for Acetylation

Lipman and his co-workers found that a coenzyme was needed for the acetylation of sulfanilamide. The coenzyme, which is commonly called as coenzyme A, accepts acetyl groups from one metabolite and then donates them to the other.

$$\text{Acetyl}-M_1 + \text{COA} \xrightarrow{\text{Apoenzyme I}} M_1 + \text{Acetyl COA}$$

$$\text{Acetyl COA} + M_2 \xrightarrow{\text{Apoenzyme II}} \text{Acetyl}-M_2 + \text{COA}$$

(where M_1 and M_2 are the two metabolities).

(4) Coenzyme for Oxidative Decarboxylation

It has been seen that the enzyme carboxylase requires thiamine pyrophosphate as a coenzyme to split pyruvic acid into CO_2 and acetaldehyde in yeast cells.

(5) Other Coenzymes

There is sufficient evidence that many other compounds act as coenzymes. Examples are glucose 1, 6—diphosphate, glyceric acid 3—diphosphate and a glucose 1—phosphate—uridine nucleotide etc.

9
Chapter

Nervous System

In mammals all body activities are either directly or indirectly under the control of the *nervous system.* Direct control is exercised by supplying nerves to the various organs and tissues, while indirect control is exercised through the endocrine glands and the circulatory system. The nervous system receives a variety of stimuli from the various parts of the body; depending on the nature of the stimuli it sends out appropriate signals to concerned organs. The stimuli which the nervous system receives are of two kinds: stimuli (sensations) like pressure, pain, heat etc., which are caused due to changes in the external conditions are called *exteroceptive sensations.* Sensations which convey the physical state of the body are called *proprioceptive sensations.* These include informations regarding the functioning of the internal organs.

Both extero-and proprio-ceptive sensations are brought to the brain and spinal cord through *sensory nerves.* Orders which are carried out in response to these sensations are conveyed to the appropriate organs through motor nerves. Although a large number of responses to exteroceptive stimuli are under the wilful control of the individual, responses to proprio-ceptive stimuli are not.

GENERAL ORGANISATION

The nervous system is divided into two parts: (1) *Central nervous system* which comprises the brain and spinal cord and (2) *Peripheral nervous system* which includes all the nerves arising out and going to the central nervous system. Functionally, the nervous system has been divided, into *somatic* (cerebro-spinal) and *automatic* (visceral) nervous systems. The latter given off two kinds of nerves, *sympathetic* and *parasympathetic.* The activity of most organs is accelerated by symapthetic nerves and inhibited by parasympathetic nerves. It must be understood, however, that all nervous activities are always controlled by the central nervous system; the peripheral nerves are merely carriers of information.

Only a very small part of the body activities are under the wilful (*voluntary*) control of the body. The centres for the control of the voluntary activities are present in the *thalamus* and *cerebral* cortex of the brain. These are therefore called the *conscious areas of the brain.* Most of the organs of the body are controlled by an automatic feedback circuit in which no conscious thinking is required. Such activities are termed involuntary and are controlled by the *autonomic nervous system.* The centres for these are present in the *medulla oblongata, pons* and *mesencephalon.*

The brain and spinal cord are divided into several parts and each part is concerned with the regulation of specific functions of the body. Fig. 9.1 gives in detail the nerve supply of the various organs of the body. This chapter is, however, mainly concerned with the functioning of the nervous system at the cellular and molecular level. The nervous regulation of individual systems has been dealt along with the discussion of that system.

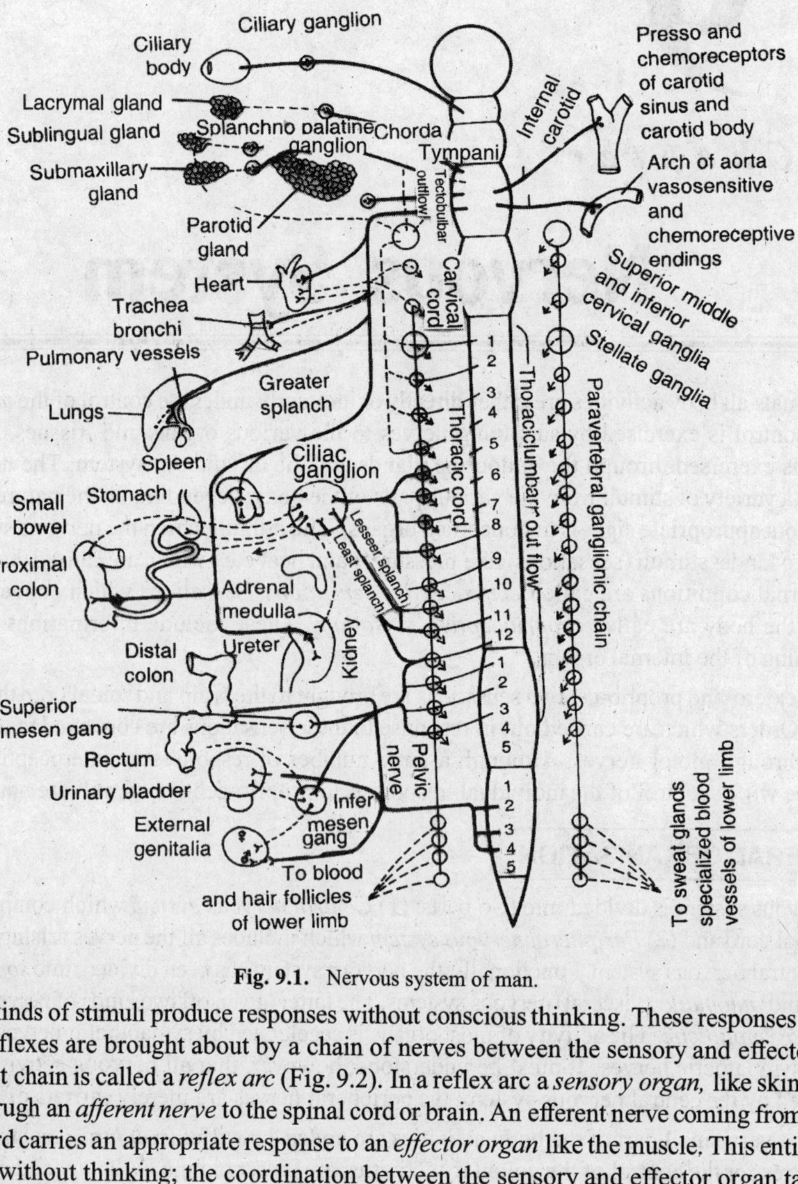

Fig. 9.1. Nervous system of man.

Certain kinds of stimuli produce responses without conscious thinking. These responses are called *reflexes.* Reflexes are brought about by a chain of nerves between the sensory and effector organs. The simplest chain is called a *reflex arc* (Fig. 9.2). In a reflex arc a *sensory organ,* like skin, sends an impulse thorugh an *afferent nerve* to the spinal cord or brain. An efferent nerve coming from the brain or spinal cord carries an appropriate response to an *effector organ* like the muscle. This entire process takes place without thinking; the coordination between the sensory and effector organ takes place through the autonomic nervous system (Fig. 9.2).

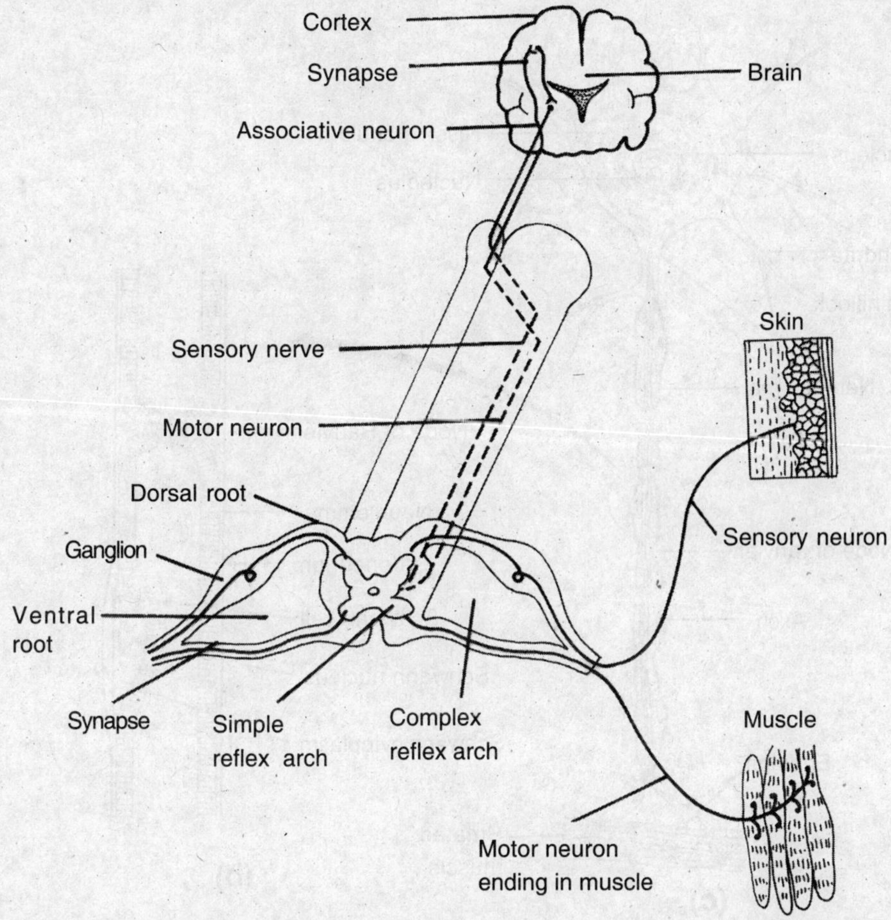

Fig. 9.2. Diagrammatic represention of a reflex arc. Two types of reflex arcs have been shown. On the left side a simple reflex arc has been shown. The impulse from a sensory nerve in the skin (afferent nerve) carries an impulse to the spinal cord through the dorsal root where it makes contact (synapse) with a motor nerve which carries it to the effector organ which may be a muscle (the skin and the muscle have not been shown on the left side). On the right side a more complex reflex arc has been shown. The sensory impulse from the skin is carried to the brain *via* the spinal cord. During this process several synapses are made. The broken lines indicate a large number of synaptic connections.

STRUCTURE OF THE NERVE CELLS

The *nerve cells* have a special structure adapted to the needs of their function. Although, these vary greatly is size and shape yet there are certain basic structures which are found in every nerve cell. The nerve cells have a *cell body* which encloses a cytoplasm containing a nucleus. The cell body gives off a number of processes. There is usually a single *axon* and a variable number of *dendrites* (Fig. 9.3).

Cell Body: A typical cell body is rounded or polygonal in shape. It contains a nucleus with a large nucleolus. Special bodies called *nissl granules* are also present in this (Fig. 9.3); these are composed of ribonucleoproteins. The nissl granules probably manufacture enzymes for the synthesis of *neurotransmitters*. Each cell body is surrounded by several special cells called *glial cells.* The membranes of glial cells surround most of the cell body except at places where the cell body makes contact with other cells.

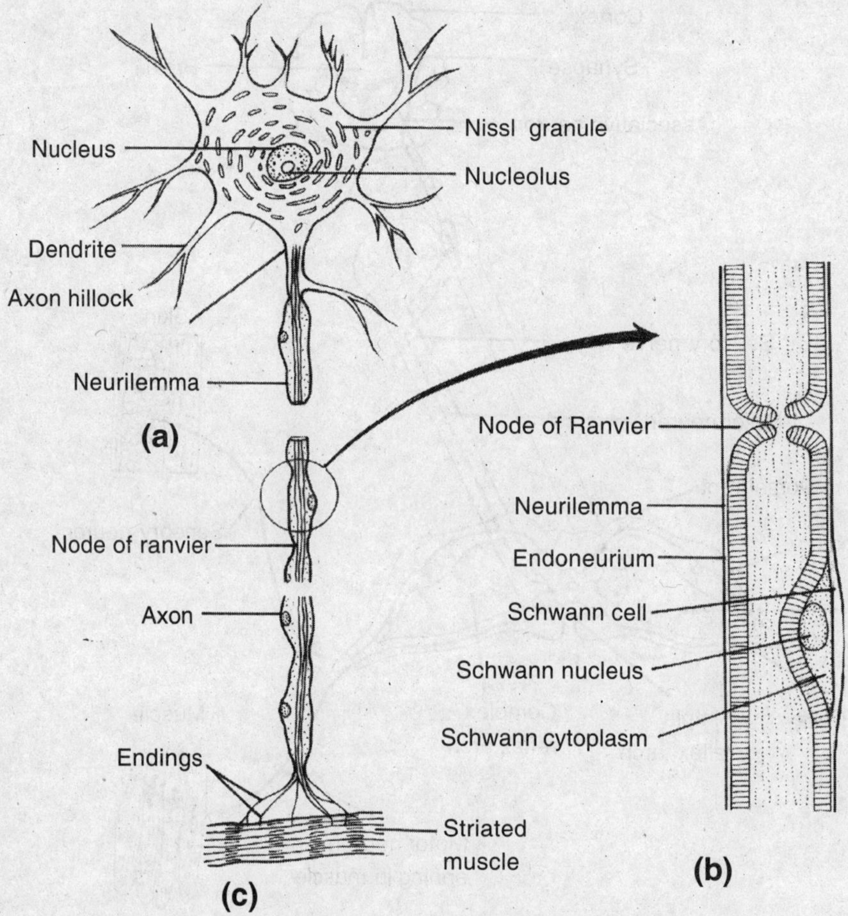

Fig. 9.3. Diagram showing a typical nerve cell. (a) a multipolar cell with a single axon and many dendrites; the axon is myelinated and is a very long process. In this diagram a part of the axon is not shown. (b) an enlargement of the boxed-in area.

Axon: The cell body gives off a number of processes. There is usually a long branch known as *axon* which is longer than the others. This arises from a special protruberance on the cell body known as the *axon hillock.* The axon is also called the *nerve fibre.* The length of the axon varies from cell to cell; sometimes it may be even more than a metre in length *e.g.* the afferent axons in the leg of man. The cytoplasm of the axon is called *axoplasm.* The axon is branched at its distal end ; the branches give off special club-shaped processes called *boutons* which makes contact with other cells (Fig. 9.20).

Dendrites: Besides the axon there are one or more shorter highly branched processes called *dendrites.* The dendrites serve to carry messages to the nerve cell. Certain types of cells *e.g.* spinal ganglion cells have only one process. Such cells are called *unipolar* cells. Depending on the number of processes (poles), nerve cells have been variously termed as *unipolar, bipolar* (*e.g.* cells of the retina) or *multipolar* (*e.g.* motor neurons of the spinal cord) (Figure 9.4).

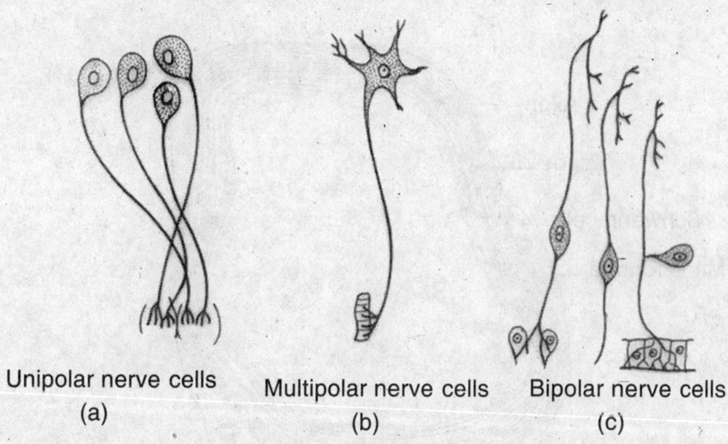

Fig. 9.4. Diagram showing various types of nerve cells.

Myelinated and non-myelinated nerve fibres: The axons, both in the vertebrates and invertebrates, are often covered with a sheath called the *myelin sheath.* Such fibres (axons) are called *myelinated fibres,* while those which are not covered are called *non-myelinated fibres.* Dendrites are always non-myelinated. Usually fibres of larger diameter are myelinated. In mammals all fibres larger than 1 μ (0.001 mm) in diameter are covered with myelin (Figure 9.5). Nerves made up of myelinated fibres are called *medulated nerves.* The structure of myelinated fibres is complicated. The axon is surrounded by the myelin sheath which may vary in structure from cell to cell. The myelin sheath is surrounded by the cytoplasm of special types of cell called *Schwann cells,* which are always closely associated with nervous tissue. In fact the Schwann cell repeatedly surrounds the axis cylinder (axon) with many concentric rings of its cell wall to form the myelin sheath of that particular region [Fig. 9.5 (a and (b)]. The myelin sheath provides both mechanical support and electrical insulation to the axon. The myelin sheath is broken at regular intervals along the lengths of the axon. The places where the myelin sheath is discontinous are called *Nodes of Ranvier* (Fig. 9.3). The structure is very important in the conduction of impulses through myelinated nerve fibres. Figure 9.3 (b) shown a single Schwann cell nucleus between two nodes of Ranvier.

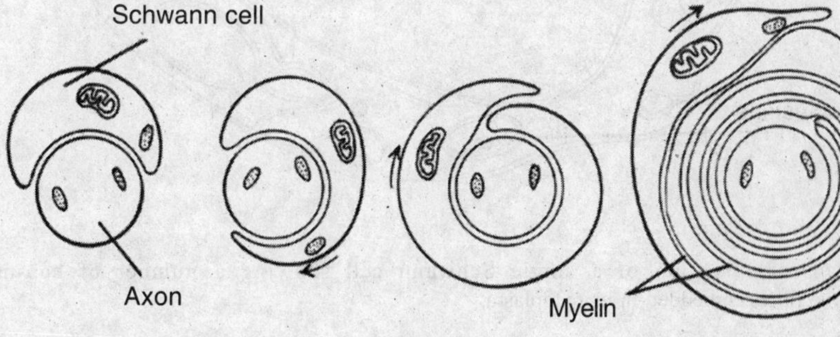

(a)

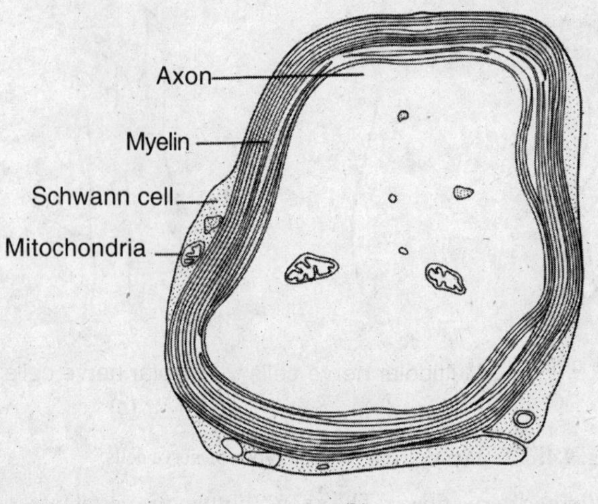

(b)

Fig. 9.5. Transverse section of myelinated nerve fibre. **(a)** diagrammatic representation of the development of a myelinated nerve fibre; the fibre develops by repeated encircling of the axon by the Schwann cell. **(b)** Cross-section of a myelinated nerve fibre as seen through an electron microscope.

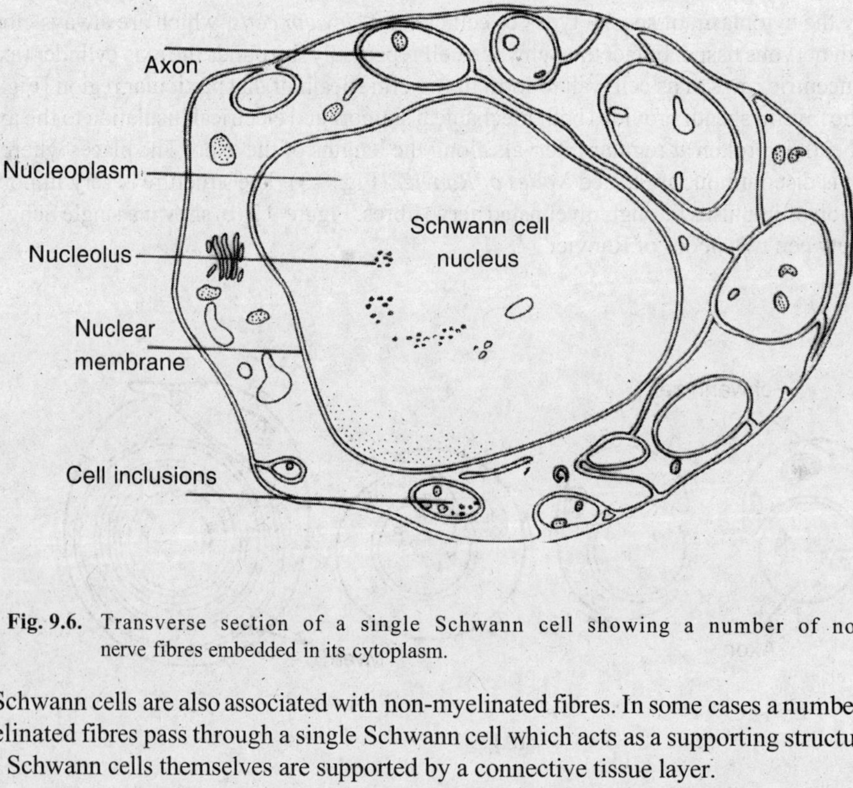

Fig. 9.6. Transverse section of a single Schwann cell showing a number of non-myelinated nerve fibres embedded in its cytoplasm.

Schwann cells are also associated with non-myelinated fibres. In some cases a number of fine non-myelinated fibres pass through a single Schwann cell which acts as a supporting structure (Fig. 9.6). The Schwann cells themselves are supported by a connective tissue layer.

The axon is covered by a structure less membrane called *neurilemma* [Fig. 9.3 (b)]. The Schwann cells lie within this membrane. The connective tissue layer surrounding the individual axon is called *endoneurium.* Several axons are bound together by another connective tissue sheath called *perineurium.* Several such bundles of axons are bound together by another connective tissue layer—*epineurium* to form a nerve (Fig. 9.7).

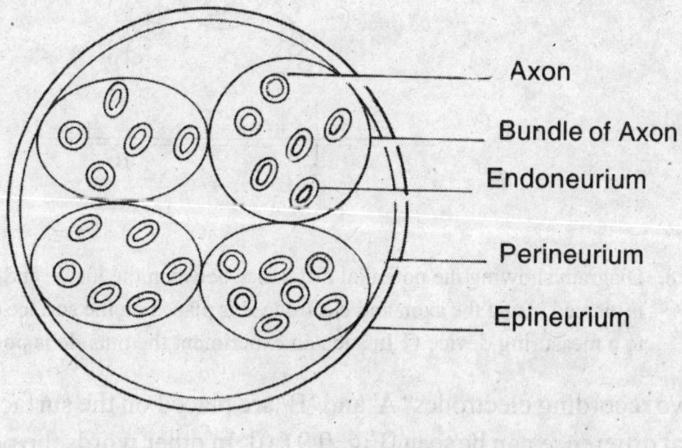

Fig. 9.7. Diagrammatic representation of crosssection of a single nerve.

Size of nerve fibres: The nerve fibres of mammals vary greatly in size. The diameter of an axon may be as large as $25\ \mu$ in some vertebrates. In human beings it ranges from $0.1\ \mu$ to $10\ \mu$. The largest mammalian nerve fibres are called 'A' fibres, those of intermediate size are called 'B' fibres and the smallest are called 'C' fibrs, 'A' fibres are further subdivided into α, β and γ fibres. This classification is based not only on size but also on conduction velocity (described later), which are closely related. The fibres of some invertebrates, especially annelids, crustaceans and molluscs are very large. The axon of *Loligo,* popularly known as *squid giant axon,* has a diameter of nearly $1500\ \mu$. Most of the fundamental research on the physiology of nervous system has been carried out on this axon.

It should be understood at this point that the cell body of every cell is confined to the central nervous system (brain, spinal cord and ganglia) only. The nerves are made up exclusively from axons. A nerve usually contains many axons. This accounts for the grey and white matter of the central nervous system. The grey colour is solely due to the cell bodies and non-myelinated processes, whereas the white matter comes from the myelinated nerve fibres as myelin has a characteristic g listening white appearance.

RESTING AND ACTION POTENTIALS

A cell membrane has a complicated structure (see Chapter 1) which endows it with specialised electrical properties enabling the cell to develop an electrical potential across the cell membrane. Although all cells develop electrical potentials the following discussion applies specially to nerve cells.

The axon is a tube like structure filled with axoplasm. If we take two fine electrodes (devices for sensing potential difference), place one of the surface of the axon, pierce the other in the axoplasm, and connect the two to a measuring device (Fig. 9.8), like a sensitive galvanometer or an oscilloscope, then we can mesure the potential difference between the inside and outside of the cell. Such measurements show a potential difference of 40.120 mV (millivolt = 1/1000 volt), the outside being postive with respect to the inside. This potential difference is called *resting potential.*

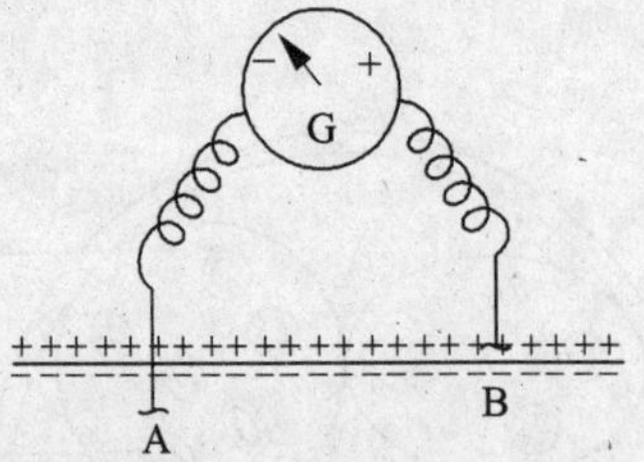

Fig. 9.8. Diagram showing the potential difference between the inside and outside of the axon. The electrode A is pierced inside the axon and electrode B is placed on the surface of the axon. These two are connected to a measuring device G. In such an experiment the outside is positive with respect to the inside.

If the two recording electrodes 'A' and 'B' are placed on the surface of the axon at two points then no potential difference can be seen [Fig. 9.9 (a)]. In other words the potential difference between two points on the surface of a resting axon is zero. Now, if a stimulus (S) is given at a point away from the electrode 'A' [Fig. 9.9 (b)] then after some time electrode 'A' becomes electronegative with respect to electrode 'B' [Fig. 9.9 (c)]. After this, the potential difference between the two electrodes again falls to zero [Fig 9.9 (d)]. Again, after a short time interval electrode 'B' becomes electronegative to 'A' [Fig. 9.9 (e)]. Finally the potential difference between the two electrodes again becomes zero [Fig. 9.9 (f)].

This experimental observation can be explained in the following manner: The brief negativity of the first electrode is due to the arrival of the impuse (stimulus) under electrod 'A' [Fig. 9.9 (c)]. In the second stage the impulse comes in between the two electrodes and neither of the electrodes can monitor it so that both electrodes shown the same potential [Fig. 9.9 (d)]. In the next stage the impulse arrives under the electrode 'B' which then becomes electronegative with respect to 'A', or in other words 'A' becomes positive with respect to 'B' [Fig. 9.9 (e)]. Finally, [Fig. 9.9. (f)] the impulse passes away, both electrodes return to their original state, the potential difference between them disappears and they record a zero potential.

The above observations suggest two very important facts: (1) when a stimulus is given it creates an *impulse* which is in the form of a brief electronegative charge created on the surface of the axon. (2) the impulse, once it has been created, travels along the length of the axon or in other words *is propagated.* The brief electrical negativity is called *action potential.* Since the action potential moves along the fibre it is also called a *propagated action potential.* A propagated action potential is, therefore, in the form of a wave of electronegativity. The action potential under a recording electrode rises over a period of time. The highest voltage which it attains is called the *spike potential* or simply *spike* (Fig. 9.10).

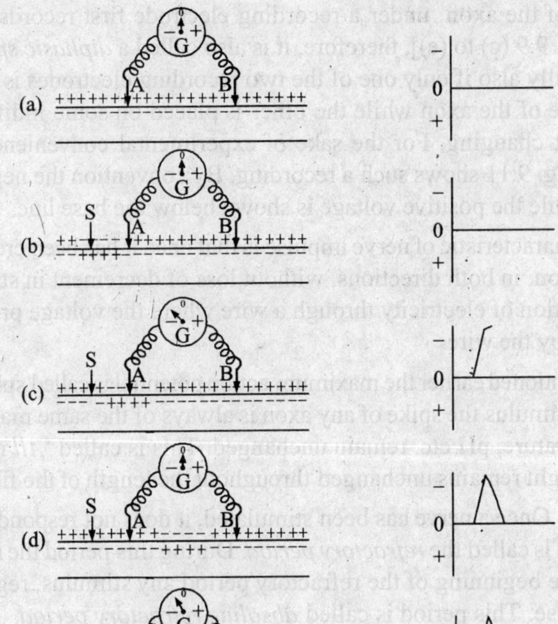

Fig. 9.9. Diagram showing the change in potential between two parts on the surface of an axon when a stimulus S is given; G-galvanometer. A and B are two recording electrodes placed some distance apart on the surface of an axon. (a) No potential difference between A and B when cell is resting. (b) When stimulus S is given, the part on the surface of the axon directly below the stimulus becomes negative on the outside and positive on the inside. (c) When the negativity reaches A the galvanometer records a negative potential from it in relation to B. (d) When stimulus is between A and B the galvanometer does not record any potential. (e) When the stimulus reaches B then B becomes negative in relation to A. (f) Finally the stimulus passes away and both A and B again have the same potential. The right hand side depicts a graph showing the potential changes; an actual recording with an osciloscope gives much the same picture.

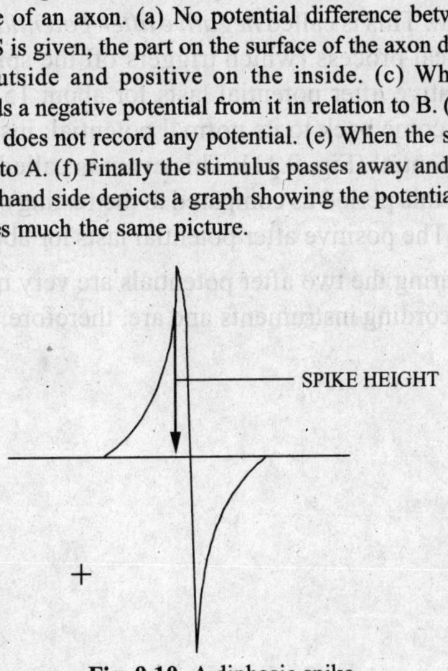

Fig. 9.10. A diphasic spike.

Since the surface of the axon under a recording electrode first records a negative and then a positive potential [Fig. 9.9 (c) to (e)], therefore, it is also called a *diphasic spike.* An impulse can be recorded monophasically also if only one of the two recording electrodes is used for measuring the changes on the surface of the axon while the other is placed on some indifferent part of the body whose potential is not changing. For the sake of experimental convenience usually monophasic recording are made. Fig. 9.11 shows such a recording. By convention the negative voltage is shown above the base line while the positive voltage is shown below the base line.

A very important characteristic of nerve impulse is that once it has been created it travels along the entire length of the axon, in both directions, without loss of decrement in strength. This is in sharp contrast to the conduction of electricity through a wire where the voltage progressively falls due to the resistance created by the wire.

All or None: As mentioned earlier the maximum action potential is called spike potential. Regardless of the strength of the stimulus the spike of any axon is always of the same magnitude, provided other conditions like temperature, pH etc. remain unchanged. This is called '*All* or *None'* principle. Not only this, the spike height remains unchanged throughout the length of the fibre.

Refractory Period: Once a nerve has been stimulated, it does not respond to a stimulus again for some time. This period is called the *refractory period.* During this period the nerve recovers from the earlier stimulus. At the beginning of the refractory period any stimulus, regardless of its strength, cannot evoke a response. This period is called *absolute refractory period.* The absolute refractory period is followed by a period where a stimulus, much higher than the normal stimulus, is required to evoke a response. This period is called *relative refractory period.* Following this the nerve fully recovers and can be stimulated again in the usual manner.

Events during spike formation: If the details of a spike are analysed (Fig. 9.11), it will be seen that a spike is made up of several distinct events. Just before start of the spike there is a small local disturbance in which the axon becomes slightly electronegative; this tiggers off the spike (a phenomenon known as *depolarisation*) which rises and falls very sharply. The spike lasts for approximately 1 m/sec. The entire period during which the spike rises and part of the period of the fall of the spike corresponds with the absolute refractory period. The latter part of the spike corresponds with the relative refractory period. Near the end of the spike the process of the disappearance of the spike (*repolarisation*) becomes slow. This is called *negative after potential.* As the negative after potential is similar in nature to the local process (which triggers off the spike) the nerve is more excitable during this period. The negative after potential lasts for about 15 m/sec. After the negative after potential the nerve does not come back to its normal potential; instead, it attains a higher positive potential than the resting potential (Fig. 9.11). This event is called *positive after potential.* As the nerve is more positive during this period as compared to the resting stage, therefore it is less excitable than a normal resting nerve. The positive after potential lasts for about 90 m/sec.

As the voltage changes during the two after potentials are very meagre, there are mostly beyond the resolving power of the recording instruments and are, therefore, normally not seen.

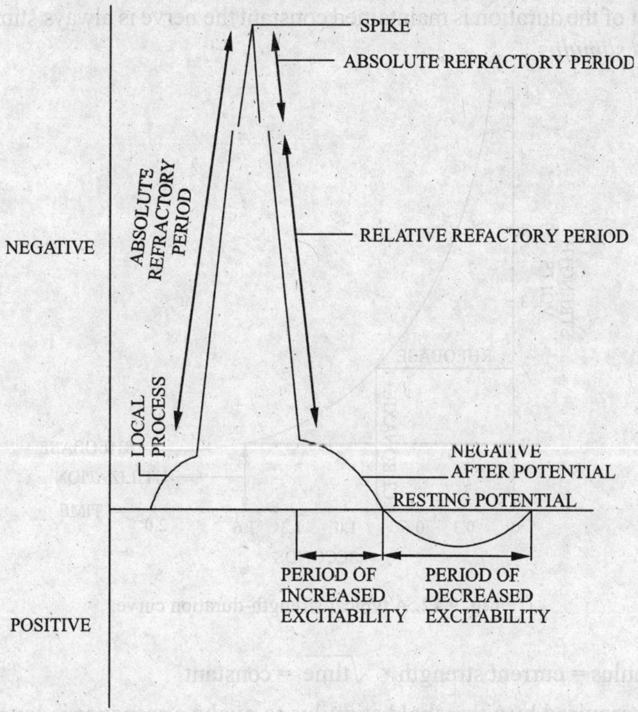

Fig. 9.11. Detailed picture of a monophasic spike. The picture is so enlarged that if the spike height was depicted in its entirety it would have been several inches above the book.

The Rate of Nerve Conduction

Once a spike has been initiated it travels along the axon at a fairly rapid rate. The rate at which the impulse travels is known as *conduction velocity*. This rate is mostly dependent upon the diameter of the fibre and the thickness of the myelin sheath. The conduction velocity of the largest mammalian fibres known as 'A' fibres can be as high as 100 metres/sec. It is about 10 metres/sec in 'B' fibres which have a smaller diameter. In 'C' fibres which have the smallest diameter the conduction velocity in only 2 metres/sec. The conduction velocity increases with increase in fibre diameter and thickness of the myelin sheath. When the product of the diameter of the fibre and the thickness of the myelin sheath is constant then the rate of conduction is also similar regardless of the fact whether the fibres are taken from squids, crustaceans, frogs, or cats.

Even though the conduction velocity of large fibres is quite high, it is obvious that spread of an impulse along the length of an axon is not simply a phenomenon of electrical conduction as the speed of electricity is high (3×18^8 M/sec).

Strength-duration Relationship

The minimum current required to stimulate a nerve is called *threshold current* or *rheobase* (Fig. 9.12). Rheobase stimulus has to be given for a long time in order to evoke a response. The time for which a current (stimulus) has to be applied in order to evoke a response is called *utilization time.* A current below rheobase strength cannot evoke a response even if it is continuously applied for a long time. If a current higher than rheobase is applied then the utilisation time is reduced. It has demonstrated that there is a definite relationship between the strength of the current and the time for which it has to be applied in order to evoke a response (Fig. 9.12). *The curve relating the stimulus strength, required to create a response, and the duration for which it has to be applied is called strength-duration curve* (Fig. 9.12). Studies on strength duration relationship show that if the product of the stimulating current and the square root of the duration is maintained constant the nerve is always stimulated. This product is called *threshold stimulus.*

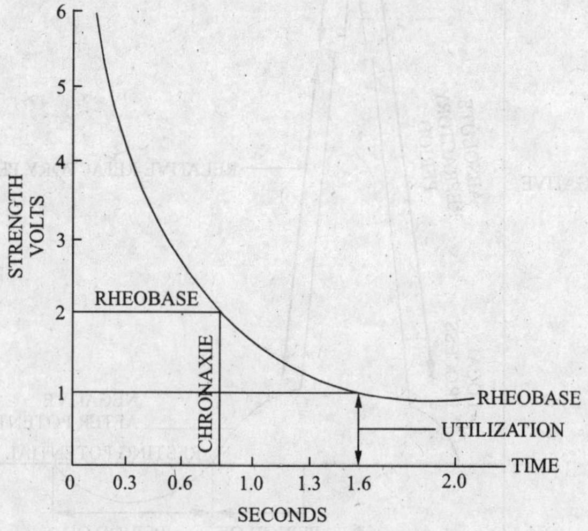

Fig. 9.12. A typical strength-duration curve.

Threshold stimulus = current strength × $\sqrt{\text{time}}$ = constant

Since the time required by a threshold stimulus to evoke a response is long, usually a stimulus twice the rheobase strength is given. The utlization time for a stimulus twice the rheobase is called *chronaxie* (Fig. 9.12).

IONIC BASIS OF NERVE POTENTIALS

It has been seen that even in a resting state there is a potential difference between the outside and inside of a cell with the outside positive with respect to the inside. This phenomenon can be explained on the basis of a physical law known as *Nernst diffusion equation.*

Nernst diffusion equation: According to this equation a potential difference is created across a semipermeable membrane if the concentration of ions to which it is permeable is different on the two sides of the membrane.

The magnitude of the potential E, which develops across the membrane, can be expressed by the following equation:

$$E = \frac{RT}{nF} l_n \frac{C_1}{C_2}$$

where,

R = gas constant (8.312 joules/degree/mol)
T = Absolute temperature
n = Valency change
F = Faraday (96,500 coulombs/gram equivalent)
l_n = Natural logarithm (2.3 × $\log_{10}$)
C_1 = Higher concentration of electrolyte
C_2 = Lower concentration of electrolyte

By substituting the values R, n, F, and taking the room temperature at 27°C (300 Absolute) and converting $\log_n$ to $\log_{10}$ we get

$$E = 59.5 \log_{10} \frac{C_1}{C_2}$$

If the concentration difference on the two sides of the membrane is tenfold then $\log \frac{C_1}{C_2} = \log 10 = 1$,

Therefore, E = 59.6 millivolts.

The *concentration gradient* across the membrane tends to force the ions from the higher concentration to the lower concentration; the *electromotive force* exerted by the charged ions opposes this tendency. E is the equilibrium potential at which these two forces are exactly balanced (Fig. 9.13).

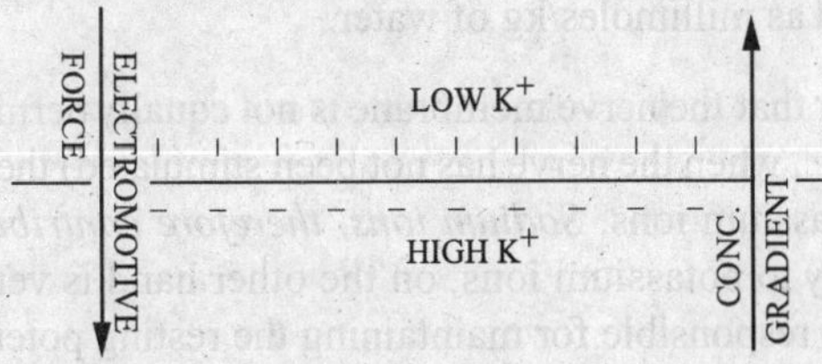

Fig. 9.13. The forces working across a cell membrane due to unequal concentration of ions.

If the potential is due to potassium ions it is written as $E_K{}^+$, if it is due to sodium ions it is written as $E_{Na}{}^+$ and so on. *The sign of the potential takes the sign of the ions on the side on which the concentration of the ions is lower.* Thus, if the concentration of potassium ions is low on the outside of the cell membrane $E_K{}^+$ will have a positive sign on the outside and negative on the inside (Fig. 9.13).

When there are several ions on the two sides of a cell membrane with unequal concentrations then the voltage developed across the membrane will depend on all these ions. It can be expressed by the following equation

$$E = \frac{RT}{nF} \ln \left(\left[\frac{P_K{}^+[K^+]\text{ in}}{P_K{}^+[K^+]\text{ out}} + \frac{P_{Na}{}^+[Na^+]\text{ in}}{P_{Na}{}^+[Na^+]\text{ out}} + \frac{P_C l^-[C^-]\text{ out}}{P_C{}^-[C^-]\text{ in}} \right] \right)$$

where E is the potential difference across the membrane $P_K{}^+$, $P_{Na}{}^+$, and PCl^- are the *permeability coefficients* for K^+, Na^+ and Cl^- respectively. The symbols 'in' and 'out' refer to the concentration of ions inside and outside the cell respectively.

The equation given above is a modification, by Hodgkin and Katz, of the *Goldman constant field equation.* This equation implies that (1) a potential will develop across a semipermeable if unequal concentration of permeable ions are present on its two sides; (2) The magnitude of this potential will depend on the ratio of the individual ions and their permeability will depend on the ratio of the individual ions and their permeability coefficients. (3) The potential developed will be the sum of the potentials created by all the ions. *It is important to note that the potential will be created by only those ions to which the membrane is permeable.*

In living cells also a situation always exists in which the concentration of ions, in the cell, and outside the cell are different. Table 9.1 shows that the concentration of K^+, Na^-, and Cl^- inside and outside are remarkably different *e.g.* K^+ is high inside and Na^+, and Cl^-, are high outside the cell. Since the nerve membrane is a semipermeable membrane a potential develops across the membrane according to the Goldman constant field equation. The sign of this potential is positive on the side where the concentration of anions is high and negative on the side where the concentration of cations is high. The actual potential is the sum of these two. This is the basis of the resting potential.

Table 9.1 Table showing the difference in ionic composition of intra and intercellular fluids from frog and squid. Similar differences have been reported in other animals also.

Tissue	*Potassium*	*Sodium*	*Chloride*
Squid giant axon	410	49	123
Squid blood	22	460	500
Frog muscle cell	125	15	1.2
Frog plasma	2.6	110	77

Note: Values expressed as millimoles/kg of water.

It is crucial to remember that the nerve membrane is not equally permeable to all ions at all times. During the resting stage (*i.e.*, when the nerve has not been stimulated) the permeability to sodium ions is about 1% of that of potassium ions. *Sodium ions, therefore contribute very little to the resting potential.* The permeability to potassium ions, on the other hand is very high, and it is, therefore, potassium which is largely responsible for maintaining the resting potential. If the concentration of K^+, outside the cell, is raised the potential falls. The permeability to Cl^- is also low so that it plays only a small part in maintaining the resting potential. If the concentration of K^+ is made equal on two sides of the cell membrane some potential can still be measured; this is mostly due to the chloride ions. Of all the ions present around the cell membrane the permeability to potassium is highest. The measured values of the resting potentials correspond very closely to the theoretical value of E_{K^+}.

The concentration of K^+ inside and Cl^- outside the cell (Table 9.1) are such that these are nearly in electrochemical equilibrium. Since the concentration of sodium is high outside, some Na^+ keeps on entering the cell even though the permeability of the membrane is low for Na^+. The Na^+ which enters in this manner is actively extruded out by a mechanism of *active transport* known as *sodium pump.*

Earlier in this chapter, it was shown that the outside of a nerve membrane is positively charged during the resting stage. This is mostly due to the low concentration of K^+ outside. It was also shown that when a nerve is stimulated the outer surface of the axon gets negatively charged for a brief moment. The *reversal of potential* from positive to negative is called *repolarisation.* The cause of the resting potential has already been explained on the basis of the Goldman constant field equation. The action potential can also be explained in a similar manner.

Origin of Propagated Action Potential

The present theory of origin and propagation of a action potential is mainly based on the work done on the squid (*Loligo*) giant axon by Hodgkin and Huxley. Briefly this theory explains the potential in the following manner:

The stimulus disturbs the delicate balance of the positive and negative ions on the two sides of the nerve membrane so that the resting potential is slightly altered; the outside potential becomes less positive than the resting stage. The disturbance, somehow, causes, the sodium permeability to rise. The increase in sodium permeability produces two things: (1) Inward movement of Na^+ since the Na^+ concentration outside is high. (2) Because of increase in Na^+ permeability the Nernst potential also becomes dependent on sodium ions which played a nominal role during the resting stage. Since the concentration of Na^+ is high outside and low inside the potential to fall. The fall in potential further increases the Na^+ permeability. The Na^+ permeability rises to its maximum in an extremely short time. During this period Na^+ enters the cell.

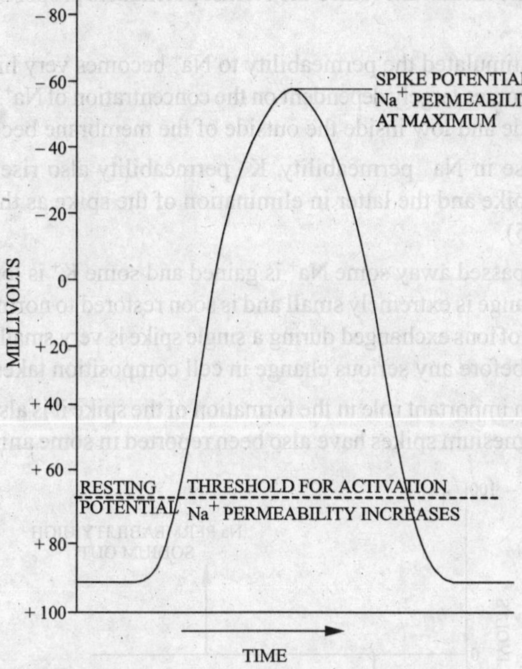

Fig. 9.14. Figure showing origin of action potential. The resting potential is about + 90 m volts. After stimualtion, due to imbalance of ions the resting potential fall to about 65 m volts. This change produces a sudden increase in Na^+ permeability. The permeability coefficient for Na^+ becomes much higher than the permeability coefficient for K^+ so that potential across the membrane is mostly due to Na^+ and the outer surface of the membrane becomes negative (–60 m volts). The total change of voltage is 90 – (– 60) = 150 m volts.

During the formation of spike the permeability coefficient to sodium ions becomes much higher than the permeability coefficient for potassium. Due to this the Nernst potential becomes more dependent on the sodium ions. The permeability to Na^+ becomes so high that it not only offsets the potential produced by K^+ but even over-shoots the resting potential and the membrane becomes negatively charged on the outside (Fig. 9.14). After the peak of the spike has been reached the Na^+ permeability falls rapidly and the membrane returns to normal.

Closely following the rise of the Na^+ permeability the K^+ permeability also rises above its normal level. The peak of K^+ permeability, however, is reached only after the spike has been formed. The fall in K^+ permeability is slower than the fall of Na^+ permeability. As soon as the Na^+ permeability starts falling the potential starts returning to normal. The rise in K^+ permeability helps in a rapid return to normal. The spike lasts for about 1 msec. It must be understood clearly that the rise in potassium permeability only helps in a rapid return to the resting potential (repolarisation). Even if there is no rise in K^+ permeability the nerve would still repolarise, although at a slower rate. The positive after potential repolarise, although at a slower rate. The positive after potential (Fig. 9.11) is caused by rise in K^+ permeability. Following positive after potential the permeability to all ions returns to normal the sodium pump starts working again and all the sodium, which entered the cell during spike formation, is pumped out. The K^+ which leaked out during the process is also brought in.

The entire phenomenon can be summarised as follows:

(1) During the resting stage the membrane is very permeable to K^+, less to Cl^- and very little to Na^+. During this stage the membrane potential is largely dependent on the concentration of K^+

on the two sides of the membrane (since the Nernst potential is proportional to the permeability coefficient).

(2) When the nerve is stimulated the permeability to Na^+ becomes very high in comparison to K^+ and the potential becomes largely dependent on the concentration of Na^+. Since the concentration of Na^+ is high outside and low inside the outside of the membrane becomes negative.

(3) Shortly after the rise in Na^+ permeability, K^+ permeability also rises. The former helps in production of the spike and the latter in elimination of the spike as the two work in opposite directions (Fig. 9.15).

(4) After the spike has passed away some Na^+ is gained and some K^+ is lost by the cell (10^{-2} M in case of K^+). This change is extremely small and is soon restored to normal by the sodium pump. Because the amount of ions exchanged during a single spike is very small, a cell can be stimulated thousands of times before any serious change in cell composition takes place.

Since Na^+ plays such an important role in the formation of the spike it is also called a *sodium spike*. Calcium, barium and magnesium spikes have also been reported in some animals.

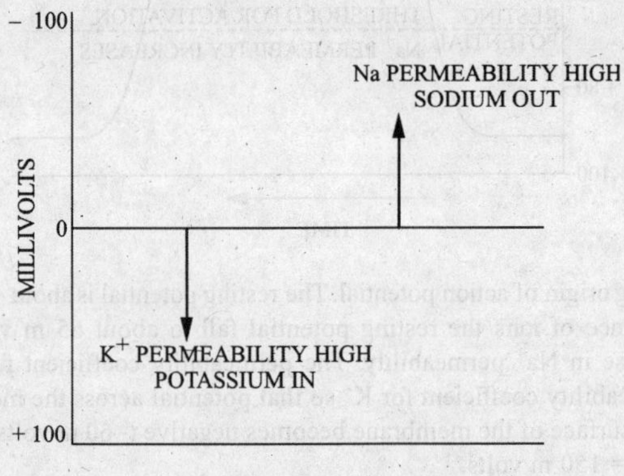

Fig. 9.15. Action potential due to change in ionic permeability (for details see text).

CONDUCTION OF NERVE IMPULSE

Only a very small portion of an axon is depolarised at a given moment and it is only in this part that the spike is seen. The conduction of nerve impulse along an axon is brough about by a progressive formation of spikes over the entire axon. This is called *spread* of the spike.

It has been shown already, that spike formation reverses the charges on the membrane. The small area which is depolarised creates a local field of negatively; local circuits are formed and positive charges move towards the negatively charged area [Fig. 9.16 (b)]. The stimulated area, therefore acts as a *sink* for positive ions on the outside and negative ions on the inside [Fig. 9.16 (b)]. This movement of ions causes an ionic imbalance in the areas adjacent to the spike. On the outer surface of the membrane the areas adjacent to the spike lose some of their positive charges to the sink and get less positively charged. Because of this the permeability to sodium in these areas increases and the membrane depolarises. This situation is very similar to the one created by an external stimulus. In this manner a wave of progressive depolarisation travels along the length of the axon. This theory has been called *local circuit theory* of the propagation of action potential.

As soon as the spike moves the nerve gets repolarised very quickly. The repolarisation is so fast that at any instant only a small length of a nerve fibre is depolarised. Since the propagation of an impulse takes place with the progressive formation of fresh spikes, whose height (magnitude) is constant the impulse travels without decrement.

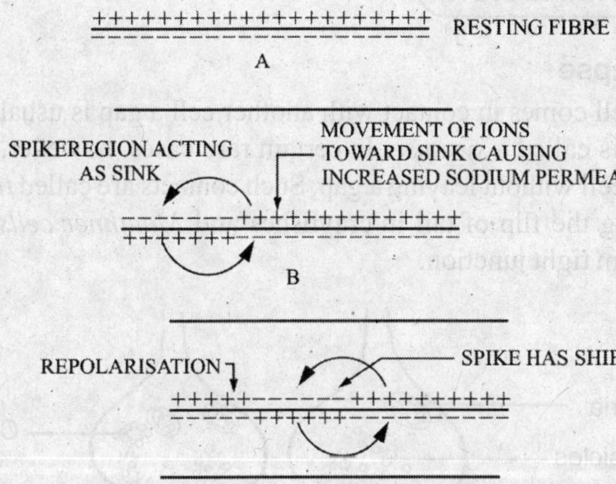

Fig. 9.16. Diagram showing local circuit theory of propagation of action potential; (a) resting cell; (b) spike has been formed at the left which acts as sink. Ions from adjacent regions start moving towards the sink so that adjacent regions becomes increasingly permeable to sodium ions; (c) original spike gets repolarised and new spike is formed in the adjacent region.

Propagation of nerve impulse in the above manner takes place only in non-myelinated fibres. In myelinated fibres the nerve is insulated from the external medium by the myelin sheath. This insulation is broken only at the nodes of Ranvier. The nerve impulse in the case of myelinated fibres skips from node as shown in Fig. 9.17. In these cases the local circuits are formed only at the nodes.

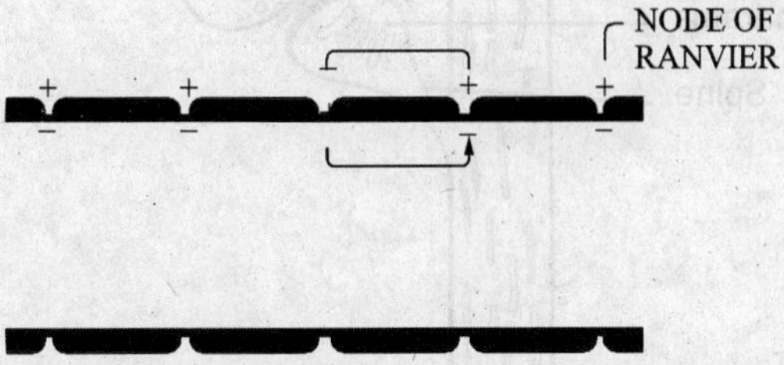

Fig. 9.17. Saltatory conduction in a myelinated nerve fibre. The spike skips from node to node because at the internodes the fibres is insulated by the myelin sheath.

This type of conduction is called *saltatory conduction* of nerve impulse. As, in this kind of conduction, the nerve impulse travels from one node to the next without travelling through the internodal region the conduction velocity is very high. The distance between two adjacent nodes is higher in fibres of larger diameter as compared to fibres of smaller diameter (Table 9.2). Because of this conduction velocity in fibres of larger diameter is rapid.

Table 9.2 Table showing distance between adjacent nodes of Ranvier in fibres of varying thickness in bullfrog neurons

Fibre diameter	*Nodal distance*
4μ	0.2 mm
12μ	1.5 mm
15μ	2.5 mm

SYNAPTIC TRANSMISSION

Structure of Synapse

Whenever, a nerve cell comes in contact with another cell a gap is usually present between the two cells; such a contact is called a *synapse.* In certain rare cases, however, a nerve cell may establish contact with another cell without leaving a gap. Such contacts are called *tight junctions* or *nexus.* The giant fibres mediating the flip of tail in crayfishes and *Mauthner cells* mediating a rapid escape response in fishes form tight junction.

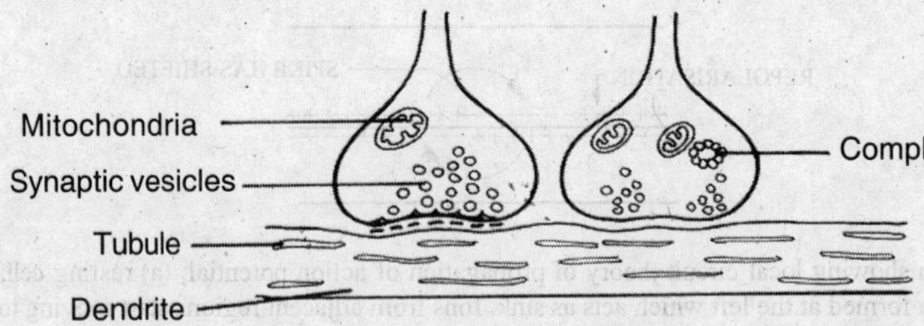

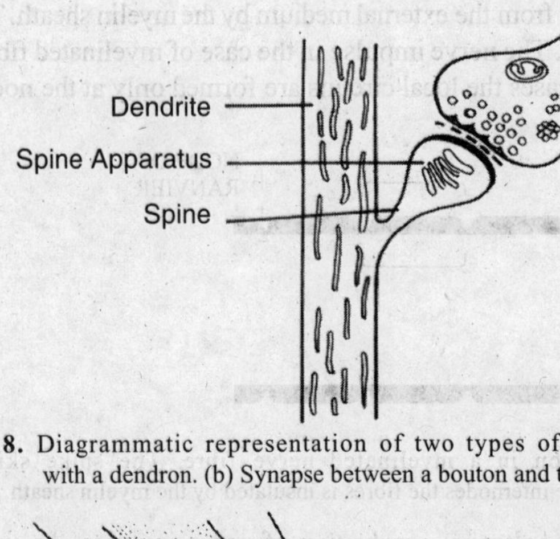

Fig. 9.18. Diagrammatic representation of two types of synapses (a) Two boutons synapsing with a dendron. (b) Synapse between a bouton and the spine of a dendrite.

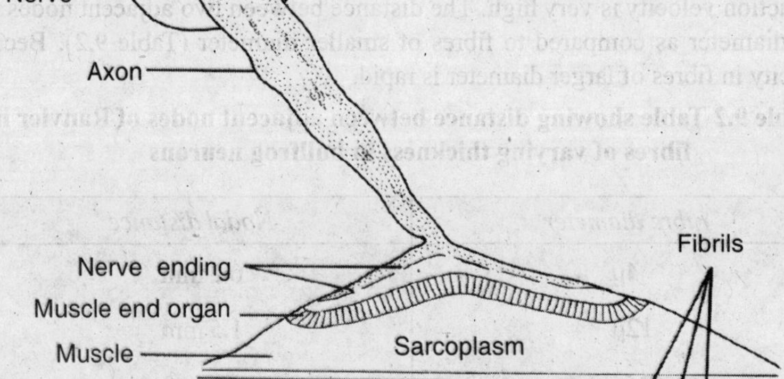

Fig. 9.19. A myoneural junction.

Synapses vary greatly in structure (Figs. 9.18 and 9.19): an axon may establish connection with dendrites, cell bodies, axons or cells of effector organs. Many times, an axon gives off a large number of offshoots which end in club-shaped protruberances called boutons (*presynaptic buttons* or *end feet*). The buttons make synaptic connections [Figs. 9.18 and 9.20 (a)] with other cells with which they come in close contact, although there is no cytoplasmic continuity. The space between the end feet and the neighbouring cell is called synaptic cleft [Fig. 9.20 (c)]. A single nerve cell may even be convered with numerous boutons arising from different cells [Fig. 9.20 (a)].

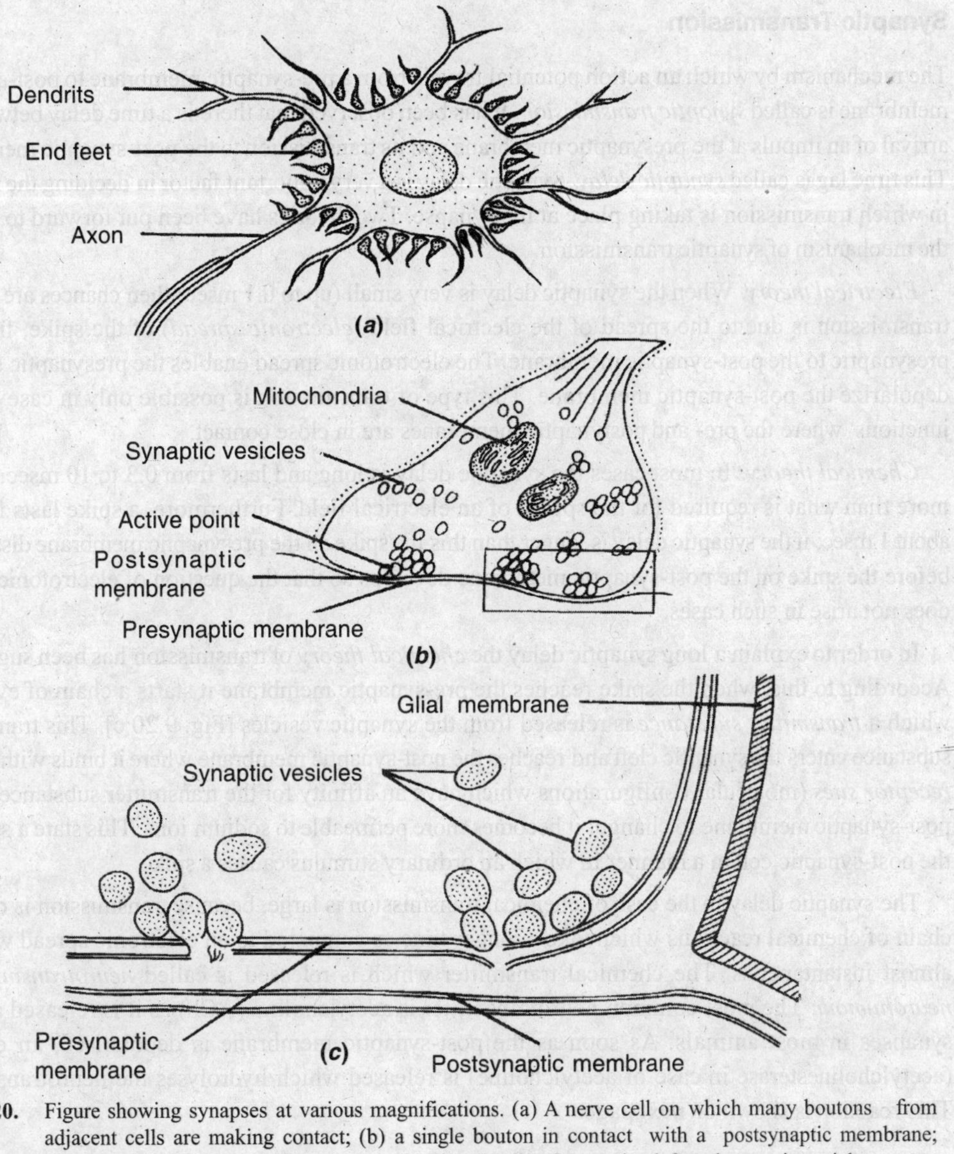

Fig. 9.20. Figure showing synapses at various magnifications. (a) A nerve cell on which many boutons from adjacent cells are making contact; (b) a single bouton in contact with a postsynaptic membrane; (c) enlargement of boxed region in (b), showing details of synaptic cleft and synaptic vesicles.

When a nerve cell forms a synapse with a muscle cell then it is termed a *myoneural* or *neuromuscular junction* (Fig. 9.19). If the structure on the two sides of the synapse is similar then the message can travel in both directions. However, when it is a dissimilar union (Fig. 9.19), the message can travel only in one direction. The cell which sends the impulse is called *presynaptic cell,* whereas the cell which receives the impulse is called *postsynaptic cell.* The boutons from the presynaptic cells make contact with some part of the postsynaptic cells. Sometimes, dendrites have special spines for making synapses with boutons [Fig. 9.18 (b)].

A typical bouton contains several structures. In addition to cytoplasmic organelles like mitochondria etc., it also has numerous small packets of vesicles called *synaptic vesicles* [Fig. 9.20 (b) and (c)]. The synaptic vesicles contain special chemicals called *neurotransmitters*

Synaptic Transmission

The mechanism by which an action potential travels from a pre-synaptic membrane to post-synaptic membrane is called *synaptic transmission.* It has been observed that there is a time delay between the arrival of an impuls at the presynaptic membrane and its transmission to the post-synaptic membrane. This time lag is called *synaptic delay.* Synaptic delay is a very important factor in deciding the manner in which transmission is taking place at the synapse. Two theories have been put forward to explain the mechanism of synaptic transmission.

Electrical theory. When the synaptic delay is very small (up to 0.1 msec) then chances are that the transmission is due to the spread of the electrical field (*electronic spread*) of the spike, from the presynaptic to the post-synaptic membrane. The electrotonic spread enables the presynaptic spike to depolarize the post-synaptic membrane. This type of transmission is possible only in case of tight junctions where the pre- and postynaptic membranes are in close contact.

Chemical theory: In most cases the synaptic delay is long and lasts from 0.3 to 10 msec. This is more than what is required for the spread of an electrical field. Furthermore, a spike lasts for only about 1 msec; if the synaptic delay is longer than this the spike on the presynaptic membrane disappears before the spike on the post-synaptic membrane develops so that the question of electrotonic spread does not arise in such cases.

In order to explain a long synaptic delay the *chemical theory* of transmission has been suggested. According to this, when the spike reaches the pre-synaptic membrane it starts a chain of events in which a *transmitter substance* is released from the synaptic vesicles [Fig. 9.20 c]. This transmitter substance enters the synaptic cleft and reaches the post-synaptic membrane where it binds with special *receptor sites* (molecular configurations which have an affinity for the transmitter substance) of the post-synaptic membrane to change; it becomes more permeable to sodium ions. This state a spike on the post-synaptic cell in a manner in which an ordinary stimulus causes a spike.

The synaptic delay in the case of chemical transmission is large, because transmission is due to a chain of chemical reactions which take a longer time as compared to an electronic spread which is almost instantaneous. The chemical transmitter which is released is called *neurotransmitter* or *neurohumour.* The most common neurotransmitter is acetylcholine (ACh) as it is released at most synapses in most animals. As soon as the post-synaptic membrane is depolarised, an enzyme (acetylcholinesterase in case of acetylcholine) is released which hydrolyses the neurotransmitter. This reaction is shown on next page.

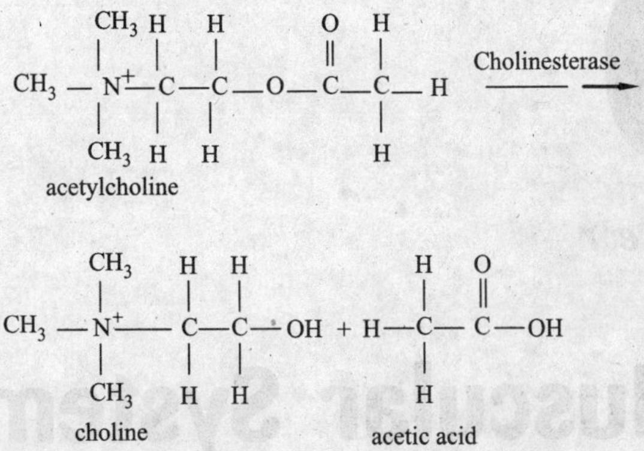

If acetylcholine is allowed to remain at the synaptic cleft the post-synaptic membrane would not repolarise and would therefore be unfit to receive a fresh stimulus. A fresh quantity of neurotransmitter is released with every impulse.

A similar mechanism is also present at the myoneural junction. In fact most of the information regarding synaptic transmission has been derived from neuromuscular junctions.

Although acetylcholine is the most important transmitter in the animal kingdom it is by no means the only one. *Noradrenaline* (*norpinephrine*) is the chemical transmitter in the postganglionic fibres of the sympathetic nervous system. 5-hydroxytryptamine (5HT) is the transmitter in many invertebrates and in the central nervous system of vertebrates also . Several identified (*e.g.* Gama amino butyric acid) and unidentified substances have been assigned with transmitter roles from time to time. The mechanism of transmission in the case of these neurotransmitters, in so far as it is understood, is similar to acetylcholine. Thus the enzymes involved in the hydrolysis of norepinephrine and 5-hydroxyliyptaume is *more amine oxidase*.

SUMMATION

Quite often a single stimulus fails to evoke a response although when a number of stimuli of the same strength are given a response is evoked. When a response is obtained by the combined effect of many stimuli the phenomenon is known as *summation.* Summation is of two kinds *temporal* and *spatial.*

Temporal summation. If an effector unit (*e.g.*, a muscle fibre) is supplied with only one nerve fibre and if a single stimulus of the nerve fibre does not evoke a response while several successive stimuli of the same strength are successful in doing so the phenomenon is known as *temporal summation.* Since the magnitude of the spike for every nerve fibre is constant the only manner in which temporal summation work is perhaps by accumulating sufficient neuro transmitter, from the successive stimuli, to depolarise the postsynaptic membrane.

Spatial summation. In spatial summation more that one nerve fibre is involved. For example, if 100 motor fibres make contact between a muscle and the spinal cord then stimulation of only one fibre may cause very little or no contraction at all. If, however, all the 100 fibres are stimulated together then there is a powerful contraction. It is possible that the stimulation of a single fibre does not liberate enough neurohumour to generate a post-synaptic potential. However, when all the fibres are stimulated together enough neurohumour to generate a postsynaptic potential is liberated.

10

Chapter

Muscular System

Mobility is one of the most distinguishing features of animals. Movement whether in cilia, muscles, nematocysts or mitotic spindle basically involves the same principles, *i.e.*, change in configuration of special types of contractile proteins. Even through most of the mechanical work done by animals is accomplished through muscles, nevertheless, for certain exceptions like calcified bone cells, nearly all cells show some movement. The muscle cells have, however, perfected the ability of movement because their entire structure and chemical composition is geared towards accomplishing motion which is brought about by the contractile proteins. The rest of the structures in muscle cells are mostly incidental and are present solely for the upkeep of the moving parts.

Large number of tissues and organs (intestine, heart, limbs) show movement of some kind, although their requirements in terms of *speed of movement* are not identical. The sleek, beautiful running movements of a race horse require swift and simultaneous contractions of the muscles of the legs, thighs and neck. On the contrary, the movements of the alimentary canal are very slow, so that while one part is in a contracted state its neighbouring parts are relaxed, causing the characteristic peristaltic movements of the intestine. The heart, on the other hand, contracts faster than the alimentary canal. Besides this, muscles of entire sections of the heart (auricle, ventricle, sinus, venosus) contract simultaneously with considerable speed. Thus, broadly speaking, there are three kinds of muscles in mammals; (i) Slow muscles, also callee *nonstriated muscles* (*visceral muscles, smooth muscles, unstriped muscles* or *involuntary muscles*). These are associated with the visceral organs like the alimentary canal, bladder etc. (ii) Fast moving muscles, called *striated muscles* (*voluntary muscles, striped muscles* or *skeletal muscles*) which are associated with the skelton. (ii) Medium-fast muscles called *cardiac muscles,* associated with the heart.

Before the detailed structure of these three types of muscle is taken up it is better to understand the relatioship between the speed of contraction and structure of muscles. For example, in long muscles such as those found in the legs the stimulus for contraction should travel over the entire length of the muscle very quickly. This means that communication from one muscle cell to next should be very rapid.

As already discussed earlier in the chapter on nervous system the spread of an impulse over the membrane of an axon is very quick.(a spike develops in 1 msec). However, when two cells come together at a synapse there is a delay in transmission [Fig. 10.1 (a)] (synaptic delay). Obviously, therefore, if there are fewer intercellular membranes (synapses) the spread of impulse would be fast. Moreover synaptic delay is lessened if the opposing membranes of two cells are very close to each other in the form of tight junctions or nexus (Chapter 9) [Fig. 10.1 (*b*)]. If there is total fusion of two cells there is no delay [Fig. 10.1 (*c*)]. This is true in the case of muscles also, nonstriated muscles have large intercellular spaces, cardiac muscles have tight-junctions and striated muscles are formed by the fusion of several cells.

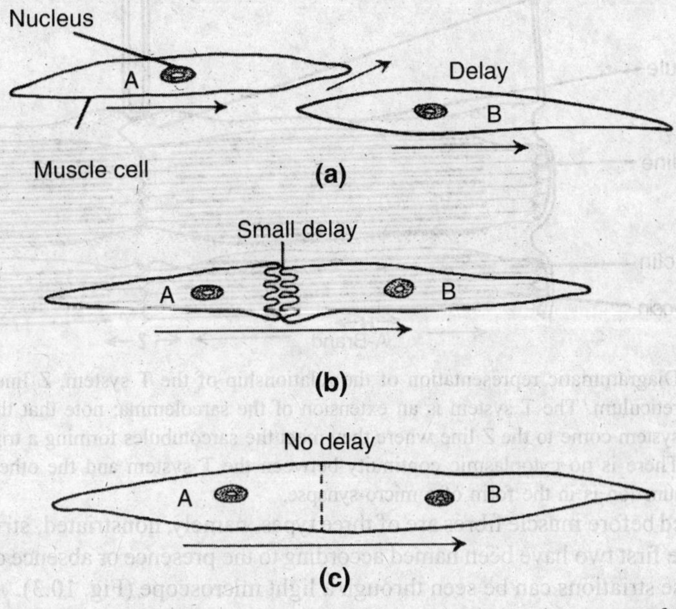

Fig. 10.1. Diagram showing the relation of speed of conduction and structure of the junction of two opposing fibres; (a) slow communication as the gap between A and B is large (*e.g* smooth muscles (b) faster communication if A and B form a tight junction *e.g.* cardiac muscle) (c) very fast communication if there are no boundaries between A and B (e.g. skeletal muscle).

STRUCTURE OF MUSCLES

A muscle cells is often referred to as a *muscle fibre* because of the fibre like elongated shape of the cell. The cell membrane of a muscle fibre is called *sarcolemma* and its cytoplasm is likewise called *sarcoplasm.* A single fibre is composed of fibrils called *myofibrils.* The spaces between adjacent fibrils are interspersed with a variable number of mitochondria termed *sarcosomes.* A single fibril is made up of many *myoflaments* (Fig. 10.5). The myofibrils and myofilaments are collectively called the *contractile elements* of the muscle. The filaments are mainly made up of four contractile proteins *actin, myosin, tropomyosin* and *troponin.* There are a large number of tubules which run through the sacroplasm. These are *sarcotubules* and they form the *sarcoplasmic reticulum.* The sarcotubules usually run parallel to the sarcomeres and contain mitochondria (sarcosomes). Beside these, the sarcolemma is also carried into the body of the muscle fibres in the form of an elaborate system of tubules called the T system. Although the tubules of the T system come very close to the sarcotubules, they never become continuous with the latter as the sarcolemma separates them. The T tubes travel up to the Z-line and make contact with the sarcotubules and Z lines forming a *triad* (Fig. 10.2).

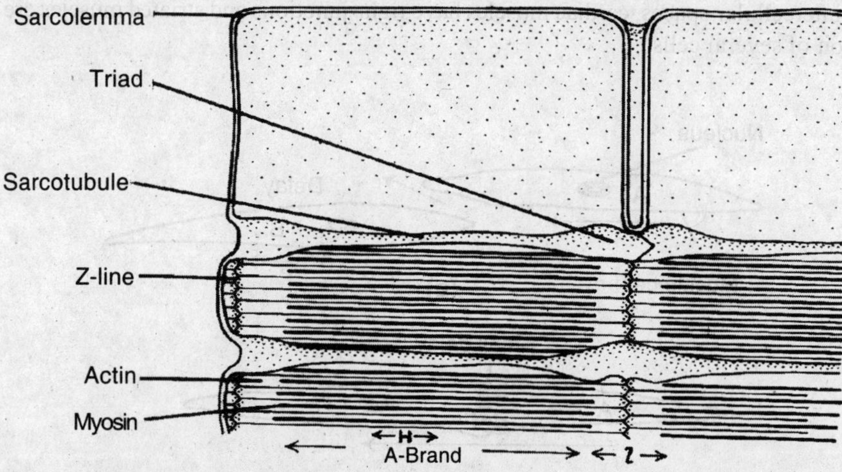

Fig. 10.2. Diagrammatic representation of the relationship of the T system, Z line and sarcoplasmic reticulum. The T system is an extension of the sarcolemma; note that the tubules of the T system come to the Z line where they meet the sarcotubules forming a triple junction (triad). There is no cytoplasmic continuity between the T system and the other structures as the junction is in the form of a micro-synapse.

As described before muscle fibres are of three types, namely, nonstriated, striated and cardiac muscle fibres. The first two have been named according to the presence or absence of cross striations on the fibre. These striations can be seen through a light microscope (Fig. 10.3).

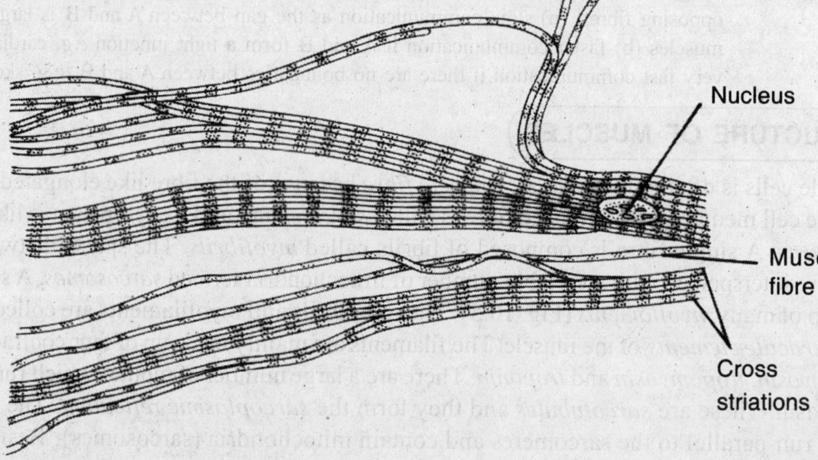

Fig. 10.3. Striated muscle fibres as seen through a light microscope. The alternate light and dark bands can be clearly seen.

Smooth or Nonstriated Muscles

The muscles of the alimentary canal, uterus, urinary bladder etc. are smooth muscles. These are non-striated, are involuntary and thus cannot be moved at will. The cells of smooth muscles are elongated, spindle-shaped structures, 15—500 μ long, 2—20 μ in diameter with a single nucleus (Fig. 10.4).

The muscle cells are interconnected with bridges. The myofibrils are arranged parallel to the longitudinal axis of the cells but do not follow any definite pattern. Under the light microscope they may show faint longitudinal striations but never transverse striations. In hollow organs (intestine etc.) these are arranged in longitudinal and circular bands. The contraction of the longitudinal bands shortens the organ while the contraction of the circular bands lengthens the organ.

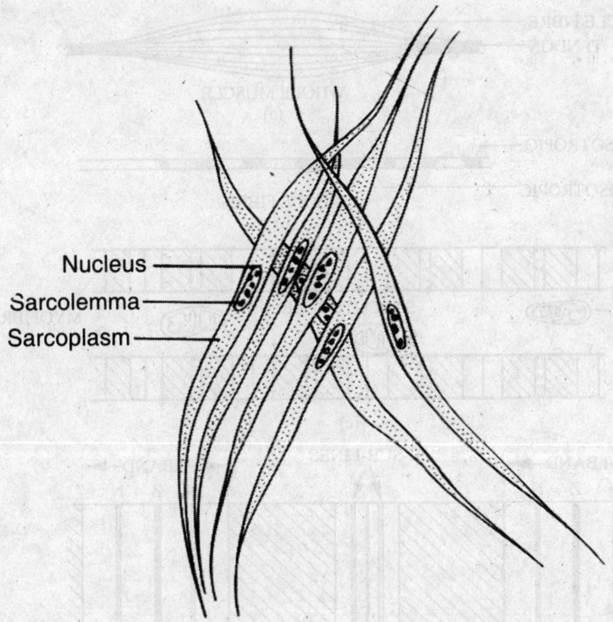

Fig. 10.4. Smooth muscle fibres.

Skeletal or Striated Muscles

Striated muscles are quick acting muscles and are found associated with the skeleton of vertebrates. They have also been reported in invertebrates in those organs which show rapid movements. These muscles are called striated or *striped muscles* because they contain bundles of myofibrils which show transversely arranged alternate light and dark bands (Fig. 10.3). A single cells of striated muscle is fairly long (1—120 mm) and contains many nuclei. The large number of nuclei indicate that it may be formed by the fusion of many cells. Fig. 10.5 (*a – d*) show a progressive enlargement of a skeletal muscle. Under the light microscope the skeletal muscle appears to have (1) dark bands of *high birefringence* (*Anisotropic* — A Band) alternating with light band of low *birefringence* (*Isotropic*—I bands). The electron microscope has revealed the detailed structure of the skeletal muscle. Fig. 10.5 (*c, d*) show a part of a single myofibril. A myofibril is transversely divided at regular intervals into several sections by membranes known as Z *lines* (*Dobies lines*). The area between two adjacent Z lines is called a sacromere [Fig. 10.5 (*d*)].

Every sacromere has two kinds of filaments—thick *myosin* filaments and *thin actin* filaments [Fig. 10.6 (a)]. One end of the actin filaments is attached to the Z line [(Fig. 10.6 a)] while the other is free. The actin filaments from two adjacent Z lines do not meet in the centre in the relaxed state. The thick myosin filaments which are present in the middle of the sacromere do not reach the Z line and are free at both ends [Fig. 10.6 (a)]. Since the actin filaments do not meet in the centre of the sacromere, this part of the sarcomere is occupied by myosin fibres alone. This area is called H *band* (Hensen's line). The areas near the Z lines where only actin filaments are present are called *I bands* (Isotropic). The entire area in which myosin filaments are present is called A band (Anisotropic). In H band only myosin filaments are present and in the I bands only actin filaments are present. A band is the region of high birefringence and appears dark. I band is the region of low birefringence and is light (Figs. 10.5 and 10.6).

Fig. 10.5 shows progressive enlargement of a striated muscle. (a) complete muscle with tendons attached at both ends; (b) a single muscle fibre showing striations; (c) a part of a single fibre showing myofibrils; mitochondria (sacrosomes) are present between the myofibrils; (d) a single sacromere showing the various regions.

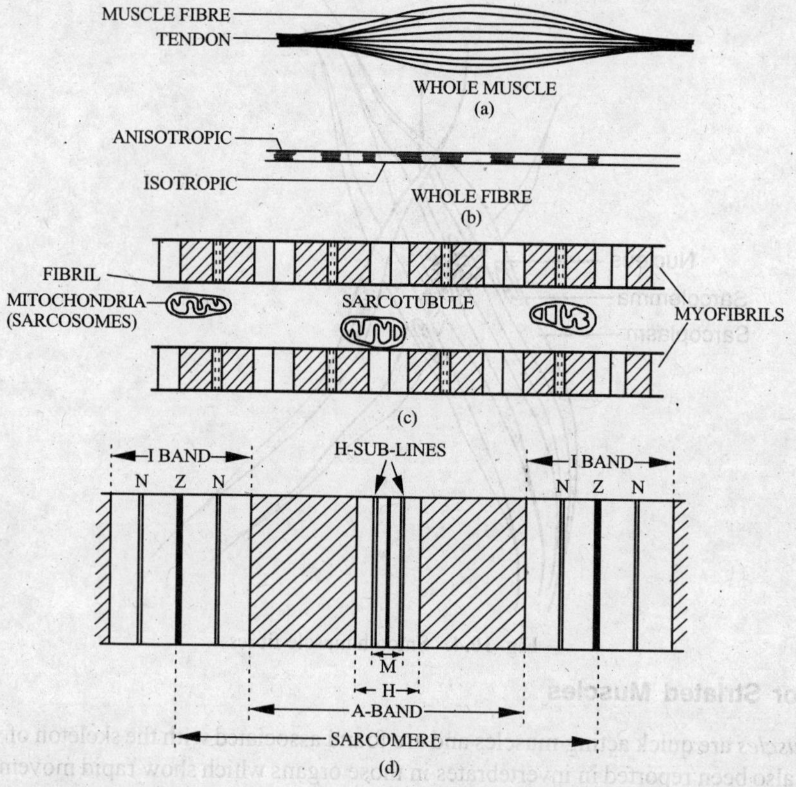

Fig. 10.5. Figure showing progressive enlargement of a striated muscle. (a) complete muscle with tendons attached at both ends; (b) a single muscle fibre showing striation; (c) a part of a single fibre showing myofibrils; mitochondria (sacrosome) are present between the myofibrils; (d) a single sacromere showing the various regions.

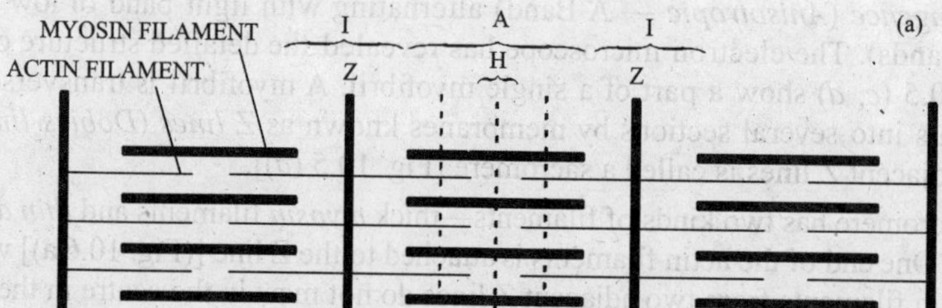

Fig. 10.6. Diagram showing the relative disposition of the actin and myosin fibres; (a) the relative arrangement of actin and myosin in adjacent sacromeres (b) cross sections through various regions of the A band.

The myosin filaments send out regularly spaced *cross bridges* towards the actin filaments. These bridges touch the surrounding actin filaments (Fig. 10.8). Another transverse band called the N band is present in the I region [Fig. 10.5 (d)]. The nature of this is not known. Similarly a line called M band is present in the centre of the H band. The nature of this is also not known.

In cross-section each myosin filament is surrounded by six actin filaments [Fig. 10.6 (b)]. The neighbouring fibrils in every fibre are aligned so that the striations are in register right across the fibre [Fig. 10.5 (c)]. Because of alignment the myosin and actin bands fall in clear lines of isotropic and anisotropic areas giving the muscle its characteristic striated appearance. In smooth muscles although both actin and myosin are present but the fibrils are not arranged in an orderly manner, hence, the striations can not be seen.

Cardiac Muscle

Vertebrate cardiac-muscles are composed of branched cells. Although, previously cardiac muscles were thought to be *syncytium* it has now been revealed by electron microscope studies that the heart muscles are made up of distinct individual cells, which *appear like a syncytium* because adjacent membranes are in very close contact with each other through special membrances are in the colse contact with each other through special structures called intercalated discs (Fig. 10.7). These discs are made up of closely approximated cells membranes. Each cell gives off several finger-like projections which closely fit into their adjacent counterparts. The intercalated discs correspond to the Z lines of striated muscles as the actin fibres are attached to these. Although heart muscles are made up of several cells, communication between cells in facilitated because of tight junctions (nexuses) in the form of intercalated discs. The heart muscle also show transverse striations like striated muscles.

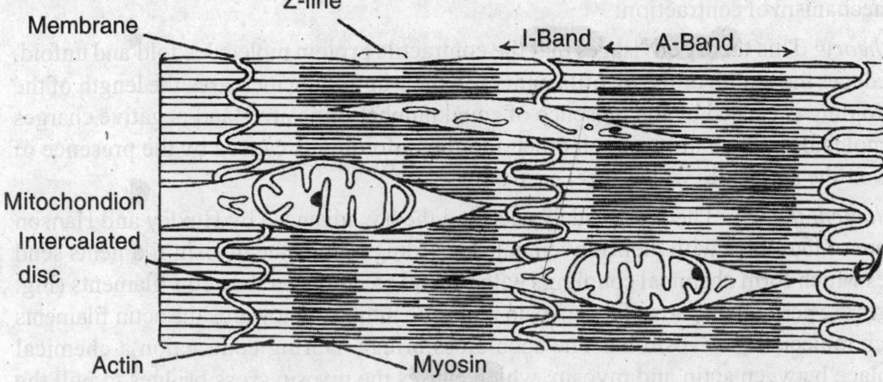

Fig. 10.7. Diagrammatic representation of a cross-section of cardiac muscle showing the intercalated discs and the striated arrangement of muscle fibres.

CHEMICAL COMPOSITION OF MUSCLE

In order to have the property of *contractility* muscle cells are provided with a special class of proteins called *contractile proteins.* The proteins make up about 25% of the bulk of muscles.

The contractile proteins are of three types—*myosin, actin* and *tropomyosin.* Besides these some other types of non-contractile proteins are also present in the muscles *e.g.*, globulins, albumins and all other enzymes necessary for cell metabolism. Lipids are present in traces, and carbohydrates mostly in the form of glycogen in amounts between 0.5 to 1%. ATP is also present in appreciable quantities. These together constitute about 5% of the muscle, the remaining 75% in water.

Myosin. The molecular weight of myosin ranges from 350,000 to 450,000, length 1500 Å and diameter about 20 Å to 50 Å. Because of its anisotropic properties it splits light waves into two mutually perpendicular vibrations (double refraction) hence it shows beautiful streaming birefringence.

Myosin also acts as an ATPase, it hydrolyses ATP into ADP and inorganic phosphate and releases energy stored in ATP. On hydrolysis with trypsin, myosin splits into two fragments of unequal size. These are called *meromyosins.* The lighter fragment in called *L-meromyosin* and heavier fragment *H-meromyosin.* The ATPase property of myosin is present in the H-meromyosin fragment.

Actin: Actin has the form of a double helix and a molecular weight of 70,000. It constitutes about 20 to 25% of the protein of mammalian skeletal muscle.

Actomyosin: When solutions of actin and myosin are mixed in the presence of salts (especially K^+) a complex of high viscosity, called *actomyosin,* is formed. The viscosity of actomyosin is higher than the combined viscosities of actin and myosin. Electron microscope studies have shown that a network is formed by the electrostatic association of the two proteins. When actomyocin is mixed with ATP in the presence of 0.6 M KCl then the thread of actomyosin contracts. The composition and particle weight of actomyosin depend upon the pH, KCl and $MgCl_2$ concentrations. Electron microscopy has shown that only the head of a myosin molecule binds with an actin molecule and ATP is hydrolysed.

Tropomyosin: Tropomyosin forms only 2.5% of the muscle protein. Its molecular weight is 60,000. It forms a complex with troponin and is part of the thin filament. This complex mediates the interaction of actin and myosin and thus regulates contraction.

MECHANISM OF MUSCLE CONTRACTION

The architecture of the muscle fibres is designed for contraction. Two theories have been advanced to explain the mechanism of contraction:

Folding theory: This theory envisages that the contractile protein molecules fold and unfold. Their folding reduces the length of the myofilaments and their unfolding increases the length of the filaments. The folding is caused by the presence of equal number of positive and negative charges on the protein molecules which attract each other and the unfolding is caused by the presence of similar changes on the molecules which repel each other.

Sliding filament theory: The most widely accepted theory, advanced by Huxley and Hanson links the structure of the fibres with their movement. According to this, the myosin filaments send out cross-bridges which form chemical couplings with special receptors on the actin filaments (Fig. 10.8). The bridges are permanent parts of the myosin molecule jutting out towards the actin filaments (Fig. 10.8). Each molecule of myosin has one such cross bridge. During contraction a chemical reaction takes place between actin and myosin, which causes the myosin cross bridges to pull the actin filaments in a *ratchet like action* for a short distance postulated to be about 10 nm. The cross bridge then fits the next receptor on actin filament and son (Fig. 10.8) till the actin filaments on the two ends of the sacromere come to meet in the centre of the H band [Fig. 10.9 c]. Further contraction crumples the actin and myosin filaments [Fig. 10.9 (d, e)]. The actin filaments as they slide towards each other shorten the length of the myofibrils so that the muscle contracts. During the process of contraction hydrolysis of ATP takes place which provides the energy for contraction. It is believed that a single pull of a cross bridge causes the hydrolysis of one ATP molecule. When the hydrolysis of ATP stops as a result of inactivation of ATPase activity of actomyosin by the removal of Ca^{++} the myosin and actin dissociate and the muscle returns to its original relaxed state.

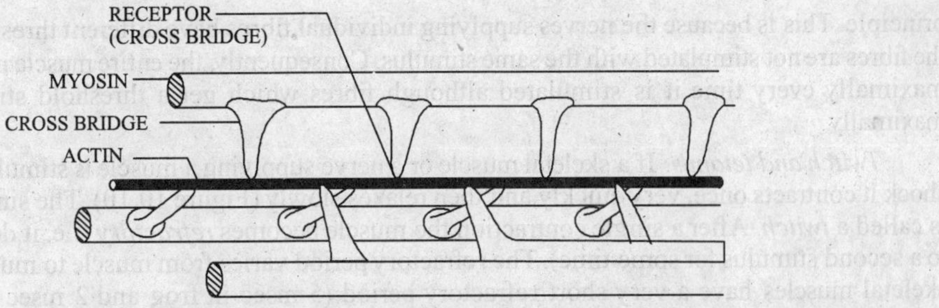

Fig. 10.8. Diagram showing cross-bridges arising out of a myosin filament which fit into actin filaments and pull the latter in a ratchet-like action during contraction. One cross-bridge fits into a receptor and the actin filament pulls it for a short distance then fits into the next and so on.

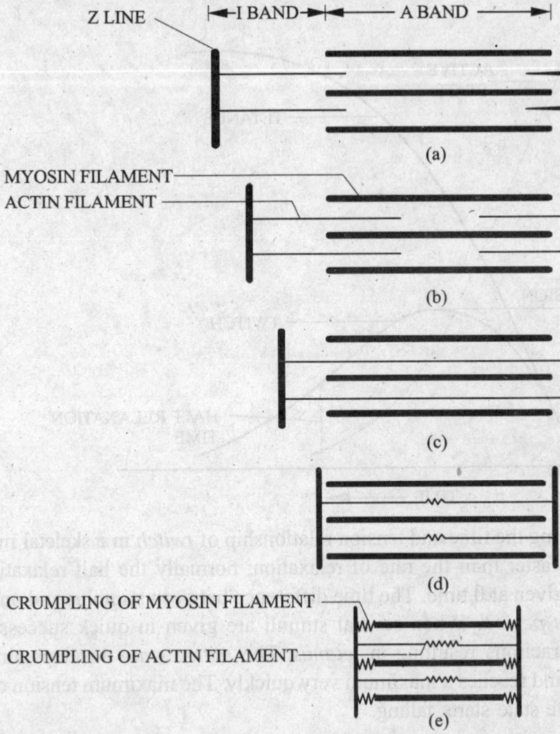

Fig. 10.9. Diagrammatic representation of muscle contraction according to the sliding filament theory; (a) relaxed muscle; b, c, d, e, progressive contraction of muscle in which the Z lines are pulled towards each other. In (c) the actin filaments have met in the centre (d) the actin filaments crumple in the middle due to further contraction (e) still more contraction produces crumpling of myosin filaments also.

MECHANICAL PROPERTIES OF MUSCLES

Muscles have two main mechanical properties, *contractility* and *elasticity*.

Contractility of Muscles

The extent and speed of contractility varies from muscle to muscle. Their different properties have been discussed below.

All or *none:* When individual muscle fibres are stimulated they either contract to their fullest extent or not at all. Thus like nerves they follow the *all* or *none* principal. However, a complete muscle which is made up of a large number of fibres does not always work on the all or none principle. This is because the nerves supplying individual fibres have different thresholds so that all the fibres are not stimulated with the same stimulus. Consequently, the entire muscle may not contract maximally every time it is stimulated although fibres which get a threshold stimulus contract maximally.

Twitch and tetanus: If a skeletal muscle or a nerve supplying a muscle is stimulated by a single shock it contracts once, very quickly and then relaxes slowly (Figure 10.10). The single contraction is called a *twitch*. After a single contraction the muscle becomes *refractory* (i.e. it does not respond to a second stimulus for some time). The refractory period varies from muscle to muscle. Generally, skeletal muscles have a very short refractory period (5 msec in frog and 2 msec in mammalian muscle). If a muscle is stimulated a second time before it has completely relaxed, but after its refractory period is over, then it contracts some more. This is called *staircase phenomenon* (Figure 10.11).

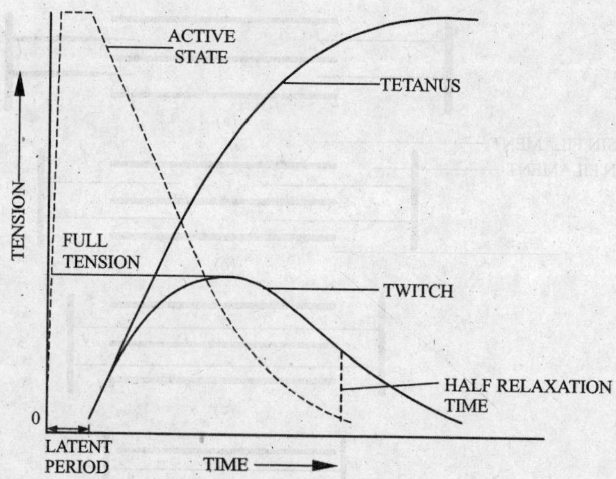

Fig. 10.10. Diagram showing the time and tension relationship of *twitch* in a skeletal muscle. The rate of contraction is faster than the rate of relaxation; normally the half relaxation time is noted. Stimulus was given at 0 time. The time difference between stimulus and onset of contraction is called *latent period.* When several stimuli are given in quick succession there is total fusion of contractions resulting in *tetanus*. The *active state* develops almost immediately after stimulus and reaches a maximum very quickly. The maximum tension of twitch develops when the active state starts falling.

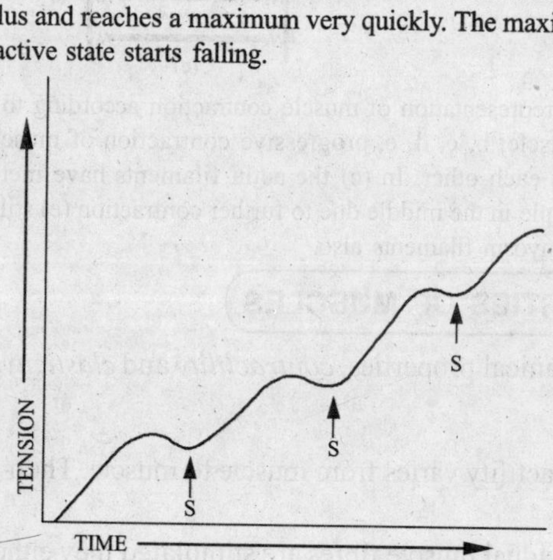

Fig. 10.11. Figure showing staircase phenomenon. The muscle is stimulated before it has completely relaxed so that extent of total contraction goes on increasing; S, stimulus.

It the stimuli for successive contractions are given in quick succession there is complete fusion (summation) of all the contractions and the muscle goes into a sustained contraction called *tetanus* (Figure 10.10). During tetanus all fibres of a muscle are contracted maximally.

The total time taken by smooth muscles for a twitch is generally long. The entire heart muscle obeys the 'all or none' principle. Since its refractory period is long it cannot be tetanised.

Isotonic and isometric contraction: The force with which a muscle contracts is called *tension.* If a muscle is stimulated, but it cannot contract because of very heavy load, it develops tension against this load. This kind of tension if called *isometric* (same length) *tension.* In this case no mechanical work is done, and all the energy produced during isometric tension is dissipated as heat. If a muscle shortens in length due to stimulus it is called *isotonic contraction e.g.,* a muscle lifting a weight contracts isotonically.

Elasticity of Muscles

A muscle is not a rigid structure. Both in stimulated and unstimulated states muscles possess certain amount of elasticity due to which their length can be increased or decreased. The property of elasticity is partly present in the contractile elements and partly in the tendons, connective tissue, and sarcolemma. Due to their elastic properties muscles resist tension. In an unstimulated muscle the resistance to stretching increases exponentially with a linearly increaseing force *i.e.*, if the force doubles the tension increases tnen times. If a load is placed on an unstimulated muscle it changes in length due to its elasticity. This elasticity is due to *parallel elastic elements* which are present in contractile elements, sarcolemma and connective tissue and are placed parallel to the contractile elements. When a load is applied to a contracted (stimulated) muscle then it also tends to lengthen due to its elastic elements lie in the contractile elements and in the tendons at the ends of the muscles.

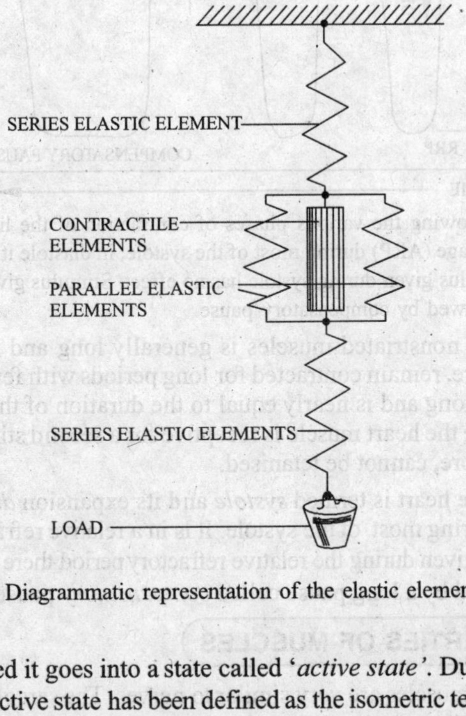

Fig. 10.12. Diagrammatic representation of the elastic elements of muscles.

Active State

When a muscle is stimulated it goes into a state called '*active state*'. During active state the ability to bear tension increases. Active state has been defined as the isometric tension which the contractile component can develop (or just bear without lengthening) at that instant. At this point it should be noted that the terms contraction and relaxation may be misleading. Contraction is usually seen as an energy-consuming active process while relaxation a mere passive process. The truth, however, is that as long as the muscle is shorter than its relaxed length (that is till nearly the end of relaxation) it is in an active state (Fig. 10.10). Contraction and relaxation merely reflect the increase and decrease of the active state. The active state develops and falls very rapidly after the muscle is stimulated. The contraction of the muscle does not develop at the same speed as the active state. In most muscles maximum tension develops after the decay of the active state has started (Fig. 10.10).

The properties of muscle described above have mostly been studied in skeletal muscles. The basic properties of smooth muscles and cardiac muscles are similar. However, there are certain differences.

Special Properties of Smooth and Cardiac Muscles

A striking property of the non-striated muscles of the digestive tract as well as the cardiac muscles of the vertebrate heart is their ability of *rhythmic contraction,* even when all nervous connections have been severed. Not only this, even pieces of alimentary canal, entire hearts or even fragments of hearts will continue to contact and relax when all nervous connnections have been cut off. This demonstrates that the ability to contract is an inherent property of the muscle itself. This is called *myogenicity*. Although myogenicity is present in the entire heart there exits a gradient of excitability in the heart. The most excitable part of the mammalian heart is the *sino-auricular node* (SA node) which acts as a pacemaker (Chapter 14).

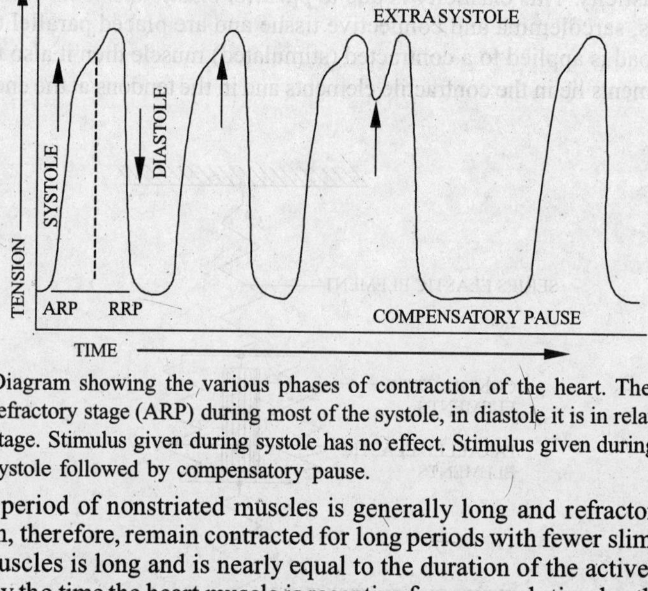

Fig. 10.13. Diagram showing the various phases of contraction of the heart. The heart is in absolute refractory stage (ARP) during most of the systole, in diastole it is in relative refractory (RRP) stage. Stimulus given during systole has no effect. Stimulus given during diatole causes extra systole followed by compensatory pause.

The twitch period of nonstriated muscles is generally long and refractory period is short. These muscles can, therefore, remain contracted for long periods with fewer slimuli. The refractory period of heart muscles is long and is nearly equal to the duration of the active state elicited by a single stimulus. By the time the heart muscle is receptive for a second stimulus the first has already died. Heart muscles, therefore, cannot be tetanised.

The contraction of the heart is termed *systole* and its expansion *diastole.* The heart is in the absolute refractory stage during most of the systole. It is in a relative refractory stage during most of the diastole. If a stimulus is given during the relative refractory period there is an additional contraction called *extra systole* followed by a long pause called *compensatory* pause (Fig. 10.13).

ELECTRICAL PROPERTIES OF MUSCLES

The electrical properties of muscles are very similar to nerves. The sarcolemma is a semipermeable membrane whose permeability can be changed by the changes on its surface. The muscle fibre maintains high *potassium* inside and high sodium outside. An unstimulated muscle maintains a resting potential of 30 – 100 mV (Fig. 10.14).

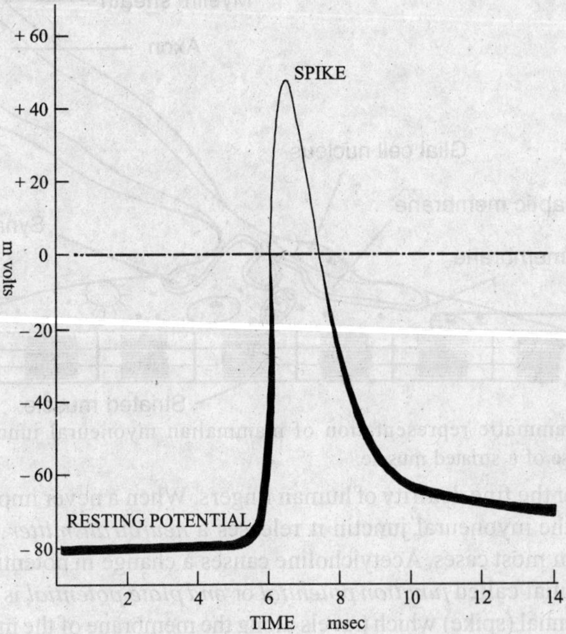

Fig. 10.14. Action potential of frog sartorius muscle. The resting potential is about – 80 mV. The peak of the spike is about + 48 mV.

Muscles can be stimulated in three ways (1) By nerves through *neurotransmitters* (Chapter 9), (2) Spontaneous stimulation due to the property of myogenicity, (3) By electronic spread from one cell to another (Chapter 9).

NERVOUS CONTROL OF MUSCLES

Nearly all muscles including myogenic muscles are supplied by nerves. In the latter, nerves have a regulatory role. Muscles are supplied both with sensory and motor fibres. The sensory nerves, send information regarding the 'stage' of the muscle (pain, fatigue etc.) to the central nervous system. Motor nerves accordingly control the activity of the muscle. Special motor nerves which prevent muscles from contracting are called *inhibitory nerves*. Nerves which control the activity of automatically contracting muscles, like visceral and heart muscles are called *regulatory nerves*. Regulatory nerves can be both *inhibitory* and *excitatory*. The point at which the axon makes contact with the muscle is called *myoneural junction* or *motor end plate*. The structure of a myoneural junction is given in Fig. 10.15. When a single axon supplies one muscle fibre with a single terminal the innervation is called *mononeural monoterminal* [Fig. 10.16 (a)]; when a single axon divides into several terminals before supplying a muscle fibre it is termed *mononeural multiterminal* [Fig. 10.16 (b)]. When a muscle fibre receives stimuli from several axons it is called *polyneuronal* innervated by it is called a *motor unit*. If a large number of motor units are present in a muscle, the muscle can contract in a finely graded manner. This is done by contracting only a discrete number of fibres at a time. When only a few motor units are activated a feeble movement is produced, whereas if all the units are activated a powerful contraction is caused. In this manner small and large contractions can be obtained from the same muscle.

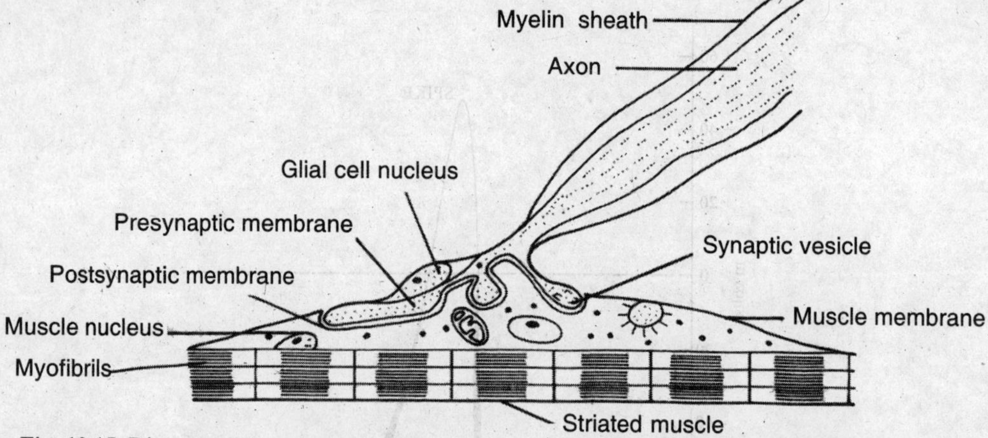

Fig. 10.15. Diagrammatic representation of mammalian myoneural junction (motor end plate) in case of a striated muscle.

This accounts for the fine dextrity of human fingers. When a never impulse travelling along a motor nerve reaches the myoneural junctin it releases a *neurotransmitter* (Chapter 9) which ia acetylcholine (ACH) in most cases. Acetylcholine causes a change in potential at the post-synaptic membrane and a potential called *junction potential* or *end plate potential* is created. This creates a propagated action potential (spike) which travels along the membrane of the muscle fibre in a manner similar to the nerve fibre.

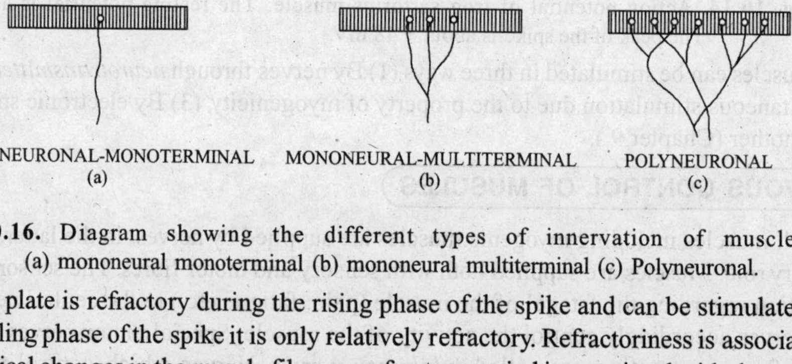

Fig. 10.16. Diagram showing the different types of innervation of muscle fibres: (a) mononeural monoterminal (b) mononeural multiterminal (c) Polyneuronal.

The end plate is refractory during the rising phase of the spike and can be stimulated again. During the falling phase of the spike it is only relatively refractory. Refractoriness is associated only with the electrical changes in the muscle fibre, no refractory period is associated with the contractile elment.

Summation

A single stimulus causes a *twitch* and repeated stimuli in quick succession cause tetanus. Both these phenomena have been described earlier. Tetanus is caused by a phenomenon called *summation.* Summation may be defined as the post-synaptic fusion of two or more pre-synaptic events (stimuli). It is obvious that summation will take place only if the succeeding stimuli are given before the previous ones have died out.

Facilitation

In some muscles it has been seen that if two (or more) stimuli of equal strength are given successively the response goes on increasing (Fig. 10.17). This phenomenon is called *facilitation.* It may be cuase by any one of the following reasons: (1) the neurotransmitter produced by the first stimulus may not be completely destroyed before the second stimulus is given. The quantity of the transmitter thus goes on increasing with successive stimuli (2) The pre-synaptic terminals may be releasing more transmitter substance with each succeeding stimulus (3). It is also possible that the activation brought about in the post-synaptic membrane by the first stimulus may not be completely neutralised before the second stimulus is given.

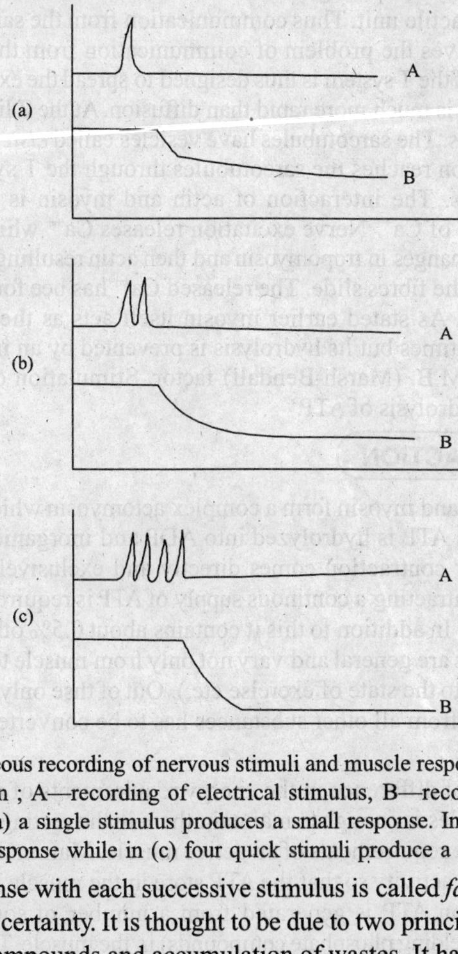

Fig. 10.17. Simultaneous recording of nervous stimuli and muscle response showing the phenomenon of facilitation ; A—recording of electrical stimulus, B—recording of muscle response. Note that in (a) a single stimulus produces a small response. In (b) two quick stimuli produce a greater response while in (c) four quick stimuli produce a still greater response.

Reduction in response with each successive stimulus is called *fatigue.* The cause of muscular fatigue is not known with certainty. It is thought to be due to two principal changes *viz.* exhaustion of high energy phosphate compounds and accumulation of wastes. It has been seen that if successive stimuli are applied at such intervals that muscle has a chance to recover and if a continuous supply of ATP is made then fatigue is delayed. Removal of waste products also delays this process.

EXCITATION-CONTRACTION COUPLING

The mechanism of the creation of a propagated action potential in muscle cell, caused by a stimulus is well understood. Once the spike is produced across the membrane of the muscle fibre (sarcolemma) it can cause the contractile elements of the fibre to move. One of the remarkable features of muscles is that they contract very quickly after nerves supplying them are stimulated. Apparently, therefore, there ought to be a rapid communcation from nerve to the sarcolemma and from the sarcolemma to the sarcomeres (contractile elements). In a number of cases the distance from the sarcolemma to the contractile elements is quite large so that the centre of fibre is far from ist surface (*e.g.* in some invertebrates the diameter of muscle fibre is up to 1 mm, which means that the centre of the cell is 500 μ away from the surface). Even in thinner fibres the centre is appreciably removed from the surface. Simple diffusion of ions is not a sufficiently fast process for accomplishing this. How then is the information travelling along the surface of the fibre communicated to the myofibrils? It has been described earlier that the sarcolemma extends into the interior of the cell body in the form of tubes forming the T system (Fig. 10.2). These tubes extend up to the Z lines forming a triad. Thus structurally there is a microsynanaptic system from the membrane to the Z lines. The Z lines form the margin of a single contractile unit. Thus communication from the sarcolemma to the contractile unit is established. This solves the problem of communication from the surface to the contractile elements. The architecture of the T system is thus designed to spread the excitation from the membrane to the Z lines at a pace which is much more rapid than diffusion. At the Z lines the T system establishes contact with the sarcotubules. The sarcotubules have vesicles called *cisternal vesicels* which contain calcium. When the excitation reaches the sarcotubules through the T system the cisternal vesicles liberate Ca^{++} at the Z lines. The interaction of actin and myosin is inhibited by troponin and tropomyosin in the absence of Ca^{++} Nerve excitation releases Ca^{++} which binds with the troponin that causes conformational changes in tropomyosin and then actin resulting in actin-myosin interaction which hydrolyses ATP and the fibres slide. The released Ca^{++} has bee found to be intimately related with the hydrolysis of ATP. As stated earlier myosin itself acts as the ATPase. Although ATP is present in themuscles at all times but its hydrolysis is prevented by an inhibitor of ATPase, present in the microsomes, called M.B. (Marsh-Bendall) factor. Stimulation of the muscle removes this factor and facilitates the hydrolysis of ATP.

ENERGY OF CONTRACTION

It has been shown that actin and myosin form a complex actomyosin which contracts in the presence of ATP. During this process ATP is hydrolyzed into ADP and inorganic phosphate. All evidences indicate that the energy for contraction comes directly and exclusively from ATP. Therefore, in order to keep the muscle contracting a continous supply of ATP is required. The mammalian muscle contains about 0.025% ATP. In addition to this it contains about 0.5% other phosphogens and about 1% glycogen. (These figures are general and vary not only from muscle to muscle but also change in the same muscle according to the state of exercise etc.). Out of thse only ATP is utilized directly for muscle contraction; energy from all other substances has to be converted into ATP before it can be utilized.

There is a great deal of difference in the energy requirements of resting and active muscles. The ATP content in the muscles is enough to maintain them in the resting state. However, it is almost insignificant for meeting the requirements of an active muscle. Mammalian muscles require 10^{-3} g/min. of ATP during instense activity so that the ATP store in the muscle lasts for less than a second. During muscular contraction ATP is generated from a number of sources: (1) There are other phosphogens (high energy yielding phosphate compounds) in the muscle. The most important amongst these is creatine phosphate in the vertebrates and arginine phosphate in the invertebrates. The ~P (high energy phosphate) of these phosphates is transferred to ADP to form ATP.

$$\text{Creatine phosphate} + \text{ADP} \xrightarrow[\text{kinase}]{\text{creatine}} \text{creatine} + \text{ATP}$$

During recovery endogenious ATP is again available and the creatine phosphate is regenerated. Creatine phosphate, however, is present only in relatively small quantities and can only provide short bursts of energy to the muscle while it is insufficient for prolonged work. In frog, for example, the phosphogens are just sufficient to produce about 50 twitches. (2) Most of the energy for muscle contraction comes from carbohydrate metabolism. During intense exercise although energy requirements may increase 1000-fold, the oxygen consumption does not increase more than 200-fold even in trained individuals like atheletes. Most of the energy, therefore, comes from anaerobic metabolism. The carbohydrates used are (a) muscle glycogen and (b) glucose. The latter is continuously brought to the muscle by blood. The anaerobic respiration of one molecule of glucose

yields two molecular of ATP. The end product of glycolysis is pyruvate (Chapter 13). Part of this pyruvate produces more ATP through the tricarboxylic acid (TCA) cycle. However, since the supply of oxygen is limited most of the pyruvate is converted to lactic acid and stored in the muscle. During rest when oxygen is available the lactic acid is reconverted into pyruvate and is oxidised thorugh the TCA cycle. (3) When very large amount of lactic acid is accumulated in the muscle, part of it diffuses into the blood which carries it to the liver where it is reconverted into glycogen through the 'Cori Cycle' depicted as under:

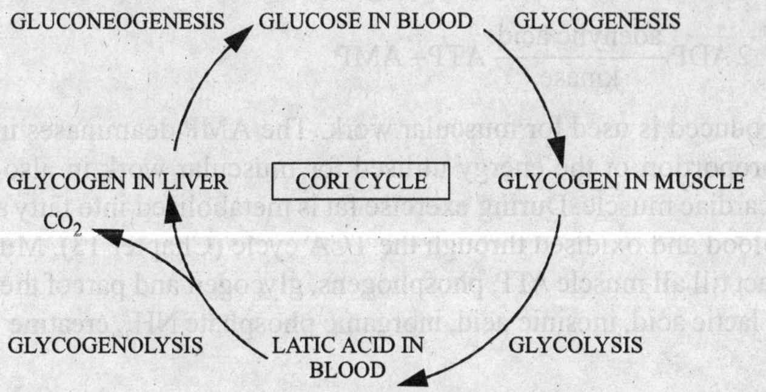

During gluconeogenesis one-sixth of the lactic acid is oxidised to provide energy for the synthesis of the remaining lactic acid into glycogen.

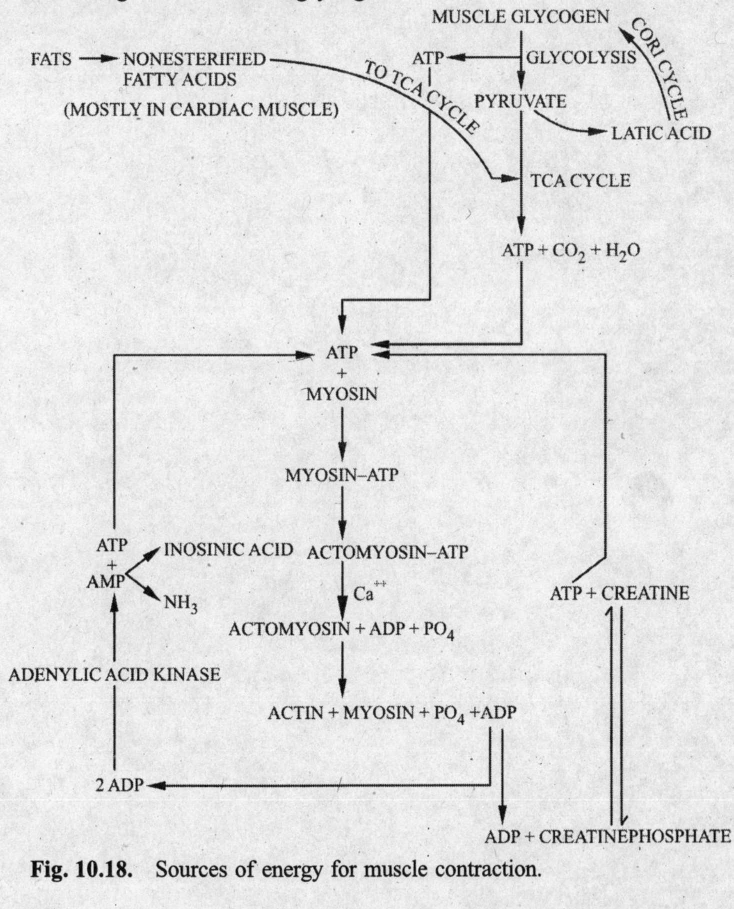

Fig. 10.18. Sources of energy for muscle contraction.

The functioning of the muscles under anaerobic conditions incurs and *oxygen debt.* After cessation of work there is increased uptake of O_2 which provides ATP required for the synthesis of glycogen from lactic acid. Since glycogen is synthesized in the liver, it is noteworthy that part of the energy for muscular work comes form the liver. (4) During the utilisation of ATP a large quantity of ADP is produced.

$$ATP — ADP + PO_4 + 7000 \text{ cals}$$

Muscles possess an enzyme *adenylic acid kinase* which works as follows:

$$2\,ADP \xrightarrow[\text{kinase}]{\text{adenylic acid}} ATP + AMP$$

The ATP produced is used for muscular work. The AMP deaminases into *inosinic acid* and NH_3. (5) A fair proportion of the energy utilised for muscular work in also derived from lipids, especially in the cardiac muscle. During exercise fat is metabolised into fatty acids which reach the muscle through blood and oxidised through the TCA cycle (Chapter 13). Muscles can , therefore, continue to contract till all muscle ATP, phosphogens, glycogen and part of the fat have disappeared and in their place lactic acid, inosinic acid, inorganic phosphate NH_3, creatine and CO_2 appear (Fig. 10.18).

11

Chapter

Hormones and Endocrine glands

Many glands in animals release their secretions through ducts. These are *exocrine glands* or *duct glands* (e.g. sebaceous, salivary, and sweat glands, etc). Other glands in the body secrete *hormones* or chemical messengers that are released directly into the blood circulation and carried to the cells on which they act. The cells influenced by a particular hormone are called *target cells* or *effector cells*. These are *endocrine glands* or *ductless glands*. Endocrine glands have a rich blood supply as the blood not only forms a vehicle for transport of the hormones, but also act as a source of the chemical raw materials from which the hormones are synthesized. *Hormones* are formed by certain cells or aggregate of cells referred to as *endocrine glands*. The role of the hormone is to direct, regulate, and coordinate activities of various organs in the body. Furthermore, there are intracellular acting substances which act as an *intermediary* for a number of hormonal effects on the cellular level. Thus, such intermediaries identified are cyclic AMP, cyclic GMP, calcium ions, diacylglycerol (DAG), and inositol triphosphate (IP3). They are referred to as the *second messengers* by E.W. Sutherland; hormones are the *first messengers*.

PROPERTIES AND FUNCTIONS OF HORMONES

Hormones are produced in very small amount by endocrine glands from nanograms (ng; 1 ng = 10^{-9}g) to milligrams per day; the secretion rate of a hormone is determined by the need of that hormone. Hormones act as *trigger substances*. They initiate biochemical reactions that may persist even after hormones are no longer detectible.

The endocrine system regulates vital body functions such as growth, somatic and sexual development, cellular oxidation, heat production, and the metabolism of water, salt, carbohydrates, fats, and proteins. In general, however, following main functions may be ascribed to the hormones.

(1) *Regulatory or homeostatic function.* Hormones can influence the composition of body fluids, rate of gaseous exchange, activity of the cardiovascular system, or central nervous system. In a normal organism the composition of the body fluids has an amazing degree of precision and constancy. Claude Bernard, the 19th century French physiologist, referred to it as the *internal fluid environment.* The concept of the precise regulation of this environment, *homeostasis,* was proposed by Walter Cannon in 1920. Hormones play a very important role in homeostasis by a mechanism called *the self adaptive negative feedback loop.* The mechanism of this regulation is explained later in this chapter.

(2) *Integrative function.* Hormones provide animals with an integrative capacity that parallels or supports the role of the nervous system. In general, however, the integrative properties of the endocrine system are slow and steady. The nervous system, on the other hand, reacts rapidly to changing environmental conditions.

(3) *Morphogenetic function.* Hormones play very important role in development right from the embryonic to the adult stage.

(4) *Permissive function.* Certain hormones require the presence (permission) of another hormone for the expression of their activity at the cellular level.

In spite of the foregoing characteristics of the endocine system, it has certain degree of dependency on the central nervous system. The inter-relationship between the two system has led to the science of *neuroendocrinology.* Such inter-relationship will be discussed in this chapter as and when need will arise.

CHEMICAL NATURE OF HORMONES

Hormones are classified into three chemical classes: (1) amines (2) peptides and proteins, and (3) steroids (Take 11.1).

The amine hormones are nitrogen–containing derivatives of the amino acid tyrosine. They include the *thyroid hormones,* (TH) namely the two iodine–containing amine hormones thyroxine (T_4) and triiodothyronine (T_3). Besides these, the adrenal medulla of each of the *adrenal glands* secretes two amino hormones epinephrine (E) and *norepinephrine* (NE) which belong to the chemical family catecholamines. *Melatonin*, secreted by the pineal gland, and prolaction release inhibiting hormone (*PIH*) one of the seven hypothalamic hormones, are also catecholamines; PIH has been identified as *dopamine* (DA).

Peptide (having three amino acids) and small proteins form majority of all hormones, secreted by *pituitary, pancreas, ovary* (secretes *inhibin*), thyroid (secretes *calcitonin*), *parathyroid* , *kidneys*, and the *gastrointestinal tract*. With one exception (PIH) all the other six hypothalamic releasing hormones are peptides. As a convention, endocrinologists refer to them all as *peptide hormones*.

Steroid hormones are produced by the *adrenal cortex* and the gonads (testes and ovaries) as well as by the placenta during pregnancy. Cholesterol is the precursor of all steroid hormones.

Peptide and catecholomine hormones are water soluble and are, therefore, carried in free form in the plasma. Their receptors are located on the plasma membrane of target cells. The signal transduction mechanisms involved for the intimate response from the effect cells are receptor operated channels, cAMP, cGMP, Ca^{2+}, IP_3 and DAG. The peptide and catecholamines are excreted or metabolized fast (minute to a few hours) by the kidneys or the liver, respectively. In contrast, the steroid and thyroid hormones are poorly soluble in water and, therefore, circulate in the blood bound to plasma proteins. Their receptors one located in the interior of the cell and the signal transduction mechanisms involve activation of genes (DNA). Excretion or metabolism of steroid and thyroid hormones, as they are bound to plasma proteins, is slow (many hours). All hormones have 24 – hour cyclical variations (circadian rhythm) in their secretory rates, which is different for different hormones. Hormone secretion is signalled by four types of inputs to endocrine cells. These are : (1) plasma ionic concentration, (2) plasma concentrations of organic materials, (3) neurotransmitters impinging on the endocrine cell, and (4) another hormone acting on the endocrine cells. However, the hormones secreted by the gastrointestinal tract are regulated by a number of inputs, such as emotional status, sight, smell and taste of food (neural regulation) as well as chemoreceptors, osmoreceptors or mechanoreceptors in the gastrointestinal wall.

Table 11.1 lists the hormones according to their chemical nature, site (endocrine gland cells) of origin, target cells or tissues, and primary functions. Fig. 11.1 shows approximate locations of the endocrine glands of man.

The hypothalamus is considered part of the endocrine system too. This is because the chemical messengers (neurohormones) released by the neuron terminals in both the hypothalamus and its extension, the posterior pituitary, do not function as neurotransmitters but enter blood circulation and carried to sites of action else where. This is the reason why hypothalamus forms a part of Table 11.1 (see Table 11.2 and Fig. 11.3 also).

A look at Table 11.1 shows that there are a large number of endocrine glands distributed throughout the body. A detailed description of these glands and the hypothalamus is given below.

Table 11.1 The Mammalian Hormones, Their Cheimical Identity, Source, Target Cells or Organs and Chief Effects

Chemical Identity and Source of Hormones	*Chief Effects*
PEPTIDES; PROTEIN HORMONES	
1. Anterior Pituitary	
Growth hormone (somatotropin, GH)	Growth via secretion of IGF-I, protein carbohydrate and lipid metabolism
Thyroid stimulating hormone (TSH, thyrotropin)	Thyroid gland
Adrenocorticotropic hormone (ACTH, corticotropin)	Adrenal cortex
Prolactin	Breast growth and milk synthesis, permissive for certain reproductive functions in the male
Gonadotropic hormones Follicle-stimulating hormone (FSH) Luteinizing hormone (LH)	Gonads (gamete production and sex hormone secretion)
β-lipotropin and β-endorphin	Largely unknown for β-lipotropin; β-endorphin, was role in memory consolidation
Melanophore-stimulating hormone (α - and β - MSH)	Melanophores; Pigment dispersal in melanophores (darkening of skin)
Lipotrophins (LPH)	Adipose tissue Lipolysis
2. Posterior Pituitary	
Oxytocin	Milk let-down uterine motility
Vasopressin (antidiuretic hormone, ADH)	Water excretion by the kidney; blood pressure
Dopamine (DA) or prolactin release inhibiting hormone (PIH)*	Prolactin secretion
Prolactin releasing factor	Prolactin secretion
3. Pancreas	
Insulin	Organic metabolism; plasma glucose
Glucagon	Breakdown of liver glycogen; elevation of blood glucose
Somatostatin	Inhibits release of GH, insulin and glucagon
Pancreatic polypeptide	
4. Gonads:	
Female: ovaries	
Relaxin	Possibly relaxation of cervix and pubic ligaments
Inhibin	FSH secretion
5. Thyroid	
Calcitonin	Plasma calcium
6. Parathyroid	
Parathyroid hormone (parathormone, PTH, PH)	Plasma calcium and phosphate

7. Kidneys	
Renin (angiotensin (II)	Aldosterone secretion; blood pressure
Erythropoietin	Erythrocyte production
1, 25-dihydroxy vitamin D_3	Plasma calcium
8. Gastrointestinal tract	
Gastrin	
Secretin	Gastrointestinal tract; liver; pancreas; gall bladder
Cholecystokinin	
Glucose-dependent insulinotropic peptide (GIP)	
Somatostatin	
9. Liver (and other cells)	
Insulin-like growth factors (IGF-I and II)	Growth
10. Heart	
Atrial natriuretic factor (ANF, atriopeptin, auriculin)	Sodium excretion by kidneys; blood pressure
11. Leukocytes, macrophages, endothelial cells and fibroblasts	
Cytokines (these include the interleukins, colony-stimulating factos, interferons, tumor necrosis factors)	Immune defenses
12. Multiple cell types	
Growth factors (e.g., epidermal growth factor)	Growth and proliferation of specific cell types, including some neurons.
13. Thymus	
Thymopoietin	T-lymphocyte function
AMINO ACID DERIVATIVS	
1. Thyroid	
Thyroxine (T_4)	Metabolic rate; growth; brain development and function
Triiodothyronine (T_3)	
2. Adrenal medulla	
Epinephrine Norepinephrine	Organic metabolism; cardiovascular function; response to stress
3. Pineal	
Melatonin	Sexual maturity; body rhythms
STEROIDS	
1. Gonads	
Male: testes	
Testosterone	Reproductive system, growth and development
Inhibin**	FSH secretion
Mullerian-inhibiting hormone (MIH)**	Regression of Mullerian ducts

2. Ovaries	
Estrogen Progesterone	Reproductive system; breasts; growth and development
3. Placenta	
Chorionic gonadotropin (CG)**	Secretion by corpus luteum
Estrogens	See ovaries
Progesterone	See ovaries
Placental lactogen	Breast development; organic metabolism
4. Adrenal cortex	
Cortisol	Organic metabolism response to stresses; immune system
Androgens	Sex drive in women
Aldosterone excretion by kidneys	Sodium, potassium and acid

Note: * DA or PIH is a catecholamine.
** Inhibin, MIH and CG are proteins.

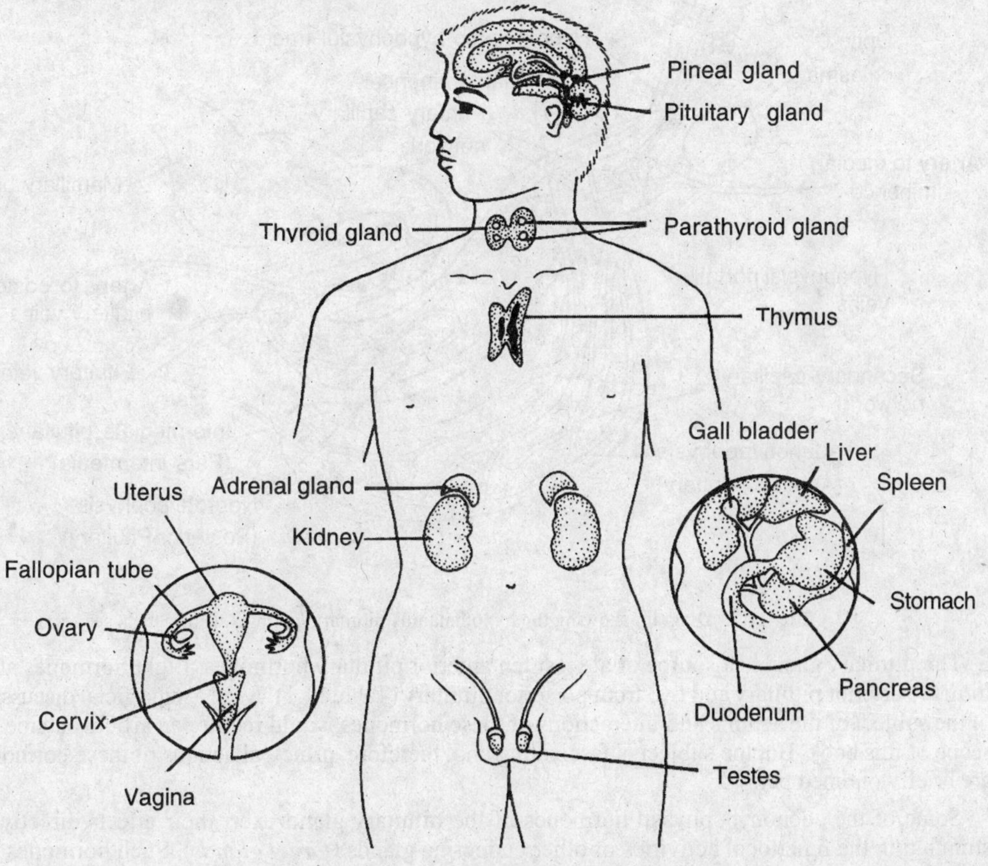

Fig. 11.1. Approximate locations of the endocrine glands of man. The encircled diagram on the right illustrates the source of gastrointestinal hormones; the diagram on the left shows the reproductive features of the female.

PITUITARY GLAND

The *pituitary gland or hypophysis cerebri* is one of the well protected glands of the body. It is unpaired organ located in the *sella turcica*, a concavity in the sphenoid bone. The gland is closely applied to the floor of the fore brain (hypothalamus) and remains attached to it by means of a stalk containing nerve fibres and blood vessels.

In adult human beings, the pituitary gland consists of two distinct adjacent lobes called the *adenohypophysis (anterior pituitary*) and the *neurohypophysis (posterior pituitary)*. The former is located in front of the latter. An adjunct to adenohypophysis is *pars intermedia* (intermediate pituitary) it lies between the adenohypophysis and the neurohypophysis. Pars intermedia is absent in certain mammals, such as the whale, Indian elephant, and armadillo.

The posterior pituitary develops from the brain and remains connected with it by the infundibular stem. It contains nerve fibres arising from the supraoptic and paraventicular nuclei of the hypothalamus, and *pituicytes*. The latter are non-neural connective tissue cells which are collectively called *neuroglia*. The nerve fibres project from the hypothalamus into the posterior pituitary where they comprise the hypothalamo-hypophysial tract. The latter passes through the median eminence (Fig. 11.2).

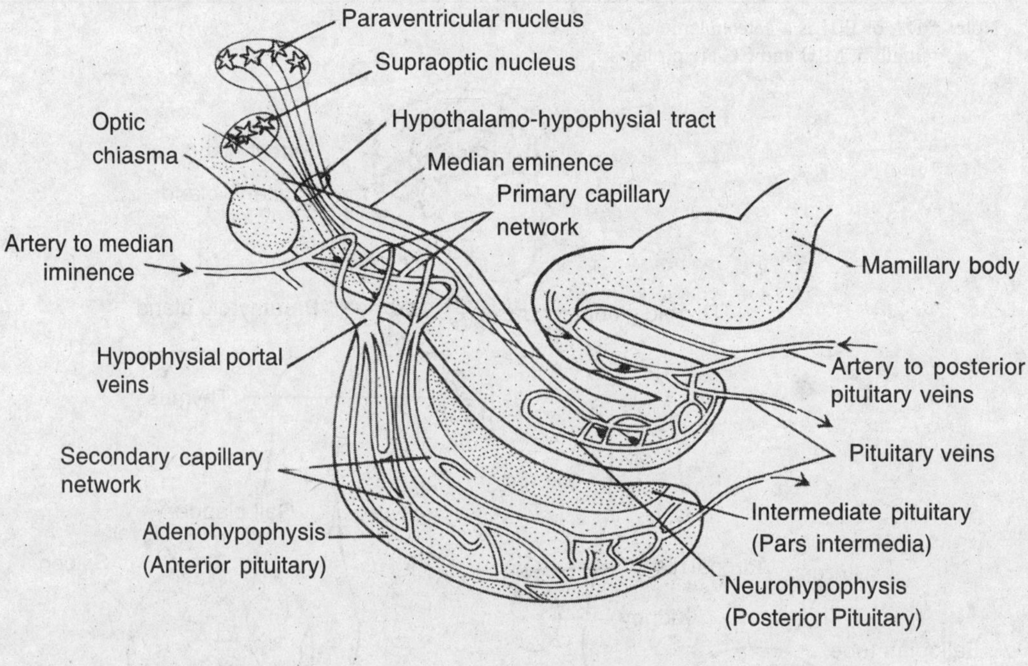

Fig. 11.2. Diagram showing the hypothalamus pituitary complex. (For details see text).

The pituitary gland is a source of at least ten anterior pituitary hormones. Eight hormones arise form the arterior pituitary and two from posterior pituitary (Table 11.1). Even a superficial discussion of the subject of the actions and interactions of these hormones would require an article outside the scope of this book. But the subject is fascinating and, therefore, principal actions of these hormones are briefly outlined below.

Some of the adenohypophysial hormones of the pituitary gland exert their effects directly by stimulating the functional activities of other endocrine glands (*target glands*). Such hormones are called *tropic* hormones. These are thyroid stimulating hormones (TSH; thyrotropin) and adrenocorlicotropic hormone (ACTH) which induce secretion of thyroid hormones, namely, thyroxine (T_4) and triiodothyronine (T_3) from thyroid gland and cortisol from the adrenal cortex of *adrenal glands*, respectively. Three other anterior pituitary hormones also stimulate secretions of another hormone but have an additional function as well. These are *follicle-stimulating hormone* (FSH) and luteinizing hormone (LH) which stimulate the secretion of the sex hormones *estrogen*, *progesterone* and *testosterone* by the gonads and, in addition, regulate the growth and development of sperm and ova. *Growth hormone* (GH) stimulates both the liver and many other body cells to secrete a growth promoting peptide hormone known as insulin-like growth factor (IGF-I, also called *somatomedin* C) and exert direcet effects on the organic metabolism of various organs and tissues. The sixth hormone is *prolactin* which does not exert influence over the secretion of a hormone by another endocrine gland. It mainly stimulates development of mammary glands and milk production by direct effect upon the breasts in females. In males, prolactin facilitates reproductive function.

This is not the end of anterior pituitary hormones as substantial evidence indicates that two more peptide hormones, namely, β-*lipotropin* and β-*endorphin* are co-secreted along with ACTH.

It may be noted here that except for FSH and LH each of the other four hormones are secrcted by distinct cell type in the anterior pituitary. FSH and LH, collectively known as ***gonadotropic hormones****, as they stimulate the gonads, are both secreted by the same cells.*

The major functions, other than those briefly stated above, of the anterior pituitary hormones are described below under the heading 'Tropic Hormones that affect other Endocrine glands'.

Mention may be made here of a reciprocal interaction present between the pituitary and its target organs called *negative feedback* mechanism. It can be explained as follows. The pituitary secretes trophic hormone which stimulates the target gland to step up the rate of secretion of its hormone. As the concentration of the target gland hormone rises in the blood, it begins to affect (feedback) the pituitary to suppress the production of the trophic hormone; this in turn reduces the secretion level of the target glands (see flow diagram below).

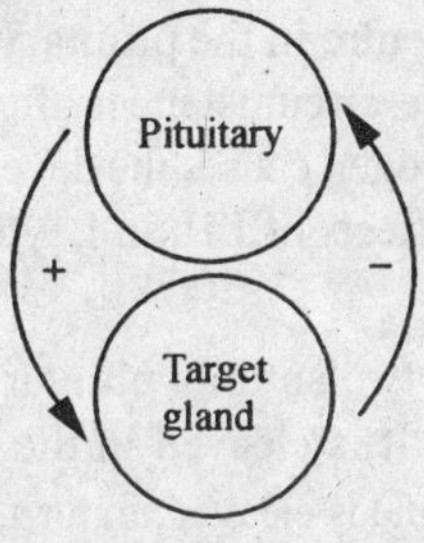

However, as the target gland hormone is utilized, metabolized, and excreted, its concentration falls in the blood. This releases the pituitary from the feedback inhibition by the trophic gland hormone, and the trophic hormone is once again secreted. In this way, a delicate balance is maintained between the secretion of the pituitary and its target gland. It may be noted here that negative feedback systems are the most common homeostatic mechanism in the body. However, there is another feedback known as *positive feedback* in which an initial disturbance in a system sets of a chain of events that increases the distrubance even further. Thus, positive feedback does not favour stability and often displaces a system away from its steady-state operating point.

Tropic Hormones that affect other Endocrine Glands

Thyroid Stimulating Hormone (TSH; thyrotropin): The anterior pituitary produces TSH which induces an increase in the rate of the breakdown of thyroglobulin, to which the thyroid hormones bind, in the thyroid gland. This results in an increased discharge of thyroid hormones, viz. thyroxine (T_4) and triio -dothyronine (T_3). The TSH also acts on fat deposits, resulting into the release of fatty acids (lipolytic effect).

In 1956 Adams and Purves described in the plasma of hyperthyroid patients a stimulatory substance which induces the discharge of organically bound iodine form rat thyroid glands. This substance is not present in the pituitary. It is known as the *long-acting thyroid stimulator* (LATS). It is an α-globulin. Another special response to hyperthyroidism is *exophthalmos* (bulging of the eye). It has been seen that highly purified TSH preparations do not induce exophthalmos, whereas crude preparation of thyroid does. The general opinion is that TSH and another hormone (*exophthalmos producing hormone*) are involved in producting exophthalmos. This hormone has not yet been obtained.

Adrenocorticotropic Hormone (ACTH): Another hormone from the anterior pituitary is adrenocorticotropic hormone (ACTH). It acts upon several sites in the body. The target it primarily influences are the two innermost zones of cell in the adrenal cortex (*zona fasciculata* and *zona reticularis*). These cell layers, as will be later seen, constitute the site of glucocorticoid synthesis and secretion. The third region—*zona glomerulosa,* the outermost layer of the adrenal cortex, which secretes aldosterone (*mineralocorticoid*) is only minimally affected by ACTH. The hormone also affects fat deposits, where it exerts a lipolytic action. This action is similar to the effect produced by GH (growth hormone), TSH, and catecholamines. The adrenal cortex is affected by ACTH in several ways: (1) causes reduction of the ascorbic acid content; (2) it converts cholesterol to glucocorticoids and increases their secretion rate; (3) it raises the level of metabolic activity of adrenal tissue by causing an increase in glucose utilization; (4) it stimulates cell division of the inner two layers and causes cortex to grow. ACTH also regenerates memory consolidation of learned experiences.

Prolactin: Yet another hormone from the adenohypophysis is prolactin. Because of its broad spectrum of effects on vertebrates in general, it is the most versatile of all adenohypophysial hormones. However, the main effect of prolactin in mammals is to stimulate breast development and *lactogenic* (production of milk). The hormone induces milk secretion provided the mammary glands have bee previously *sensitized* by adequate amounts of estrogen and progesterone. In some women at parturition (chilbirth), the prolaction centre in the pituitary gland is destroyed by some disease. The breasts fail to secrete milk under these circumstances. In addition, prolactin also stimulates hair maturaiton, growth, and along with androgen causes the development of secondary male sex characters. Prolactin potentiates the stimulatory effects of LH on Leydig cells and of testosterone and its target cells.

Lipotropin and β-Endorphin: These two anterior pituitary hormones are synthesized as part of the large pro-opiomelanocortin molecule. It is cleaved in the cell to form ACTH and other peptides, including β-lipotropin. The latter molecule is cleaved, in turn, to form the powerful opioid β-endorphin. Their functions are largely unknown but some possible roles are stated here. The secretion of both lipotropin and endorphin is increased during stress. Endorphin is a potent endogenous opiate and plays a role in mediating both relief of pain and mood alteration in stress. It also decreases appetite. An interesting role of endorphin is transferring short-term memory to long-term form, i.e. in *memory consolidation.*

Growth Hormone (GH): has widespread effects on the metabolism of protein, carbohydrate and lipids. It also enchances the effect of GH, insulin and glucocorticoids. The primary physiological effects of GH are to stimulate growth and protein metabolism. Only excess of GH does produce abnormalities of lipid and carbohydrate melabolism Growth hormone's major growth promoting effect is stimulation of mitosis is its many target tissues. Thus, maturation and mitosis of the chondrocytes in the epiphyseal plates widen the plates providing more cartilaginous material for bone formaton. (GH has no effect on fetal growth). Excess of GH during childhood produces *giantism*, whereas deficiency produces *dwarfism*. In adults, after epiphyseal closure, the bones cannot be lengthened; this produces disfiguring of bones and overgrowth of other organs known as *acromegaly*.

It is important to note here that GH exerts its mitosis-stimulating (*mitogenic*) effect *not directly* but *indirectly* through the mediation of a chemical messenger called *insulin-like growth factor 1 (IGF-1*;

also known as *somatomedin C*) whose synthesis and release are induced by GH. The IGF-1 is mainly synthesized in the liver and secreted into the blood, which carries it to its target cells. Consequently, IFG-1 is categorized as a hormone. IGF-1 is also secreted by cells other than those in the liver. In such cases, it acts in these sites as a paracrine or autocrine.

The interaction of GH and IGF-1 in exerting mitogenic effect on the epiphyseal plates is as follows: GH stimulates prechondrocytes to differentiate into chondrocytes; these cells begin to secrete lGF-1and, as a result, the chondrocytes undergo mitosis. A similar interaction also occurs between GH and IGF-1 for stimulating growth in the target tissues of GH; GH stimulates precursor cells to differentiate and to secrete IGF-1 which , on its turn, stimulates mitosis in the differentiated cells.

In addition to its specific growth promoting effect on mitosis via IGF-1, GH also causes mobilization of non-esterified fatty acids from fat depots. It inhibits glucose utilization by muscle tissues and decrease the sensitivity of animals to insulin. The first of these actions is called *ketogenic* effect and the latter two are *diabetogenic* actions. Such effects are referred to as anti-insulin effects. Current studies on GH action point to a role in *transcription* and *translation* steps leading to protein biosynthesis. It has been shown that GH treatment causes an increase in liver-RNA polymerase, increasing tRNA and mRNA formation. Injections of GH can partially restore the lost ability of ribosomes to incorporate amino acids into proteins. All these actions lead to increased protein synthesis and thus stimulate growth.

Melanophore Stimulating Hormone (MSH): In mammalian pituitaries MSH (also known as intermedin) is present in two forms as α-and β-MSH. The former has greater biological activity. The hormone is used in colour adaptation in lower vertebrates. The animal becomes pale on a light background and dark on a dark background. This reguation is effected by certain cells containing the black pigment *melanin.* The role of and α-and β-MSH in mammals is not clear. Some mammals, *viz.* Indian elephant and the whale do not possess pars intermedia. In man this hormone increases melanin synthesis in the melanocytes and thus the skin darkens.

This is not the end of the list of adenohypophysial hormones, as evidence indicates that several more hormones are produced. Substantial evidence shows that another group of hormones, the *lipotrophins* should be added to the list of hormones already described. The lipotrophins (chemically polypeptides exist in two forms, viz., β-lipotrophins (β-LPH) and γ-lipotrophins (γ LPH). As the very name of hormones indicates, LPH has lipolytic activities, that is they have a fat-mobilizing action, which is similar to the lipolytic effects of ACTH. Very little is known about the role of lipotrophins in vertebrates.

Existence of certain lesser adenohypophysial hormones has also been reported, *viz.*, *diabetogenic hormone,* which produces a diabetic state in normal animal by inhibiting the action of insulin, and by diminishing the production of insulin by the pancreas. The list could further be extended by the addition of several more hormones, namely, *pancreatrophic, glucostatic, glycotrophic,* and *ketogenic* hormones. However, concrete proof to support their existence in the anterior pituitary is wanting.

The Posterior Pituitary and its Hormones

The posterior pituitary releases two specific hormones, *oxytocin* and *vasopressin.* The posterior pituitary is formed of nerve fibres arising from the supraoptic and paraventricular nuclei of the hypothalamus (Fig. 11.2). The hormones are stored in small vesicles accumulated at the axon terminals. The two hormones are transmitted from the axon terminal to the capillary beds in the posterior pituitary. In response to generation of action potential in the axon as a result of synaptic input, the hormones are first liberated from the nerve terminals into the blood capillaries and are carried away by the blood returning to heart to be distributed to the entire body. The transmission of the hormones is facilitated through fenestrated lining of the vessel, a feature typical of capillaries in regions of active absorption and secretion (see Fig. 11.4). Each hormone is secreted independently from its own hypothalamic nucleus. The supraoptic nucleus secretes vasopressin and the paraventricular nucleus secretes oxytocin.

It is obvious from the above that the term posterior hormones is a misnomer since the hormones are actually synthesized in the hypothalamus, of which the posterior pituitary is merely an extension.

Vasopressin constricts arterioles, and, therefore, causes marked rise in blood pressure. The effect is similar to epinephrine (adrenaline). The effect of the latter, which is an amine, wears off in a few minutes, whereas the effect of the former, which is a polypeptide, lasts much longer. Vasopressin also brings about a reduction in the urine volume by causing kidney tubules to withhold more water. This is the reason why vasopressin is also *called antidiuretic hormone* (ADH). The absence of vasopressin in mammals results in *diabetes insipidus*. This is a condition in which large volumes of water are consumed and then lost in the urine without elevation of sugar in the urine.

Oxytocin causes contractions in the smooth muscles. It is important only in a pregnant and lactating female. The hormone is released from the posterior pituitary during the later part of pregnancy. It starts the process of birth by causing contraction of the uterine muscles. It is also responsible for the ejection of milk. Commercial form of oxytocin is frequently used to induce labor.

The Hypothalamus and Pituitary Gland

As already said, vertebrates have evolved two major systems serving the functions of integration and adaptation—the nervous and endocrine systems. The development of links between these two organizations enables the activities of endocrine tissues to be functionally related not only to the fluxes (changes) in the internal environment but also to the external world. The latter comprises the changing conditions of light, temperature, sound, smell, contact, stresses, etc.

To some extent the activities of the anterior putuitary are regulated by the secretory products of the organs which it controls (feedback mechanism). However, the feedback mechanism alone is not adequate to adapt the animal to changes in the external environment. Some link of the anterior pituitary with the external environment via the nervous system is, therefore, necessary. This link between the nervous system and the pituitary gland is established through the hypothalamus (Fig. 11.2).

The *posterior pituitary* is an outgrowth of the hypothalamus and is a neural tissue. In contrast to the neural connections between the hypothalamus and posterior pituitary (Fig. 11.2), there are no neural connections between the hypothalamus and the anterior pituitary. There is however, an un usual blood-vessel connexion. The capillaries in the *median eminence* into the *hypothalamic-pituitary portal vessels.* It may be reminded that the term 'portal signifies vessels that connect distinct capillary heds. The portal vessel pass down the stalk connecting the hypothalamus and pituitary and enter the anterior pituitary capillaries. Thus, the portal vessels offer a local route in blood flow directly from the hypothalamus to the anterior pituitary. Its should once again be pointed out here that the posterior pituitary capillaries, into which vasopressin and oxytocin are secreted, immediately join the main blood stream. In contrast the capillaries in the median evince of the hypothalamus do not join the main blood stream but empty into the portal vesells, which provide a short direct route to the anterior pituitary.

There is therefore, a neural connexion between the hypothalamus and the anterior pituitary too, via the hypothalamo-pituitary portal vessel. In fact, the secretion of anterior pituitary hormones is regulated by still other hormones produced by hypothalamic neurons. These hormones are called *hypothalamic releasing hormones.*

HYPOTHALAMIC RELEASING HORMONES

There are seven major hypothalamic releasing hormones, each secreted by a particular group of hypothalamic neurons. They regulate the release of one or more of the anterior pituitary hormones. Figure 11.4 summarizes the releasing hormones and the classical anterior pituitary hormones they influence. *Each releasing hormone is named according to an anterior pituitary hormones whose secretion it controls.* The "+" and "–" symbols in the figure denote the stimulatory and inhibitory effects, respectively, of the releasing hormones on the anterior pituitary hormones. It is evident from the figure that two of the releasing hormones do not cause release of anterior pituitary hormones but inhibit their release. Therefore, the term "releasing hormone" is misleading. To avoid this awkward situation, some endocrinologists refer to the entire group as "**hypophysiotropic hormones**" rather than releasing hormones. The reader may once again be reminded here that these are blood-borne hormones which are introduced into the anterior pituitary via the hypothalamo- pituitary portal capillaries in the median eminence. As discussed already, the hypothalamic releasing hormones are called neurohormones.

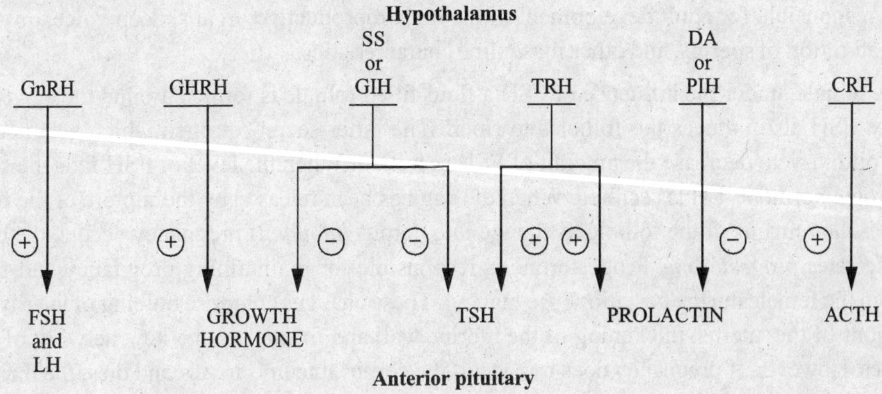

Fig. 11.3. Hypothalamic-anterior-pituitary system. Major hypophysiotropic hormones and ther effects on the anterior pituitary.

Table 11.2 lists the hypothalamic hormones and their effect on anterior pituitary hormones.

Table 11.2 Major hypothalamic releasing hormones and their effects on anterior pituitary hormones

Hypothalamic releasing hormones	*Effect on anterior pituitary hormones*
Gonadotropin releasing hormone (GnRH)	Stimulates secretion of LH and FSH
Growth hormone releasing hormone (GHRH)	Stimulates secretion of GH
Somatostatin (SS; also known as growth hormone release inhibiting hormone, GIH)	*Inhibits* secretion of GH and TSH
Thyrotropin releasing hormone (TRH)	Stimulates secretion of TSH and prolactin
Dopanine (DA, also known as Prolactin release inhibiting hormone PTH)	*Inhibits* secretion of prolactin
Prolactin releasing hormone (PRH)	Stimulates secretion of prolactin
Corticotrapin releasing hormone (CRH)	Stimulates secretion of ACTH

Chemically, all hypothalamic hormones are peptides except PIH which is dopamine (a catecholamine). As stated above, a releasing hormones may influence one or more anterior pituitary hormones. Thus, TRH stimulates the secretion of both TSH and prolactin. But the PRH acts only on prolactin. CRH is a major stimulator of ACTH release, but vasopressin from the posterior pituitary may also stimualte ACTH release.

Tropic Hormones in Reproduction

There are two types of hormones which affect the reproductive system, viz, *luteinizing hormone* (LH) and *follicle-stimulating hormone* (FSH). These two together are referred to as *gonadotrophic* hormones because they stimulate in both sexes and are secreted by the same cells. The major function of FSH and LH are shown in the following flow diagram.

In human beings, the gonads of the young individual remain dormant until the age of about 10-14 years, when they are stimulated by gonadotropic hormones of the anterior pituitary. At this age for reasons not yet understood, the adenohypophysis begins to secrete the FSH and LH. In the male, the FSH promotes the formation of the sperm spermatogenesis in the semniferous-tubule of sertoli cells. These cells recreate a protein hormone inhiblin which inhibit FSH secretion of the testes. The LH when it is acting in the male, stimulates the function of the *interstitial cells of* Leydig. These specialized cells are located between the seminiferous tubules. The interstitial cells in turn, produce *testosterone,* which is responsible for adult development of the male reproductive organs, deep voice, growth of beard, maturation of sperms, and other masculine characteristics.

In the female, under the influence of FSH a fluid-filled follicle is formed around the egg cells in the ovary. FSH also induces the follicles to ripen. The latter secret estrogen which acts upon the anterior pituitary to decrease the amount of FSH secretion. When the level of FSH falls the second gonadotropic hormone-LH is secreted. When the egg has been released by the rupture of the follicle LH causes the remains of the follicle to change into corpus *luteum*. If pregnancy occurs, the corpus luteum secretes *progesterone*. Progesterone is responsible for maintaining pregnancy and for the changes in the female during gestation (pregnancy). These changes comprise quieing of the rhythmic contractions of the uterus, thickening of the uterine wall and formation of a new network of blood capillaries. However, if pregnancy does not occur, the corpus luteum shrivels and dies; the thickened uterine epithelium sloughs off (*menstruation*), a new egg follicle ripens and the cycle is repeated. For further details, see chapter 16.

Although FSH and LH are the principal gonadotrophins of pituitary origin, three other hormones HMG, PMS, and HCG, of non-pituitary origin are in common use. HMG (*human menopausal gonadotropin*) isolated from the urine of menopausal women (those women in whom menstruation has stopped; this usually happens after the age of 45) shows both FSH-like and LH like activity. PMS (*pregnant mares serum ganadotropin*) is formed in the uterine endometrium and accumulates in the blood because it cannot be excreted by the kidneys—because PMS has the tendency to aggregate in electrolyte solutions which cannot pass through the renal membrane. It is more FSH-like but with small LH-like action. HCG—*human chorionic ganadrotropin,* appears in women about seven days after conception and a peak in about six weeks. It has both LH and FSH like activities.

The Neuroendocrine System

Hypothalamic releasing hormones have been classified as *neurohormones.* These are unlike *neurohumors* (for instance acetylcholine, nor-epinephrine serotinin, etc.) which are released at nerve endings and activate the adjacent nerve bodies. The neurohormones are released into the blood and activate cells some distance from their point of release. The neurosecretory cells have at their terminals two different sized granules, large ones composed of neutrosecretory substance and small ones (*synaptic vesicles*; Fig. 11.4), containing neurohumoral substance. It is suggested that the afferent nerve stimuli activate the release of neurohumor, which in turn, facilitate the discharge of neurohormone.

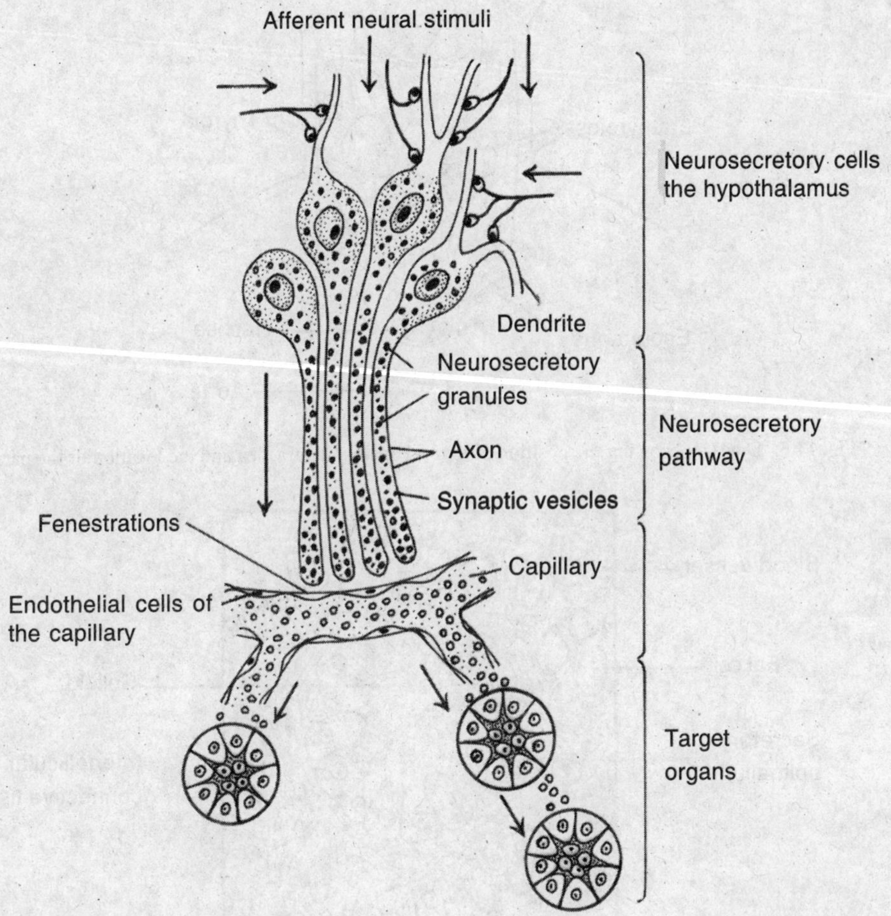

Fig. 11.4. Diagram of neurosecretory cells in the median eminence showing four nerve terminals on a blood capillary.

The neurons that produce releasing hormones serve as an intermediate structure between the nervous system and the endocrine system. They are considered as *transducer cells,* converting neural signals into hormonal signals. The circuit by which sensory information is converted from neural into hormonal signal, and finally into a physiological response is called a *neuroendocrine reflex.* The significance of the systems is obvious; it allows environmental stimuli perceived in the central nervous system to exert its effects on the endocrine system.

In conclusion, therefore, it may be said that the pituitary gland is an essential link in the neuroendocrine sytem. Moreover, since the pituitary is subservient to the nervous system and some of the other endocrine glands, it is misleading to call the pituitary as the master gland of the body as was the practice in yester years.

THYROID GLAND

The *thyroid gland* is located in the lower part of the neck at about the level of the larynx and upper part of the trachea (Fig. 11.5). It is the largest endocrine gland in the body. The thyroid gland consists of two lobes connected by a narrow bridge (isthmus) across the ventral surface of the trachea. The functional units of the gland are called *follicles*. These are separated by *interfollicular connectives tissue* containing numerous capillaries. Each follicle consists of a single peripheral layer of cuboidal cells surrounding a cavity filled with a colloid protein secretion called *thyroglobulin* (Fig. 11.6).

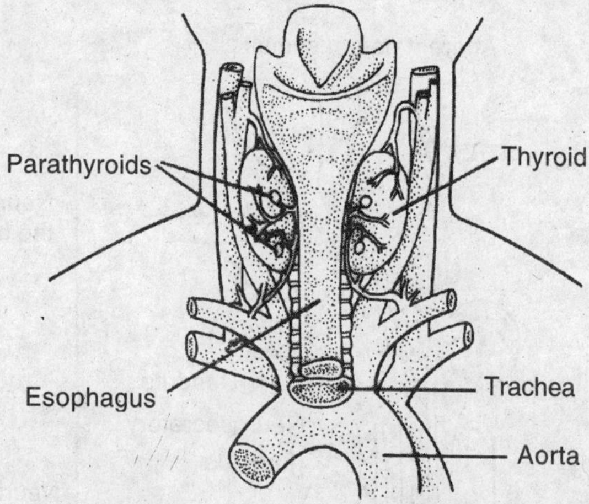

Fig. 11.5. Diagram showing the position of thyroid against the trachea and the location of the parathyroid.

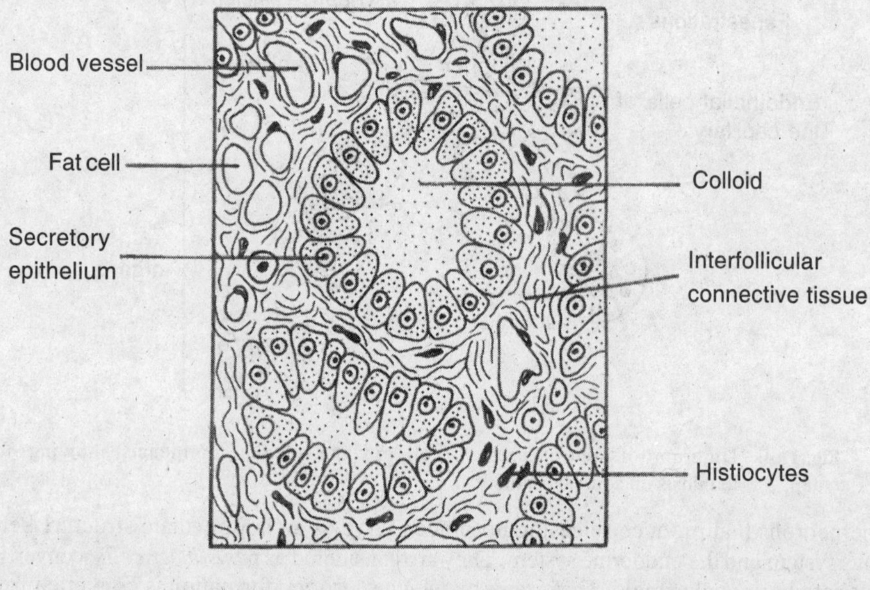

Fig. 11.6. Histological features of the thyroid gland.

Thyroid Hormones

The thyroid gland secretes three hormones, namely, thyroxine (T_4), triiodothyronine (T_3) and calcitonin. T_4 and T_3 are collectively known as *thyroid hormones (TH). Calcitonin (a peptide hormone) will not be included in subsequent references to thyroid hormones.*

Iodine is an essential component of both T_4 and T_3. The gastrointestinal tract converts most of the ingested iodine in food to iodide (ionized form) which is transported via blood to the thyroid follicular cells. Inside the follicular cells, the iodide is oxidised and converted back to iodine, which is then used, along with tyrosine, for the synthesis of T_4 and T_3. The two hormones are then secreted into the follicle cavity wherein they bind to thyroglobulin. The latter is, therefore, a storage form of thyroid hormones. *This is probably the only example to an endocrine gland that stores its products outside of the cells.* Hormone secretion occurs by enzymatic splitting of T_4 and T_3 from the thyroglobulin and the entry of the free hormones into the blood. T_4 form the bulk of secretion and T_3 is secreted in smaller amounts. However, most target tissues and other cells first convert T_4 to T_3 by removing one of the iodine atoms stores from T_4. It may be noted here that the normal thyroid gland stores several weeks supply of thyroid hormones bound to thyroglobulin.

A variety of factors influence the secretion of the thyroid gland. These include the anterior pituitary, dietary components such as iodine, temperature, the adrenal glands, and stress. Physiological effects of the thryoid hormones are widespread. They can be categorised in two general groups: (1) metabolic effects, which include *calorigenesis* (increased oxygen consumption and heat production; this is a major contributor to the maintenance of the body temperature and an important determine of basal metabolic rate (BMR) in homeotherms), regulation of water and ion transport and the regulation of intermediary metabolism, and (2) growth promoting developmental actions, which include influences on the growth rate of tissues of homeotherms, and the control of protein synthesis. As such the thyroid hormones are permissive for the growth – promoting effects of GH.

The thyroid hormone regulates the metabolic rates of most of the adult tissue—skeletal muscle, heart, liver, kidney, etc. But it does not normally affect the lungs, lymphatic system, gonads and accessory organs, nervous tissue, skin, and smooth muscles.

Hypothyroidism (decreased thyroid activity) causes extracellular retention of sodium, chloride, and water; as such the blood volume is considerably reduced. TH plays a crucial role in development of central nervous system during fetal life and the first few months after birth. Hypothyroidism in infants results in *cretinism.* A *cretin* is a sexually immature dwarf with the mentality of an imbecile. A hypothyroid man exhibits sluggishness and poor mental function. Conversely, hyperthyroidisms makes a man jittery and hyperactive. Thyroid hormones accelerate the uptake of glucose in the blood. This causes liver glycogen stores to diminish and the blood sugar rises. Hypothyroidism appears to delay sexual development in the young male. In females it impairs the ovaries and causes the menstrual cycles to cease and become irreguar. Hypothyroidism also increases the size of the heart and reduces the output of the blood.

During periods of iodine deficiency, the pituitary gland secretes larger amounts of TSH. This causes an increase in size (hypertrophy) and number (hyperplasia) of the thyroid follicular cells. This increase in size partially makes up for the deficient activity of the cells, causing *goiter*. Goiter can also be produced by blocking the synthesis of thyroid hormones by drugs. Such agents are called *goitrogens* (antithyroid drugs). Some of the goitrogens are the sulphonamides, thioureas and the thiocyanates.

Calcitonin (CT): The third hormone of the thyroid is calcitonin which is involved in the regulation of calcium levels in the blood. In 1961 Copp, Davidson, and Cheney identified a *calcium lowering factor* as an impurity in commercial parathyroid extracts. They called this *hypocalcemic factor as calcitonin* (CT). Later, evidence was presented that in mammals the calcium lowering principle originated in the thyroid, rather than in the parathyroid. CT is synthesized in parafollicular C-cells (calcitonin cells)present in the connective tissue around thyroid follicular cells.

Calcitonin is a peptide hormone and is very much distinct from T_4 and T_3. CT is associated with calcium and phosphate metabolism. It acts by causing a transfer of calcium from blood into bone, by increased calcification of the bones or by diminished decalcification, or by both mechanisms. This function of CT is opposed by the hypercalcemic hormone produced by the parathyroid. Calcium homeostasis will be discussed in detail under the parathyroid gland.

PARATHYROID GLAND

The *parathyroid glands* are small, yellowish and rounded bodies located postero-laterally on the surface of the thyroid gland (Fig. 11.5). Typically, four glands are displayed, although more or fewer numbers may also be observed. They are composed of cords of secretory epithelial cells, separated by small number of fat cells (Fig. 11.7). The secretory epithelial cells are composed of two types of cells, relatively small *principal cells* (*chief cells*), and the *oxyphil cells*. The *chief cells* are the source of the *parathyroid hormone* (PTH) or parathormone (PH) and are rich in glycogen. The hormone is a peptide. The oxyphil cells are rich in mitochondria but lack glycogen.

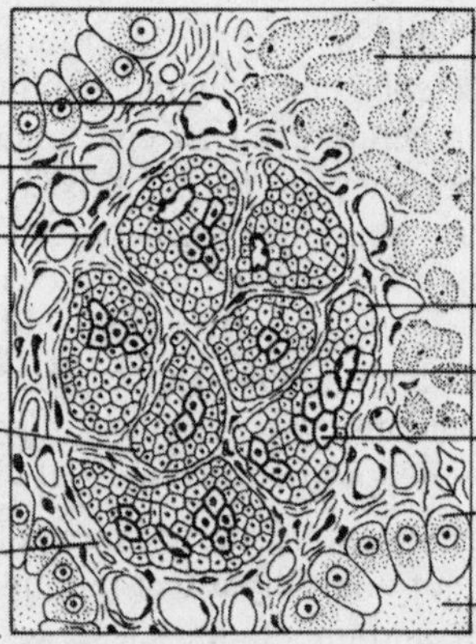

Fig. 11.7. A section of thyroid and parathyroid glands. Note that the parathyroid gland lies near the surface and is surrounded on three sides by the thyroid follicles.

The *parathyroid hormone* functions to regulate the normal level of calcium in the blood and body fluids. Oversecretion of PTH causes withdrawal of calcium from the bones and thus increase the calcium level in the various body fluids. In this condtion, excretion of calcium *via* the urine is increased, and it may be deposited in the kidney, bladder or soft tissue (specially in the large arteries). Withdrawal of calcium from the bones weakens them and may cause pronounced deformity. Undersection of PTH is accompanied by a low blood calcium level which may produce *tetany* characterized by severe muscular spasms and convulsive seizures. However, any condition that tends to reduce ionized calcium in body fluids can cause tetany (such as alkaloids, resulting from over-medication with bicarbonate of soda or other alkali), Tetany can be relieved by administration of either PTH or calcium.

One of the most important constituents of internal environment is the calcium ion. An optimum calcium ion concentration is necessary for the normal transmission of nerve impulses, the contraction of the muscle, coagulation of blood, the fertilization of egg, and the formation of the bones. Therefore, a precise endocrine control mechanism has been evolved to ensure a stable concentration of this essential ion. The three hormones of importance in *calcium homeostasis are vitamin* D, PTH and CT.

The term vitamin D denotes a group of closely related chemicals that regulate absorption of ingested calcium from the intestine. Vitamin D_3 is formed by the action of ultraviolet radiation (usually form sunlight) on 7-dehydrocholesterol found in the skin. Vitamin D_3 can be called a hormone since it was made in the skin (which is an organ). However, because man no longer wanders nude and/or seminude, vitamin D_3 is now no longer available from the skin. Therefore, man is dependent on dietary source of vitamin D. This is the second form of vitamin ingested in food. It is for this reason it has been classified as a vitamin. There is hardly much difference between the structures of ingested vitamin D and D_3. Therefore, no distinction, will be made between the two in the following description.

The plasma vitamin D_3 is metabolized by addition of hydroxyl group first in the liver and then in the kidneys to 1, 25-dihydroxy vitamin D_3 (abbreviated 1, 25- $(OH)_2$ D_3), which is the active form of vitamin D. It may be noted here that 1, 25-$(OH)_2$ D_3 is really a hormone, not a vitamin, since it is made in the body. The significant role of 1, 25 – $(OH)_2$ D_3 is to stimulate the intestine to actively absorb the calcium ingested in food. Consequently, vitamin D deficiency leads to decreased intestinal calcium absorption, resulting in decreased plasma calcium. The blood concentration of 1, 25- $(OH)_2$ D_3 is subject to physiological control. The control mechanism operates when the second hydroxyl group is added to vitamin D_3 in the kidney. The enzyme which catalyzes this step is stimulated by PTH. Thus, 1, 25 – $(OH)_2$ D_3 and PTH overlap in function and work synergistically (reinforce each other's activity).

The secretory activity of the parathyroid is controlled by the concentration of the calcium ion itself. When the concentration of calcium ion falls, the glands are stimulated to secrete more hormone, which enhances the production of $1, 25-(OH)_2 D_3$ (1, 25-dehydroxyvitamin D_3).

PTH acts on the intestinal cells to increase the absorption of calcium. It acts on the cells of the kidney tubules to increase resorption of calcium from the filtered urine. These processes contribute to the restoration of plasma calcium levels to normal. PTH also acts upon the bone cells to convert them to osteocalastic cells, which breakdown the non-exchangeable bone and release the stored calcium. All these effects lead to an increase in the concentration of calcium in the internal environment. PTH increases excretion of phosphates through urine and, thus, diminishes the amount of phosphate in the internal environment.

Under normal conditions calcium is excreted primarily in the feces and phosphate primarily in the urine. $1, 25-(OH)_2D_3$ reduces the excretion of each by promoting their retention in the body. When vitamin D is lacking, there is a declinc in the total amount of calcium and phosphate in the internal environment. When this occurs in children, the bone matrix protein no longer becomes calcified and eventually the bone stops growing, bends and breaks. This is the condition known as rickets.

PANCREAS

The *pancreas* is both an *exocrine* and *endocrine* gland. The former quality result in the formation of many digestive enzymes and the latter in the regulation of carbohydrate metabolism. It lies transversely below the stomach between the curve of duodenum and spleen (Fig. 11.1). Pancreas is a compact and lobulated organ. The exocrine function of the pancreas is accomplished by epithelial cells grouped into hollow spheres (*acini*); Fig. 11.8. These are drained by ducts and open into the upper duodenum via the pancreaticoduodenal ducts.

Scattered throughout the pancreas are small islets (groups of specilized cells) which secrete the hormone—*insulin* and *glucagon.* The specialized cells are called *islets of* Langerhans (Fig. 11.8). The islets contain at least four distinct cell types: *A (or alpha), B (or beta) D, and F* cells, each of which secrete a different hormone. Insulin is formed in the B-cells and glucagon is formed in the A-cells. There are two other pancreatic hormone *somatostatin* and *pancreatic polypeptide* by the D cells and F cells, respectively, but the functions of these hormones are not known. B-cells are more abundant; A cells are less abundant and tend to be arranged about the periphery of the islets. Tail of the pancreas lacks A-cells.

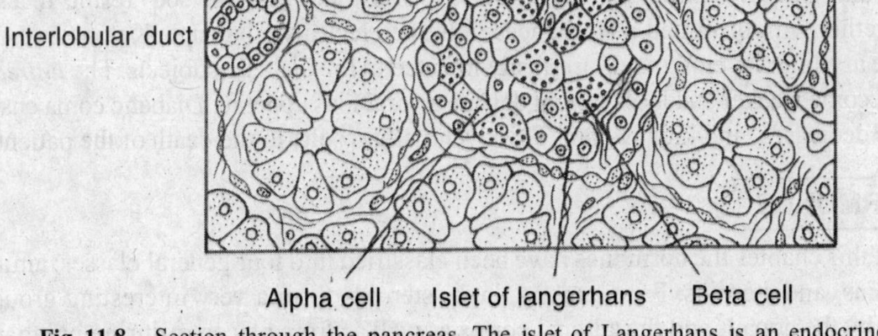

Fig. 11.8. Section through the pancreas. The islet of Langerhans is an endocrine gland whereas the surrounding acinar tissue forms an exocrine gland.

The *insulin* molecules secreted by the B-cells are extremely complex consisting of 51 amino acids. Insulin has manifold influences in the body. It acts directly or indirectly to affect many kinds of biochemical processes. An overall effect of the hormone is to facilitate the utilization of glucose at the cellular level. It also prevents excessive breakdown of glycogen stored in the liver and muscles. It has also important regulatory actions on the metabolism of fats and proteins. Insulin favours fat synthesis (lipogenesis) from glycerol acetate units. These are formed by the degradation of glucose. Insulin also inhibits lipolysis (antilipolytic).

Insulin facilitates the movement of amino ocids in the cells. Thus, it promotes a positive nitrogen balance, and favours the synthesis of protein (proteogenesis). At the same time it also acts as an anti-proteolytic (discourages excessive breakdown of tissue protein) agent. This action is similar to the action of GH and testosterone. Insulin is, in all respects, an anabolic hormone.

Five factors regulate the secretion of insulin: (1) hyperglycemia (increased blood sugar levels), (2) carbohydrate-rich food in stomach and duodenum, (3) high-protein meal, (4) glucagon and (5) activation of parasympathetic neurons. Both epinephrine and non-epinephrine inhibit the secretion of insulin. Epinephrine is more effective of the two catecholamines. However, the catecholamines are not involved in the normal glucose homeostasis. Glucagon is a much more effective regulator of glucose release from the liver than is epinephrine.

It may be noted here that insulin-activated receptor is a protein kinase enzyme that catalyzes its own phosphorylation (using ATP). This phosphorylation further activities the receptor's kinase activity. How the activated receptor then influences the cells ultimate response to insulin (also termed as signal transduction mechanism) is still unclear. *However, insulin is one hormone that does not utilize the adenylate cyclase-cAMP second messenger system. In fact, insulin, inhibits adenylate cyclase, thereby decreasing intracellular cAMP concentration.*

Glucagon is a straight chain polypeptide consisting of 29 amono acids. Its secretion is stimulated by hypoglycemia (decreased blood sugar levels). The actions of glucagon are opposed to that of insulin. Glucagon, unlike insulin, uses the cAMP second messenger system to influence its target cells. Glucagon produces hyperglycemia. Acting in the liver, it stimulates glycogenolysis (glycogen breakdown), gluconeogenesis (production of glucose from non-carbohydrate sources) and ketone synthesis Glycogenolytic effect of glucagon is similar to the action of epinephrine and ACTH. Glucagon also produces lipolysis in adipose tissue.

The normal blood glucose of fasting adult persons is 80 mg/100 ml. Values of blood glucose ranging between 150-200 mg/100 ml in a fasting individual indicate a disorder of the body known as *diabetic state.* It is produced by the inhibition of insulin secretion from B-cells of pancreatic islets. Some of the symptoms of diabetes are: rise in blood sugar and appearance of sugar in urine (glycosuria). The diabetic person passes large amounts of urine (polyurea) and consequently, is thirsty and consumes much fluid (polydipsia). Since tissues cannot utilize glucose normally, even though they need fuel, the diabetic individual is constantly hungry and tends to eat excessively (palyphagia). There is an increase in urinary elimination of nitrogen; the continued loss of water from the body results in a state of dehydration together with profound disturbances in tissue electrolyte balance. Ketone bodies in blood and urine are at a high level. Wounds are easily infected in the diabetic subjects. The *untreated* person gradually become weaker and loses weight despite a voracious appetite. Diabetic coma ensues, the plasma volume decreases, and kidneys fail. This eventually results in the death of the patient.

STERIOD HORMONES

In the beginning of this chapter the hormones have been classified into four general classes: amines, peptides and proteins, and steroids. From among these, steroids form a very interesting group of hormones secreted by the gonads and the adrenal cortex as well as from placenta during pregnancy.

Because the different steroid hormones are closely related with each other, a brief chemistry, biosynthesis, and activity-relationship are outlined here.

The most commonly occurring animal steroid is *cholesterol*. Cholesterol is in fact a major constituent of all normal tissue. An 80 kg man has about 240 g of cholesterol. The name cholesterol is derived from the fact that it is a common constituent of human gallstones and is deposited in the bile (chole) duct. It is a metabolic precursor of the five major classes of steroid hormones, they are shown in Table 11.3.

Table 11.3 Cholesterol and Steroid Hormones

Groups	*Principal Compound*	*Principal Source*
Sterol	Cholesterol	All tissues and cells
Estrogen	Estradiol	Ovary: follicles
Androgens	Testosterone	Testes: interstitial cells
Progrestogen	Progesterone	Ovary: Corpus luteum
Mineralocorticoid	Aldosterone	Adrenal cortex: zonaglomerulosa
Glucocorticoid	Cortisol	Adrenal cortex: zona fascicularis

The *estrogens* are steriods produced by the developing ovarian follicles. The most active naturally occurring estrogen is *estradiol* 17β. The *androgens* are produced by the interstitial cells of the testes. *Testosterone* is the principal androgen in mammalian testes. The principal representative of *progestogen* is *progesterone.* It is formed by the *corpus luteum.* (*Progresterone delays ovulation; this property of the hormone is used to develop orally active synthetic contraceptive agents—pills, which prevent pregnancy*). The *adrenocortical steroids* are basically progesterone derivatives. The names *mineralocorticoid* and *glucocorticoid* given to adrenal steroids refer to the two distinctive type of biochemical activity. *Cortisol* is the most prevalent natural glucocorticoid. *Aldosterone,* on the other hand, is a principal mineralocorticoid.

As regard the biosynthesis of steroid hormones, the pioneering work of R. Schoenheimer, D. Rittenberg, and K. Birch (Nobel prize winners in medicine and physiology, 1964) established that most tissues can make cholesterol from *acetate* (CH_3COO) as precursors. The machinery for building cholesterol are the gonads, adrenals, liver and skin. The serves of steps from cholesterol to other steroids produced in the adrenals, ovaries and testes can be summarized as under.

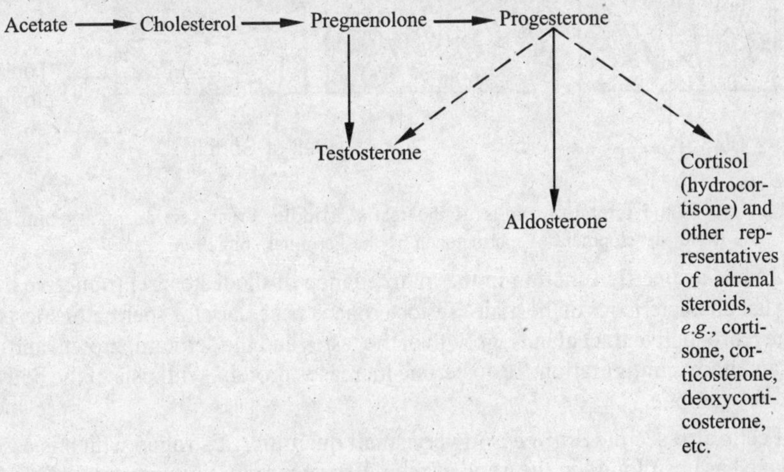

As shown above, the pathway for testosterone biosynthesis is from cholesterol through *pregnenolone* or progesterone. Estrogen formation proceeds primarily from androgens. Therefore, numerous tissues, including the ovary, placenta, and adrenals have the ability to convert testosterone to estrogenic derivatives. The adrenal steroids also originate from cholesterol *via* pregnenolone and progesterone.

The metabolized steroids are excreted by the kidneys, with the consequent appearance of urinary 17-ketosteriods. The adult male excrete 8-30 mg of 17-ketosteroids/day, whereas the normal adult female only about 5-20 mg. A sharp three-fold increase in production and excretion accompanies puberty.

THE SEX HORMONES

Testes

The *testes* are paired glands enclosed in the *scrotal sacs* (Figs. 11.1 and 16.1). (*However, in a number of mammals the testes remain in the abdomen throughout life, e.g., elephant, whale, dolphin, etc*). The testes not only produce *sperms* but also secrete the male sex hormone *testosterone.* Each testes is composed of long *seminiferous tubules* containing sperms in various stages of development. The heads of theofsperms are embedded in *Sertoli cells* where they complete their maturation. Sertoli cells secrete two protein hormones: *inhibin and Mullerian inhibiting hormone* (MTH). Outside the seminiferous tubules are masses of cells the *interstitial cells of* Leydig (Fig. 11.9). These are the sources of *androgens: testosterone* and very small amount of *estrogens.*

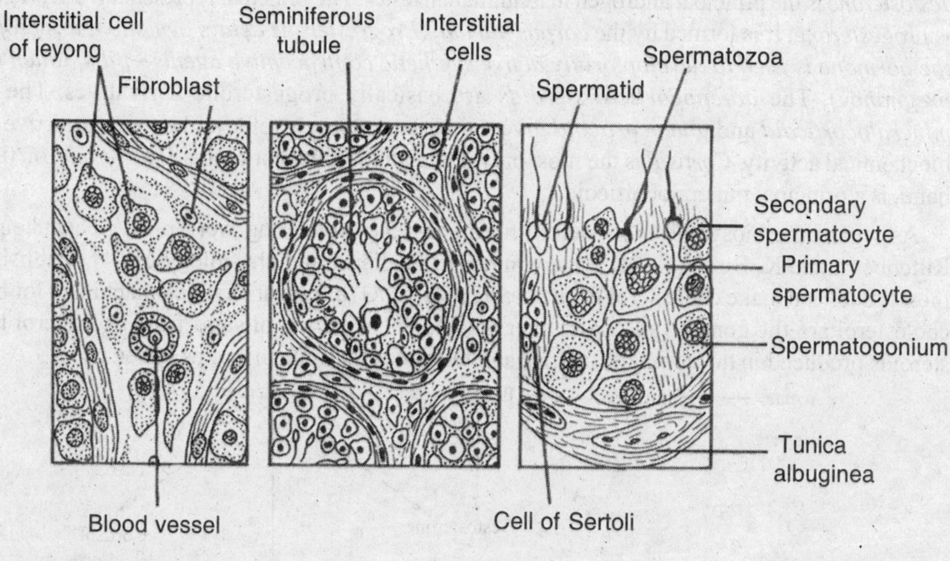

Fig. 11.9. Left, the interstitial cells of the testis. Middle, cross-section of a semniferous tubule. Right, development of spermatozoa by the germinal epithelium.

Testosterore has a specific function in the maintenance of the testes, reproductive tract, and all secondary sexual characteristics of the male. Testosterone is necessary for spermatogenesis, continued function of the reproductive tract glands, growth of the testes and the scrotum, growth and distribution of body hair, and body configuration. Testosterone increases protein synthesis in the body and hence it is an *anabolic* hormone.

Interstitial cells also secrete *estrogens* in very small quantities. Estrogen which is essentially the female hormone has been found in the urine of men. Estrogens can cause reversal of sex in the male with the formation of an *ovo-testes* or even complete feminization.

The two protein hormones, *inhibin* and *Mullerian inhibiting hormone* (MIH) secreted by the Sertoli cells peroform two very important functions. It has already been described that FSH acts only on Sertoli cells. The secretion of FSH is inhibited by inhibin. Mulberian inhibiting hormone is secreted by Sertoli cells of the fetal testes which causes Mullerian duct system to degenerate, and testosterone from the Leydig cells causes the Wolffian ducts to differentiate into epididymis, was deferens, ejaculatory duct, and seminal vesicle. (*It may be pointed out here that prior to the functioning of the fetal glands, the primitive reproductive tract includes a double genital duct system — Wolffian ducts and Mullerian ducts, and a common opening for the genital ducts and urinary system to the outside. In developing fetus, with the differentiation of testes by the presence of testes — determining factor (TDF) on Y-chromosome, only the wolffian ducts persist, whereas in females the Mullerian ducts persist and transform into uterus and uterine tubes because of the absence of MIH. Moreover, absence of testosterone in the ovary of the female fetus causes the regression of Wolffian ducts. Therefore, there is no duct system in females; instead, tissues give rise to vagina and female external genitalia rather than male structures).*

Hormone LH from the anterior pituitary acts only on the Leydig cells to control the production and release of testosterone. The secretion of LH is inhibited by testosterone by a feedback loop.

Ovaries

The *ovaries* are small oval structures located in the upper part of the pelvic cavity, one on each side of the uterus (Figs. 11.1 and 16.3) Ovaries are composed of *follicles* in various stages of development varying from *primary* to mature (*Graafian*) follicles. Distributed among the follicle cells is the interstitial tissue (*stroma*) of the ovary. In adult mammals the ovary contains degenerating follicles, *corpora lutea* (formed after ovulation), and *corpora albicans* (the residue of corpora lutea)-Fig. 11.10. The ovaries produce two hormones *estrogens* and *progesterone.* Two other peptide hormones, *inhibin* and *relaxin* are also secreted by the ovaries.

Estrogens (a group of steroid hormones, namely, estradiol, estrone and estriol of which estradiol is the major estrogen secreted by ovaries) are produced during follicular growth by steroid-secreting cells among the fibroblasts of *theca interna.* After the release of the ovum it is also secreted by the remnant of the follicle *corpus luteum.* It accounts for the increased uterine growth, and growth of the vagina at puberty. In also stimulates the development of secondary sex organs such as breasts, myometrium, etc., distribution of body fats (for instance, the development of the hip pads), and thickening of the uterine cavtity's lining in preparation for embedding of the fertilized ovum. Estrogen also acts in the repair of the endometrium (inner mucous membrane lining of the uterus) following menstruation. Over-secretion of estrogen causes menstrual cycle irregularities and atrophy or under-development of the breast and the uterus.

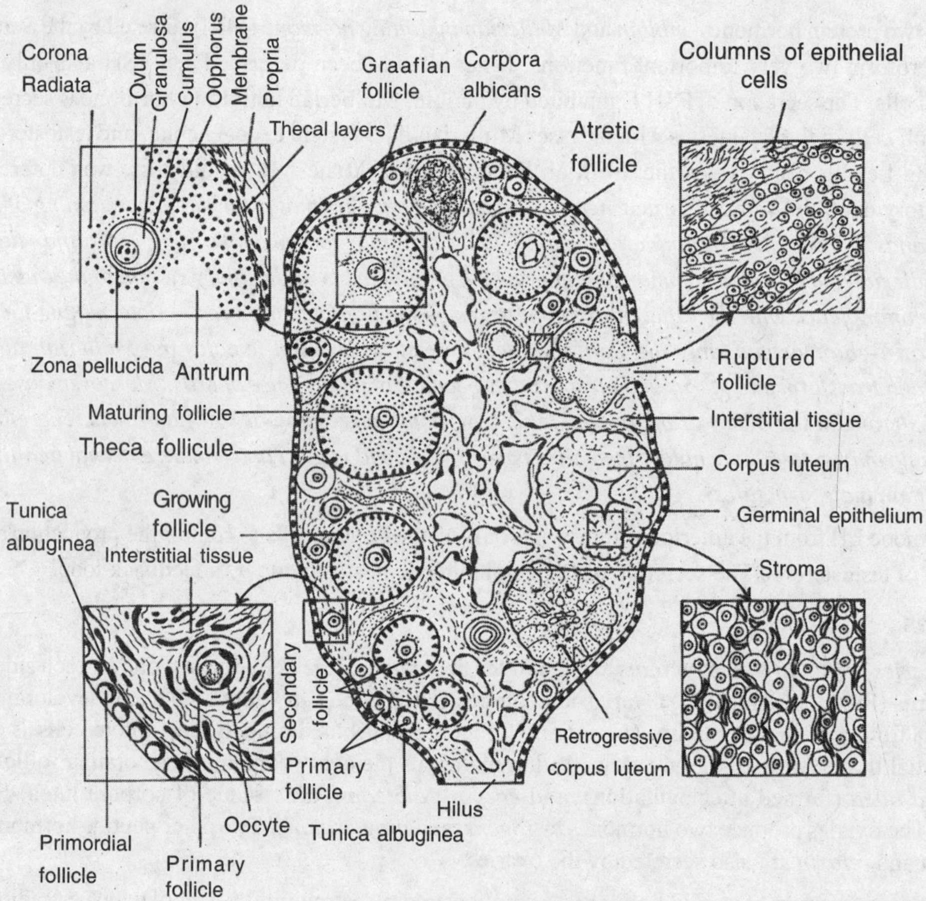

Fig. 11.10. Diagram of composite mammalian ovary. Ordinarily all these stages would not be seen at any one time during the cycle. Progressive stages in differentiation of Graafian follicle shown on the top the mature follicle may become atretic (right) or ovulate and undergo leutenization (bottom left).

Progesterone is secreted by the *corpus luteum* (a recurring structure associated with menstrual cycle) and by the placenta during pregnancy. Progesterone plays a number of important roles in the reproductive life. It speeds the movement of the ovum through the fallopian tubes, and prepares the uterine endometrium for implantation. It also inhibits contractility of the uterine myometrium, thus allowing the pregnancy to continue. Decreased secretion of progesterone cause abortion in pregnant women, and menstrual irregularities in non-pregnant women. Progesterone also prevents the ovary form releasing additional ova during pregnancy. It is also essential for the development of mammary glands. However, it requires three hormones for the final secretion of milk. In the presence of estrogens which causes lactiferous duct growth, progesterone causes the development of glandular portions—the alveoli. Production of milk requires prolactin as well as ACTH, and GH.

The inhibin hormone secreted by the *granulosa cells* of *ovarian follicles* inhibits FSH secretion (as in males), where as the secretion of both FSH and LH is inhibited by estrogen (unlike males where testosterone mainly inhibits LH). *Relaxin* is mainly secreted by the *corpus luteum.* The major function of this hormone is cervical softening (cervix becomes soft and cervical glands extensible by the break up of collagen fibres during delivery of the infant (*parturition*) by the mother. Indeed, estrogen and prostaglandins, too, help in this process.

Curiously, ovaries also produce *testosterone.* But it is converted into estrogen. If over-production of testosterone occurs, it may cause development of masculine characters in females, such as deep voice and beard (*hirsutism*). Testosterone may cause *sex reversal* in females with the formation of *ovotestes* or even complete transformation to a male.

Mention may be made here of two other hormones, namely, *chorionic gonadotropin* (CG) and *placental lactogen*. The former is a protein hormone secreted by trophoblastiç cells of blastocyst and placenta. It maintains secretary activity of corpus luteum during first three weeks of pregnancy. Placental lactogen is produced by placenta and had effects on breast development and organic metabolism (similar to both prolactin and growth hormone) in the mother.

ADRENAL GLAND

In the mammals, the *adrenal glands* are oval masses embedded in fatty tissue resting near the anterior poles of the kidneys. The adrenal gland is composed of an outer *cortex* and inner *medulla* (Fig. 11.11). The cortex is divided into three layers which secrete five own steroid hormones, namely, *dehydroepiandrosterone, androstenedione, corticosterone, cortisol*, and *aldosterone*. The medulla is responsible for the synthesis of epinephrine (E; adrenaline) and norepinephrine (NE; noradrenaline) which are catecholamines.

Adrenal Cortex

The outermost cortical layer, the *zona glomerulosa,* consists of columnar cells responsible for the secretion of *aldosterone*—the mineralocorticoid because it effects salt (mineral) metabolism on kidneys handling of sodium, potassium and hydrogen ions. Zona glomerulosa possesses enzymes for the conversion of corticosterone to aldosterone, but not to corticosterone. The next two layers are *zona fosciculata* and *zona reticularis* (Fig. 11.11). The latter surrounds the medulla. The cells of the two inner layers of the cortex synthesize the *glucocorticoids* namely, cortisol and corticosterone. The name glucocorticoids refers to the role of these hormones as they affect metabolism of glucose and other organic nutrients. Cortisol is far the more important than corticosterone. Synthesis of the latter is under the control of the adenohypophysial ACTH by the negative feedback control.

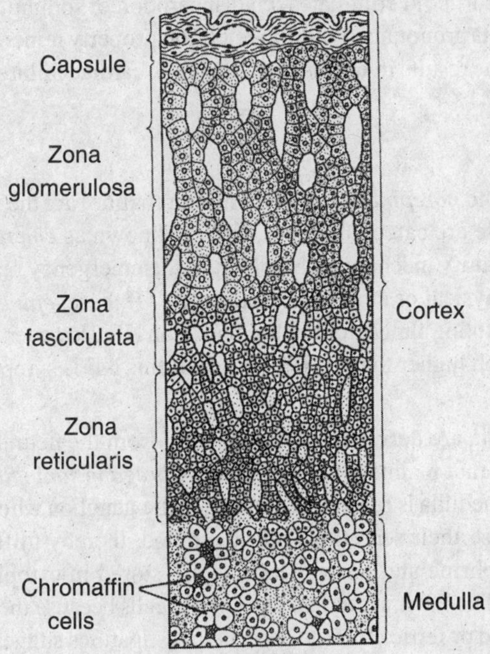

Fig. 11.11. Histological structure of the adrenal gland.

Dehydroepiandrosterone and androstenedione are androgens (similar to testostenone produced by Leydig cells in the testes). The two adrenal cortical androgens share similar roles with testosterone, but the adrenal androgens are much less potent and have very little physiological significance in the male. They do however, play roles in the female in 'sexual drive' or sexual desire which is maintained beyond the menopause.

The *glucocorticoids* reduce amino acid uptake in many tissues. They also have a catabolic action on proteins. The amino acids derived from the protein degradation serve as substrates for gluconeogenesis rather than protein synthesis. This causes rise in blood sugar level resulting in *steroid diabetes*.

In some instances hypersecretion of corticosteroid results in a reversal of the synthetic process. During this androgens are synthesized from the steroids leading to *adrenal virilism* in women and precocious sexual development in children. The masculanization process in the female is accompanied by growth of beard, atrophy of breasts, and inhibition of ovulation.

Glucocorticoids such as *cortisone* and *hydrocortisone* also possess sodium retention activity. Removal of the adrenal gland or Addison's disease (adrenal insufficiency) leads into a profound disturbance of electrolytes and water balance. Sodium is excreted and potassium is retained. There is also an immense loss of water from the circulation. This results in fall of plasma volume (hemoconcentration), blood flow, body temperature, oxygen uptake and blood pressure (by decreased cardiac output). Urea and plasmaproteins increase in the blood.

Glucocorticoids also have inflammation-inhibiting effects. As a result, corticoids are also related to *immunizing mechanisms*. This is achieved by the release of plasma γ-globulins from lymphoid tissue which provide the substrate for the formation of *antibodies*. Glucocorticoids have anti-allergic activities. Allergy is often accompanied by an increase in the number of eosinophils in the circulating blood. The corticoids reduce eosinophils.

Aldosterone—the mineralcorticoid stimulates renal absorption of sodium, hydrogen, ammonium and magnesium ions. Due to its pronounced sodium retention property it increases water absorption. The effect of excess aldosterone results in edema, hypertension, cardiac failure and extreme muscular weakness.

Adrenal Medulla

Epinephrine (E;adrenaline) and *norepinephrine* (NE; noradrenaline) are the two hormones secreted by the *adrenal mudulla.* These are catecholamines, and are known as *emergency hormones.* They constitute the principal regulatory mechanism involved in the emergency reactions that premit the animal to mobilize to meet physical or emotional challenges. Of these, *epinephrine* is the principal hormone of the *medulla* constituting almost four times more than NE. However, under basal conditions, the plasma level of NE is much higher than E. The source of this NE is sympathetic neurons not the adrenal medulla.

Cells of the adrenal medulla are derived embryologically from the neural crest tissue. They are *modified postganglionic cells* that maintain contact with the *preganglionic fibres* of the sympathetic nervous system. The adrenal medulla is really like a sympathetic ganglion whose cells do not send out nerve fibres but instead release their secretions into the blood, thereby fulfilling a criterion of an endocrine tissue. Both epinephrine and norepinephrine are stored in cytoplasmic granules of the *chromaffin cells* (Fig. 11.11). The latter are called chromaffin cells because they give colour reactions when treated with chromic acid or ferric chloride. Recent investigations suggest that epinephrine and norepinephrine are formed by different cells of the medulla.

As in the case of oxytocin and vasopressin, release of epinephrine and norepinephrine is under the direct neural control. Stimulation of the splanchnic nerve (a sympathetic nerve) causes a secretion of these hormones. On the other hand, cutting the nerve or application of adrenergic blocking agents greatly reduces their secretion.

Epinephrine exerts its effect on carbohydrate metabolism. There is a rapid rise both in blood sugar and lactate at the expense of liver glycogen and muscle glycogen. The hyperglycemic effect thus opposes insulin actions. Epinephrine inhibits insulin secretion and stimulates glucagon secretion. Epinephrine also increases the free fatty acid levels of blood by promoting lipolysis in adipose tissue. The overall effect of the availability of these oxidisable substances is to provide a quick source of glucose so that energy release is increased. These responses constitute *flight or fight* responses under stressful situations.

Epinephrine causes a redistribution of blood in the body by constricting some vessels and dilating others so that the blood supply to the skin is decreased while the blood supply to skeletal muscles and heart is increased.Sinoatrial node is activated by it which increases heart rate. The animal is thus better prepared to face environmental stresses. The brain and the lungs are relatively unaffected.

It is interesting to note that hypofunction of the adrenal medulla is unknown. However, hyperfunction may occur. It is induced by certain tumors. This increases the plasma level of epinephrine and norepinephrine many times and results in persistent hypertension.

THYMUS GLAND

The *thymus* is a flat, pinkish grey bilobed gland located in the chest behind the sternum (Fig. 11.1). It is large in the fetus and early childhood, but stops growing after puberty; thereafter, it begins to atrophy, and replaced by fatty tissue.Prior to atrophy, the thymus consists mainly lymphocytes. These lymphocytes are liberated into the circulation and migrate to peripheral lymphoid organs such as lymph nodes, spleen and tonsils. Thymus also contains endrocine cells that secrete a group of hormones, known collectively as *thymosin* (or *thymopoietin*). The regulatory function of the hormone on lymphocytes of thymic origin is poorly understood.

As the lymphocytes mediate specific immune responses, it is worthwhile to briefly describe how these cells arise and the organs and tissues in which they come to reside and their important role in defenses against foreign matter (*antigen*); see Table 11.4. There are three major classes of lymphocytes: the *lymphocytes* or *T cells*, *B – lymphocytes* or *B cells* and the *natural killer cells* or NK cells. The T lymphocytes are produced by the thymus and their production is completed early in life. They are carried by the blood to the lymphoid organs. All generations of lymphocytes that subsequently arise from the T cells by mitosis in the peripheral lymphoid organs are identical to the parent cells, that is, they all will be T cells. The maturation process occuring in the thymus for all T cells is not identical and, as a result, a number of subclasses of T cells exist. Based on their functions these are called *cytotoxic, helper*, and *suppressor T cells*. The cytotoxic are suppressor T cells contain a protein, known as T_8, in their plasma membrane, whereas T_4 protein is found in the helper T cell plasma membrane. Consequently the corresponding T cells are termed as T_8 and T_4 cells. The B lymphocytes are produced and mature in the bone narrow and then carried by the blood to the peripheral lymphoid organs. B cells also undergo mitosis and their offspring are identical to B cells. They differentiate into *plasma cells* during immune response. In addition to T and B cells, there is third class of lymphocytes, also referred to as "*third population*". This class has several subclasses, but the most important one is called natural killer cells. The origin of these (third population) is still not clear.

Table 11.4 Summary of Lymphocyte mediating Immune Responses

Lymphocytes	*Site produced*	*Functions*
B cells	Bone marrow	Differentiate into antibody-secreting plasma cells
T cells Cytotoxic T cells	Thymus	Directly attacks specific antigen-bearing cells, major killer of viruses and tumor cells.
Helper T cells		Enhances antibody production, secret lymphokines (a protein) which enhances cytotoxic T- and NK-cell function
Suppressor T cells		Inhibits antibody production and cytotoxic T cell function
"Third population" NK cells	Unknown	Bind nonspecifically to cells bearing foreign antigens and kill them directly

A description and discussion of immune response is beyond the scope of this chapter. Readers are advised to consult books on Immunology.

THE PINEAL GLAND

The *pineal gland* is a minute white structure, an outgrowth from the roof of diencephaon, deeply embedded between the cerebral hemispheres (Fig. 11.1). It secretes a hormone *melatonin* which is *N-acetyl-5-methoxy-tryptamine* synthesized from *serotonin.* Melatonin does not fit in the definition of hormones. It is not certain that metatonin reaches target cells via the blood which is an essential criterion for classification as a hormone. In fact, melotonin acts as a paracrine and autocrine. It is therefore called a *candidate hormone*. The function of pineal gland and melatonin is largely unknown in humans. The secretion of melatonin is stimulated by sympathetic postganglionic neurons and it undergoes a 24-hour cycle, being high at night and low during the day. The possible relationship of melatonin to "winter depression", its ability to reduce symptoms of jet-lag, and its possible role in the onset of puberty are conjectural and being currently probed. All these suspected functions ascribe the pineal acting as a "biological clock".

GASTROINTESTINAL HORMONES

The hormones that control the gastrointestinal system are secreted mainly by endocrine cells scattered all over the epithelium of the stomach and small intestine and not formed into organs like other endocrine glands. Each endocrine cell is exposed to the lumen of the gastrointestinal. At this surface, various chemical substances present in the chyme stimulate these endorcine cells to release the hormones from the opposite, i.e., the blood side of the cells. In some cases the secretions of these cells have been detected in the lumen of the gastrointestinal tract and act as paracrines.

A number of hormones are seemingly involved in the process of digestion, but only four can qualify the criterion of hormones. These hormones are *secretin*, *cholecystokinin* (*CCK*), gastrin, and glucose insulinotropic peptide (GIP) Secretin, a peptide hormone, is secreted by anterior duodenal mucosa when acid chyme enters the duodenum. The hormone stimulates pancreas to secrete bicarbonate into the small intestine. Cholecystokinin, secreted by duodenum, is a peptide hormone that regulates gastric motility and secretion, gall bladder contraction, and pancreatic enzyme secretion. Cholecystokinin represents an example of a hormone acting directly upon muscular tissue. Gastrin, another peptide hormone, is secreted by antral region of stomach when the stomach is distended with food. The hormone stimulates gastric acid secretions. The glucose insulinotropic peptide, the fourth major peptide hormone, also called a gastric inhibitory peptide. GIP, is an intestinal hormone; it stimulates insulin secretion in response to glucose and fat in small intestine. *Somatostatin* is yet another hormone suspected to be secreted by stomach, but its role is unclear.

RENAL HORMONES

The kidneys are the source of two hormones: erythropoietin and 1, 25-dihydroxyvitamin D_3. Kidneys are also the site of secretion of an important enzyme – renin which, in fact is synthesized and secreted by the *juxtaglomerular apparatus* (JGA) a renal structure consisting of macula densa and granular cells. Renin splits off a small polypeptide *angiotensin* I from a large plasma protein *angiotensinogen* synthesized by the liver. Angiotensin I then undergoes further cleavage to form angiotensin II. This conversion is mediated by an enzyme called *converting enzyme* found in luminal walls of capillary endothelial cells of lungs. Therefore, conversion of angiotensin I to angiotensin II occurs in the lung capillaries. Angiotensin II, therefore, qualifies as a hormone. Renin acts as a rate limiting factor in angiotensin II formation. This system is called *renin-angiotensin system*. The output of renin decreases by an increase in renal pressure and by increased levels of plasma Na^+ and angiotensin II. The latter indirectly affects kidney function by increasing aldosterone secretion by the adrenal cortex which leads, in turn, to an anti-diuretic effect (increased Na^+ reabsorption). This is referred to as the renin-angiotensin-aldosterone system (Fig. 11.12).

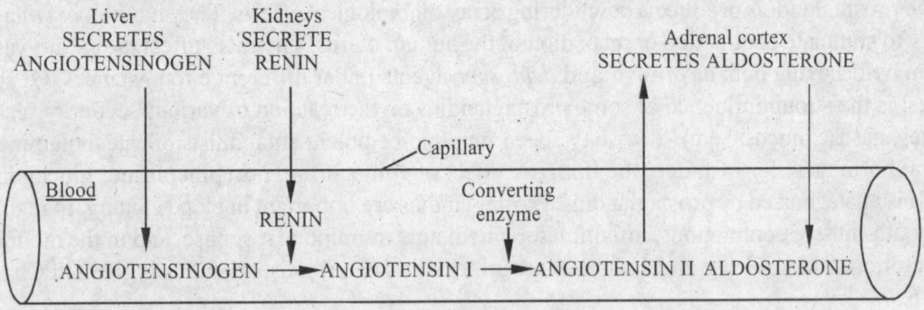

Fig. 11.12. Renin-angiotensin-aldosterone system. Converting enzyme present on the surface of capillary endothelial cells, specially in the lungs. Renin acts as rate-limiting factor in the system.

Erythropoietin is a hormone secreted by the cells in the kidneys which directly acts on the bone marrow to stimulate the proliferation of precursor cells and their maturation into erythrocytes. The stimulus for increased erythropoietin secretion results from decreased oxygen delivery to the kidneys. As a result of increased erythropoietin secretion, the plasma erythropoietin concentration, erythrocyte production, and the oxygen-carrying capacity of the blood cell increases, i.e., oxygen delivery to the tissues returns towards normal. It may be noted here that although iron, folic acid and vitamin B_{12} are essential for normal erythrocyte production, but it is erythropoietin which constitutes the signal that regulates production rate of erythrocytes.

As already discussed, 1, 25-dihydroxyvitamin D_3 or 1, 25-$(OH)_2$ D_3 hormone is formed by kidneys and is active form of vitamin D. The hormone stimulates the intestine to actively absorb the calcium ingested in food. This restores plasma calcium toward normal.

Day after day, new hormones are being identified and characterized and their names and functions are added to those already described in this chapter. The site of origin, names and possible roles of such hormones/factors are very briefly described below.

The liver and many other body cells secrete peptide hormones called *insulin-like growth factors* (IGF-I and IGF-II), when stimulated by a rise in plasma GH. Both IGF-I and II promote growth in target cells. A peptide hormone known as *atrial natriuretic factor*. (*ANF*) is secreted by atrial cells of the heart in response to atrial distension. ANF causes decreased renal sodium reabsorption. *Interleukin 1(IL-1), interleukin 2 (IL-2), and interleukin 6* are secreted by *macrophages* and other cells. They all are proteins- IL-1 and interleukin 6, secreted by macrophages and other cells, mediate activation of helper T cells and also mediate several systemic acute phase responses, including fever. IL-2, secreted by activated helper T cells, causes proliferation of activated helper T, cytotoxic and NK cells. *Monokines* and *tumor necrosis factor* (TNF) are protein messengers are secreted by monocytes and macrophages, act as local and hormonal inflammatory mediators. *Nerve growth factor* , a peptide hormone secreted by multiple cell types, stimulate growth and differentiation of some neurons.

PROSTAGLANDINS

A chapter on endocrine glands and hormones will be incomplete without a brief description of prostaglandins. *Prostaglandins* are *active lipids* (20-carbon hydroxy fatty acids), one of a group of eicosanoids, that occur in many mammalian tissues, *viz*. lung, thymus, brain, spinal cord, kidney, iris, menstrual fluid and placenta. However, one of the richest sources of prostaglandins is the *semen*. They are produced from the fatty acid *arachidonic acid* which is present in phospholipids of the plasma membrane. The formation of prostaglandins is catalyzed by the enzyme *cycloxygenase*. Prostaglandins then act locally as paracrines and autocrines. Whether they can act as hormones by being carried by blood, remains an unresolved problem.

The prostaglandins produce a bewildering array of biological effects. They have an *oxytocin-like ability* to stimulate contraction or relaxation of the human uterus. They also affect the cardio-vascular system by behaving both as *pressor* and *depressor* agents under different circumstances. Of special interest is the strong influence of some prostaglandins on the reaction of various secretory reactions (kidney, gastric mucosa, etc). They have been found to be potent antagonists of catecholamines and several hormones, for instance, the lipolytic effect of epinephrine, norepinephrine, glucagon, and ACTH is anatagonized by prostaglandins. Prostaglandins are important in blood clotting, in regulation of smooth-muscle contractions, inflammatory mediators in immune response, and in the modulation of neurotransmitter release and action. More information about prostaglandins is given in Chapter 4 and 16.

CHEMICALS AS INTERCELLULAR MESSENGERS IN HOMEOSTATIC SYSTEMS

The regulatory or homeostatic function of hormones need to be explained in some more detail because these, along with other chemical agents, not only act as intercellular messengers but also for local response, as described below. Most homeostatic systems require cell-to-cell communication via *chemical messengers* which, on their turn, bind to specific *receptors*, located on the cell's *plasma membrane*. The *messenger-receptor* complex results in *receptor activation* which causes the cell to respond to the messenger. How does the information, contained in the messenger-receptor combination in the plasma membrane, get relayed to cell's internal components will form a part of this discussion. Such relay mechanisms called *signal transductions* and the *signal transduction pathways*. Their use will also be explained by taking into account the role of *G proteins, second messengers, protein kinases, adenylate cyclase, phospholiipase* and other enzymes. Hopefully this discussion will avoid later confusion while understanding cell-cell signalling of hormones.

Essential to homeostasis is the ability of cells to communicate each other through chemicals as *intercellular messenger* (Fig. 11.13a, b). There are four categories of such chemical messengers: (1) *hormones* which are the products of endocrine glands, with the blood acting as the delivery system to transport the hormones to the target cell, (2) *neurotransmitters* which are released by nerve cell endings, diffuses through the extracellular fluid separating two nerve cells and etc. upon the second neuron of affect effector cells, (3) *neurohomones* which are the product of nerve cells secreted into the bloodstream to act on cells elsewhere in the body, and (4) paracrines which are synthesized and released by cells, diffuse into the extracellular fluid and exert their effect on cells and tissues near their secretion site. Paracrines are specifically secreted for the regulation of local target cells (hence sometimes referred to as "local hormones") One may be tempted to include neurotransmitters into paracrine, *but by convention neurotransmitters are not paracrines.*

All the four chemical messengers as described above, act intercellulary. But there is one class of local agent that is *not* an intercellular messenger. The general term for such a substance is *autocrine. In fact, an autocrine acts upon the very cell that secreted it.*

It must be remembered here that a neuron, endorcine gland cell, or another cell type may all secrete the same chemical messenger. Thus, a particular messenger may sometimes be called a neurotransmitter, as a hormone, and as a paracrine or autocrine.

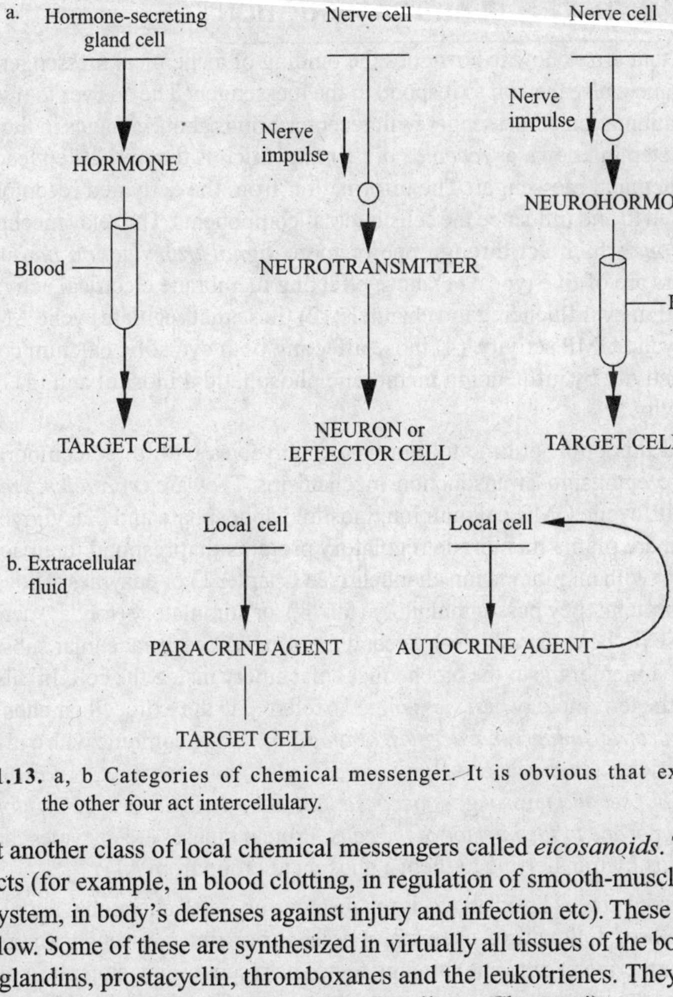

Fig. 11.13. a, b Categories of chemical messenger. It is obvious that except for autocrines, the other four act intercellulary.

There is yet another class of local chemical messengers called *eicosanoids*. These expert a wide variety of effects (for example, in blood clotting, in regulation of smooth-muscle contraction, in the reproductive system, in body's defenses against injury and infection etc). These will be described in chapters to follow. Some of these are synthesized in virtually all tissues of the body. The eicosanoids include prostaglandins, prostacyclin, thromboxanes and the leukotrienes. They all are unsaturated fatty acids that have a carbon ring at one end (For Details are Chapter 4).

Cell surface receptors resolve the question that arises now is how does an intercellular chemical messenger act on a target cell? This is accomplished by binding of the messenger on its target cell to specific protein molecules of the target cell. These molecules are known as *receptors*. The messenger forms a "*ligand*" and the receptor a "*binding site*". Receptors are located more commonly in the plasma membrane. *But in the case of steroid hormones, vitamin D_3, retinoic acid and thyroid hormones* *which can easily pass through the phospholipid plasma membrane, their receptors are in the interior of the target cell.* A single type of receptor can produce different responses in different cell types by a molecular mechanism of "switched on" by the chemical messenger binding to it. Further, a single cell contains many different receptor types for different chemical messengers. Receptors are subject to physiological regulation by their own chemical messengers. This includes *down-regulation* and *up-regulation.*

Down-regulation results in the decrease of the number of receptors in the presence of high concentration of extracellular chemical messenger. Up-regulation occurs with the cell to have many more receptors, when exposed to very low concentration of extracelluar chemical messenger. These two phenomena of down-regulation and up-regulation are made possible as a result of continuous degradation and synthesis, respectively, of recepotrs.

PATHWAY OF INTRACELLULAR TRANSDUCTION

The final question that arises now is how does the binding of a chemical messenger to a receptor on the plasma membrane cause the cell to respond to the messenger? The answer to this query has been discovered. The combination of messenger with receptor brings about changes in the conformation of the receptor. This step in known as *receptor activation* which is the initial step leading to the cell's responses to the chemical messenger. The information from the activated receptor, on the plasma membrane, is relayed to and influence the cells internal components. The relay mechanisms are called *signal transuductions* which act through one or more *signal transduction pathway.* The plasma membrane receptors are of five types: (1) those effecting membrane electrical activity and cytosolic calcium concentration by influencing ion channels, (2) those that regulate cyclic AMP formation (3) those regulating cyclic GMP activity, (4) those affecting both cytosolic calcium concentration and protein kinase C activity by influencing membrane phosphatidyl inositol and (5) those possessing protein kinase activity.

It is desirable at this point of time to familiarize ourselves with three components which play significant role in receptor signal transduction mechanisms. They are *G proteins*, *second messengers* (namely cyclic AMP, cyclic GMP, calcium ion, inositol triphosphate and diacylglycerol and *protein kinases*). G protein are plama membrane regulatory proteins that respond to an activated receptor and, in turn, interact with membrane ion channels (see Chapter 1) or enzymes. They are of two types, viz., G_i protein G_s protein; they posses inhibitory (for "i") or stimulatory (for "s") action, respectively, on adenylate cyclase (to be discused later). Second messengers are intracellular substances that serve as relay from plasma membrane to the biochemical machinery inside the cell. In this context, it may be pointed out that the terminology *first messenger* is referred to the extracelluar chemical messengers (viz., *hormones, neurotransmitters, paracrines, or autocrines*) that combine with a plasma-membrane receptor or an electric message that arives at the cell's plasma membrane. *It may be noted here that the G proteins are not second messengers because they do not themselves pass information directly from the plasma membrane to the interior of the cells.* Protein kinases are enzymes that phosphorylate certain other proteins by transferring to them a phosphate group from ATP.

The first type of receptor, as listed above, is the receptor-operated channel in the plasma membrane that is opened or closed following a chemical messenger or first messenger (namely, a hormone, neurotransmitter, paracrine or autorine) binding to it. The receptor may either be a part of the protein complex constituting the ion channel (see chapter 1) or it may affect the channel through the mediation of G protein in the membrane. The ion channel is specific for different ions and the diffusion of ion or ions is due to an electric signal. The movement of calcium ions from the extracellular fluid into the cytosol due to electrochemical gradient has a great significance when calcium acts as a second messenger; this will be discussed later in this section.

The second type of receptors generate the second messenger cyclic AMP (cAMP; 3, 5 –adenosine monophosphate; Fig. 11.14) in the cytosol. The cyclic AMP second messenger system is depicted in Fig. 11.15. The combination (hormone, neurotransmitter, or paracrine) with its specific receptor allows the receptor to bind to the membrane G protein termed G_s proetin (subscript "s" denotes stimulatory action). This protein in turn binds to the membrane enzyme *adenylate cyclase*, activates it and induces it to catalyze transformation of ATP to cyclic AMP. (The catalytic site of adenylate cyclase faces the cytosol). The cAMP activates an enzyme known as *cAMP-dependent protein kinase* which transform inactive protein kinase into active protein kinase. The active protein kinase then phosphorylates inactive specific proteins (or enzymes) to active protein by transferring phosphate group to them. This triggers biochemical reactions leading to the cell's response to the first messenger. *The cAMP, thus, acts as a second messenger*. As shown in Fig. 11.15, the action of cAMP is terminated by its breakdown to AMP by the action of an enzyme cAMP *phosphodiesterase*. (Fig. 11.16)

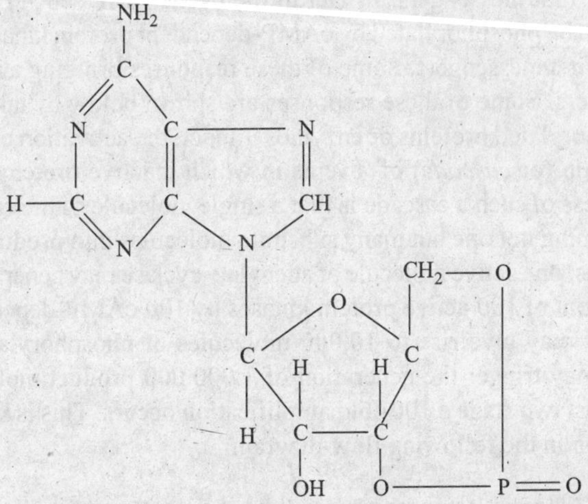

Fig. 11.14. Cyclic AMP (cAMP) in which the phosphate group is bonded internally to form a cyclic molecule; it is generated from ATP by the enzyme adenylcyclase. (see Fig. 11.16)

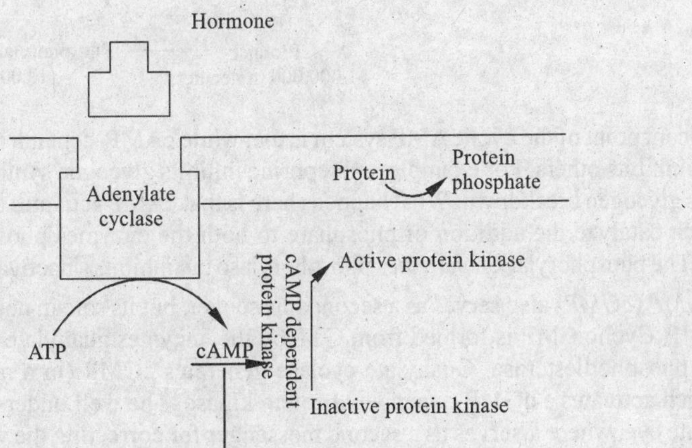

Fig. 11.15. The role of adenylate cyclase and cAMP mediating the action of a hormone. The hormonal action plasma results in an alteration in the plasma membrane. The hormone binds to membrane G_s protein which then binds to adenylate cyclase. The latter gets activated and indces transformation of ATP into cAMP.

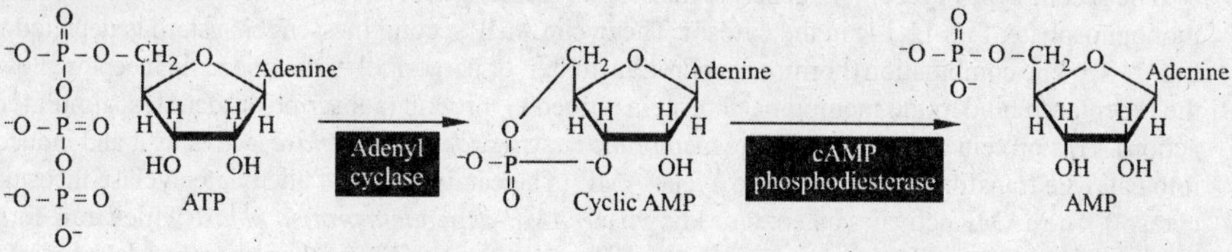

Fig. 11.16. Cyclic AMP is synthesized from ATP by adenyl cyclase (also called adenylate cyclase) and degraded to AMP by cAMP phosphodiesterase.

The activation of protein kinase by cAMP is an event common to a great variety of biochemical responses to as many first messengers. In fact in the cyclic AMP seocnd messenger system larger number of proteins can be phosphorylated by cAMP-dependent protein kinase bringing about specific resources to specific first messengers. Some of these responses bringing about specific responses to specific first messengers. Some of these responses are shown below by taking random examples of eight different phosphorylated proteins or enzymes. In fact, the activation of adenylate cyclase by G_s protein *initiates* a chain (or *cascade*) of events in which inactive proteins are converted to active proteins. The usefulness of such a cascade is that a single molecule of the enzyme adenylate cyclase is capable of transforming not one but many substrate molecules into products. Thus, for example, if there are 100 substrates, one active molecule of adenylate cyclase may generate 100 cAMP molecules, leading to the formation of 100 active protein kinases by 100 cAMP-dependent kinases. These 100 active protein kinases may give rise to 10,000 molecules of phosphorylated proteins or enzymes which on their turn, may trigger the generation of 1,000,000 product molecules. In this particular example, during the last two stage a 100-fold amplification occurs. This is known as *amplification of cAMP system* as shown in the following flow-diagram;

Messenger-Receptor (1 molecule) ⟶ Active adenylate cyclase (1 molecule) ⟶ cAMP (100 molecules) ⟶ Active Protein Kinase (1 molecules)

↓

Product (1,000,000 molecules) ⟵ Phosphorylated Protein/Enzyme (10,000 molecules)

A significant point of the cyclic AMP system is that while cAMP-dependent kinase activates certain enzymes, it inhibits others. For example, epinephrine inhibits glycogen synthesis but at the same time it stimulates glycogen breakdown. What happens here is that cAMP activates cAMP-dependent protein kinase which catalyze the addition of phosphate to both the enzymes phosphorylase and glycogen synthetase. The phosphorylation *activates* phosphorylase but inhibits (inactivates) glycogan synthetase.

Cyclic GMP (*cGMP*) also serves as a second messenger, but its role in not as clearly understood as that of cAMP. Cyclic GMP is formed from GTP by the enzymes guanylate cyclase and degraded to GMP by a phosphodiesterase. Guanylate cyclase generates cGMP (in a manner similar to that of cAMP) which activates a cGMP-dependent protein kinase. The well understood role of cGMP is in the vertebrate eye, where it serves as a second messenger for correcting the visual signals received as light to nerve impulses. In brief absorption of light by retinol (a small molecule) associated with rhodopsin activates the G protein-coupled rhodopsin. The G protein then stimulates cGMP phosphodiesterase, to leading to a decrease in intracellular concentration of cGMP.

Calcium ion (Ca^{2+}) functions as a second messenger in a large number of responses to first messengers. It is known that Ca^{2+} is maintained at a very low concentration in the cytosol, by using Ca-ATPase pumps and by secondary active calcium counter-transport with Na^+. When a stimulus is provided by a first messenger, the cytosolic concentration of Ca^{2+} increases. This triggers a sequence of reactions leading to the cell's response. However, this is not the only mechanism for increasing cytosolic Ca^{2+}. Such an increase in calcium ion concentration also occurs by the movement of calcium from the endoplasmic reticulum and other cell organelles (Fig. 11.17). The chain of events starts with the activation of the receptor by first messenger. The receptor stimulates a plasma membrane enzyme *phospholipase C (PLC)* via the G protein. The enzyme phospholipase C catalyzes the breakdown of a plasma membrane phospholipid called *phosphatidylinositol bisphosphate* (PIP_2), localized to the cytosolic suface of the membrane (see Chapter 1), to inositol *triphosphate* (IP_3) and *diacylglycerol (DAG)* as shown in Fig. 11.17. The IP_3 enters the cytosol and acts upon the membranes of the endoplasmic reticulum to cause movement of calcium from this organelle into the cytosol. Fig. (11.17).

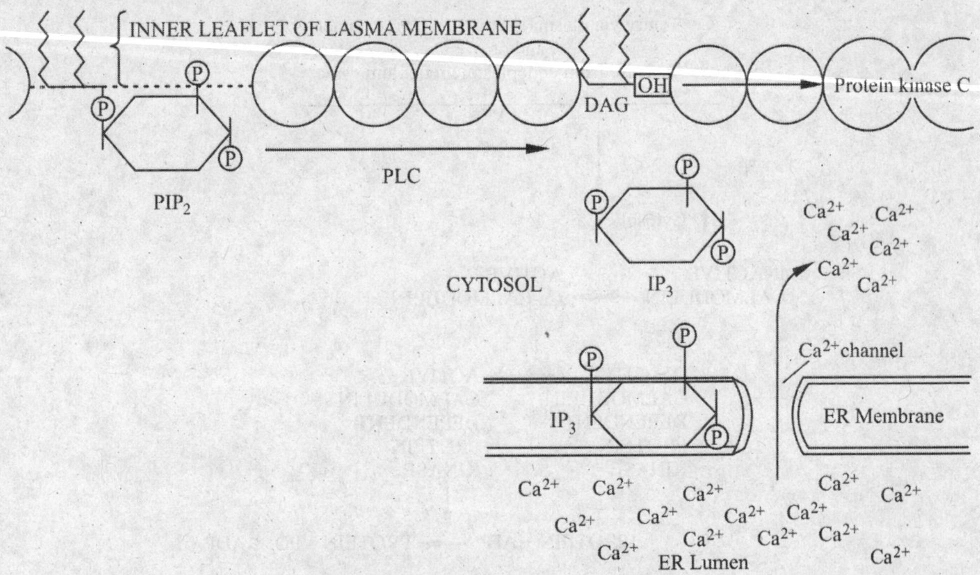

Fig. 11.17. *Hydrolysis of PIP_2.* Phospholipase C (PLC) catalyzes the hydrolysis of phosphatidylinositol 4, 5-bisphosphate (PIP_2) to yield diacylglycerol (DAG) and inositol trisphosphate (IP_3). Diacylglycerol activates members of the protein kinase C family and IP_3 signals the release of Ca^{2+} from endoplasmic recticulum (ER).

It may be noted here that in the above chain of reactions IP_3 is acting as second messenger to mobilize yet another second messenger calcium. As shown in Fig. 11.17, the two mechanisms for increasing cytosolic calcium by opening plasma membrane channels on the outer surface and releasing calcium form the endoplasmic reticulum are triggered simultaneously by the same messenger.

The next question which arises is how does increased cytosolic calcium concentration elicit cells response? This is accomplished by the ability of calcium to bind specifically to various cytosolic proteins. One of these proteins is calmodulin (Fig. 11.18) which is a component of almost all cells in an inactive state. The calcium ions bind to calmodulin and activate the inactive form. Further action of calmodulin is analogous to cAMP because many proteins affected by active calmodulin are protein kinases. Activation or inhibition of *calmodulin-dependent protein kinases* leads, through phosphorylation, to activation or inhibition of protein or enzymes in the cells final response to the first messenger.

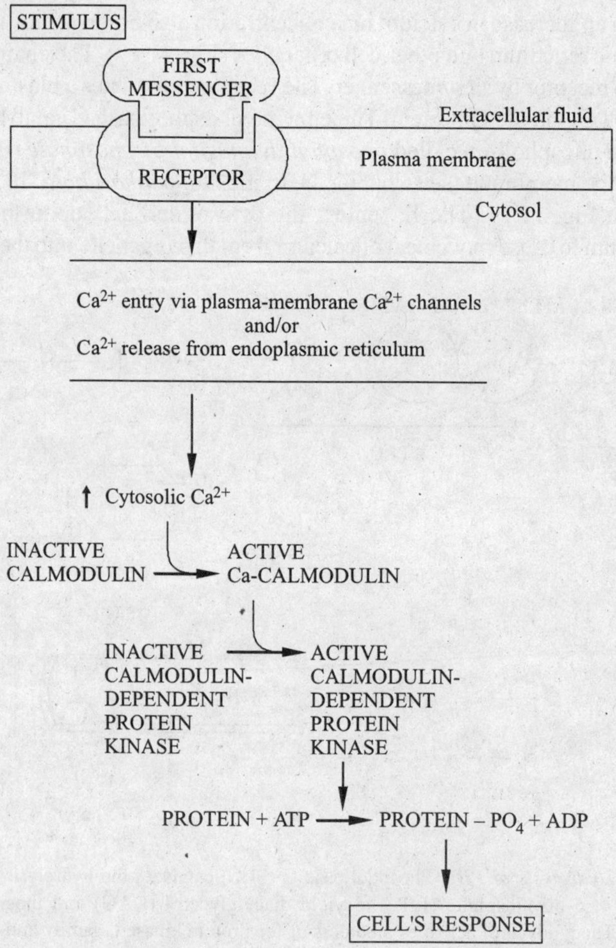

Fig. 11.18. Calcium, calmodulin, and the calmodulin-dependent protein kinase system. The mechanisms for increasing cytosolic calcium concentration and how does this induces the cell's response.

Diacylglycerol (*DAG*) activates another protein called *protein kinase C* (Fig. 11.17). The enzyme then phosphorylates a large number of other proteins. It must be emphasized here that protein kinase C is an entirely different protein kinase than those dependent on cAMP, cGMP and calmodulin. Further, the proteins that are phophnorylated by protein kinase C are also different. Moreover, a first messenger that triggers the generation of IP_3 and DAG causes both the activation of protein kinase C (mediated by DAG) and also, an increase in cytosolic calcium concentration (mediated by IP_3's release of calcium from the endoplasmic reticulum). Thus, both the pathways may interact in an additive or synergestic group.

In contrast to the G-protein-coupled receptors that activate protein kinase via cAMP, cGMP, calcium-activated calmodulin and DAG, the cell membrane receptors are directly linked to intracellular enzymes, that is, the receptors themselves become active protein kinases. These are receptor – protein-tyrosine- kinases which phophorylate their substrate proteins on tyrosine residues. First messenger involved in growth and development (such as insulin) act through these active protein kinases.

A summary of the effects of the five second messengers described above is given below:

Second messengers	*Effects*
Cyclic AMP	Activates cAMP-dependent protein kinase
Cyclic GMP	Activates cGMP-dependent protein kinase
Calcium ion	Activates calmodulin
Inositol triphosphate (IP_3)	Releases calcium from endoplasmic reticlum
Diacylglycerol (DAG)	Activates protein kinase C

As already noted , signaling molecules (first messengers) act by binding to receptors commonly located on the plasma membrane of the target cells. But some receptors are *intracellular proteins* located in the cytosol or the nucleus . The *steroid hormones* (sex steroids producted by gonads, testostesrone, estrogen and progesterone; corticosteroids produced by adrenal glands; glucocorticoids and mineralocorticoids), vitamin D_3, thyroid hormone and retinoic acid are the signaling or messenger molecules for the intracellular receptors. These messenger are able to diffuse across the plasma membrane. Once inside the cell, they bind to intracellular receptors, which directly stimulate transcription of their target genes as in the case of steriod hormones (Fig. 11.19). It may be mentioned here that the receptor binds DNA in either the presence or absence of hormone; a classic example of this is the thyroid hormone receptor which acts as a repressor in the absence of the hormone, but hormone – binding converts it to an activator that stimulates transcription of thyroid hormone-inducible genes (Fig. 11.19).

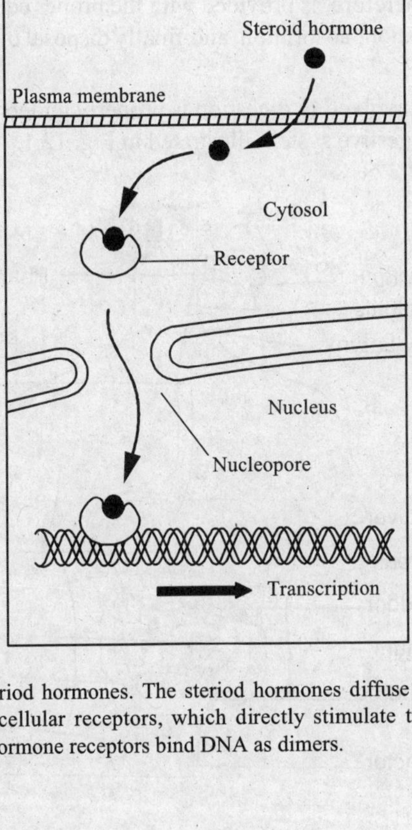

Fig. 11.19. Action of steriod hormones. The steriod hormones diffuse across the plasma membrane and bind to intracellular receptors, which directly stimulate transcription of their target genes. The steriod hormone receptors bind DNA as dimers.

12
Chapter

Digestion and Absorption

Only a few components of the diet such as salts, certain vitamins, water and other small water soluble molecules are directly absorbed by the digestive tract. These, however, form only a small fraction of the diet. The major constituents of the diet are relatively complex molecules. These are carbohydrates, proteins and fats etc. which can not be absorbed unless they are broken down into simpler compounds. The digestive system is responsible for the intake of food, its alteration to a simple and absorbable form during its passage through the different regions of the alimentary canal and finally its absorption into the blood stream. The fragmentation of food into simpler forms is called *digestion.* This is put into effect by a number of *hydrolytic enzymes* secreted by the alimentary canal and associated glands.

The digestive system, therefore, is provided with the proper equipment for the transport of food, secretion of enzymes, digestion, absorption, and finally disposal of non-utilizable compounds of the food.

Before the mechanism involved in digestion is properly understood, it is essential to understand the gross anatomy of the digestive system illustrated in Fig. 12.1.

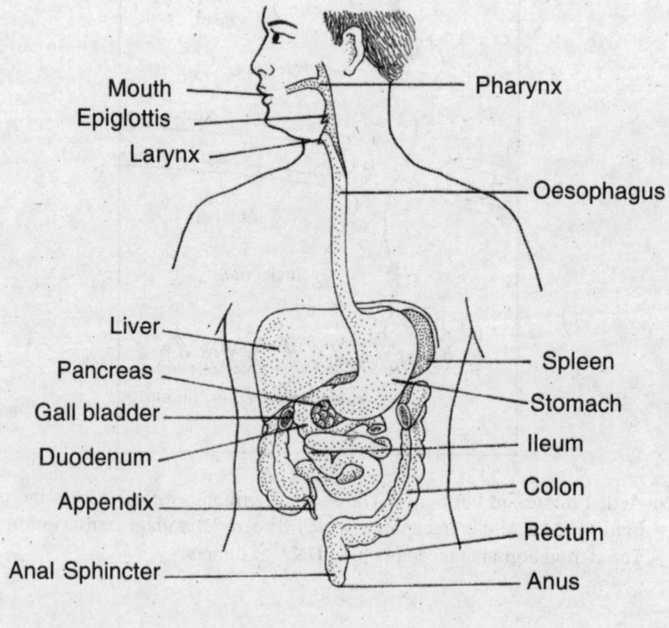

Fig. 12.1. Diagrammatic representation of the gastro-intestinal assembly.

ORGANISATION OF GASTROINTESTINAL TRACT

Gross Structure

The mouth is followed by a muscular tube called *pharynx.* The latter leads to a spacious chamber, the *stomach,* through a narrow tube known as *oesophagus.* The passage from the oesophagus to the stomach is guarded by a valve known as the *cardiac sphincter.* The stomach is followed by *small intestine* which can be divided into three parts. The distal end of the stomach opens through *pyloric sphincter* into the *duodenum* which is first part of the small intestine. The ducts from the liver and pancreas pour their secretions into the duodenum. The middle part of the small intestine is called the *jejnum,* and the last *ileum.* The last section of the alimentary canal is known as *large intestine.* Ileum opens into the first part of large intestine, the *colon* through another valve known as *ileo-colic valve.* A *vermiform appendix* is present at the site of ileo-colic valve which is vestigial in human beings. The colon leads into the last tubular part of large intestine—*rectum.* The latter, whose distal end is guarded by *anal spincters,* opens outside through the anus.

The wall of the entire alimentary canal is typically made up of four layers. These are the *mucosa, submucosa, muscularis* and the *serosa* (Fig. 12.2 A).

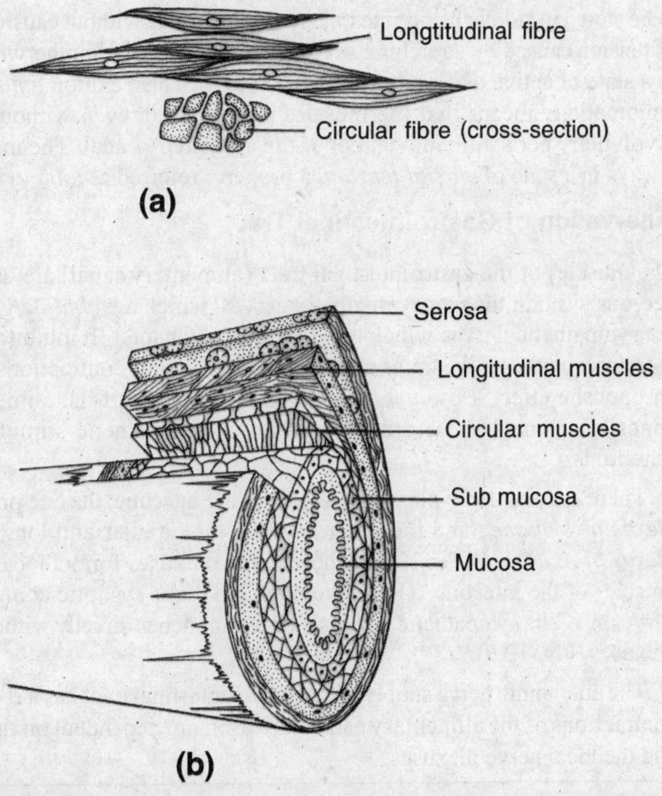

Fig. 12.2. (A) Microscopic appearance of intestinal smooth muscles; (B) Diagrammatic representation of arrangement of layers of the intestinal wall

The innermost layer of the alimentary canal is known as *mucosa.* It is made up of a single layer of stratified or columnar epithelial cells. Throughout its lengths, the epithelium is provided with *gland cells*. The epithelium rests upon a thin membrane known as *basement membrane*. The *submucosa* contains blood and lymph vessels, and a nerve plexus held together by loose connective tissue. The *muscularis* (Fig. 12.2 A and B) consists of an inner *circular* and an outer *longitudinal* layer of muscles. This layer is also provided with nerves forming another plexus. The *serosa* is the outermost layer of the alimentary canal. It is continuous with mesenteries and contains blood and lymph vessels, and sometimes adipose tissue. In oesophagus, however, the serosa is absent and the outermost cost is made up of fibrous tissue known as *fibrosa*.

Properties of Intestinal Smooth Muscles

The muscular is layer of gastro-intestinal tract is made up largely from smooth muscle fibres. The muscle fibres are arranged in circular and longitudinal layers (Fig. 12.2 A and B). The circular muscle fibres run circumferentially whereas the longitudinal muscle fibres run parallel to the longitudinal axis of the gut. Contraction of the circular muscle fibres reduces the diameter whereas that of longitudinal muscles shortens the length of the alimentary canal. The simultaneous contraction of these two muscle layers brings about the characteristic intestinal movements called *peristalsis*. This movement mixes the food as well as propels it forwards. Normally, the food moves only towards the rectum. Under abnormal conditions the food may be propelled towards the mouth by *antiperistaltic movements* causing *vomiting*.

The smooth muscles of the alimentary canal are *plastic* in nature, as the tension exerted on the muscles due to stretching disappears within a very short period. This allows large quantities of food to be stored in stomach and intestine for long periods without causing any damage. The disappearance of tension caused by stretching saves energy which would otherwise be required to keep the muscles in a state of active tension (stretch). The muscles also exhibit *automaticity* and *tonicity* (Chapter 9). Automaticity means that the muscles may contract even without nervous stimuli. This results in involuntary peristaltic movement of the alimentary canal. The muscles of the alimentary canal are always in a state of *partial tension,* a property referred as *tonicity*.

Innervation of Gastrointestinal Tract

The muscles of the gastrointestinal tract (alimentary canal) are under the control of the autonomic nervous system; the parasympathetic nerves, which are *cholinergic* (Chapter 9) accelerate whereas the sympathetic nerves which are *adrenergic* (Chapter 9) inhibit their contractions. Acetylcholine, therefore, increases the tonus and magnitude of active contraction. Adrenalin, on the other hand, has an opposite effect. However, the valves (sphincters) of the alimentary canal are innervated in an opposite manner. These are relaxed by parasympathetic stimuli and contracted by sympathetic sensations.

There are two nerve plexuses present in the intestine; the one present in the sumucosa is called the *plexus of Meissner,* and the other between the circular and longitudinal muscles is known as the *plexus of Auerbach* (Fig 12.3). These nerve plexuses form a local system of conduction within the muscles of the intestine. These two plexuses make synaptic connections with the parasympathetic nerve only. The sympathetic nerve sends out branches directly without making connections with these plexuses (Fig. 12.3).

The autonomic nerve supply of the gastro-intestinal tract has a regulatory function only. The normal contractions of the alimentary canal, however, are dependent on the automaticity of smooth muscles and the local nerve plexuses.

MOVEMENTS OF THE GASTROINTESTINAL TRACT

The functions of different parts of the alimentary canal are accomplished by a variety of movements peculiar to the objectives of that region and the local structures. These movement are under the control of nervous system and hromones produced by gastrointestinal tract.

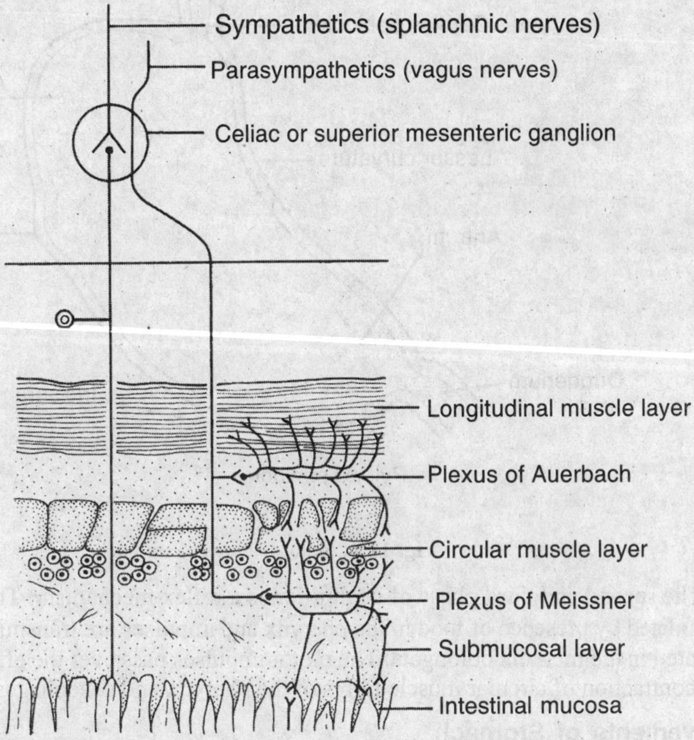

Fig. 12.3. Diagrammatic representation of the innervation of intestine.

Movement of Mouth and Pharynx

The mouth and pharynx perform two functions—that of *mastication* and *deglutition* respectively.

Mastication: In this act the lower jaw is moved upwards, downwards and laterally. A combination of these movements allows the food particles to be finely ground by the teeth. The tongue and cheeks help in the act of mastication as these keep the food between the teeth. During the process of mastication, saliva is poured in the mouth by salivary glands. The saliva helps to moisten and lump the food particles together to form a *bolus*. As saliva contains the enzyme amylase, it also hydrolyses starch into glucose and maltose.

Deglutition: Deglutition is the process of swallowing. It is accomplished in three stages: the first stage is voluntary and the other two are involuntary.

In the *first* stage the food is brought on the dorsal depressed surface of the tongue by the joint action of lips, cheeks and the tongue itself. Elevation and backward movement of the tongue then forces the food into the pharynx. In the *second* stage, food passes from pharynx into the oesophagus. The pharynx is the common passage for food and air, but during the act of swallowing the inspiratory and expiratory movement are suspended by reflex action. Fluids pass rapidly from pharynx to oesophagus, but the solids and semi-solids are pushed into the oesophagus by strong constrictor muscles of the pharynx. In the *third* phase the liquid passes directly from oesophagus to stomach, but the solids and semi-solids are forced by peristaltic actions. This is accomplished by relaxation of muscles immediately ahead the bolus and simultaneous contractions of the muscles directly behind it. Such successive peristaltic movements slowly push the bolus into the stomach.

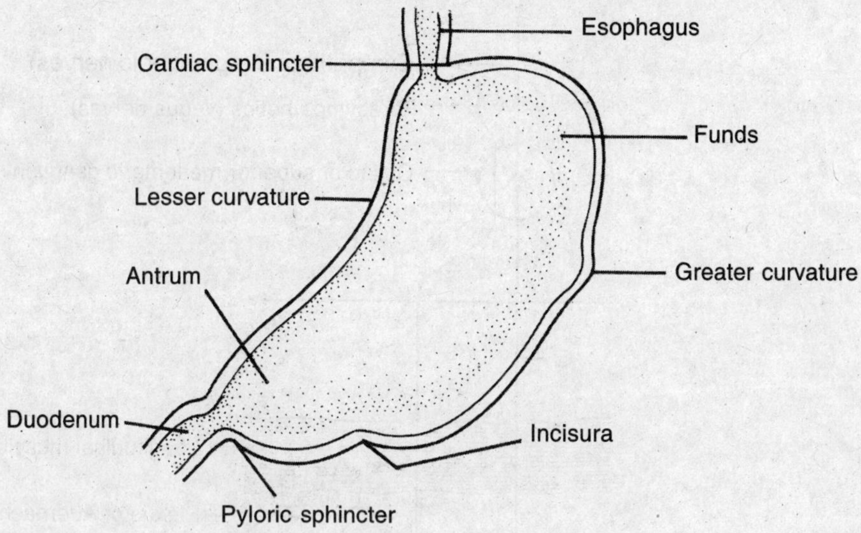

Fig. 12.4. Major features of stomach.

The second and third phase of deglutition is a reflex mechanism. The sensory nerve endings are stimulated by presence of food in the pharynx and impulses are transmitted to the *deglutition centre* situated in the medulla oblongata. The motor impulses pass over the efferent nerves and bring about the contraction of circular muscle fibres of the pharynx and oesophagus.

Movements of Stomach

Fig. 12.4 shows the various parts of stomach. The walls of any empty stomach are in a state of near collapse; however, the *fundus* may be distended with gas. In this condition, there is little or no activity in the stomach. As the food enters and fills the stomach , the muscles stretch and the volume is adjusted according to the contents. To start with, the peristalsis begins at the *incisura* (Fig. 12.4). Later, the activity of the stomach increases; and the origin of the peristaltic movements progressively moves towards the fundus. In case of very high activity, the entire stomach is seen actively engaged in peristaltic activity. During such high activity, the fundus of stomach is tonically contracted and the stomach takes the form of a a cylindrical tube. The peristaltic movements propel the food into the *duodenum.*

The movement of the stomach besides being under the control of autonomic nervous system, are also regulated by certain hormones. The secretion of the hormones of the gastro-intestinal tract depends on the chemical composition of food. For instance, the secretion of *enterogastrone,* a hormone produced by duodenal and intestinal mucosa, is regulated by the fat content of the food. It is produced when fat-rich food enters the stomach. This hormone inhibits the muscle contractions of the stomach, hence, the food is retained for longer periods, if the amount of enterogastone is high. The quantity of the latter is proportional to the quantity of fat taken with food. During hunger, the stomach undergoes a type of sweeping peristalsis which are called *hunger contractions.* Although these contractions stimulate hunger, they have nothing to do with appetite as they quickly disappear when a small amount of food is taken.

Movements of Intestine

Intestine generally shows three-type of movements:

(1) Segmenting movements, which are rhythmic contractions producing churning and mixing effects. This movement is not under the control of nervous system.

(2) Pendular movement, which probably produce slight mixing effects and are not very important in man.

(3) Peristaltic movements, which can be initiated by mechanical, electrical of chemical stimuli. However, the most powerful stimulus is the presence of food in the intestine. When the food reaches the duodenum, regular peristaltic movements are initiated. At intervals powerful peristaltic contractions take place which extend thought the entire length of the intestine. This is called *peristaltic rush* and it moves the food for a considerable distance in the intestine. The peristaltic rush causes the relaxation of the *ileo-colic sphincter* so that part of the bolus is pushed into colon. The frequency of the peristaltic rush is increased if an irritant poison or a purgative is taken. The villi of the intestine (Fig. 12.6) are also in a state of constant movement. Such movement are due to contractions of smooth muscle fibres within each villus.

The peristaltic rush also pushes the bolus in the colon every time food is eaten or when it reaches the duodenum. This is due to a phenomenon known as *duodenocolic* or *gastrocolic reflex.* These reflexes often cause the urge to defecate, especially immediately after taking meals.

Vomiting

Vomiting is a protective device as by this process harmful material from the alimentary canal is ejected out through the mouth. The stimulus for vomiting may arise from a number of places, *viz.*, the receptors present in the mucosa of stomach; stimulation of pharynx; and noxious psychic factors like pain and isgust. The centre for vomiting is present in the medulla of brain. The stomach plays a passive role during vomiting. The act of vomiting is brought about by a complex series of movement of the respiratory and abdominal muscles. It can take place even after all the nerves to stomach and intestine are excised.

SECRETIONS OF THE GASTROINTESTINAL TRACT

The secretions of the gastrointestinal tract are of two types: (1) *mucus,* which protects the wall of gastro-intestinal tract from being digested and (2) the *enzymes* and allied substance which are effective in chemical digestion of food. The latter include saliva, gastric juice, intestinal juice, pancreatic juice and bile. These juices are secreted only when they are needed by the system. The control of secretion of juice is well regulated and is discussed below.

Mucus

Mucus, a mucoprotein, is secreted by the entire gastro-intestinal tract. When the food enters the alimentary canal, muscus is secreted which forms a covering over the intestinal mucosa. This protects the intestine from being digested by its own juices.

Salivary Secretions

Saliva is secreted mainly by three pairs of glands, all located in the mouth. These are the *submaxillary glands* (under the angle of jaw bone), *sublingual glands* (under the tongue) and the *parotid glands* (in front of the internal opening of ears). In addition to these glands, there are numerous small buccal glands situated in the mucosa of mouth which also secrete saliva to keep the mouth moist.

The saliva has a specific gravity of about 1.0007, a pH of about 6.6 and is composed of approximately 99.5% water, 0.2% inorganic and 0.3% organic matter. The inorganic matter consists of phosphates, carbonates, bicarbonates and chlorides of sodium, potassium, calcium and magnesium. The organic matter consists of *mucin* and *enzymes.* The former gives saliva its viscous property which helps to lubricate the food. The *enzymes* include amylase, acid phosphatase, aldolase and cholinesterase. Of these, amylase is the most potent enzyme which acts on starch and breaks it to glucose and maltose.

The secretion of saliva is exclusively controlled by autonomic nervous system; no hormone has been found to interfere with the production of saliva. The *inferior* and *superior* salivary nuclei, situated in the medulla of brain control the secretions of salivary glands. Parasympathetic stimulation produces *profuse* and *watery* saliva, whereas sympathetic stimulation produces *sparse, thick* and *mucinous* secretion. Saliva secretion is mostly a reflex action; the smell, very sight or even thought of food causes the secretion of saliva. Oral stimulation, caused by chewing the food increases saliva production. Even, the presence of food in the mouth causes salivation. Smooth food *increases* the saliva production whereas rough food reduces it. Normally, man secretes about 1—1.5 litres of saliva in 24 hours, but great individual variations occur.

Secretion of Gastric Juice

The stomach secretes large amount of gastric juice which may exceed even 2 litres per day. Rate of secretion is high when food is present in the stomach. However, even when the stomach is empty, there is a continuous but slow secretion of gastric juice. As a result, the stomach always contains about 50 ml of fluid; this is called *residium*. The gastric juice is strongly acidic and is a mixture of the substances secreted by three different types of cells present in the stomach lining (Fig. 12.5). These are: (1) *parietal* (*border*) *cells,* which produce hydrochloric acid, (2) *mucus cells,* which produce mucus and (3) *chief cells* (*zymogenic*), which produce pepsin and other digestive enzymes (Fig. 12.5). The hydrochloric acid produced by parietal cells has a strength of approximately 0.10 N and a pH of about 1. The secretion of mucus and chief cells dilute HCl in the stomach. This raises the pH of gastric contents from *one* to 1.5—2.0. The hydrochloric acid is produced by selective absorption of hydrogen and chloride ions from the blood. The reaction is a reversal of neutralization of sodium bicarbonate with HCl in blood.

$$\underset{\text{(blood)}}{Na_2CO_3} + \underset{\text{(blood)}}{NaCl} \rightleftharpoons \underset{\text{(gastric juice)}}{HCl} + \underset{\text{(blood)}}{NaHCO_3}$$

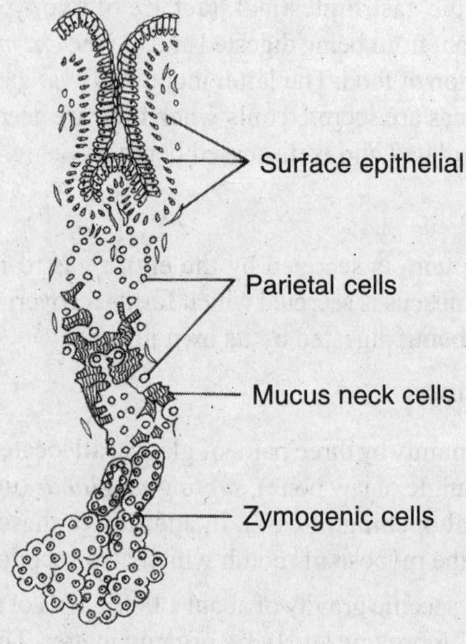

Fig. 12.5. Transverse section of stomach showing the parietal mucus and chief cells.

After the ingestion of a heavy meal, large amount of HCl is produced by parietal cells. This increases the alkalinity of blood due to excess formation of $NaHCO_3$. The kidneys excrete the extra bicarbonate in the form of sodium salts. This is referred as *alkaline tide* and imparts a cloudy appearance to the urine caused by the precipitation of urinary phosphates (see Chapter 15).

Gastric juice contains about 99% water, 0.5 % HCl, 0.2% inorganic salts and 0.3 organic matter. The inorganic salts are the same as found in the blood. The organic matter consists mainly of *mucin, pepsin* and a rather ineffective enzyme, the *gastric lipase*.

The secretion of the gastric juice is controlled by three agencies; the *brain,* the *stomach* itself and the *duodenum*. These three have been referred as *cephalic, gastric* and *intestinal* phase of gastric secretion. More often than not, these three phases occur in the sequence given above. The cephalic phase is purely nervous and is mediated by vagus stimulation, the gastric phase is both hormonal and physical whereas the intestinal phase is purely hormonal. In contrast to the secretion of salivary gland, which is controlled purely by the nervous system, the nervous system in the case of gastric secretion plays only a minor role.

Sight, smell taste or even throught of food stimulate the cephalic centre to secrete the gastric juices. The *cephalic phase* of secretion is controlled by vagus nerve.

The *gastric phase* of secretion is linked with the distention of stomach rather than the presence of food. In addition, protein-rich foods like meat, egg etc., stimulate the formation of a hormone *gastrine* (Fig. 12.7) secreted by the mucosa of stomach. The gastrin, thus produced, is absorbed into the blood stream and then carried to the gastric cells, which are finally stimulated to secrete the gastric juice.

The *intestinal phase* of gastric secretion is brought about when the food enters the duodenum. This phase is also under the control of gastro-intestinal hormone. When fat-rich food enters the duodenum, the intestinal mucosa secretes a hormone known as *enterogastrone* (Fig. 12.7) which enters the blood stream. Through the blood stream it reaches the stomach where it inhibits the secretion of stomach.

SECRETION OF INTESTINAL JUICE

Intestinal juice is secreted by two types of glands (1) *glands of Brunner* (duodenal gland) present in the submucosa of duodenum (Fig. 12.6). They secrete mucin and a weak proteolytic enzyme. (2) *Crypts of Lieberkuhn* (intestinal glands) which are tubular glands present throughout the small intestine.

The intestinal juice (succus entericus) contains about 99% water, 0.5% inorganic salts and 0.5% organic matter. The inorganic salts are similar to those described for saliva. However, in this, $NaHCO_3$ is present in large amounts which makes the intestinal juice alkaline (pH 7.6 to 8.5). The organic matter consists of mucin and several enzymes which have been discussed separately.

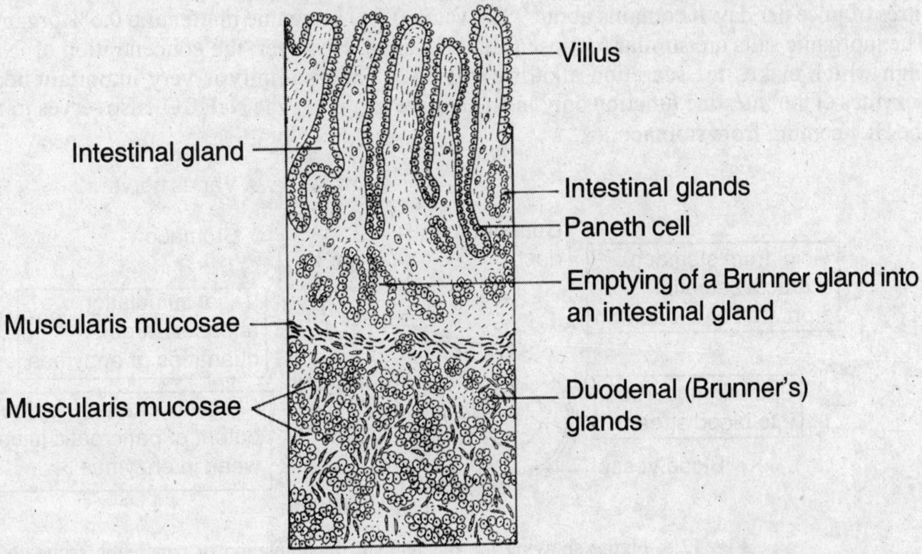

Fig. 12.6. Transverse section of duodenum showing the intestinal villi and glands of Brunner.

The secretion of intestinal juice is more or less continuous but it is increased remarkably by the secretion of a hormone, *enterocrinin* (Fig. 12.7), produced by the intestinal mucosa. The secretion of *enterocrinin* is increased by mechanical pressure on intestinal wall. The presence of food provides such stimulus. The nerves to intestinal glands are in general inhibitory.

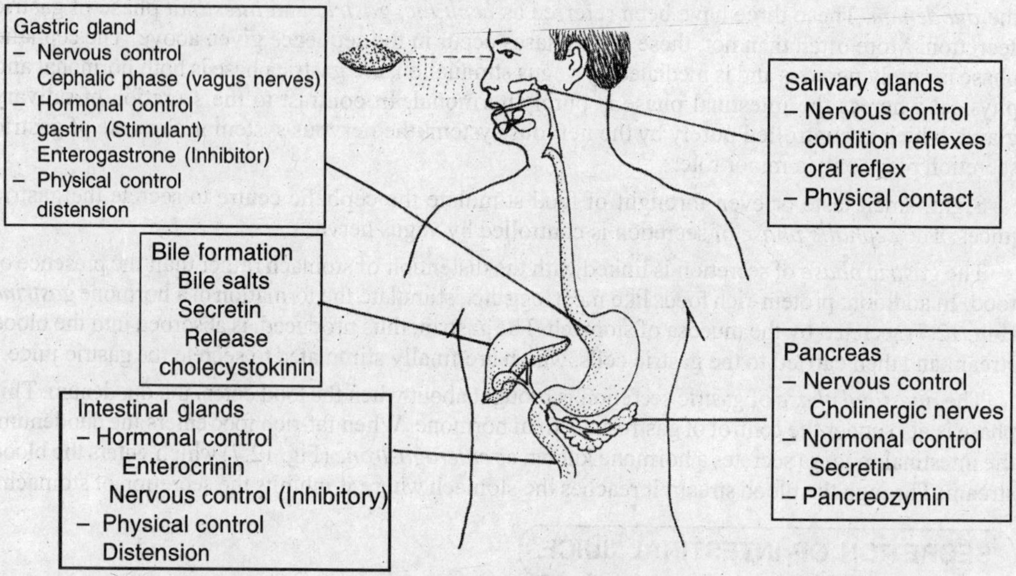

Fig. 12.7. Figure showing the factors influencing the secretions of gastro-intestinal tract.

The various factors influencing the secretion of gastrointestinal tract are summarised in Fig. 12.7.

Secretion of Pancreatic Juice

The *pancreas* is a large gland situated between the duodenum and stomach (Fig. 12.8). The pancreatic secretions are drained through the *pancreatic duct* which join the *bile duct* coming from the liver. The two form the *common bile duct* which open in the duodenum. (In about 29% population the pancreatic duct opens independently into the duodenum). The pancreas normally produce 1 to 1.5 litres of juice per day. It contains about 99% water, 0.5% inorganic matter and 0.5% organic matter. The inorganic salts are similar to those found in saliva. However, the concentration of $NaHCO_3$ is high which makes the secretion alkaline (pH 8.0). The alkalinity is very important because the enzymes of the intestine function only in an alkaline medium. The $NaHCO_3$ also serves to neutralize the HCl coming from stomach.

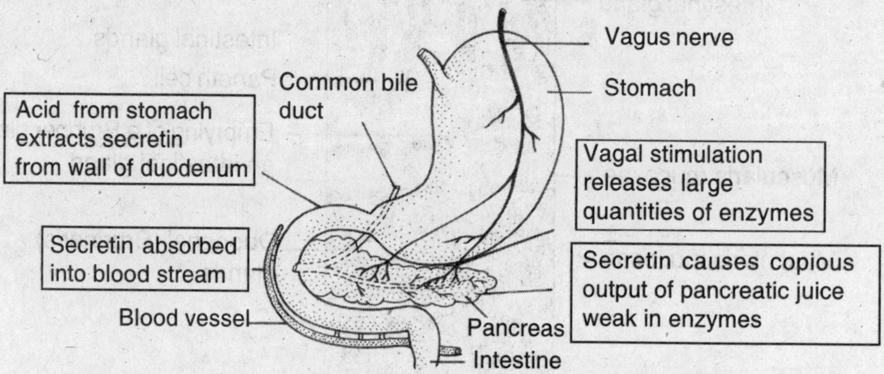

Fig. 12.8. Figure showing the mechanism for regulation of pancreatic secretion.

The secretion of pancreatic juice is both under nervous and hormonal control. Parasympathetic (vagus) stimulation increases the secretion of enzyme-rich pancreatic juice. The parasympathetic innervation is *cholinergic* because it is blocked by atropine.

When acidic chyme from the stomach reaches the intestine, the HCl in the chyme stimulates the duodenal mucosa to produce a hormone called *secretin.* This is absorbed into the blood stream. The secretion then acts on the pancreas which produces *thin* pancreatic juice; this contains salt and water but no enzymes. (In addition, secretion also stimulates liver to some extent). Simultaneously, another hormone known as *pancreozymin* (Fig. 12.7) is also produced by intestinal mucosa which works on the zymogen granules of the acinous cells of the pancreas and stimulates the secretion of pancreatic juice which is *copious* and *rich* in enzymes. The juice produced by pancreozymin is similar in nature to that produced by parasympathetic stimulationof pancreas.

Secretion of Bile

Liver continuously secretes a juice called *bile*. Bile is collected by a number of *hepatic ducts* which join together to form the *common hepatic duct.* This immediately joins the duct from the gall bladder (cystic duct) and continues towards the intestine as the *bile duct.* The opening of the bile duct into the intestine is guarded by a sphincter known as *sphincter of Oddi* (Fig. 12.9). When sufficient pressure builds up in the bile duct, the bile backs into the gall bladder. In the gall bladder extra water is absorbed from bile. Hence, it becomes sufficiently concentrated. It is stored in this form until needed for the digestive processes.

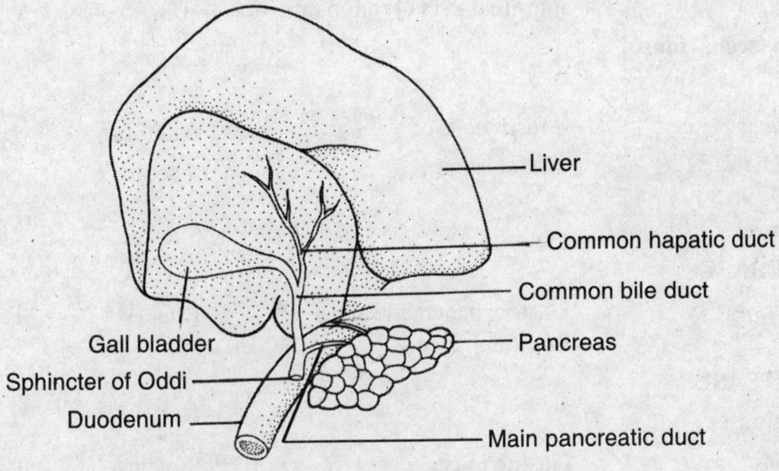

Fig. 12.9. Figure showing the anatomical arrangement of biliary system and pancreatic ducts.

Gall bladder bile contains about 86% water and 14% salts. The salts are made up of 8% bile salts (which are sodium salts of glycocholic and taurocholic acid), 2% bile pigments and mucin, 3% cholesterol and other lipids, and about 1% inorganic salts. Bile has an amber colour due to the presence of the pigement, *bilirubin,* but on exposure it changes to green colour (the amber pigment *bilirubin*is converted to the green pigment *biliverdin*). Bile is alkaline (pH 7.7 —8.6) due to the presence of large concentration of bile salts. The output of bile by the liver is about 1 litre per day.

The release of bile is under the control of a hormone *cholecyst okinin* (Fig. 12.7), which is produced by the mucosa of *upper intestine,* when fats and acid enter the intestine. The hormone then enters the blood stream and causes the simulataneous contraction and relaxation of sphincter of Oddi. Thus, bile is poured in duodenum. Bile salts have strong detergent action, which is important for absorption of fat.

Although the release of bile is mostly controlled by cholecystokinin, it has been shown that the vagus nerve also exercises a control on its opening. Stimulation of the vagus nerve causes *spasm* in the sphincter of Odii. If the vagus nerve is cut, the gall bladder empties slowly.

DIGESTIVE ENZYMES

The enzymes concerned with digestion may be conveniently grouped according to the major food substance digested by them. Thus, *carbohydrases* are the enzymes which act on *carbohydrates: lipases* and *esterases* are those which act on *fats* and *proteases* are those which act on *proteins.*

A brief classification of digestive enzyme is presented in Table 12.1.

Table 12.1 Digestive enzymes, their occurrence, substrates and end products

Name of the Enzymes	*Occurrence*	*Substrates*	*End products*
CARBOHYDRASES			
Polysaccharidases			
Amylase	Salivary pancreatic and intestinal and intestinal juice.	Starch	Maltose
Cellulase	Absent in man; but produced by bacteria found in the gastro-intestinal tract of ruminants	Cellulose	Glucose
Oligosaccharidases			
α-glucosidase	Mucus cells of the intestine	Maltose	Glucose
β-galactosidase	Mucus cells of the intestine	Lactose	Glucose and Galactose
Invertase		Sucrose	Glucose and Fructose
ESTERASES			
Lipases	Gastric, pancreatic and intestinal juice	Fats	Fatty acids, Glycer-ides
PROTEASES			
Proteinases			
Pepsin	Gastric juice	Proteins, Collagen and Nucleo-proteins	Proteoses, Peptones and Polypeptides
Trypsin	Pancreatic juice	Proteosis, Peptones and poly-peptides	Polypeptides and Amino acids
Polypeptidases	Pancreatic and intestinal juice	Polypeptides and Amino acid	Amino acids

The digested food material is mostly absorbed in the small intestine through many million small *villi* (Fig. 12.9). The villi are small finger-like projections of the mucus membrane. Each villus contains an arteriol which is connected to a venule through a network of capillaries. The venule ultimately empties in the portal vein. There is also a blind lymph vessel in each villus called the *lacteal* (Fig. 12.10). During digestion and absorption the villi are in a constant movement controlled by the *p exus of Meissiner* and a hormone known as *villikinin* (see Chapter 11).

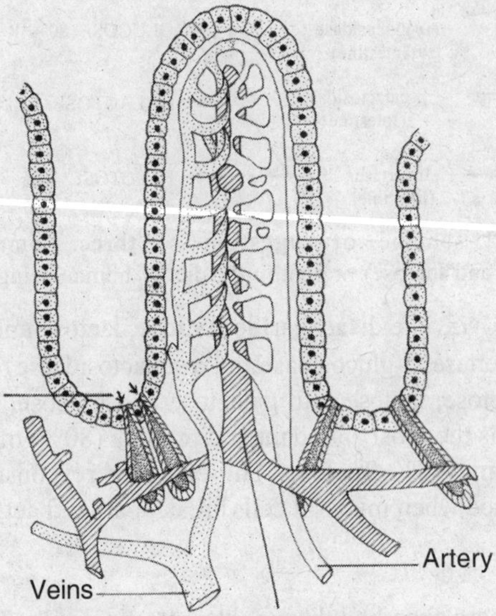

Fig. 12.10. Figure showing the arrangement of blood vessels and lacteal in a villus of intestine.

DIGESTION AND ABSORPTION OF CARBOHYDRATES

Digestion

In carbohydrate digestion the complex carbohydrate molecules are broken down into monosaccharides by the addition of water. This process is known *hydrolysis*.

The common carbohydrates of the diet are strach, cellulose, glycogen, lactose, sucrose, maltose, glucose etc. Of these glucose is directly absorbed in the blood. The human body does not possess enzymes necessary for cellulose digestion. Hence, the cellulose present in the foods of plant origin is passed out undigested in the faeces. Only ruminants are able to digest cellulose with the help of symbiotic mircro-organism present in the rumen.

The outline of the digestion of the remaining carbohydrates has been schematically presented in Figure 12.11.

The first step in the digestion of *starch* is its conversion to maltose with the help of amylase. A large amount (40%) is digested in the mouth by *slivary amylase*. When the chyme enters the stomach, part of the undigested starch is hydrolysed by the HCl present in the gastric juice. The total digestion is accomplished in intestine where it is exposed both the pancreatic and intestinal amylase of which the latter plays only a minor role.

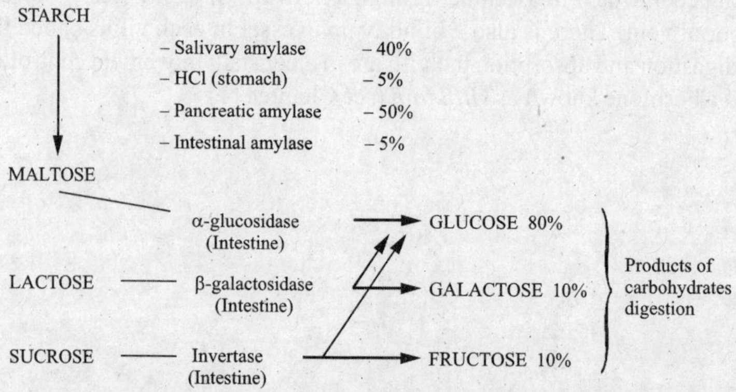

Fig. 12.11 General scheme of digestion of three common carbohydrates (starch, lactose and sucrose) present in the diet of human beings.

The other carbohydrates, *viz.*, the disaccharide sucrose, lactose and maltose are digested in the intestine by the enzymes invertase, β-glucosidase and α-galactosidease respectively. Sucrose is broken down into glucose and fructose; lactose into galactose and glucose, and, maltose is splitted into glucose. Evidently glucose is the most abundant and product (80%) of carbohydrate digestion. The others are 10% galactose and 10% fructose. The enzymes responsible for starch digestion are intracellular enzymes released when mucosal cells break apart and act at pH 7—8.5.

Absorption

The digested carbohydrates are absorbed through the intestinal villi. There are two mechanisms of absorption in the intestine; (1) *diffusion* and (2) *active absorption*. Fructose is transported by simple diffusion in response to a concentration gradient by the absorptive cells of the villi. Glucose and galactose, on the other hand, are actively transported. For this they firstly combine with a carrier substance. As this process requires energy, the carrier compound is first *Phosphorelated* with ATP. A probable mechanism of active transport has been given in Chapter 13.

Notwithstanding the earlier statement mode about the digestion of disaccharides, it must be understood at this point that the rate of digestion of sucrose and lactose is quite low; the rate of absorption in the intestine, on the other hand, is quite high. This means that the intestine must absorb these two disaccharides from some other route also. It is now believed that these two are taken up directly by the mucosal cells where they are converted into monosaccharides by the intracellular carbohydrates. Sucrose is mainly absorbed in the jejunum in man. The former belief, that all the dietary carbohydrates are first broken down into monosaccharides, is therefore, untendable.

Of the three monosaccharides (Fig. 12.11), glucose is immediately transported throughout the body fluids. Fructose and galactose however, are converted into glucose in the liver and then transported to different parts of the body. Thus, the end product of carbohydrate digestion is glucose. The concentration of glucose in blood is about 90 mg/100 ml.

After the ingestion of heavy meal the glucose level in portal veins may rise to fold. Some of the excess glucose is taken up by tissues for oxidation, but most of it is taken up by muscle and liver for storage as glycogen. This process is facilitated by the hormone *insulin* (Chapter 11). When blood glucose level falls, the glycogen is reconverted to glucose under the influence of a hormone *glucagon* and transferred to blood (glycogenolysis) (Fig. 12.12). Hence, liver is an efficient buffer organ for blood glucose level as when its (glucose) level rises, it is stored and when the glucose level falls it is again compensated.

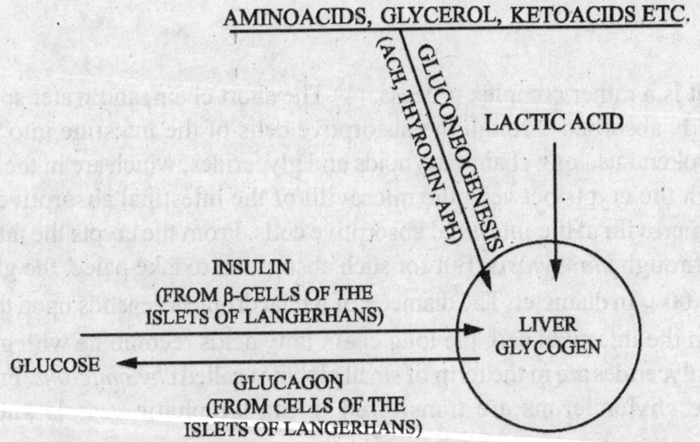

Fig. 12.12 Diagrammatic representation of the reversible equilibrium between liver glycogen and blood glucose. It also shows gluconeogenesis by liver.

When the blood glusose level falls very low and is not adequately compensated by liver glucogen, then glucose is synthesized from proteins and to some extent from fats. This is known as *gluconeogenesis*. This is a very important phenomenon as it provides glucose to the blood even during starvation because the amount of glycogen present in the liver is not sufficient to maintain the blood glucose for more than 24 hours. When glucose is not available, cells are able to utilise fats for energy. The ganglion cells of the brain are unable to utilize fats, and without glucose they die quickly. Gluconeogenesis is, therefore, very important in maintaining the blood glucose level even during starvation.

DIGESTION AND ABSORPTION OF LIPIDS

Digestion

The digestion of fat, like carbohydrates, is also a hydrolytic process in which the lipids are broken into fatty acids, glycerol and glycerides (Table 12.1). Fats are digested by the enzyme lipase secreted in the stomach, pancreas and intestine. The outline of fat digestion has been given in Fig. 12.13.

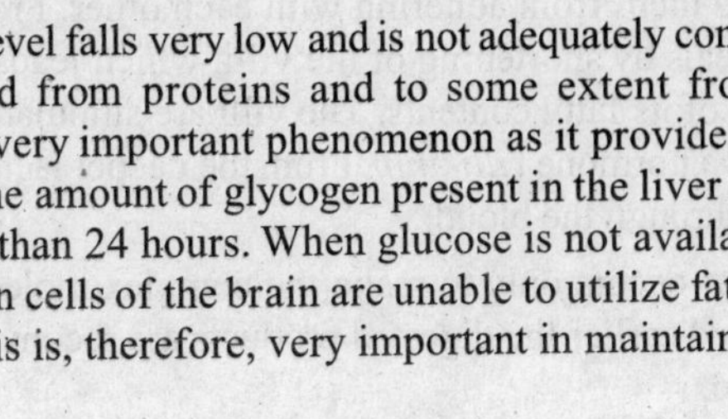

Fig. 12.13. Diagrammatic representation of digestion of fats presents in the diet.

There is no buccal digestion of fat, however, small part of the fat is digested in the stomach with the help of gastric lipase. Most of the fat digestion takes place in small intestine. When the fat enters the small intestine, they are acted upon by the bile salts simultaneously released from liver. As the bile salts have a detergent action, they emulsify the fat into small globules—known as *micelles*. These offer a greater surface area for lipases to act. These are further degraded by intestinal pancreatic lipases into the end product of fat digestion—fatty acids, glycerol and mono and diglycerides (Chapter 4). In the fat digestion, only 50% fat is completely digested to fatty acid and glycerol. The remaining 50% are glycerides (Fig. 12.13).

Absorption

The absorption of fat is a rather complex process. (1) The short chain, and water soluble long chain fatty acids are directly absorbed through the absorptive cells of the intestine into the portal blood system. (2) The unbroken fats, long chain fatty acids and glycerides, which are in the form of micelles are absorbed through the crypts between the microvilli of the intestinal absorptive cells. From the crypts between the microvilli of the intestinal absorptive cells. From the crypts the fats are transferred into the cytoplasm through *pinocytosis*. But for such absorption to take palce, the globules must not be more than 0.5 to 1.00 μ in diameter. The diameter of the fat droplets depends upon the emulsification of fat by bile salts. In the intestinal cell, the long chain fatty acids recombine with glyceride to form *triglycerides*. The triglycerides are in the form of small globules called *chylomicrons*. From the intestinal absorptive cells, the chylomicrons are transferred to the lymphatic vessels known as *lacteals* (Fig. 12.10).

Immediately after entering the lacteals these chylomicrons absorb proteins on their surface and are known as *chyle*. This prevents them from adhering with each other. From the lacteals the chyle is transported to the deeper lacteals by shortening of the villi, which leads to the compression of the central lacteal and evacuation of its fatty contents. The villi are stimulated mechanically by certain constituents of the food and by a hormone *villikinin*. From the deeper lacteals the fat is transported to liver of fat tissue of the body through the blood.

In brief, the digestion of fat occurs only to make the complex molecule diffusible through the intestinal villi, because as stated earlier the digested products are reconverted again into the natural fats as were present in the food.

DIGESTION AND ABSORPTION OF PROTEINS

Digestion

During digestion the proteins (like that of carbohydrates and fats) are also the hydrolysed. The end-products of protein digestion, are *amino acids* (Fig. 12.14).

Digestion of protein is initiated in the stomach by the action of the enzyme *pepsin*. Pepsin is secreted by the chief cells of stomach in an inactive form, *pepsinogen*. The latter is converted autocatalytically into pepsin between **pH** 5.5 in the presence of H^+ and pepsin itself. The H^+ is provided by HCl secreted by the *parietal cells*. Although pepsinogen is converted into pepsin below pH 5.5, optimum pH for its activity lies between 1 and 2.

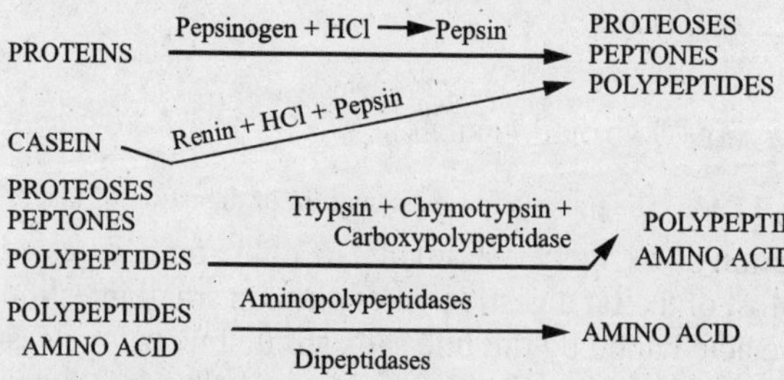

Fig. 12.14. Schematic representation of digestion of proteins present in the diet.

Pepsin attacks the peptide linkages adjacent to the aromatic amino acids and breaks the protein into *proteoses*, peptones and *polypeptides*. The proteoses are the largest fragments of the protein molecules, peptones are smaller and polypeptides are smallest containing very few amino acids.

The milk of the diet is clotted by another gastric enzyme known as renin. The coagulated caesin (a protein found in milk) is then acted upon the pepsin. Renin is found in large quantities in infants.

When the semi-digested proteins (proteoses, peptones and polypeptides) reach the small intestine, they are acted upon by *trypsin* and *chymotrypsin* which are simultaneously poured from pancreas. Trypsin is secreted as *tropsinogen* which is activated to trypsin by an intestinal enzyme *enterokinase*. Trypsin attacks peptide bonds between the *carboxyl groups* of *arginine* and *lysine*, and the *amino groups* of other amino acids. In addition to this, trypsin also acts as an esterase and hydrolyses numerous esters and fatty acids. The pH optimum for trypsin is between 7.0 to 9.0. There are at least four chymotrypsin; they are secreted as chymotrypsinogen and are activated between pH 7.5 to 9.0 in presence of trypsin. Chymotrypsins are major proteolytic enzyme of the pancreas. They hydrolyse the peptide bonds involving the L-isomers of aromatic amino acids. These also act on amides and esters of amino acids.

L. Ala — L. Arg — L. Lys — L. Try — L. Phe
(Aromatic)
Trypsin Chymotrypsin

The two enzymes (trypsin and chymotrypsin) thus break the proteose, peptones and polypeptides into small fragments. The polypeptides thus formed are finally hydrolysed by peptidases which break the terminal bonds and releases the individual amino acids. These are secreted by intestine and pancreas.

In addition to the dietary proteins of which has been described above, large amount of protein is passed into the alimentary canal as gastro-intestinal secretion. The quantity is formidable and exceeds the normal intake of dietary protein. About 140 g of secretion and about 25 g of gastro-intestinal debris pass through the gastro-intestinal tract every day. This is digested and absorbed along with the dietary proteins.

Absorption

At the end of protein digestion although most of the proteins are converted to amino acids a part is left in the form of small polypeptides. In general peptides having a molecular weight of 200 to 300 are poorly absorbed and pass out in the faces. Some dipeptides are *intracellularly* digested; these enter the intestinal cells and are hydrolysed there. The free amino acids are absorbed both by *diffusion* and *active transport*. The active transport involves oxidative phosphorylation. The transport of neutral amino acid and sodium go together; but there are separate systems for basic and acidic amino acids. The entire mechanism of absorption of product of protein degestion, however, has not yet been clearly understood.

From the absorptive cell of the intestine, the amino acids enter the blood from where these go to the liver; very little enters the lacteals.

Very little quantity of proteins is also absorbed by *pinocytosis*. This does not have any nutritional value. In certain cases, some of the unchanged proteins may act as *antigens* for producing *antibodies*. When sufficient quantities of antibodies are formed, further absorption of protein produces a reaction in the form of *ashes* or *asthmatic* conditions. An individual is then said to be *allergic* to a particular protein.

A common stock or *metabolic pool* of free amino acids is maintained in blood, liver and tissues, from the amino acids absorbed in the gastro-intestinal tract and those obtained for the break down of body proteins. Part of these is utilized for building body tissues, hormones and enzymes, etc., and the remainder degraded to form nitrogenous wastes (Chapter 15).

The body proteins are extremely labile and are continuously renewed e.g., half of the protein of liver and muscle is renewed every 10th and 160th days, respectively. The synthesis of protein from amino acids is under the control of anabolic hormones such as *testosterone* and *GH*. Breakdown of proteins to amino acids, on the other hand, is controlled by catabolic hormones such as *ACTH*. The protein metabolism in the body, therefore, is in a state of constant *flux*. Nearly all cells of the body have a free amino acid pool from which they are capable of synthesizing proteins (Chapter 5). When the amino acid level of one part of the body falls, amino acids from other parts are mobilized and transferred to the deficient organ.

DEFECATION

Digestion and absorption of food is nearly completed by the time it reaches the hind-gut. In the latter, ions and water is absorbed specially by the caecum and ascending colon. The large intestine secretes a thick mucoid substance which acts as lubricant to facilitate the passeage of faecal matter through the colon. The faecal matter is stored in the large intestine for about 36 hrs. during which it is acted upon by a number of bacetria which impart the faeces their characteristic foul smell.

Defecation is controlled by the inner and outer sphincter muscles which encircle the anus. The inner sphincter is involuntary whereas the outer sphincter is voluntary. The expulsion of faeces is brought about by contraction of diaphragm and skeletal muscles of the abdominal walls. This raises the abdominal pressure even up to 200 mm of Hg which forces the faeces out.

The frequency of defecation is variable from person to person. In some people defecation even after 4—5 days is common and they have no problem. Hence, it is difficult to define *constipation*. But when constipation occurs, it can produce nausea, halitosis, headache, general dullness, etc.

There is a strong belief that constipation is a gift of civilization as it may also result from emotional disturbances.

13

Chapter

Respiration

RESPIRATION

Respiration originally stood for taking up oxygen from air and giving off carbon dioxide in exchange. It was inherent in this concept that during respiration oxygen was utilized for oxidation of carbon and hydrogen in the lungs. The oxidation was supposed to provide "animal heat"— the energy that makes life possible. Blood was considered to be the vehicle for the transport of the "heat" produced in the lungs. This view was discarded after the realization that the lungs serve only as a device for the exchange of gases between the blood and the environment, and that the oxygen was actually utilized at the cellular level to oxidise the foodstuffs (carbohydrates, lipids, and proteins) to carbon dioxide and water with the concomitant release of energy. The energy thus obtained is utilized to sustain life. Respiration, therefore, essentially consists of two phases. The first phase, consisting of gaseous exchange between the blood and the environment via the lungs, is referred to as *external respiration*; the second phase, consisting of the oxidative energy-yielding reactions of the cells, is called as *internal respiration* (Fig. 13.1).

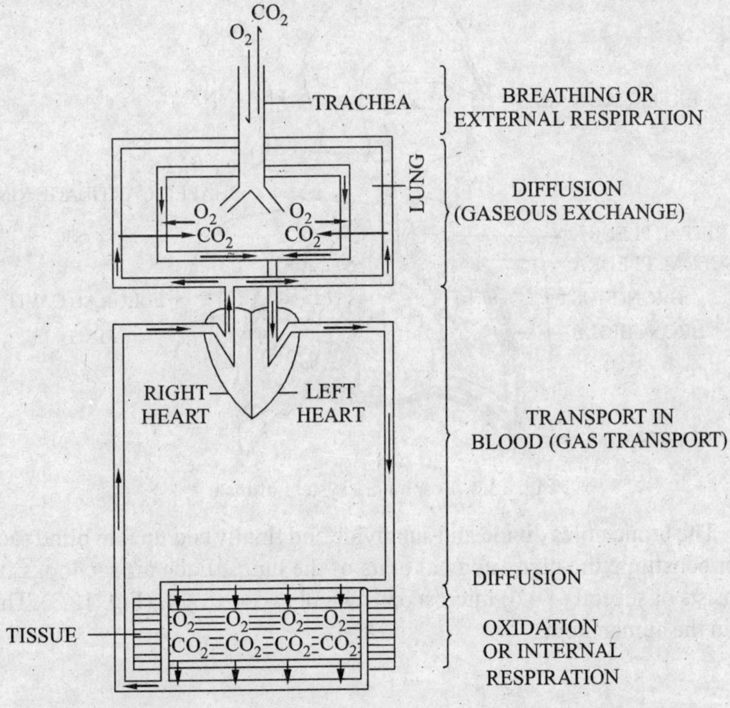

Fig. 13.1. Diagrammatic representation of exchange of gases in the body.

EXTERNAL RESPIRATION

Organs of Respiration

The respiratory system consists of the air passages, lungs and associated organs.

Respiratory Passage: It consists of the air passages (Fig. 13.2) of the *nasal canals, pharynx, larynx, trachea, bronchi*, and *bronchioles*. Air enters through the *nostrils* which lead into two internal openings called *internal nares;* the latter open into the pharynx.

Situated between the pharynx and the trachea is a cartilaginous chamber, known as the *larynx;* it is associated with the production of sound. One of the cartilages of the larynx is called the *epiglottis.* The latter serves as a lid which automatically closes the opening of the larynx during the act of swallowing.

The *trachea* is a tubular structure about 4.5 inches long and having a diameter of about 1 inch. It extends from the larynx downward through the neck into the *thorax* where it divides into two branches — the *right* and the *left bronchi*. The trachea is provided with a series of 15 -20 *C*-shaped cartilage rings (Fig.13.2) *incomplete dorsally*, which prevent the trachea from collapse.

The right and left bronchi enter the lungs of their respective sides. Immediately upon entering the lungs, they divide and subdivide into progressively smaller and narrower *bronchial branches*. Because the right lung consists of three and the left lung of two primary lobes, the right bronchus initially divides into three and the left bronchus into two main branches— one for each lobe of the lung.

The bronchial tubes have the same type of cartilaginous rings as the trachea. As the bronchi divide into smaller and narrower tubes, the cartilage rings become fewer in number. When the smaller bronchial tubes attain a diameter of 1 mm or less, the cartilage is entirely lacking and the tubes are referred to as *bronchioles*.

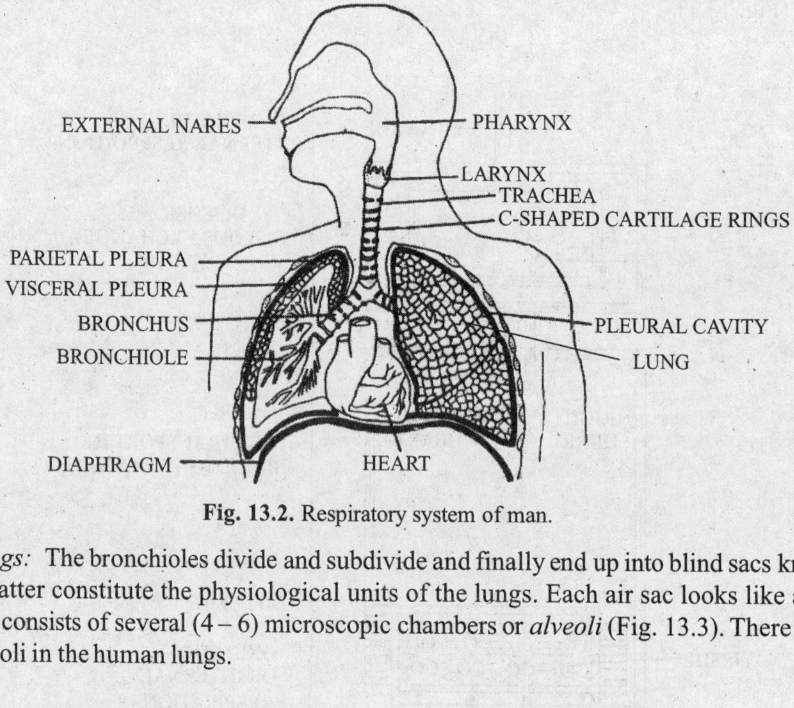

Fig. 13.2. Respiratory system of man.

The Lungs: The bronchioles divide and subdivide and finally end up into blind sacs known as *air sacs*. The latter constitute the physiological units of the lungs. Each air sac looks like a cluster of grapes and consists of several (4 – 6) microscopic chambers or *alveoli* (Fig. 13.3). There are about a billion alveoli in the human lungs.

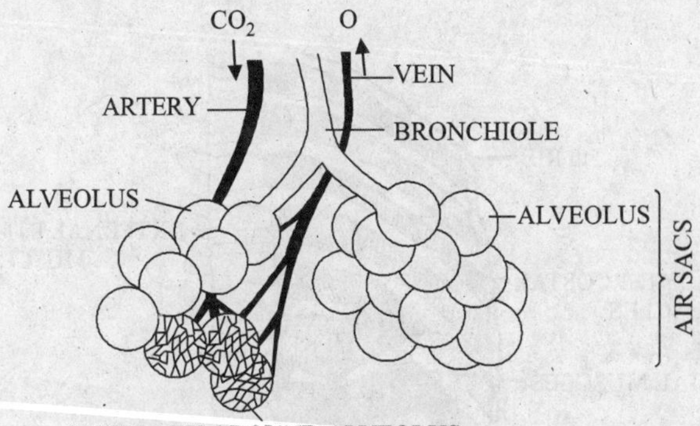

Fig 13.3. Diagram showing the interrelation between the alveoli of the lungs and the blood vessels.

The respiratory tract is lined by ciliated columnar epithelial cells. The cilia beat synchronously propelling mucus and foreign bodies outwards towards the main bronchi and trachea from which they are finally cleared out by coughing. The walls of the pulmonary or lung alveoli, on the other hand, are simply composed of a single layer of cuboidal non-ciliated epithelial cells. The alveoli are surrounded by a rich network of capillaries (Fig. 13.3). These capillaries receive the blood supply from the *pulmonary artery* which is returned through the *pulmonary vein*. The exchange of gases between the air and blood takes place in the alveoli.

The alveolar wall is permeable only to gases, but not to plasma in the capillaries. Furthermore, the blood pressure in the capillaries is only 10 mm Hg which is not enough for the ultrafiltration of blood. Therefore, the alveoli are kept dry. However, if the capillary blood pressure becomes high, fluid escapes into the alveoli and causes a condition known as *pulmonary edema.*

Associated Organs: The *thorax* (synonymous with chest) which contains the lungs is bounded by the rib cage. On the lower side of the thorax, however, the lung is bounded by a dome-shaped sheet of skeletal muscle — the *diaphragm* (Fig. 13.2).

Within the thorax the lungs are enclosed in the *pleural cavities*. The pleural cavity is a membranous double-walled structure; the inner wall adjacent to the lungs is known as the *visceral pleura,* whereas the outer wall is called the *parietal pleura.* There is a thin layer of *serous fluid* between the two pleuras. This prevents friction of the lungs during breathing movements. When there is accumulation of excess fluid between the pleural membranes, the pleuras get inflamed and cause a disease known as *pleurisy*.

Mechanism of Breathing (Ventilation)

Breathing is the process by which air is taken in and expelled out of the lungs. The former process is known as *inspiration* and the latter *expiration.* Normally, a man breathes 15-20 times a minute. Breathing movements are caused by the alternate expansion and contraction of the chest cavity. When the chest expands, its volume increases so that the pressure within the chest cavity falls below the atmospheric pressure. Air from outside enters the lung under pressure (inspiration). When the chest cavity contracts, its volume is reduced and the air is forced out (expiration). The change in volume of the chest cavity is brought about by the alternate contraction and relaxation of the diaphragm and *external intercostal muscles* present between the ribs (Fig. 13.4).

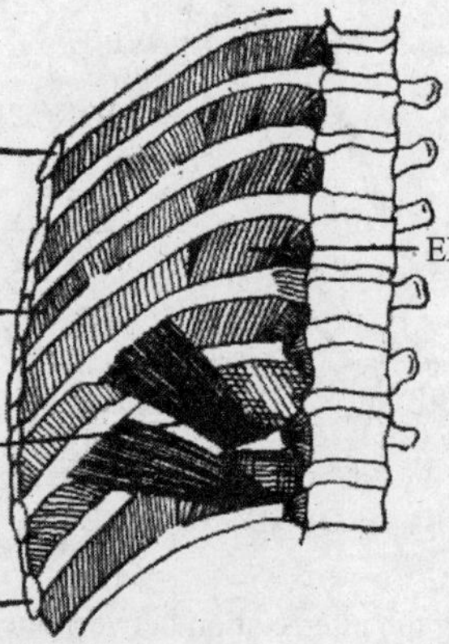

Fig. 13.4. The muscles associated with the ribs; the relation between the internal and external intercostal muscles can be clearly seen.

Enlargement of chest cavity is attained by the contraction of the dome-shaped *diaphragm;* it flattens and descends in the abdominal cavity. Simultaneously, the contraction of the external inter-costal muscles raises the ribs and the breastbone upwards and forwards. These two movements together decrease the pressure within the chest cavity. Because of the decreased pressure, air flows into the lungs and initiates the inspiratory phase of breathing. Subsequent relaxation of the diaphragm and the external intercostal muscles is responsible for a decrease in chest volume, and a consequent increase in the internal pressure on the lungs. This causes the air to be expelled out to initiate the expiration phase of breathing.

Respiratory movements are primarily under the control of the *autonomic nervous system.* However, a person can *voluntarily* speed up or slow down or even stop the rate and depth of breathing movements for a limited period of time.

Lung Capacity: The lung is a spongy organ which always contains some air in it. This air, which is about 1200 ml, cannot be removed from a living individual and is known as the *residual volume*. In addition to the residual volume, another 1000 ml, known as the *expiratory reserve volume* of air is always present in the lungs. This, however, can be forcefully exhaled. Thus, a man breathing normally always contains about 2200 ml (residual volume+expiratory reserve volume) of air in the alveoli which keeps them in an expanded state. This is known as the *functional residual capacity.* During a forceful inspiration a healthy person can inspire a maximum of 3000 ml of air. This is known as the *inspiratory capacity.* However, under resting conditions only 500 ml of air is exchanged. This is known as the *tidal volume*. The remaining 2500 ml is designated as *inspiratory reserve volume.*

An individual who has inspired fully has about 5200 ml of air in the lungs; 3000 ml of inspired air, 1000 ml of expiratory reserve volume, and 1200 ml of residual volume. As this is the maximum amount of air which the lungs can hold, it is known as the *total lung capacity.* The sum of the inspiratory capacity (3000 ml) and the expiratory reserve volume (1000 ml) together are known as the *vital capacity,* which is about 4000 ml. Of the tidal volume (500 ml), only 360 ml reaches the alveoli and the remaining 140 ml is trapped as dead air in the air passages. A diagrammatic representation of the lung capacity is given in Fig. 13.5.

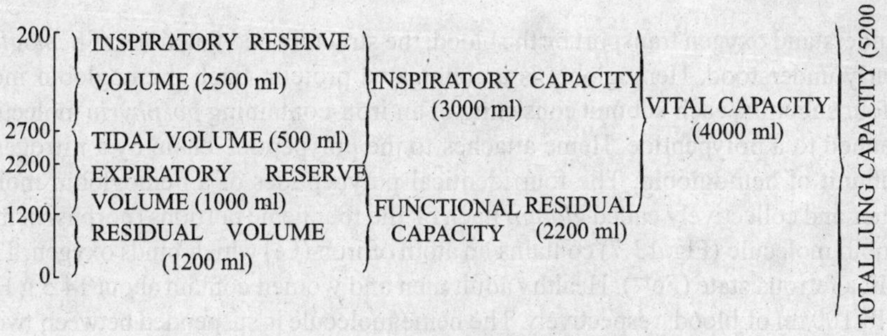

Fig. 13.5. Total lung capacity.

Gaseous Exchange

Oxygen of the inspired air diffuses into the blood through the walls of the alveoli and the capillaries. Likewise, carbon dioxide from the blood diffuses into the alveoli. The diffusion of the gases depends upon the difference in partial pressure across the capillary membrane.

In a mixture of gases, each gas exerts its own pressure. The total pressure of dry air in the alveoli is 713 mm Hg, of which oxygen constitutes about 14%. The partial pressure of oxygen (P_{O_2}) in alveoli, therefore, is approximately 100 mm Hg ($P_{O_2} = 713 \times 0.14 = 92.82$). The partial pressure of carbon dioxide (P_{CO_2}) of the alveolar air is 36 mm Hg. The P_{O_2} and P_{CO_2} of venous blood is approximately 40 and 46 mm Hg, respectively. Therefore, the difference in pressure of 60 mm Hg allows the passage of oxygen from the pulmonary alveoli to venous blood which is thus oxygenated. Similarly, there is difference of pressure of carbon dioxide between the venous blood and alveoli. The difference is about 10 mm Hg, and is sufficient to cause the diffusion of carbon dioxide from venous blood to alveoli (Fig. 13.6). After gaseous exchange, blood has a P_{O_2} of 100 mm Hg and P_{CO_2} of 36 mm Hg.

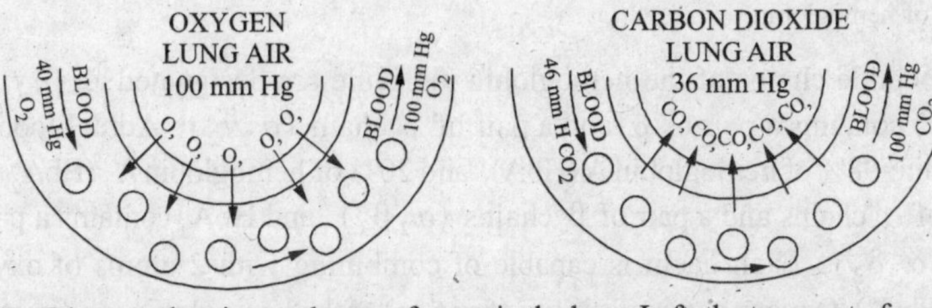

Fig. 13.6. Diagram showing exchange of gases in the lung. Left, the transport of oxygen from the lung to the blood due to higher partial pressure of O_2 in the lung. Right, the transport of CO_2 from the blood to the lung due to higher partial pressure of CO_2 in the blood.

The diffusion of nitrogen, which forms a considerable portion of the atmospheric air, does not arise because the pressure of nitrogen is the same in blood as in the pulmonary alveoli.

GAS TRANSPORT

Oxygen is relatively insoluble in plasma (about 0.3 ml/100 ml). Therefore, only part of it is carried in solution in plasma; most of the oxygen is transported by *hemoglobin,* a respiratory pigment present in the erythrocytes. Most of the carbon dioxide, on the other hand, is transported by the plasma in the form of soluble bicarbonates.

Oxygen Transport

In order to understand oxygen transport by the blood, the structure and functions of h*emoglobin (Hb)* must be clearly understood. Hemoglobin is a conjugated protein. Each hemoglobin molecule is made up of four subunits, each subunit consisting of an iron-containing *porphyrin* molecule known as *heme* attached to a polypeptide. Heme attaches to the polypeptide chain by a nitrogen atom to form one subunit of hemoglobin. The four identical polypeptides of a hemoglobin molecule are bound together and collectively called *globin*. Each of the four heme portions (porphyrin molecules) of a hemoglobin molecule (Fig. 13.7) contains an atom of iron (Fe) which binds oxygen. The iron in the heme is in a ferrous state (Fe^{++}). Healthy adult men and women contain about 14.5 g Hb/100 ml and 13.5 g Hb/100 ml of blood, respectively. The heme molecule is suspended between two residues of histidine with its iron atom directly linked to one of the histidine residues.

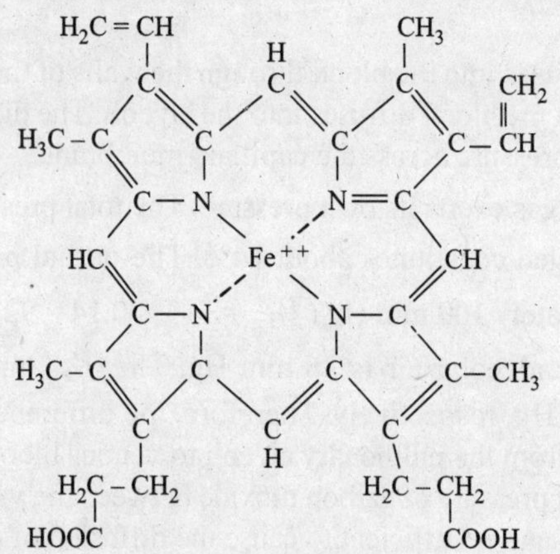

Fig. 13.7. Heme-porphyrin; the iron-containing porphyrin molecule that forms the oxygen-binding portion of hemoglobin.

The four polypeptide chains of the hemoglobin molecule are designated α, β, γ and δ. *Fetal hemoglobin* (*HbF*) contains a pair of α and a pair of γ chain $(\alpha_2\gamma_2)$. Adult blood contains two types of hemoglobin: 98% of hemoglobin A (HbA), and 20% of hemoglobin A_2 (HbA_2). Hemoglobin A contain a pair of α chains and a pair of β chains $(\alpha_2\beta_2)$, and HbA_2 contains a pair of α and a pair of δ chains $(\alpha_2\delta_2)$. Each haem is capable of combining with 2 atoms of oxygen to form *oxyhemoglobin* (HbO_2). Oxygen carrying ability (fixation) of the hemoglobin is a function of the iron atom. The association of hemoglobin with oxygen is *unstable*. The interconversion $Hb + O_2 \rightleftharpoons HbO_2$ depends on the partial pressure of oxygen P_{O_2}. Oxyhemoglobin is formed when P_{O_2} is high and gives up oxygen when P_{O_2} is low. It must be borne in mind that this reaction is not an oxidation-reduction as the iron remains in the ferrous state (Fe^{++}) in both hemoglobin and oxyhemoglobin. The oxidation product of hemoglobin and oxyhemolgobin is distinctly different from oxyhemoglobin and is called as *ferrihemoglobin* (Fe^{+++}) or *methemoglobin.*

As one atom of iron combines with one molecule of oxygen, one molecule of hemoglobin containing four atoms of iron will, therefore, combine with four molecules of oxygen. One molecule of hemoglobin with molecular weight 64,500 d will carry four gram moles (32 × 4 grams) of oxygen. As one gram-mole of oxygen at STP (standard-temperature and pressure) has a volume of 22.4 litres, therefore four gram-moles of oxygen will have a volume of 22.4 × 4 litres. One gram of hemoglobin will therefore carry 1.34 ml of oxygen $\left(\frac{22.4 \times 4}{66500} = 0.00134 \text{ litres, or } 1.34 \text{ ml}\right)$. This is called the *respiratory capacity* of the blood. Assuming the amount of hemoglobin to be 14.5 g/100 ml blood, it is obvious that 100 ml of blood will carry 19.43 ml (14.5 × 1.34) in the form of HbO_2. In addition, 0.393 ml is carried in the plasma in soluiton. Therefore, the total amount of oxygen carried in 100 ml of blood will be approximately 20 volumes per cent. This is called the *total respiratory capacity* of the blood. This represents the maximum capacity (100 per cent saturation) of the hemoglobin for oxygen when the blood is fully oxygenated. However, under normal conditions arterial blood exposed to the alveoli of the lungs is not 100% oxygenated with an oxygen tension of 100 mm Hg; it is usually 98% saturated.

Carboxyhemoglobin: Hemoglobin can combines with gases other than oxygen. For example, it combines reversibly with carbon monoxide ($CO + Hb \rightleftharpoons COHb$) forming carboxyhemoglobin. Carbon monoxide is known for its toxic effect in causing what is known as carbon monoxide poisning, and is an essential component of coal gas and exhaust fumes of automobiles. The affinity of hemoglobin for carbon monoxide is about 300 times high as compared to oxygen. A very little of carbon monoxide (even 0.1% in the immediate surrounding of a person) will bind most of the hemoglobin. As a result, it deprives the blood of the ability to transport oxygen and the affected person may die of *anoxia* (lack of oxygen). Carbon monoxide of carboxyhemoglobin can be displaced under very high pressures of oxygen. Therefore, a person suffering from carbon monoxide poisoning can be revived by placing him into a high pressure oxygen chamber (*hyperbaric* oxygen).

Hemoglobin-Oxygen Equilibrium

Four factors, viz., partial pressures of oxygen and carbon dioxide, temperature, H^+ concentration, and 2, 3-diphosphoglyceraldehyde (2, 3-DPG) influence the equilibrium $Hb + O_2 \rightleftharpoons HbO_2$.

P_{O_2} **:** The quantity of oxygen which can be held by hemoglobin depends on the partial pressure of oxygen P_{O_2}. The relationship between the two can be graphically represented by a curve called *oxygen-hemoglobin dissociation curve* (Fig. 13.8). The curve is *sigmoid* (S-shaped) in nature, and depicts the salient features concerning the uptake and dissociation of oxygen by hemoglobin at varying partial pressures of oxygen. The essential feature of the oxygen-hemoglobin dissocation curve is that

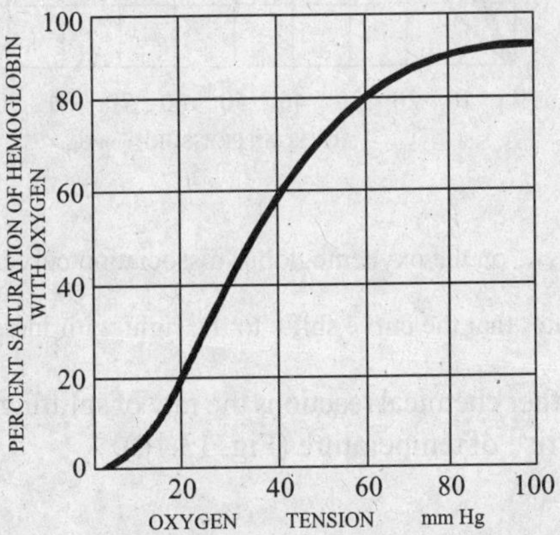

Fig. 13.8. Graph showing the relationship between the partial pressure of oxygen and the formation of oxyhemoglobin– 'Oxyhemoglobin-dissociation curve'.

hemoglobin takes up oxygen when the partial pressure of the latter is high; oxyhemoglobin (HbO_2) dissociates when P_{O_2} is low. The hemoglobin is completely oxygenated (approximately 98% saturation) by an oxygen tension of 100 mm Hg. However, when the oxygen tension falls below 60 mm Hg the oxyhemoglobin dissociates rapidly. This is indicated by the steeper slope of the curve at low oxygen tensions (Fig. 13.8). When the P_{O_2} is zero, oxyhemoglobin loses all its oxygen. The hemoglobin gets completely oxygenated (98% saturation) in the alveoli of the lungs where the oxygen tension is 100 mm Hg and, later, donates the oxygen to the tissues where the oxygen tension is low and where oxygen is needed most.

$\boldsymbol{P_{CO_2}}$: Partial pressure of carbon dioxide also determines the equilibrium $Hb + O_2 \rightleftharpoons HbO_2$. The dissociation of oxyhemoglobin at four different tensions of carbon dioxide is depicted in Fig. 13.9. It shows that oxyhemoglobin releases oxygen more rapidly at higher concentrations of carbon dioxide. In effect this rapid dissociation of HbO_2 is due to the acidity produced by the carbon dioxide. Thus, a fall in pH of blood (below 7.4) shifts the oxygen-hemoglobin dissociation curve to the right. This is known as the *Bohr effect*. Bohr effect is of great biological importance. The cells and tissues of the body respiring at a rapid rate have high levels of carbon dioxide. This causes a rapid release of oxygen in these tissues. Conversely, the uptake of oxygen and the formation of oxyhemoglobin in the lungs is facilitaed by the removal of carbon dioxide by exhalation.

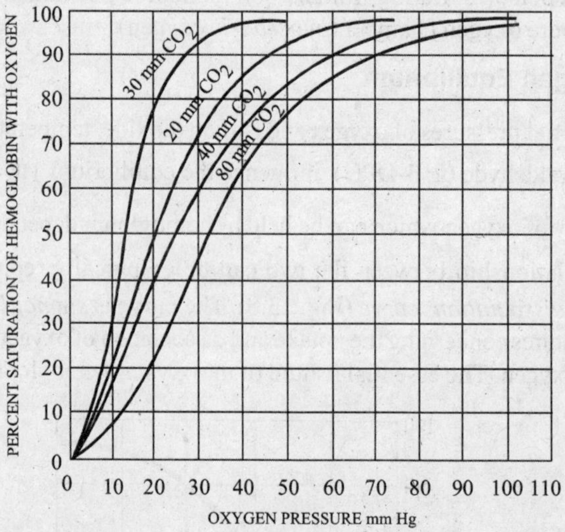

Fig. 13.9. Effect of P_{CO_2} on the oxyhemoglobin-dissociation curve. The graph is drawn at four different P_{CO_2} ; note that the curve shifts to the right with increase in P_{CO_2} .

Temperature: Like all other chemical reactions the rate of splitting of oxyhemoglobin to hemoglobin and oxygen increases with rise of temperature (Fig. 13.10).

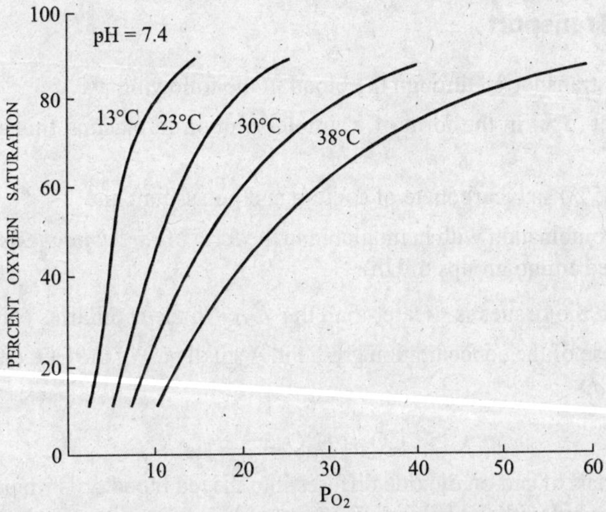

Fig. 13.10. Effect of temperature on the oxyhemoglobin-dissociation curve. The graph is drawn at four different temperatures; note that the curve shifts to the right with increase in temperature.

Thus, an active tissue which has a relatively high P_{CO_2}, low pH, and raised temperature sets free more oxygen from hemoglobin for its respiratory processes.

H^+ concentration: The H^+ concentration in plasma also modifies the equilibrium $Hb + O_2 \rightleftharpoons HbO_2$. At low oxygen tensions oxyhemoglobin liberates oxygen more easily in the presence of hydrogen ions than in their absence. This fact is of great physiological importance in the tissue capillaries where the H^+ concentration is high and oxygen tension low; therefore, more oxygen is liberated from the oxyhemoglobin.

The question arises as to what is the mechanism by which greater P_{CO_2}, hydrogen ions, and temperature influence hemoglobin's affinity for oxygen? Carbon dioxide and hydrogen ions do so by combining with the Hb and altering its molecular configuration. Thus, these effects are forms of allosteric modulation. An elevated temperature also decreases hemoglobin's affinity for oxygen by altering its molecular configuration.

Yet another factor which influences the oxygen-hemoglobin dissociation curve is the 2, 3-*diphosphoglyceraldehyde* (2, 3-DPG) concentration in the erythrocytes. DPG is synthesized in the erythrocytes from 1, 3-*diphosphoglyceraldehyde* with the help of enzymes *disphosphoglyceromutase.* DPG *stabilizes* the hemoglobin; therefore, its affinity for oxygen is greatly reduced and oxyhemoglobin rapidly releases oxygen; the oxygen dissociation curve shifts to the right. The synthesis of DPG is stimulated by hypoxia and high altitudes. This gives the additional amount of oxygen needed during these conditions.

In summary, an increase in blood P_{CO_2}, H^+ concentration, and DPG causes the dissociation curve to shift to the right which means that at any given P_{O_2}, hemoglobin has less affinity for oxygen. In contrast, decrease in any of these factors causes the dissociation curve to shift to the left, which indicates that, at any given P_{O_2}, hemoglobin has a greater affinity for oxygen.

Finally, it should be obvious that the amount of oxygen transported in blood will be dependent on the hemoglobin content of blood; blood containing less hemoglobin will carry less oxygen to the cells and tissues. This happens in *anemic* persons.

Carbon Dioxide Transport

Carbon dioxide is transported through the blood in the following ways:

(1) a small amount (5%) in the form of a simple solution in plasma; this amount is more than dissolved O_2;

(2) a major part (85%) as bicarbonate of sodium and potassium; and

(3) about 10% in combination with hemoglobin in the form of *carbaminohemoglobin* (combination of CO_2 with free amino groups of Hb).

When the P_{CO_2} in the tissues is greater than the P_{CO_2} in the capillaries carbon dioxide diffuses into capillaries because of the concentration gradient. A small part of the CO_2 dissolves in the plasma to form *carbonic acid.*

$$CO_2 + H_2O \rightleftharpoons H_2CO_3$$

However, the major part of carbon dioxide diffuses into the red blood cells where it is converted very rapidly into carbonic acid with the help of an enzyme *carbonic anhydrase*. In fact, the speed of the reaction is about 1,300 times greater in the erythrocytes than in the plasma. The carbonic acid dissociates into bicarbonate and hydrogen ions.

$$H_2O + CO_2 \underset{\text{anhydrase}}{\overset{\text{carbonic}}{\rightleftharpoons}} H_2CO_3 \rightleftharpoons HCO_3^- + H^+$$

The hydrogen ions thus liberated are taken up by the protein molecules of the blood in exchange for metal ions (K^+ and Na^+). Diffusion of carbon dioxide from respiratory tissues into blood, therefore, leads to the formation of bicarbonates of potassium and sodium. Carbonic anhydrase also catalyses the splitting of the H_2CO_3 to H_2O and CO_2 in the lungs. This results in a rapid release of carbon dioxide during expiration.

When the carbon dioxide concentration is high, some of it entering the erythrocytes from the tissues reacts with amino group ($-NH_2$) of hemoglobin to form *carbaminohemoglobin.*

$$HbNH_2 + CO_2 \underset{\text{Low } P_{CO_2}}{\overset{\text{High } P_{CO_2}}{\rightleftharpoons}} HbNHCOOH$$

The reaction is very rapid although no enzyme is required. Low carbon dioxide concentration leads to the dissociation of carbaminohemoglobin to CO_2 and hemoglobin. Carbaminohemoglobin formed at the site of high P_{CO_2} gives up carbon dioxide when it reaches the lungs where the P_{CO_2} is low.

Plasma proteins, like hemoglobin, also transport some CO_2 in the form of *carbamino compounds.*

$$\underset{\text{(protein)}}{Pr\,NH_2} + CO_2 \underset{\text{Low } PCO_2}{\overset{\text{High } P_{CO_2}}{\rightleftharpoons}} \underset{\text{(carbamino compound)}}{Pr\,NHCOOH}$$

In summary, it can be said that both oxygen and carbon dioxide compete for hemoglobin in order to form oxyhemoglobin and carbaminohemoglobin, respectively. The concentration of oxyhemoglobin rises when the P_{CO_2} is low and P_{CO_2} is high. Similarly, the concentration of carbaminohemoglobin rises when P_{O_2} is low and CO_2 is high. This means that oxygen saturation of hemoglobin affects its CO_2 content. This is known as *Haldane Effect.* Thus, in the arteries where the P_{CO_2} is low, the blood is able to carry more oxygen as compared to the vein where the P_{CO_2} is high.

The transport of hydrogen ions between tissues and lungs may be very briefly described also. Most of the hydrogen ions, generated in the erythrocytes form carbonic acid during blood passage through tissue capillaries bind to *deoxyhemoglobin.* This reaction is facilitated because deoxyhemoglobin formed, as oxygen unloads from oxyhemoglobin, has a high affinity for hydrogen ions. Deoxyhemoglobin is abbreviated HbH rather than Hb to denote its binding with H^+. In effect, the reaction is:

$$HbO_2 + H^+ \rightleftharpoons HbH + O_2$$

In this manner only a small number of hydrogen ions generated in the blood remain free. This explains why the acidity of the venous blood is only slightly greater than that of arterial blood. The pH of arterial blood is 7.4 and that of venous blood is 7.36. (*It must be remembered here that pH goes down as H^+ concentration goes up*). The binding of hydrogen ions to deoxyhemoglobin is reversed as the blood through the lung capillaries and the ions combine with bicarbonate and yield CO_2 and H_2O.

Despite the change in carbon dioxide content of the blood, outlined above, the pH of blood is held remarkably constant at about 7.4. This is achieved by the *buffering mechanism* present in the blood (see Chapter 7).

RHYTHMICITY AND CONTROL OF BREATHING

Breathing is caused by the rhythmic contraction and relaxation of the diaphragm and the external intercostal muscles of the chest. These muscles consist of skeletal muscles and cannot contract unless stimulated to do so by nerves. The movement of these muscles is brought about by the generation of *efferent* and *afferent nerve* impulses. The efferent nerve impulses are generated at *respiratory centres* in the anterior *medulla oblongata* and posterior *pons* (part of the brain stem just above the medulla); the afferent form the *pulmonary stretch receptors* (Hering-Breuer *receptors*) in the lungs.

The *respiratory centre* consists of three functional units, *viz.*, the *pneumotaxic* or *pontine centre* in the rostral-most part of the pons, the *expiratory centre* in the lower and middle pons, and the *inspiratory centre* in the ventro-lateral medulla. These units are connected in such a manner that the inspiratory centre both excites and inhibits the expiratory centre, the expiratory centre only inhibits the pneumotaxic centre, and the pneumotaxic centre inhibits the inspiratory centre (Fig. 13.11).

The *inspiratory centre* (cluster of neurons) sends out *efferent* nerves (phrenic nerves to the diaphragm and intercostal nerves to the intercostal muscles which cause the contraction of these muscles. This accounts for the *inspiratory phase* of the breathing cycle. The inspiratory centre simultaneously excites the expiratory centre. This in turn stimulates the pneumotaxic centre; the latter then inhibits the inspiratory centre. This stops the discharge of its inspiratory impulses through the efferent nerves. As a result of this, the diaphragm and the external intercostal muscles relax and the expiratory phase of breathing cycle commences. It is obvious from the above discussion, therefore, that *inspiration is an active process while expiration passive.* The expiratory centre, nervertheless, is also connected with efferents to the diaphragm and external muscles, which can bring about active expiration.

The inspiratory centre sends out continuous impulses so that if all nerves except the efferents are severed, a continuous inspiration (apneusis) occurs.

The rhythmicity of breathing is also influenced by *afferent impulses* generated by the *pulmonary stretch receptors* (Hering-Breuer *receptors*) of the lungs. The stretch receptors discharge impulses when the lungs inflate. These impulses travel upwards in the *vagus nerves* and inhibit the inspiratory centre. This is known as *Hering-Breuer reflex*. The Hering-Breuer receptors act like "circuit breaker" and are stimulated only when they are fully stretched. This takes places at the height of the inspiratory phase when the lungs reach a certain threshold tension. The time delay between the onset of inspiration and the simulation of Hering-Breuer receptors allows successful completion of the inspiratory phase.

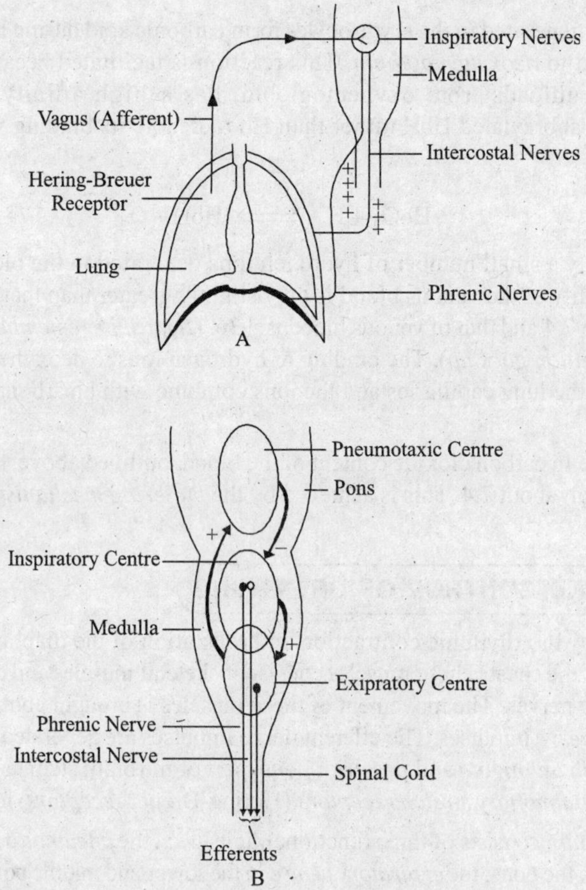

Fig. 13.11. Diagrams showing regulation of respiratory rhythm. (A) The Hering-Breuer mechanism; (B) The Pontine mechanism.

However, even if the vagus nerves are cut, the breathing continues to be rhythmic, although it comes deeper and slower. Hering-Breuer reflex, therefore, is not the only mechanism that causes rhythmicity of breathing. The respiratory centres along with the Hering-Breuer reflex make the rhythm in breathing more reliable.

Voluntary control of breathing is also mediated through the same nervous mechanism as involuntary breathing. In the former case the higher centres of the brain, cerebral cortex, send out impulse to the respiratory centres and control their activity.

Strong emotions, fright, excitement, and other mental factors also have a striking influence on the breathing process. Choking and coughing are reflex acts of protective value which are usually caused by stimulation of nerve endings in the lining of the respiratory passages by chemicals or foreign particles. Choking and coughing generally result in expulsion of the irritating substances.

Factors Affecting Respiratory Centre

Several chemical and physical factors influence the respiratory centre and, therefore, the rate and depth of the breathing process or *ventilation*. There are several inputs to the medullary inspiratory neurons, but the most important for the involuntary control of ventilation are from peripheral chemoreceptors and central chemoreceptors. The former are located in the neck at the bifurcation of the common carotid arteries and in the thorax on the arch of the aorta, called the carotid and aortic bodies (Fig. 13.12). The carotid bodies contain epithelial-like cells and neuron terminals in close contact with arterial blood. Afferent nerve fibres arising from these terminals pass to the brain stem, where they synapse with medullary inspiratory neurons. The peripheral chemoreceptors are sensitive to changes in the arterial P_{O_2}, H^+ concentration, and P_{CO_2}. The central chemoreceptors are located in the medulla and provide inputs to the medullary inspiratory centre. They respond to the P_{CO_2}-dependent changes in the H^+ concentration due to the reaction:

$$CO_2 + H_2O \underset{\text{anhydrase}}{\overset{\text{Carbonic}}{\rightleftharpoons}} H_2CO_3 \rightleftharpoons HCO_3^- + H^+$$

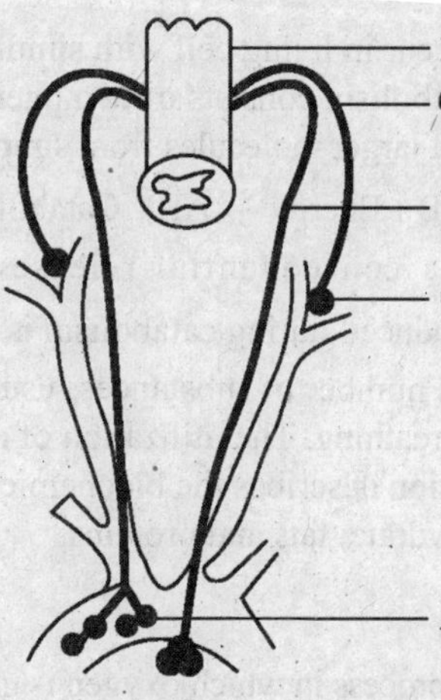

Fig. 13.12. Peripheral chemoreceptors for the involuntary control of ventilation.

The other important factors are: (1) the P_{CO_2}, the H^+ concentration of the fluid surrounding the nerve cells, and (3) the P_{CO_2}. Amongst these the P_{CO_2} in the blood is perhaps the most important factor as it influences the other factors also.

The respiratory centre is highly sensitive to the partial pressure of carbon dioxide in arterial blood which is normally 40 mm Hg. Even a slight rise in the P_{CO_2} excites the respiratory centre in the brain resulting in an increased rate and the depth of the breathing cycle. An increase of 1.5 mm Hg in the P_{CO_2} of arterial blood, for example, results in the doubling of the rate of ventilation. Similarly, increased acidity of the blood also has an excitatory effect on the inspiratory centre. Changes in venilation due to changes in arterial P_{CO_2} are brought about by both peripheral and central chemoreceptors. A rise in the concentration of the level of lactic acid or keto acids causes an increase in the rate and depth of breathing.

Small changes in the oxygen concentration do not affect rate of breathing. However, low P_{O_2} excites the inspiratory centres through carotid and aortic bodies chemoreceptors. At high altitudes of 10,000 feet or more the oxygen pressure of atmosphere is greatly reduced. At these altitudes the rate of breathing increases. This increased rate of breathing cannot be ascribed to an increase in arterial carbon dioxide. At still higher altitudes where the P_{O_2} in the atmosphere is so low that even increased rate of breathing cannot compensate for the insufficiency of oxygen; consequently, *altitude sickness* develops.

Body temperature and *blood pressure* also influence the breathing rate. A rise in the blood temperature during fever or exercise stimulates breathing, whereas a lowered body temperature has the opposite effect. A decrease in blood pressure causes a decrease in breathing rate.

A very interesting phenomenon which should not escape mention is the initiation of breathing in a newborn baby. Several factors are in fact, involved in this process. There is a rise in P_{O_2} and fall in P_{O_2} of the arterial blood of the baby when the connexion between the baby's circulation and that of the mother is severed at the time of the birth. This causes sudden excitation of the inspiratory centre and the breathing process is initiated. In case the breathing does not start, stimulation of skin receptors, for example by slapping the buttocks of the newborn, also serves to stimulate the respiratory centre in a reflex manner.

INTERNAL RESPIRATION

A large number of chemical reactions occur in living cell with simultaneous energy changes. These are collectively called *metabolism*. Metabolism consists of two phenomena: (1) *anabolism* and (2) *catabolism*. Anabolism is the synthesis of larger molecules from simpler ones. This process generally consumes energy (*i.e., endergonic*; $A + B + \text{Energy} \rightarrow AB$). Catabolism is the degradation of larger molecules into smaller ones with a consequential release of energy (*i.e., exergonic*; $AB \rightarrow A + B + \text{Energy}$). The energy produced during catabolism is utilized for anabolic processes. During catabolism, oxidation of a large number of substances, usually nutrients, takes place. The oxygen for oxidation is obtained from breathing. The utilization of oxygen within the living cells is known as *internal respiration*. This section describes the biochemical pathways by which energy is obtained from the breakdown of carbohydrate, fats and proteins.

Oxidation-Reduction

In a classical sense oxidation involves a process in which oxygen is gained, and reduction is a process where hydrogen is gained. In all cases where oxgyen is gained, there is loss of an *electron*. Similarly, when hydrogen is added there is gain of an electron. According to the modern concept, all reactions which involve the loss of an electron are termed *oxidation*, while those which involve the gain of an electron are called *reduction*. This definition, therefore, includes those reactions also where there is no involvement, of oxygen and hydrogen. *All biological oxidations are simply reactions in which there is transfer of electrons.*

$$Fe^{++} \underset{\text{Reduction}}{\overset{\text{Oxidation}}{\rightleftharpoons}} Fe^{+++} + \text{electron}$$

In the above reaction when *ferrous* iron is changed to *ferric* iron, it is an oxidation reaction because it involves a loss of an electron, whereas when ferric is changed to ferrous, it is a reduction as it involves the gain of an electron. Because oxidation and reduction go hand in hand, such reactions are known as *redox reactions*.

Chemical reactions, however, do not generally involve free electrons or atoms. A compound, thus, can only be oxidised if an equivalent amount of another compound is simultaneouly reduced. Redox reactions can, therefore, be represented in the following way, where A and B are the two compounds taking part. The subscripts denote their oxidised and reduced forms.

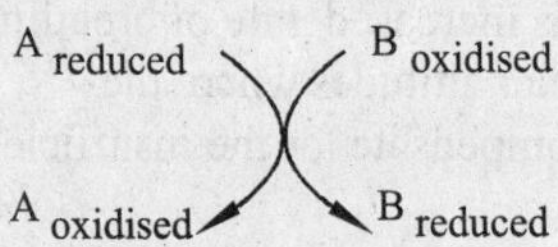

Electron Transport in Biological Oxidations

Redox reactions involving electron transfer occur in almost all biological oxidations. One very important aspect of such reaction is the oxidation of *carbohydrates*, *proteins*, and *lipids* by cells, which yields a considerable amount of *energy*. The energy is used for the mainfold activities of the body.

In most biological oxidations, hydrogens (electrons) are picked up from molecules which are being catabolized, and transferred to specific enzymes. In all living organisms there is always a series of enzymes which participate in the stepwise transfer of electrons. These enzymes are *dehydrogenases, flavoproteins,* and *cytochromes.* These three sets of enzymes are found in all animals. The dehydrogenases either have NAD or NADP as *coenzymes*. Flavoproteins, similarly, have FMN and FAD as coenzymes (see Chapter 8). The dehydrogenases and flavoproteins of different animals differ only in their apo-enzyme (protein) as the coenzymes are always the same (see Chapter 8). There are several cytochromes, which are protein pigments, and are dealt later.

Dehydrogenases remove hydrogen atoms from the substrate, and get reduced in the process. Coenzymes NAD or NADP of the dehydrogenases act as the hydrogen acceptors.

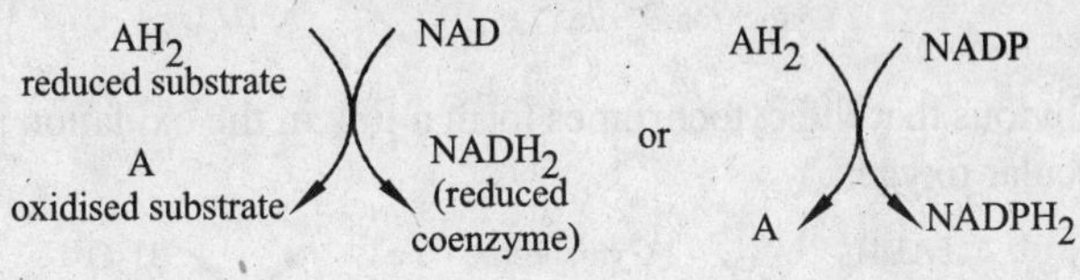

The dehydrogenases involved are highly specific both for the substrate and the coenzyme. The reduced form of the coenzyme NAD or NADP at physiological pH is in the *ionized state* (see Chapter 8). Therefore, the above reaction may be represented as

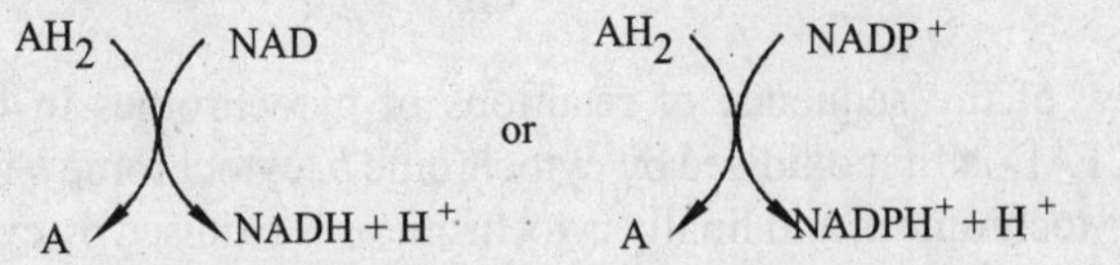

Both NAD^+ and $NADP^+$ are present in very small quantities in the cells. In order not to exhaust their supply in the cells, NADH and NADPH are reoxidised back to NAD^+ and $NADP^+$. This is achieved with the help of either coenzyme FMN or FAD which can undergo reversible reduction and oxidation.

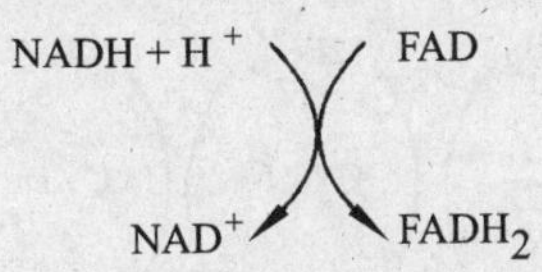

Therefore, the overall reaction for the transfer of 2H from the substrate AH_2 to NAD^+ and then to FAD is as under:

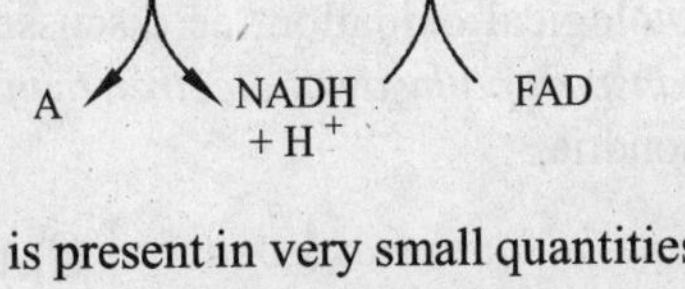

However, like NAD^+, FAD too is present in very small quantities in the cell. Therefore, to keep a steady supply of it, $FADH_2$ is oxidised to FAD with the help of *cytochromes*. In all, there are five

cytochromes; these are b, c_1, c, a, and a_3. The cytochromes participate in redox reactions which involve a change in the valency of the iron atom held in the in the centre of the porphyrin system (see heme).

$$Fe^{+++} \underset{-e}{\overset{+e}{\rightleftharpoons}} Fe^{++} \qquad (e = \text{electron})$$

The cytochromes oxidise the FAD in the following manner and *two molecules of cytochrome are required for each molecule of FAD*

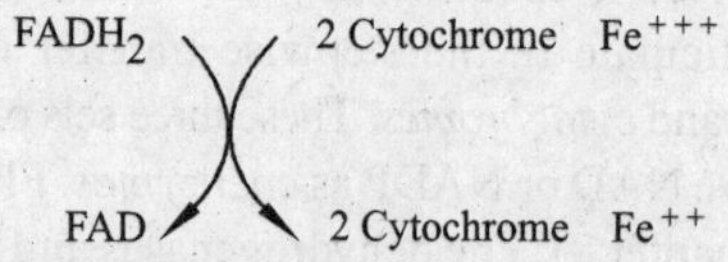

The cytochromes are then reoxidised by molecular oxygen:

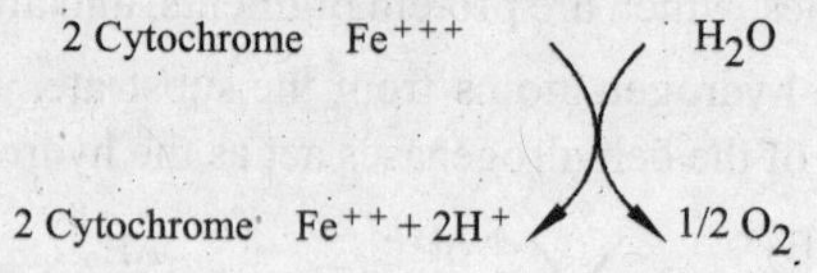

It is, therefore, obvious that the cytochromes form a link in the oxidation process by which FAD is oxidised by molecular oxygen.

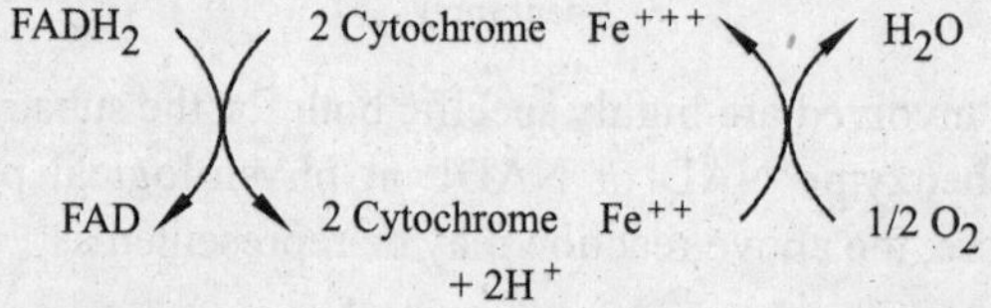

The general view of the sequence of reactions of cytochromes in oxidation of the FAD is diagrammed below. FAD is first oxidised by cytochrome b, cytochrome b is oxidised by cytochrome c, cytochrome c by cytochrome a, and finally cytochrome a is oxidised by cytochrome a_3. Cytochrome a_3 carries out the final stage of transferring the electrons to molecular oxygen and combining it with the hydrogen ions liberated in the electron transport chain (see Krebs' Cycle; Fig. 3.14) to form water. This is the only stage in internal respiration at which oxygen is needed.

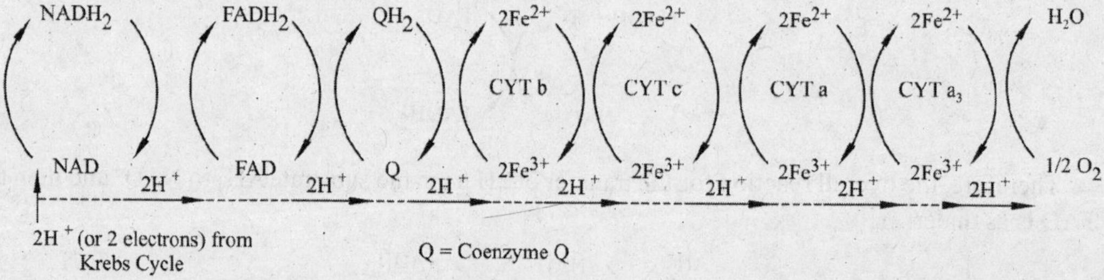

The electron transport in biological oxidation, as discussed above, is also referred to as the *respiratory chain* or *electron transport pathway* or *electron transport system,* and is the main energy transforming system of mitochondria.

Compounds generally known as *ubiquinones* (*coenzyme Q*) also take part in the electron transfer between FAD and cytochrome b. Similarly, cytochrome c_1 participates between cytochrome b and c (see Fig. 13.15). It has been shown that some of the stages in the oxidation chain can be inhibited by certain "poisons". These act by combining with the specific groups in the chain responsible for electron transfer. For instance, *amytal* acts between NAD^+ and FAD, and antibiotic *actinomycin A* between cytochromes *b* and *c*, and *cyanides* inhibit cytochrome a_3.

Oxidative Phosphorylation

In order to maintain a steady state, cells must have a continuous supply of energy. This energy is released in small amounts during the transfer of electrons in the electron transport system. *Phosphorylation* is able to temporarily "store" these small amounts of energy as ATP (adenosine triphosphate). The latter is formed from oxidation in the presence of suitable enzyme system. This phenomenon is called *oxidative phorphorylation.*

Adenosine Diphosphate (ADP) and ATP are the diphosphate and triphosphate of a naturally occurring nucleotide *adenosine-5-monophosphate* (AMP). The nucleotides are formed from adenine, ribose, and phosphate groups attached to the 5 position. AMP, ADP and ATP are interconvertible as illustrated below:

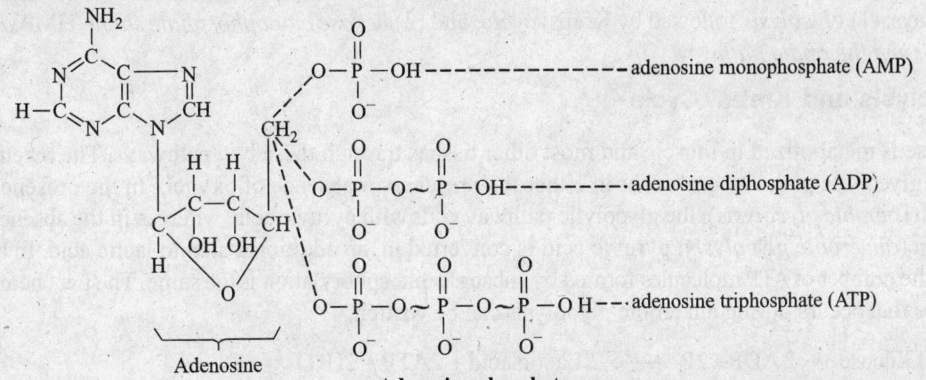

Adenosine phosphate

Adenosine triphosphate (ATP) is rich in chemical energy because it contains two "high energy" bonds (~) between the terminal and second, and between the second and the third phosphate groups. These are *ester linkages* but differ from the ordinary ester bond (–) in possessing high levels of chemical energy. Each of these two high energy bonds contains abut 7,000 calories per mol of ATP in contrast to the 2,000 to 3,000 calories of the ordinary phosphat-ester bond. In fact, when ATP is hydrolysed, the removal of the terminal phosphate group is accompanied by the release of 7,000 calories/mol (or 7 kcal/mol) of energy (ATP + $H_2O \rightarrow$ ADP +Pi + 7 Kcal/mol). It may be noted here that 7 kcal of energy are released when 1 mol (6×10^{23} molecules) of ATP are hydrolyzed, and not just 1 molecule).

It has been found that during the transfer of every $2H^+$ from NAD^+ to the end of the oxidation chain, 3 molecules of ATP are formed. This can be represented in a simplified manner as follows, where A is the substrate being oxidised:

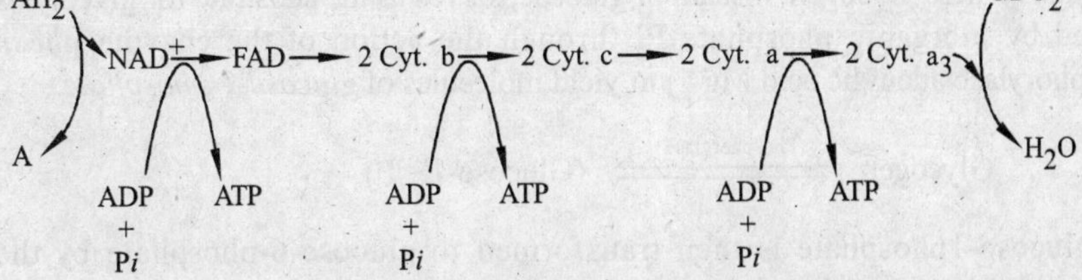

The steps in the electron transfer mechanism concerned with the release of energy for the formation of ATP have been identified. The first molecule of ATP is produced in the reaction between NAD^+

and FAD, the second molecule between the cytochromes *b* and *c*, and third molecule of ATP is produced in the reaction connecting cytochromes *a* and a_3.

From the foregoing description of ATP formation, it follows that electron transport system will only function as long as it is supplied with ADP and P*i* which it can convert to ATP. Moreover, the release of energy during the oxidation process is gradual and stepwise. *It must be emphasized here that cells use ATP not to store energy but rather to transfer it. ATP is an energy-carrying molecule that transfers relatively small amounts of energy from high energy fuel molecules.*

In higher animals, *carbohydrates, fats,* and *proteins* are used to provide *substrates* for this energy-yielding electron (hydrogen) transfer. Of these three nutrients, carbohydrate is the most important source of energy.

CARBOHYDRATE METABOLISM

Monosaccharides in general and those related with glucose in particular are the most common carbohydrates to be metabolized in the cell. *Glucose* may be metabolized by one of the following two pathways: (1) *glycolysis* followed by Krebs' *cycle,* and (2) *hexose monophosphate shunt* HMP *shunt* or *phosphogluconate pathway.*

Glycolysis and Krebs' Cycle

Glucose is metabolized in muscle and most other tissues through these two pathways. The reactions in the glycolytic pathway can occur in either the presence or absence of oxygen. In the presence of oxygen (*aerobic glycolysis*), the glycolytic pathway ends with pyruvic acid, whereas in the absence of oxygen (*anaerobic glycolysis*) pyruvic acid is converted in an additional step to lactic acid. In both cases the number of ATP molecules formed by substrate phosphorylation is the same. The net chemical change that occurs during anaerobic glycolysis can be written:

$$\text{Glucose} + 2\,\text{ADP} + 2\text{P}_i \longrightarrow 2\text{ Lactic acid} + 2\text{ATP} + 2\text{H}_2\text{O}$$

It must be noted here that conversion of glucose to pyruvic acid also occurs during aerobic condition. Additional energy is gained under aerobic conditions by oxidative phosphorylation in the mitochondria by the Krebs' cycle (also called the *citric acid* or *tricarboxylic acid or TCA cycle*).

There are special cases in which glycolysis supplies all of a cell's ATP. For example, erythrocytes contain the enzymes for glycolysis but have no mitochondria. All of their ATP production occurs by glycolysis.

Glycolysis: Glycolysis is also known as anaerobic respiration or **Embden-Meyerhof** pathway. This takes place in the cytoplasm of the cell; the different enzymes required are present in the form of soluble molecules. Glycolysis takes place in following *steps*, and as shown in Fig. 13.13.

1. Glucose is phosphorylated to *glucose-6-phosphate* by reaction with a molecule of ATP which itself is converted into ADP. The reaction is Mg^{++} dependent, and is catalysed by the enzyme *glucokinase (hexokinase).*

   In animals where glycogen, instead of glucose, serves as the substrate for glycolysis, it is first cleaved by inorganic phosphate Pi through the action of the enzyme *phosphorylase* (phosphorylase adenylic acid Mg^{++}) to yield molecules of *glucose-1-phosphate.*

$$\text{Glycogen} \xrightleftharpoons{\text{Phosphorylase}} (\text{Glucose-1} \sim \text{P})_n$$

   The glucose-1phosphate is then transformed to glucose-6-phosphate by the enzyme *phosphoglucomutase (phosphoglucomutase* Mg^{++}*).*

$$\text{Glucose-1} \sim \text{P} \xrightleftharpoons{\text{Phosphoglucomutase}} \text{glucose-6} \sim \text{P}$$

Glucose-6-phosphate should, therefore, be considered as the starting point of glycolysis. It may either be derived from phosphorylation of glucose or phosphorylation of glycogen.

2. The glucose-6-phosphate so formed undergoes an *isomeric transformation* to *fructose-6-phosphate*. This reaction is catalyzed by an enzyme called *glucose phosphate isomerase (phosphohexose isomerase).*
3. The fructose-6-phosphate combines with a further molecule of ATP to form *fructose 1, 6-diphosphate*. This reaction is Mg++ dependent, and is catalyzed by an enzyme *phosphofructokinase (phosphohexokinase).*

   The above three steps are preparatory steps; the first and third are endergonic as they consume 2 ATP.
4. The six-carbon fructose-1, 6-phosphate splits into two 3-carbon molecules (*trioses*), viz., *3-phosphoglyceraldehyde* and *dihydroxyacetone phosphate.* This reaction is mediated by an enzyme *aldolase*. The two triose phosphates are isomers and are interconvertible in the presence of an enzyme *triose phosphate isomerase (isomerase).* However, only the dihydroxyacetone phosphate gets converted into 3-phosphoglyceraldehyde; *this results in the formation of two molecules of 3-phosphoglyceraldehyde from fructose 1, 6-phosphate.*

   From this stage there are, obviously, two 3-carbon molecules taking part in the subsequent steps of glycolysis.
5. The 3-phosphoglyceraldehyde thus formed is oxidised to 1, 3-diphosphoglyceric acid with the help of the enzyme *glyceraldehyde-3-phosphate dehydrogenase (3-phosphoglyceric aldehyde* dehydrogenase) with the addition of inorganic phosphate Pi. This is an oxidation-reduction reaction in which there is transfer of electrons. The energy released during this transfer is utilized to bind the Pi to the 3-phosphoglyceraldehyde with a high energy bond. Therefore, the 1, 3-diphosphoglyceric acid has two ~ P. These ~ P are later utilized for obtaining ATP. The dehydrogenase with NAD+ as coenzyme gets reduced to NADH which later on provides ATP.

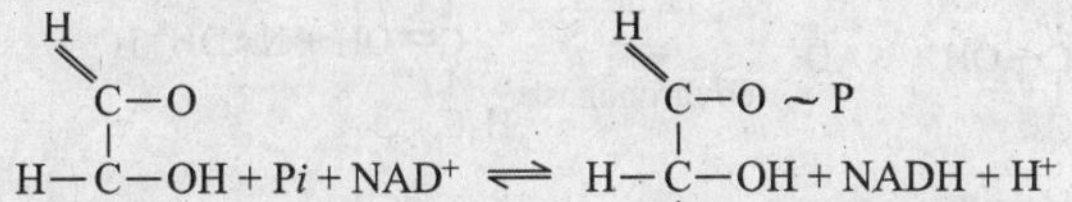

3-phospoglyceraldehyde 1, 3-phospoglyceric acid

6. In the presence of a transferring enzyme called *3-phosphoglyceric transphosphorylase* (phosphokinase), one of the high energy phosphates is transferred (dephosphorylated) from 1, 3-diphosphoglyceric acid to ADP to form ATP and *3-phosphoglyceric acid (PGA).*
7. The 3-phosphoglyceric acid is converted into *2-phosphoglyceric acid*, with the migration of the phosphate from 3 to 2 position. This transformation is carried out by an enzyme *phosphoglyceromutase*.
8. Another successive reaction catalyzed by the enzyme *enolse (phosphopyruvate hydratase)* transforms 2-phosphoglyceric acid to *phosphoenol pyruvic acid* (PEPA) to 2-phosphopyruvic acid-enol form by removal of one molecule of water.
9. Dephosphorylation of phosphoenol pyruvic acid yields one further ATP, and one molecule of *enol-pyruvic acid.* This reaction is accomplished in the presence of an enzyme *pyruvic kinase* (or *ATP-phosphoenolpyruvic transphosphorylase);* phosphoenol pyruvic acid gives up its phosphate group to ADP to form enol-pyruvic acid. The latter spontaneously converts itself into keto-pyruvic acid which is oxidised in Tricarboxylic Acid Cycle or Krebs' cycle.

It is important to remember that in step 4 a hexose derivative fructose-1, 6-diphosphate is split into two triose monophosphates and, finally, two 3-phosphoglyceraldehydes are obtained. In steps 5 to 9, the dehydration of only one of the 3-phosphoglyceraldehyde is described in which one pyruvic acid and two ATP are formed. The other 3-phosphoglyceraldehyde also undergoes similar dehydration to produce one pyruvic acid and two ATP. Thus, when one molecule of glucose undergoes glycolysis two pyruvic acid and 4 ATP are produced. Out of these two ATP are consumed in steps 1 and 3. Therefore, when one molecule of glucose undergoes anaerobic metabolism there is a net gain of 2 ATP. The two NADH produced in step 5 yield a further 6 ATP during oxidative phosphorylation.

Conversion of Pyruvic Acid to Lactic Acid or Ethyl Alcohol during Anaerobic respiration (Oxidation of NADH): Under normal conditions the pyruvic acid formed during glycoloysis is further metabolized through Krebs' cycle to carbon dioxide and water. In *muscles,* which usually respire aerobically, however, a situation may arise when glycolysis does not stop at pyruvic acid. This occurs during vigorous exercise when oxygen supply is inadequate. In this case muscles obtain extra energy by reduction of pyruvic acid to lactic acid (Fig. 13.13). NADH reduces pyruvic acid to lactic acid in the presence of an enzyme *lactic dehydrogenase.* NADH itself gets oxidised to NAD^+.

The NADH utilized in the above reaction is obtained from the oxidation of 3-phosphoglyceraldehyde (*step* 5 of glycolysis). Thus the coenzyme NAD^+, by accepting hydrogens or electrons from 3-phosphyoglceraldehyde and then donating them to pyruvic acid, undergoes an alternate reduction and oxidation in the anaerobic respiratory pathway. This process is shown below:

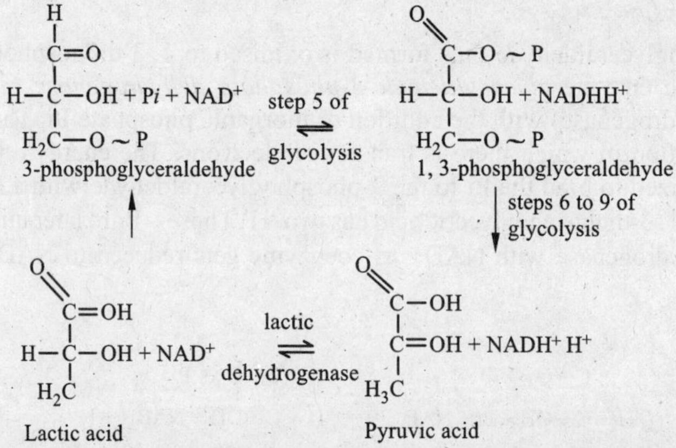

In many *microorganisms* (yeast, etc), under limited oxygen supply, the pyruvic acid is converted to *ethyl alcohol* and *carbon dioxide* instead of lactic acid. This occurs in *two steps. First,* pyruvic acid is decarboxylated to *acetaldehyde. Second,* acetaldehyde is reduced by NADH in the presence of an enzyme *alcohol dehydrogenase* to ethyl alcohol (Fig. 13.13).

It must be emphasized here that if one of the specific enzyme is missing or inhibited (by poisoning, etc.), the entire anaerobic pathway is brought to a standstill. Moreover, many of the enzymes are metalloproteins and they require a specific divalent metal ion (Mg^{++}). Yet another interesting feature of glycolysis may be pointed out here. Glucokinase (hexokinase) and phosphofructokinase (two of the initial enzymes in the glycolytic pathway) act as key control regulatory elements of the glycolytic pathway. These enzymes are inhibited by high levels of ATP. Inhibition of phosphofructokineae results in an accumulation of glucose-6-phosphate, which in turn inhibits glucokinase. Thus, when the cell has an adequate supply of metabolic energy in the form of ATP, the breakdown of glucose is inhibited.

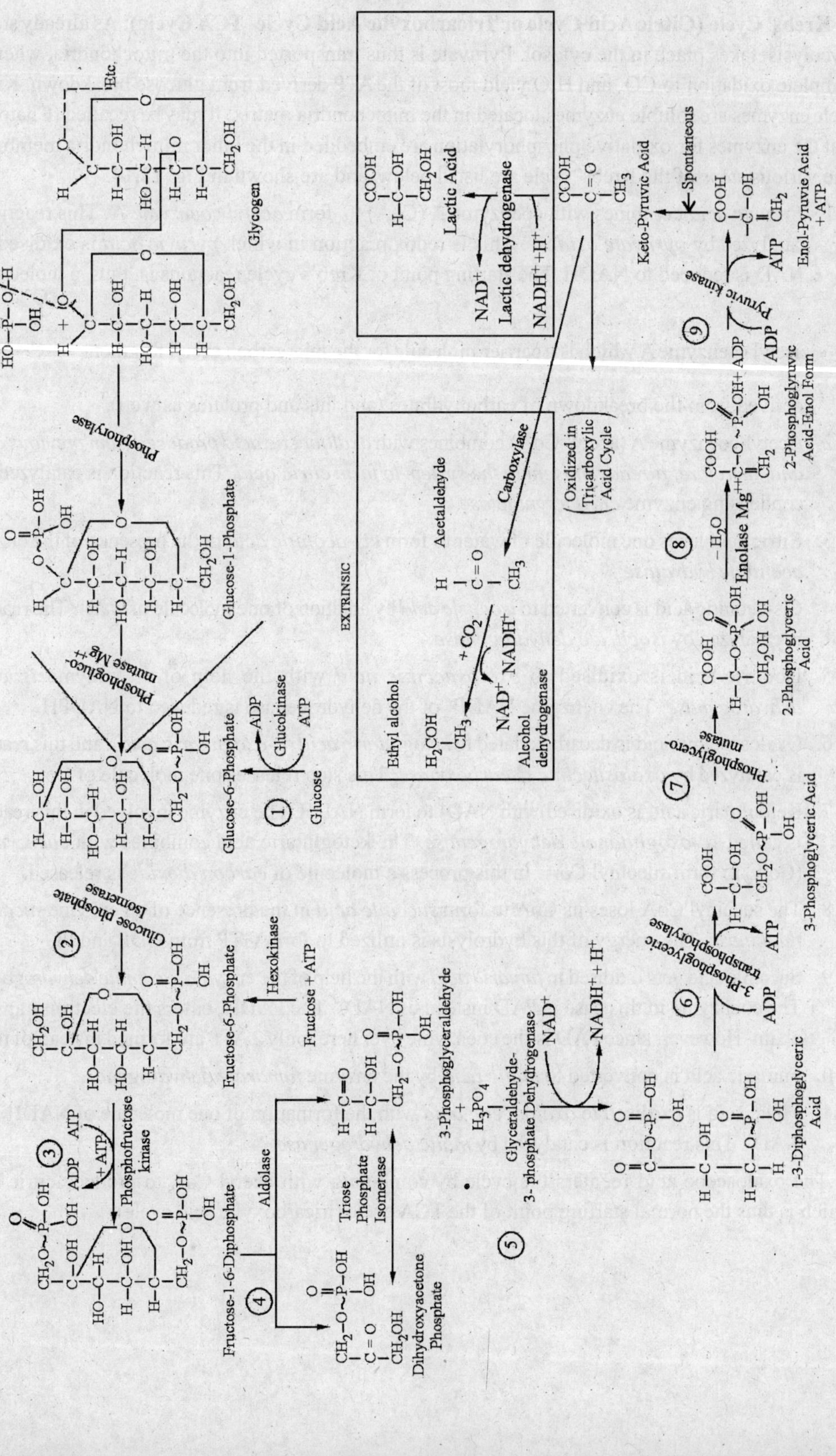

Fig. 13.13. Glycolysis (Embden-Meyerhof pathway) or anaerobic decomposition of glucose.

Krebs' Cycle (Citric Acid Cycle or Tricarboxylic Acid Cycle- TCA Cycle): As already stated, glycolysis takes place in the cytosol. Pyruvate is thus transported into the mitochondria, where its complete oxidation to CO_2 and H_2O yield most of the ATP derived from glucose breakdown. Krebs' cycle enzymes are soluble enzymes located in the mitochondria matrix. It may be recalled (Chatper 1) that the enzymes for oxidative phosphorylation are embedded in the inner mitochondria membrane. The various *steps* of the Krebs' cycle are listed below, and are shown in Fig. 13.14.

1. *Pyruvic acid* combines with coenzyme A (CoA) to form *acetyl coenzyme* A. This reaction is catalyzed by *pyruvate oxidase*. This is redox reaction in which *pyruvic acid* is oxidised and NAD is reduced to NADH. The starting point of Kreb's cycle reactions is, thus, a molecule of acetyl coenzyme A which is a carrier molecule for the two-carbon acetyl fragment $-\overset{\overset{\displaystyle O}{\|}}{C}-CH_3$ derived from the breakdown of carbohydrates (and fats and proteins as well).
2. Acetyl-coenzyme A (acetyl-CoA) combines with *oxaloacetic acid produced from previous cycle and, therefore, already present in the system to form citric acid.* This reaction is catalyzed by a condensing enzyme *citrate synthetase.*
3. Citric acid loses one molecule of water to form *cis-aconitic acid* in the presence of the enzyme *aconitate hydratase.*
4. Cis-aconitic acid is converted to *isocitric acid* by addition of one molecule of water. The reaction is catalyzed by *isocitrate dehydrogenase*.
5. Isocitric acid is oxidised to *oxalosuccinic acid* with the help of an enzyme *isocitric dehydrogenase*. The coenzyme $NADP^+$ of the dehydrogenase is reduced to NADPH.
6. Oxalosuccinic acid is decarboxylated to *ketoglutaric acid (α-oxoglutaric acid)*, and this reaction is catalyzed by *oxalosuccinic decarboxylase*. This step releases one molecule of Co_2.
7. Ketoglutaric acid is oxidised with NAD^+ to form NADH. The enzyme involved in this reaction is called α *-oxoglutanate dehydrogenase*. The ketoglutaric acid combines with *coenzyme* A (CoA) to form succinyl CoA. In this process a molecule of *carbon dioxide* is released.
8. The succinyl CoA loses its CoA to form *succinic acid* in the presence of an enzyme *succinate thiokinase*. The energy of this hydrolysis is utilized to form GTP from GDP and Pi.
9. Succinic acid gets oxidised to *fumaric acid* with the help of the enzyme *succinate dehydrogenase*. The coenzyme in this case is FAD instead of NAD^+. The $FADH_2$ enters the electron transport chain. However, since FAD is the coenzyme over here, only 2 ATP are formed instead of three.
10. Fumaric acid is converted to *malic acid* by the enzyme *fumarate dehydrogenase.*
11. Malic acid is oxidised to *oxaloacetic acid* with the formation of one molecule of NADH from NAD^+. This reaction is catalyzed by *malic dehydrogenase*.

The oxaloacetic acid reenters the cycle by combinnig with acetyl CoA to produce citric acid, which is thus the normal starting point of the TCA cycle (tricarboxylic acid cycle).

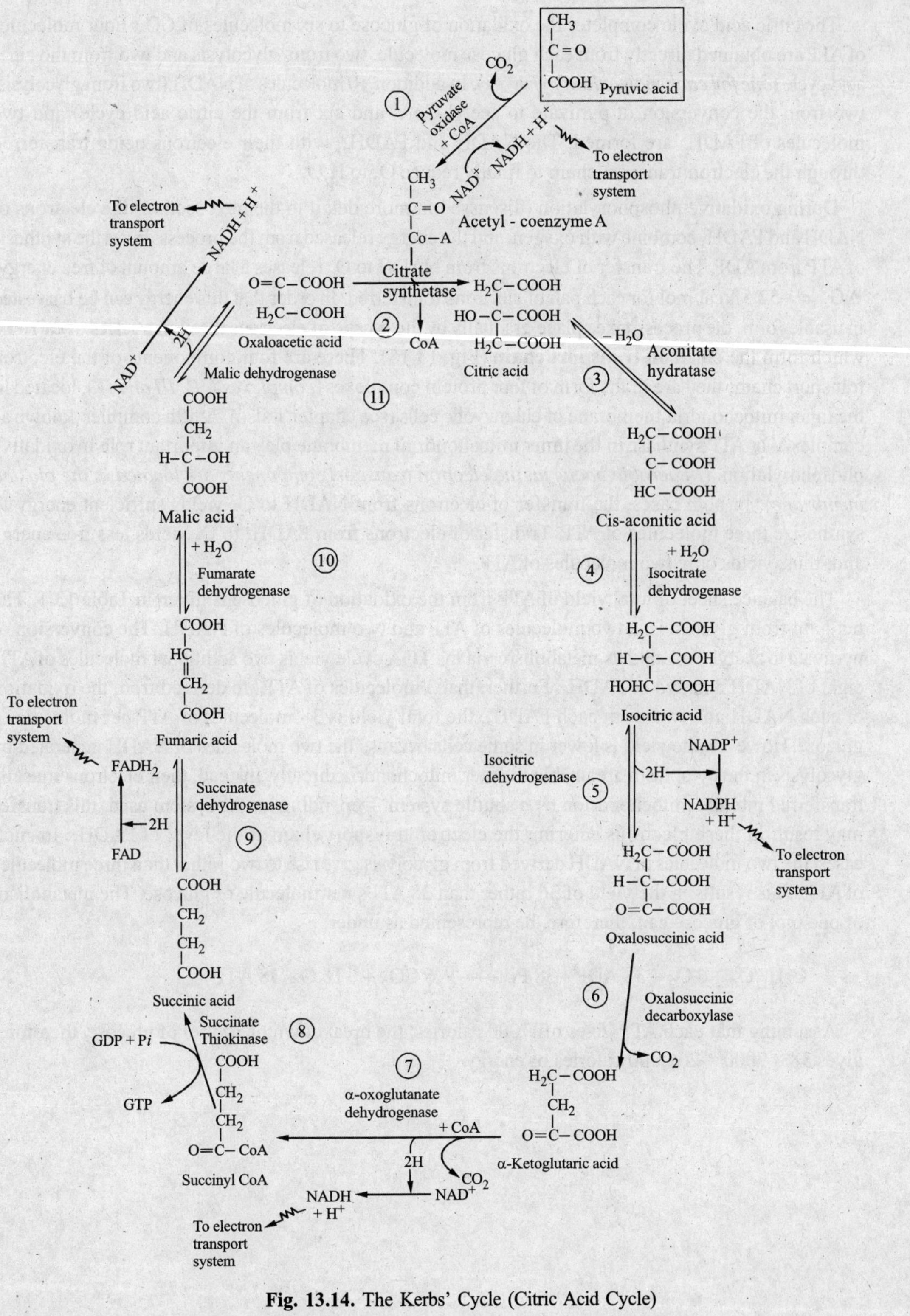

Fig. 13.14. The Kerbs' Cycle (Citric Acid Cycle)

The citric acid cycle completes the oxidation of glucose to six molecules of CO_2. Four molecules of ATP are obtained directly from each glucose molecule: two from glycolysis and two from the citric acid cycle (*one for each molecule of pyruvate*). In addition 10 molecules of NADH (two from glycolysis, two from the conversion of pyruvate to acetyl CoA, and six from the citric acid cycle) and two molecules of $FADH_2$ are formed. The NADH and $FADH_2$, with their electrons being transferred through the electron transport chain to finally reduce O_2 to H_2O.

During oxidative phosphorylation (discussed in more detail in the next section) the electrons of NADH and $FADH_2$ combine with oxygen, and the energy released from the process drives the synthesis of ATP from ADP. The transfer of electrons from NADH to O_2 releases a large amount of free energy: $\Delta G^{0'} = -52.5$ kcal/mol for each pair of electrons transferred. In order that this energy can be harvested in usable form, the process takes place gradually by the escape of electrons through a series of carriers, which form the **electron transport chain** (Fig. 13.15). There are four components of the electron transport chain; they are in the form of four protein complexes (*complexes I, II, III and IV*), located in the inner mitochondria membrane of eukowyolic cells (see chapter 1 also). A fifth compiler known as complex V or ATP synthase in the inner mitochondrial membrane play an important role in oxidative phosphorylation. (*In aerobic bacteria, the electron transport components are located in the plasma membrane*). In both cases, the transfer of electrons from NADH to O_2 yields sufficient energy to synthesize three molecules of ATP. Transfer of electrons from $FADH_2$ to O_2 yields less free energy and, thus, yields only two molecules of ATP.

The balance sheet of total yield of ATP from the oxidation of glucose is given in Table 13.1. The net gain from glycolysis is two molecules of ATP and two molecules of NADH. The conversion of pyruvate to acetyl CoA and its metabolism via the TCA cycle yields two additional molecules of ATP, eight of NADH and two of $FADH_2$. Further, that 3 molecules of ATP are derived from the oxidation of each NADH and two from each $FADH_2$, the total yield is 38 molecules of ATP per molecule of glucose. However, the yield is lower in some cells because the two molecules of NADH generated by glycolysis in the cytoplasm are unable to enter mitochondria directly. Instead, their electrons must be transferred into the mitochondrion by a shuttle system. Depending on the system used, this transfer may result in these electrons entering the electron transport chain at the level of $FADH_2$. In such cases, the two molecules of NADH derived from glycolysis give rise to two rather than three molecules of ATP. This results in the yield of 36 rather than 38 ATPs per molecule of glucose. The metabolism of one mol of glucose can, therefore, be represented as under:

$$C_6H_{12}O_6 + 6\,O_2 + 38\,ADP + 38\,Pi \longrightarrow 6\,CO_2 + 6\,H_2O + 38\,ATP$$

Assuming that each ATP gives off 7000 calories, the breakdown of 1g mol of glucose, therefore, gives 38 × 7000 = 266,000 calories of energy.

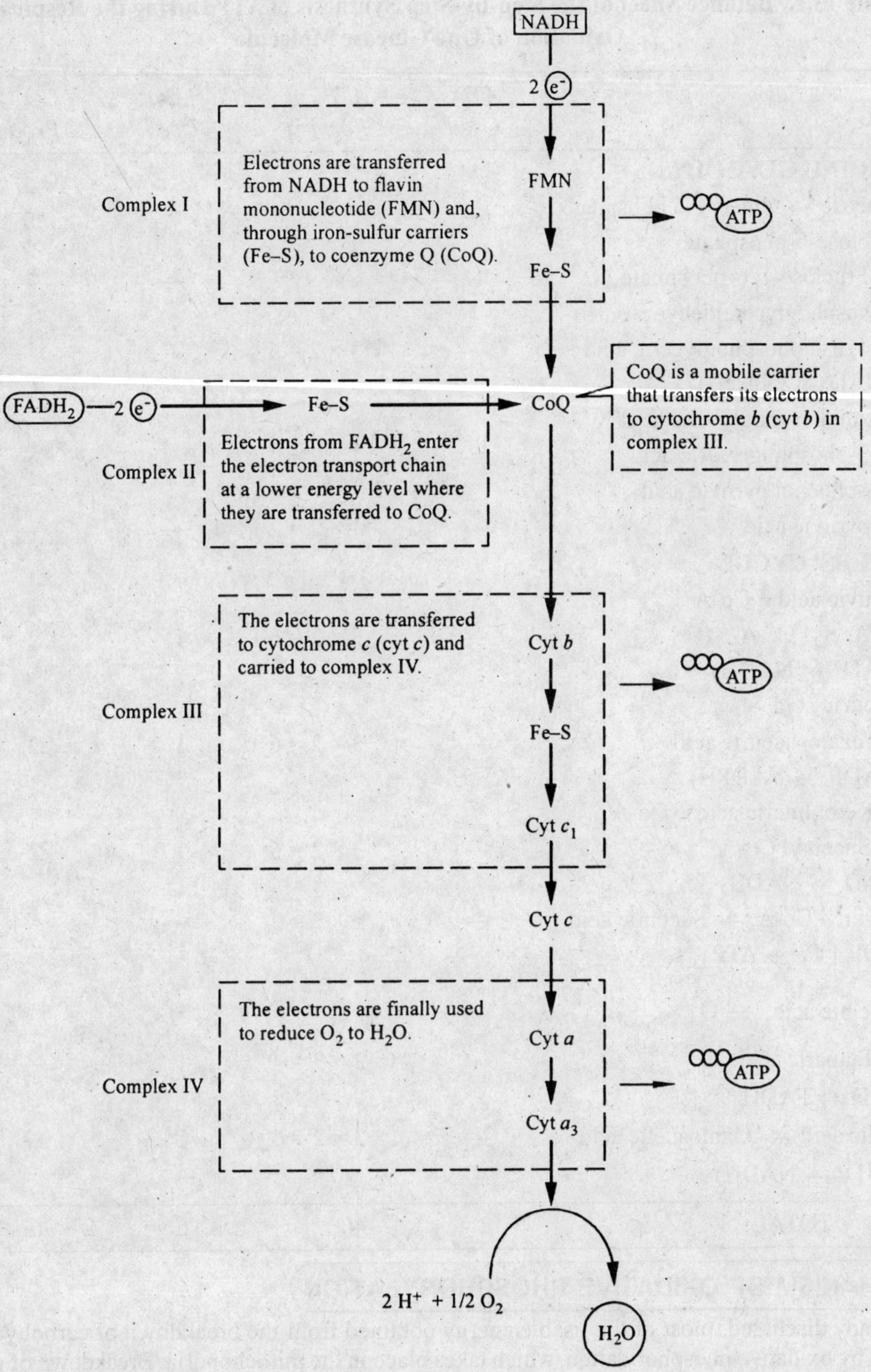

Fig. 13.15. The electron transport chain. Electrons from NADH and $FADH_2$ are transferred to O_2 through a series of carriers organized into four protein complexes in the mitochondrial membrane. The free energy derived from electron transport reactions at complexes I, III and IV is used to drive the synthesis of ATP.

Table 13.1. Balance Sheet of the Step-by-Step Synthesis of ATP During the Respiratory Oxidation of One Glucose Molecule

Steps	*ATP Used*	*ATP Formed*	*$1/2O_2$ Used*	*CO_2 Produced*
DURING GLYCOLYSIS				
Glucose → Glucose-6 phosphate	1	—	—	—
Fructose-6-phosphate → Fructose-1, 6-phosphate	1	—	—	—
3-phosphoglyceraldehyde → 1, 3-diphosphoglyceric acid (2NAD → 2NADH)	—	$3 \times 2 = 6$	—	—
1, 3-diphosphoglyceric acid → 3-phosphoglyceric acid	—	$1 \times 2 = 2$	—	—
Phosphoenol pyruvic acid → pyruvic acid	—	$1 \times 2 = 2$	—	—
KREBS' CYCLE				
Pyruvic acid + Co-A → Acetyl Co-A (NAD → NADH)	—	$3 \times 2 = 6$	1	2
Isocitric acid → oxalosuccinic acid (NADP → NADPH)	—	$3 \times 2 = 6$	1	2
α-Ketoglutaric acid + Co-A → Succinyl Co-A (NAD → NADH)	—	$3 \times 2 = 6$	1	2
Succinyl Co-A → Succinic acid (ADP + P*i* → ATP)	—	$1 \times 2 = 2$	—	—
Succinic acid $+\frac{1}{2}O_2$ → Fumaric acid ($FAD \rightarrow FADH_2$)	—	$2 \times 2 = 4$	1	—
Malic acid → Oxaloacetic acid (NAD → NADH)	—	$3 \times 2 = 6$	1	—
TOTAL	2	40	6	6

MECHANISM OF OXIDATIVE PHOSPHORYLATION

As already discussed, most of the usable energy obtained from the breakdown of carbohydrate is derived by oxidative phosphorylation, which takes place in the mitochondria. Breakdown of glucose by glycolysis and the TCA cycle yields a total of 4 molecules of ATP, ten molecules of NADH, and two molecules of $FADH_2$. Electrons from NADH and $FADH_2$ are then transferred to molecular oxygen, coupled to the formation of an additional 32 to 34 ATP molecules by oxidative phosphorylation. Electron transport and oxidative phosphorylation are critical activities of protein complexes in the inner mitochondrial membrane, which eventually serves as the major source of energy.

The Electron Transport Chain

During oxidative phosphorylation, the electrons derived from NADH and FAD combine with O_2 and the energy released from the oxidation/reduction reactions is used to drive the synthesis of ATP from ADP. As already stated, the transfer of electrons from NADH to O_2 is very energy-yielding reaction with $\Delta G°' = -52.5$ kcal/mol for each pair of electrons transferred. To be harvested in usable form, this energy must be produced gradually by the passage of electrons through a series of carriers which constitute the electron transport chain. The carriers are organized into four protein complexes (*called complexes I, II, III, and IV*) in the inner mitochondrial membrane. A fifth protein complex (*complex V* or *ATP synthase*) then serves to couple the energy-yielding reactions of electron transport to ATP synthesis. Transport of electrons from NADH and $FADH_2$ are described below.

Transport of Electrons from NADH

Electrons from NADH enter the electron transport chain in complex I (Fig. 13.16). These electrons are initially transferred from NADH to flavin mononucleotide (FMN) and, through an iron-sulphur carrier, to coenzyme Q (also called ubiquinone). This is an energy-yielding process with $\Delta G°' = -16.6$ kcal/mol. The coenzyme Q, which is a lipid soluble mobile carrier, carries the electrons from complex I through the membrane to complex III. In complex III, electrons are transferred from cytochrome *b* to cytochrome *c*, which is another energy-yielding reaction with $\Delta G°' = -10.1$ kcal/mol. Cytochrome *c* (a peripheral membrane protein bound to the outer face of the inner mitochondrial membrane) where they are finally transferred to O_2 ($\Delta G°' = -25.8$ kcal / mol).

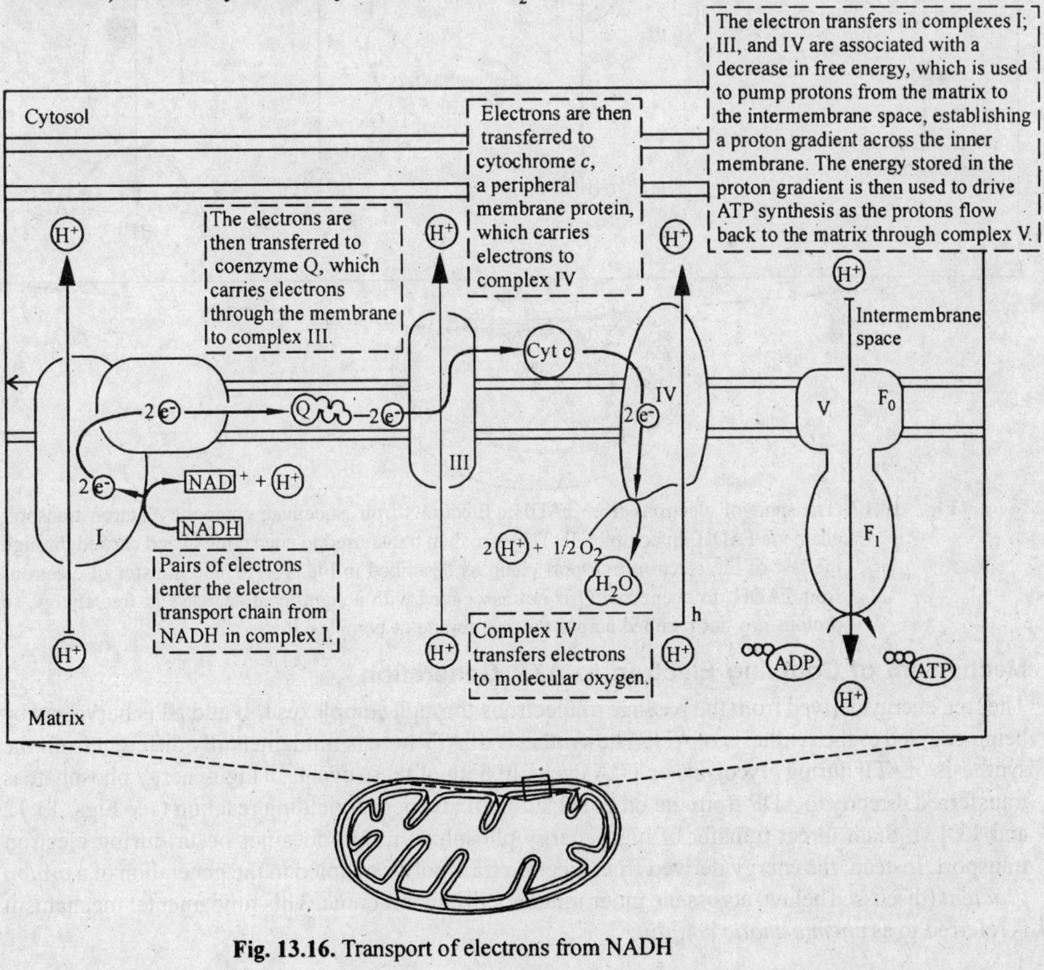

Fig. 13.16. Transport of electrons from NADH

Transport of electrons from $FADH_2$

A protein complex II receives electrons from the TCA cycle intermediate succinate (see Fig. 13.13). These electrons are transferred to $FADH_2$ (and not to NADH) and then to coenzyme Q (Fig. 13.17). From coenzyme Q, electrons are transferred to complex III and then to complex IV. *In contrast to transfer of electrons from $NADH_2$ to coenzyme Q is not associated with a significant decrease in free energy and, therefore, not coupled to ATP synthesis.* Consequently, the passage of electrons from $FADH_2$ through the electron transport chain yields energy only at complexes III and IV.

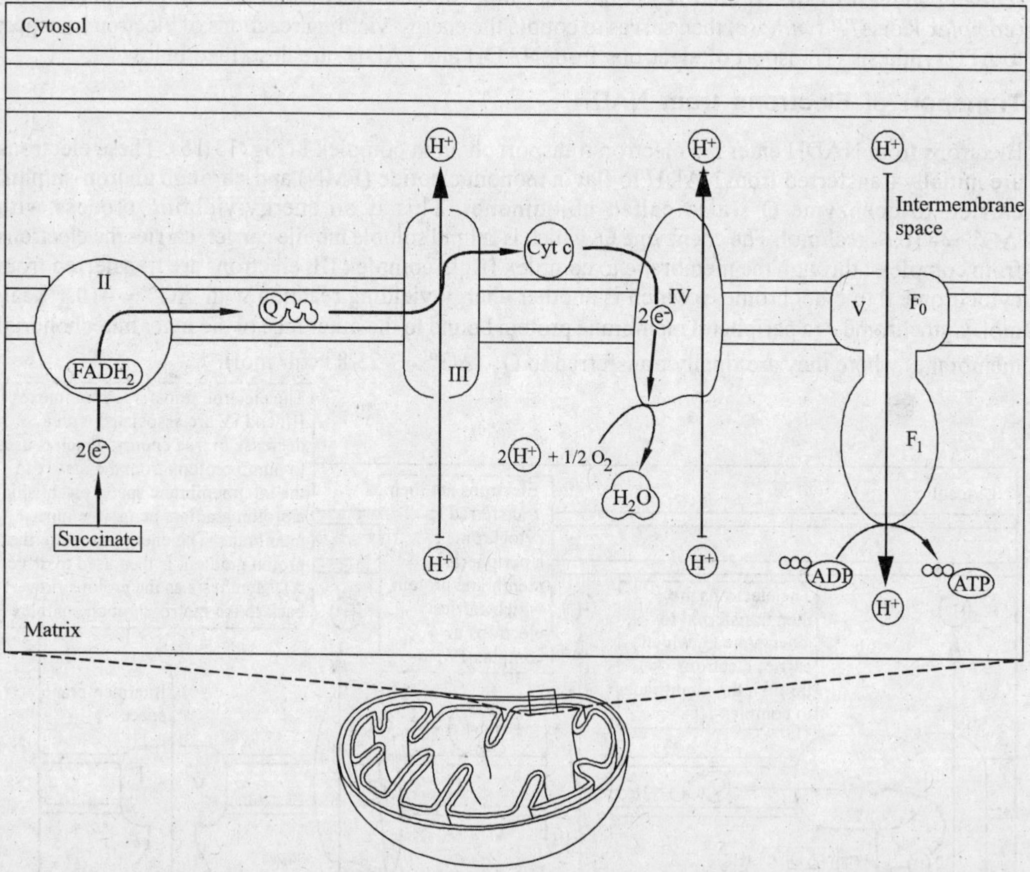

Fig. 13.17. Transport of electrons from $FADH_2$. Electrons from succinate enter the electron transport chain via $FADH_2$ in complex II. They are then transferred to coenzyme Q and carried through the rest of the electron transport chain as described in Fig. 13.16. The transfer of electrons from $FADH_2$ to coenzyme Q is not associated with a significant decrease in free energy, so protons are not pumped across the membrane at complex II.

Mechanism of Coupling Electron to ATP Generation

The free energy derived from the passage of electrons through complexes I, II and III is harvested by being coupled to the synthesis of ATP. The synthesis of ATP here is fundamentally different from the synthesis of ATP during glycolysis or TCA cycle. In both of these events, a high-energy phosphate is transferred directly to ADP from the other substrate of an energy-yielding reaction (see Figs. 13.12 and 13.13). Such direct transfer of high-energy phosphate group does not occur during electron transport. Instead, the energy derived from electron transport is coupled to the generation of a *proton gradient* (discussed below) across the inner mitochondrial membrane. This fundamental mechanism is referred to as *chemiosmotic coupling*.

Chemiosmotic Coupling: The mechanism of coupling electron transport to ATP generation is explained by the *chemiosmotic theory* proposed by *Peter Mitchell*. His work was recognized with a Nobel Prize in 1978. He explained that ATP is generated by the use of energy stored in the form of proton gradient across biological membranes. *It may be noted here that chemiosmotic theory not only operates in mitochondria but also in chloroplasts and in bacteria.*

As shown in Figs. 13.16 and 13.17, electron transport through complexes I, III and IV is coupled to the transport of protons (H^+) out of the matrix of the mitochondria. Thus, the energy-yielding reactions of electron transport are coupled to the pumping of protons from the matrix to the intermembrane space (space between the outer and inner mitochondrial membranes), which establishes a proton gradient across the inner membrane. Complexes I and III each pump four electrons per pair of electrons transferred, while two protons per pair of electrons are transferred across the membrane by complex IV. In total, therefore, ten protons are transferred across the membrane for each pair of electrons derived from NADH, while six protons are transferred per pair of electrons derived from $FADH_2$ (Figs. 13.16 and 13.17). Proton pumping plays a significant role of converting the energy derived from the *oxidation/reduction reactions* of electron transport to the potential energy stored in a proton gradient.

Because protons are electrically charged particles, the potential energy stored in the proton gradient is both electric and chemical in nature. The electric component corresponds to the voltage difference across the inner mitochondrial membrane; the matrix of the mitochondrion negative and the inter-membrane space positive. The corresponding free energy is gives by the equation

$$\Delta G = -F\Delta V$$

where F is the Faraday constant and ΔV is the membrane potential. The additional free energy corresponding to the difference in proton concentration across the membrane is given by the equation

$$\Delta G = RT \ln \frac{[H^+]_i}{[H^+]_0}$$

where $[H^+]_i$ and $[H+]_0$ refer, respectively, to the proton concentration inside and outside the mitochondria.

In metabolically active cells, protons are pumped out of the matrix such that the proton gradient across the inner membrane corresponds to about one pH unit or a tenfold lower concentration of protons within the mitochondria (Fig. 13.18). The pH of the mitochondrial matrix is therefore about 8, compared to the neutral pH 7 of the cytoplasm and intermembrane space. This gradient generates an electric potential of approximately 0.14 V across the membrane, with the matrix negative. Both the pH gradient and electric potential drive the protons back into the matrix from the cytoplasm, so that they combine to form an *electrochemical gradient* across the inner mitochondrial membrane. This coresponds to a ΔG of approximately – 5 kcal/mol per proton.

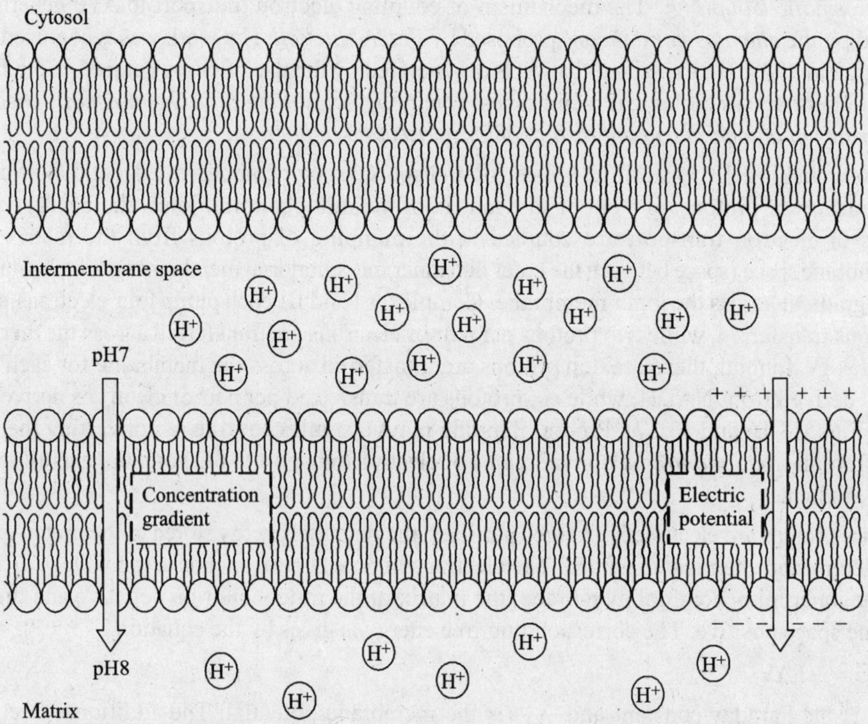

Fig. 13.18. *The electrochemical nature of the proton gradient.* Since protons are positively charged, the proton gradient established across the inner mitochondrial membrane has both chemical and electric components. The chemical component is the proton concentration, or pH, gradient, which corresponds to about a tenfold higher concentration of protons on the cytosolic side of the inner mitochondrial membrane (a difference of one pH unit). In addition, there is an electric potential across the membrane, resulting from the net increase in positive charge on the cytosolic side.

As the phospholipid layer is impermeable to ions, protons cross the membrane only through protein channels. This restriction allows the energy in the electrochemical gradient to be harvested and converted to ATP as a result of the action of the fifth complex (complex V or ATP synthase) in oxidative phosphorylation (Figs. 13.16 and 13.17).

ATP synthase (complex V) has two structurally distinct parts: F_0 and F_1 (Figs. 13.16 and 13.17). The F_0 spans the inner membrane and provides a channel through which protons flow back into from the intermembrane space to the matrix. The energetically favourable return of the protons to the matrix is coupled to ATP synthesis by the F_1 subunit. The F_1 subunit catalyzes the synthesis of ATP from ADP and phosphate ions (P*i*). The free energy required for ATP synthesis corresponds to a ΔG of approximately + 12 kcal/mol, so the flow of approximately three protons back across the membrane is needed to drive the synthesis of each molecule of ATP. Because ten protons are pumped across the membrane per pair of electrons derived from NADH, the oxidation of one molecule of NADH leads to the synthesis of approximately three molecules of ATP. However, the oxidation of $FADH_2$, which results in the transfer of six protons, drives the synthesis of only two ATP molecules. The free energy contributed to the gradient at each of the three complexes that transfer protons across the membrane (viz. complexes I, III and IV) is thus sufficient to drive the synthesis of the ATP molecule.

Five significant facts of the Krebs' cycle and oxidative phosphorylation may be pointed here. These are: (1) the CO_2 formed in the Krebs' cycle is the major waste product of metabolism; (2) the two gases O_2 and CO_2, which enter and leave the body by way of the lungs, take part in reactions that are exclusively located in the mitochondria; (3) the O_2 that appears in CO_2 is derived not from molecular oxygen but from oxygen atoms in the carboxyl groups of the Krebs' cycle intermediates; (4) molecular oxygen, in fact, ends up in water molecules after combining with hydrogen during oxidative phosphorylation; and (5) water is a product *not* of the Krebs' cycle *per se*, but of oxidative phosphorylation.

Hexose Monophosphate Shunt (HMP Shunt)

An alternate route of carbohydrate metabolism is the *hexose monophosphate shunt,* also known as the *pentose oxidative pathway* or *phosphogluconate pathway.* Like the anaerobic pathway, the enzymes of the pentose oxidative pathway are also present in the cytoplasm of the cell. The pathway is cyclic and consists of several enzymic reactions.

As shown in Fig. 13.19, the first reaction of the pathway involves the phosphorylation of glucose by ATP in the presence of an enzyme *glucokinase* to form *glucose-6-phosphate.* The latter instead of entering the glycolytic pathway enters the pentose oxidative pathway. This is, therefore, known as hexose monophosphate shunt or HMP shunt. The glucose-6-phosphate is enzymatically dehydrogenated by *glucose-6-phosphate dehydrogenase,* having coenzyme $NADP^+$, to form 6-phosphogluconic acid.

$$\text{Glucose}-6-\text{P}+\text{NADP}^+ \xrightleftharpoons[\text{dehydrogenase}]{\text{Glucose-6-phosphate}} 6-\text{phosphogluconic acid}+\text{NADPH}+\text{H}^+$$

This is followed by an oxidative decarboxylation to yield an unstable *pentose phosphate* derivative—*ribose-5-phosphate.* This reaction again involves $NADP^+$ which is a part of the enzyme *6-phosphogluconate dehydrogenase.*

$$6-\text{phosphogluconic acid}+\text{NADP}^+ \xrightleftharpoons[\text{dehydrogenase}]{6-\text{phosphogluconate}} \text{Ribose-5-P}+\text{NADPH}+\text{H}^+$$

In both reactions, $NADP^+$ acts as the electron or hydrogen acceptor becoming reduced to NADPH. In subsequent steps the unstable ribose-5 phosphate is converted into a variety of other pentoses, such as *xylulose-5-phosphate* through *ribulose-5-phosphate* by means of *epimerase* enzymes. These pentoses then undergo a successive series of other enzymatic steps involving isomerization, cleavage, transfer and rejoining of the resulting active fragments to form a number of different phosphorylate intermediate products. These intermediaries include phosphate esters consisting of seven, four, three, and six carbon sugar molecules. These are *sedoheptulose-7-phosphate, eythrose-4-phosphate, glyceraldehyde-3-phosphate,* and *fructose-6-phosphate,* respectively (Fig. 13.19).

At the end of the completion of each cycle of HMP shunt, which starts with a single molecule of glucose, one carbon dioxide, two NADPH molecules, and several smaller sugar derivatives are produced. Thus, for six rotations of the cycle, starting with six molecules of glucose, the pathway ends up with five molecules of glucose phosphate, six molecules of carbon dioxide and twelve molecules of NADPH (Fig. 13.20). In summary, therefore, a complete breakdown of glucose to carbon dioxide and hydrogens (hydrogens in the form of NADPH) occurs without the participation of aerobic respiration.

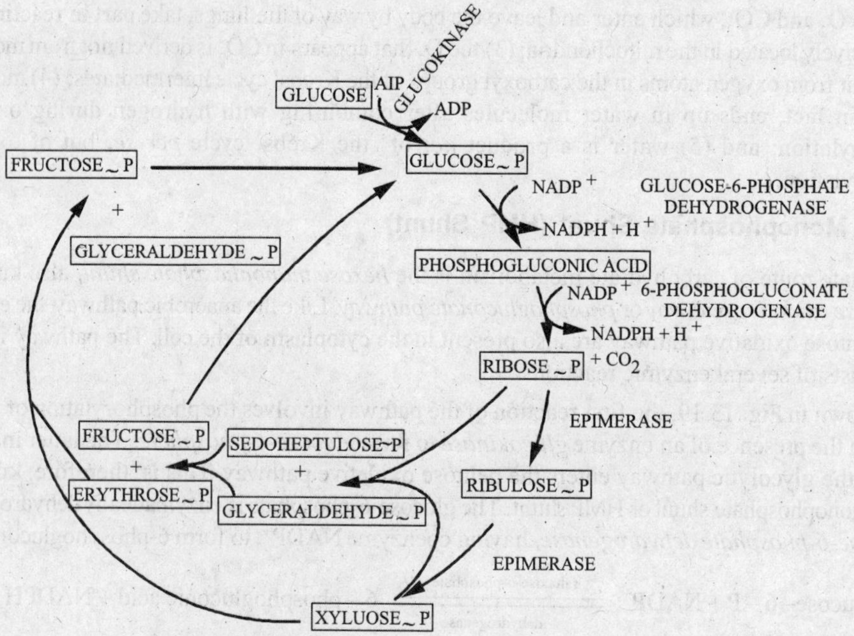

Fig. 13.19. The Hexosemonophosphate shunt.

In terms of energy yield, NADPH is finally oxidized like NADH by the electron transport system. This would then yield three molecules of ATP for each NADPH molecule. In total, therefore, 36 ATP molecules are formed in the complete respiration of a glucose molecules by the HMP shunt. This energy yield is almost equal to that produced during glycolysis and Krebs' cycles. *However, evidence indicates that the NADPH formed during the HMP shunt is not necessarily used in the electron transport system; it is actually consumed for the synthesis of fats and steroids in the liver and adrenal cortex, respectively. This is because the HMP shunt mostly occurs in the above two organs. The HMP shunt also provides a source of pentoses for nucleic acid synthesis.*

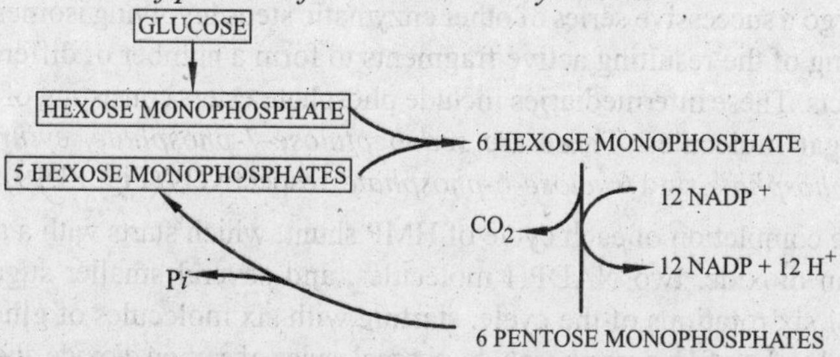

Fig. 13.20. Diagram showing the reduction of 12 NADP+ to 12 NADPH during the HMP shunt. Note that during this process only 36 ATP are formed as compared to 38 ATP during glycolysis and Krebs' cycle.

LIPID AND PROTEIN METABOLISM

Glucose or its polymer glycogen is mostly used as the primary source of energy. However, lipids and proteins can also be utilized for obtaining energy. *The basic mechanism for obtaining energy in case of lipids and proteins is also the same as that of carbohydrates because these also utilize the Krebs' cycle for the final release of energy. The following account of the metabolism of lipids and fats outlines the step which leads to the formation of acetyl CoA which is the starting point of TCA cycle.*

Lipid Metabolism

Lipids are metabolised mainly in the liver. Fats (triglycerols) are the major storage form of lipids. The first step in their utilization is their breakdown into glycerol and free fatty acids. The various steps are outlined below, and in Fig. 13.21.

1. The *fatty acid* form a thioester with CoA. This reaction is mediated by the enzyme *fatty acid thiokinase.* This initial step is accompanied by the breakdown of ATP to AMP and two P*i*. Since the terminal phosphates of both and ATP and ADP are high-energy bonds, the energy released in this reaction is equivalent to that released when two ATP are hydrolyzed to two ADP and two P*i*. Thus, although one ATP is used in the reaction, the energy cost is equivalent to two ATP.
2. Two hydrogen atoms are removed and FAD is reduced to $FADH_2$; this step is mediated by *fatty acyl coenzyme dehydrogenase.*
3. An enzyme, *enoyl hydratase,* then brings about the addition of one molecule of water.
4. There is a further loss of hydrogen, and NAD^+ is converted to NADH. This reaction is mediated by an enzyme α *-hydroxy-acyl dehydrogenase.*
5. An enzyme, β *-oxoacyl thiolase*, catalyzes the reaction, leading to the production of *acetyl CoA* plus the thioester of the *remaining fatty acid chain.*
6. This fatty acid chain now re-enters the cycle, being combined with CoA but two carbon atoms shorter than before. Each cycle removes a *two-carbon unit* from the fatty acid. Each round of oxidation also yields one molecule of NADH and one of $FADH_2$. This process continues step by step until the molecule of fatty acid has been utilized in these reactions. This is commonly called as the β- *oxidation helix.*

This β- *oxidation helix* was worked out by Knoop. During the β- oxidation helix the *carboxylic acid* is first converted to the corresponding *keto acid (step 1).* In the final step (step 5) two carbons are split from the fatty acid to form acetyl CoA. The split in the fatty acid takes place at the carbon at the β- position relative to the carboxyl group of the fatty acid. Each acetyl CoA thus obtained enters the TCA cycle and each two-carbon actyl fragment $-\overset{\overset{\displaystyle O}{\|}}{C}-CH_3$ producing two molecules of CO_2 and, ultimately, 12 molecules of ATP.

How much ATP, then, is formed as a result of total catabolism of a fatty acid? Most fatty acids in the body contain 14 to 22 carbon atoms, but 16 and 18 being most common. The breakdown of one 16-carbon saturated fatty acid palmitate thus yields 7 molecules of NADH, 7 molecules of $FADH_2$ and 8 of acetyl CoA. In terms of ATP generation, this yield corresponds to 21 molecules of ATP derived from NADH (3 × 7), 14 ATP from $FADH_2$ (2 × 7), and 96 from acetyl CoA (8 × 12), that is, 131 ATPs are generated. Since energy equivalent of 2ATP was used to start the process, the net gain is 129 ATPs per molecule of a 16-carbon fatty acid. An 18-carbon fatty acid, thus, produces 146 molecules of ATP. When these yields are compared with the net gain of 38 ATPs per molecule of glucose, the advantage of lipids over polysaccharides as energy storage molecules is obvious. If the difference in molecular weights of the fatty acid and glucose is taken into account, it turns out that the amount of ATP formed from the breakdown of a gram of fat is about 2.5 times greater than the ATP produced from the breakdown of a gram of carbohydrate. This means that if an average person stored the fuel as carbohydrates rather than as fat, his or her body weight would have to be approximately 30 per cent greater in order to store the same amount of usable energy. Consequently, the person would consume more energy moving this extra weight around. Thus, a major step in fuel economy occurred with animals evolving the ability to store fuel as fat.

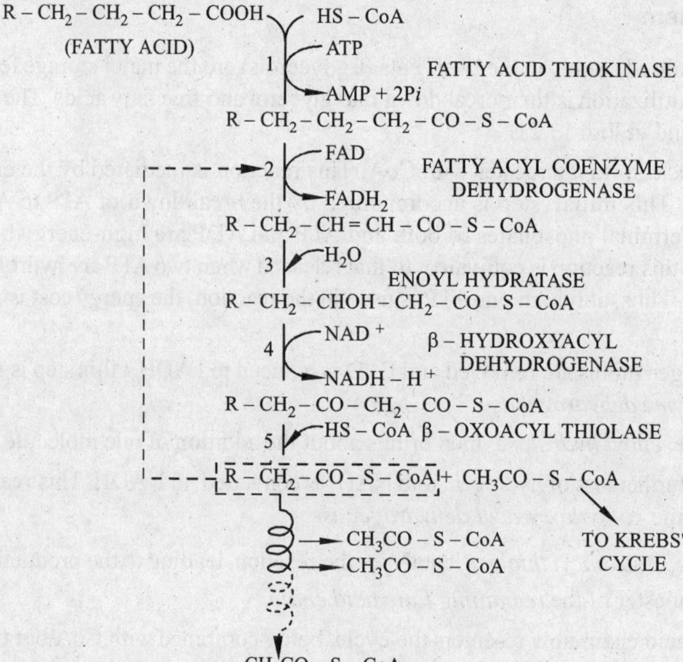

Fig. 13.21. The metabolism of fatty acids.

Protein Metabolism

The proteins are first broken down into amino acids. The first step into amino acid catabolism is their conversion into corresponding keto acids. The conversion may take place in either of the three ways described below:

(1) The *amino acid* is deaminated with the help of *L-amino acid oxidase* which has FAD as the coenzyme. This is a process of oxidative deamination.

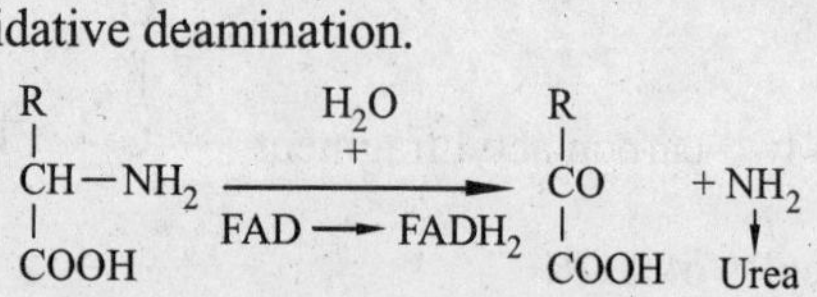

The above enzyme is usually found in the liver and the kidneys but its concentration is so low that it is almost insignificant.

(2) In the case of glutamic acid the deamination is effected through an enzyme *glutamate dehydrogenase* which has NAD$^+$ as a coenzyme. In this process α- ketoglutaric acid (α- oxoglutaric acid) is formed.

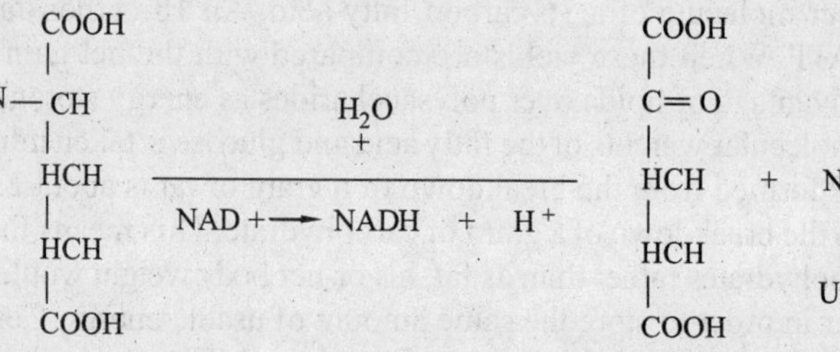

The action of glutamate dehydrogenase is a reversible reaction and is an important link between carbohydrate and protein metabolism; it converts α- *ketoglutaric acid* (an intermediate of the citric acid cycle) *to the key amino acid glutamic acid.*

(3) Most amino acids, however, are converted to keto acids through a process known as *transamination;* an amino group (–NH_2) is transferred from a donor amino acid to keto acid under the influence of enzymes known as *transaminases* or *amino transferases.* During this process the donor amino acid becomes a keto acid and the recipient keto acid becomes an amino acid.

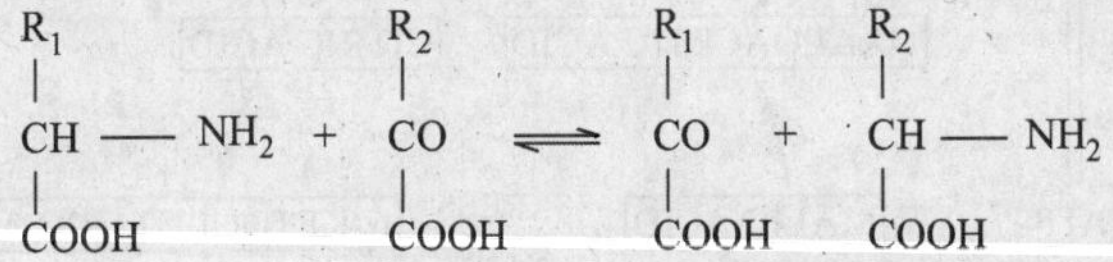

Amino acid (donor) | Keto acid (recipient) | Keto acid | Amino acid

Although most amino acids may act as donor, only three keto acids, *viz.*, α- ketoglutaric acid, oxaloacetic acid, pyruvic acid, can act as a recipient which are readily available from the Krebs' cycle. The amino acids formed from these keto acids are glutamic acid, aspartic acid, and alanine, respectively.

Fig. 13.22 illustrates the oxidative deamination of the amino acid glutamic acid and the transamination of amino acid alanine. It may be noted that the keto acids formed intermediates either in the Krebs' cycle (α- keto glutaric acid) or glycolytic pathway (pyruvic acid). Thus, these keto acids, when formed, can be metabolized to produce CO_2 and form ATP or they can be used as intermediates in the synthetic pathway leading to the formation of glucose.

As already mentioned, transamination results in the formation of a variety of keto acids. The keto acids thus formed do not have the same fate, as evident from the following: (a) Some of these are converted into glucose or glycogen in the liver. Most amino acids known as ketogenic or antiketogenic amino acids give rise to keto acids which are converted into glucose or glycogen. This is further metabolized to yield energy. (b) Other amino acids known as ketogenic amino acids yield keto acids which pass through various stages and get converted into acetoacetic acid which finally enters in TCA cycle. (c) There are still other amino acids which do not belong to either of the two categories; they can give rise to both glucose and keto bodies (see Chapter 2).

In summary, the three major classes of nutrients, namely, carbohydrates, lipids and proteins, can enter the Krebs' cycle through some intermediate and, thus, all the three can be interchangeably used as a source of chemical energy for the synthesis of ATP by oxidative phosphorylation (Fig. 13.22). This process is termed *intermediary metabolism.* Glucose can be converted to fat or some amino acid by way of some common intermediates such as pyruvic acid, α- ketoglutaric acid, and acetyl CoA. Fatty acids cannot be converted into glucose. Fatty acids can be used to synthesize proteins of the keto acids used to form amino acids. It can be said that the carbon backbone of living chemical substance is in a constant flux; it may be the skeleton of an amino acid at one moment, part of a lipid at another (Fig. 13.22).

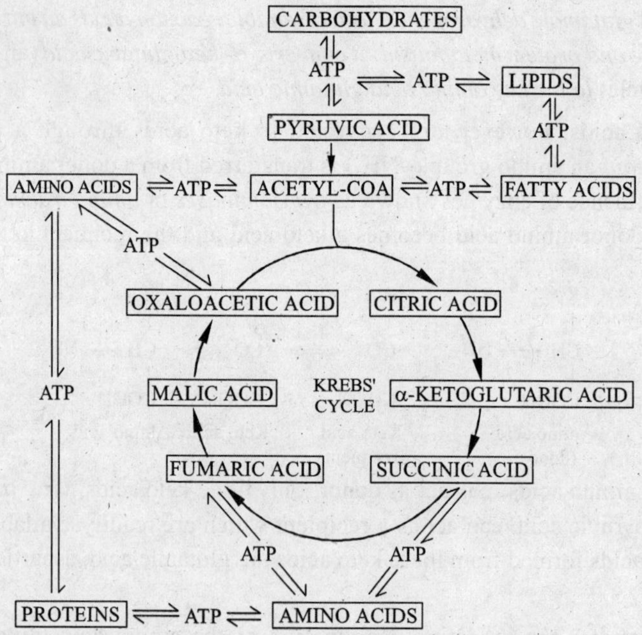

Fig. 13.22. Flow diagram showing the steps involved in the breakdown of amino acids, carbohydrates and lipids.

14

Chapter

Blood and Circulation

Blood performs several vital functons of the body, hence it has been frequently referred to as the *seat of the soul*. It distributes nutrients, oxygen, detoxicating compounds, and maintains a uniform body temperature. It is a carrier of hormones which control and integrate the activities of the various organs. It conveys antibodies and cells that protect the body from invading pathogens and also transports drugs administered to the body so that they may produce their therapeutic effects on specific sites of ailment. The blood also collects the waste material generated by the body's metabolic activities.

The science dealing with the study of blood is referred to as *hematology*.

The blood can carry out its functions efficiently only if it is kept circulating in the body. This is effected by the heart which acts as a pump, and an intricate system of distributing pipes – the blood vessels, *viz.*, arteries, veins and capillaries. The heart and the blood vessels together form the *cardiovascular system*.

BLOOD

Blood is a fluid tissue. In human beings there is about 90 ml of blood per kilogram of body weight. However, this volume depends on the age and health of the individual. Blood has a specific gravity of 1.05 to 1.06, and a pH of 7.4. It is made up of fluid matrix called *plasma*, and several types of cells termed *formed elements* or blood corpuscles. These two components separate when blood is centrifuged. The plasma forms 54% of the blood volume and the cells 45%. The latter is referred to as the *hematocrit* or *packed cell volume* (PCV). The plasma contains many inorganic and organic substances. The formed elements consist of red blood cells (RBC) or *erythrocytes, white blood cells* (WBC) or *leucocytes,* and *platelets or thrombocytes.*

Plasma

Blood plasma is the straw-coloured fluid obtained when all formed elements are removed from blood either by centrifugation or sedimentation. The ratio of plasma to formed elements is held relatively constant by the homeostatic controls of the body chemical etc. The composition of plasma is given in Table 14.2

The plasma contains inorganic salts with a dominance of chloride (Cl^-), sodium (Na^+), and calcium (Ca^{++}) ions, and lesser quantities of potassium (K^+) and magnesium (Mg^{++}) ions. Smaller concentrations of bicarbonate, sulphate, and phosphate, and traces of iron, copper, and iodine are also present.

The plasma also contains glucose which is used as a principal energy source in cellular metabolism. A variety of other substance, such as amino acids, vitamins, lipids (neutral fats and cholesterol), hormones, urea, uric acid, and other metabolite waste products are also present in the plasma. In addition, it contains dissolved oxygen, carbon dioxide; and nitrogen. Plasma also reflects certain abnormal conditions occurring in cells and tissues of the body. For example, within few hours following a heart attack, there is a noticeable increase in the level of the enzymes transaminase and lactic dehydrogenase in the plasma arising from the affected heart muscles. This has proved to be of significant value to a clinician in the diagnosis of heart attacks.

The plasma of man contains about 70 g proteins/1. Some of these proteins possess enzymic activities (*e.g.,* phosphatases), while other fractions show no activity at all. Plasma proteins are classified according to their molecular weights and electrical properties. In order of decreasing molecular weight these are: β*-lipoprotein* (MW : 1,300,000), *fibrinogen* (MW; 350,000), α-*lipoprotein* (MW: 200,000), γ*-globulin* (MW: 155,000), β-globulin (MW: 90,000), and *serum albumin* (MW: 70,000). The functions of the plasma proteins can be summarized as follows: (1) Lipoproteins carry lipids and also transport lipid-soluble vitamins and steroids hormones. (2) They are amphoteric compounds (also ampholytes) because they act both as acids and bases. At the pH of human blood (pH 7.4) proteins behave as acids and combine with cations such as Na^+ and form a buffer; they are, however, weak buffers. (3) Being macromolecules plasma proteins create a *colloidal osmotic pressure* in the blood which is higher than the colloidal osmotic pressure in the interstitial fluid. This favours the passage of interstitial fluid into the plasma. Finally (4) plasma proteins, prothrombin and fibrinogen participate in blood clotting.

Formed Elements (Blood corpuscles)

To formed elements consist of the *erythrocytes, leucocytes,* and the *thrombocytes* (Fig. 14.1)

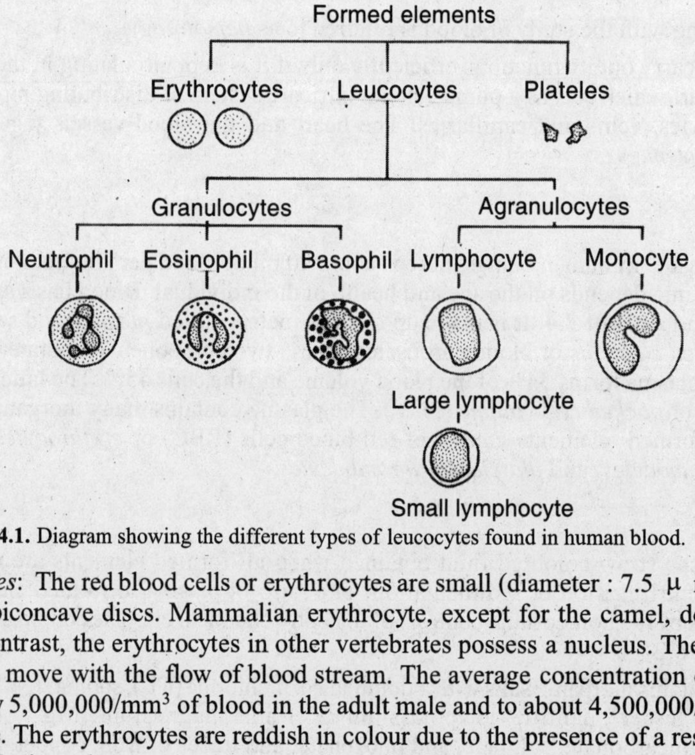

Fig. 14.1. Diagram showing the different types of leucocytes found in human blood.

Erythrocytes: The red blood cells or erythrocytes are small (diameter : 7.5 μ in man and 9.0 μ in elephant) biconcave discs. Mammalian erythrocyte, except for the camel, does not possess a nucleus. In contrast, the erythrocytes in other vertebrates possess a nucleus. The erythrocytes are immobile and move with the flow of blood stream. The average concentration of erythrocytes is approximately 5,000,000/mm^3 of blood in the adult male and to about 4,500,000/mm^3 in the adult human female. The erythrocytes are reddish in colour due to the presence of a respiratory pigment *hemoglobin.* Each erythrocyte contains an estimated 200 to 300 million molecules of hemoglobin. The latter serves to transport oxygen and, to some extent, carbon dioxide, to and from the lungs. The details of gaseous transport by hemoglobin have been dealt in Chapter 13. Hemoglobin is easily removed when the cell disintegrates (or lysed) by exposure to a hypotonic medium, or water. The resulting structure of the erythrocyte is called a "ghost" which consists of network of protein and lipids.

The average life span of an erythrocyte in man is about 120 days; the old erythrocytes are continuously breaking down and new ones are formed. The erythrocytes are destroyed by the phagocytic macrophages found in the reticuloendothelial system (in bone marrow, spleen, liver, etc.). The porphyrin section of the hemoglobin is catabolized and forms bile pigments (bilirubin and biliverdin). However, most of the iron is retained in the bone marrow to be again incorporated into new erythrocytes.

Leucocytes: The leucocytes are nucleated cells, and are somewhat larger than the erythrocytes: the white blood cells are also less abundant than the red blood cells. There are 5,000 to 9,000 leucocytes/mm^3 of blood. An increase in the number of leucocytes above 10,000/mm^3 of blood is known as *leucocytosis,* and a fall below 4,000 is referred to as *leucopenia.* Leucocytosis is caused by violent exercise, infection, hemorrhage, and a high protein diet. Leucopenia results from malnutrition, deficienty of vitamin B_{12}, folic acid, and pyridoxine.

The leucocytes are classified into *granulocytes* and *agranulocytes* depending on the presence or absence of *granules* in the cytoplasm (Fig. 14.1).

Granulocytes are divided into three basic types, viz., *neutrophils, eosinophils,* and *basophils.* This classification of the granulocytes is based on the nuclear structure and affinity of the cytoplasmic granules to dyes (acid or basic). In the neutrophils the cytoplasm is filled with fine uniformly distributed granules which do not show any affinity to either acid or basic dyes, and hence the name neutrophil. Their nucleus is polymorphic divided unequally in 3–5 lobes and separated by deep indentations or by narrow strands of chromatin; therefore, neutrophils are also referred to as *polymorphocytes*. These form 65-70% of the total leucocytes. The size of each neutrophil ranges from 10 – 12μ. The eosonophils are loaded with granules which are relatively bigger in size than in the neutrophils. The granules show affinity to acid dyes, such as eosin; the latter imparts them an intense red colour, and hence the name eosinophil. These are also referred to as *acidophils*. The nucleus is bilobed and the two lobes are connected by a thin strand of chromatin. The eosinophils form 2 – 4% of the total number of leucocytes. In the basophil the cytoplasm contains fewer granules of basophilic nature; the granules take an intense blue colour with basic dyes. The nucleus in the basophil is irregularly polymorphic. The basophils have the least representation in the total leucocyte count; the normal percentage is only 0.5.

Agranulocytes are of two types, *viz., lymphocytes* and *monocytes*. The structure of the nucleus of arganulocytes is not polymorphic and is, therefore, different from that of the granulocytes. Lymphocytes fall into the categories: *small* and *large* lymphocytes. The nucleus of the small lymphocyte is large and it almost occupies the entire cell leaving only a peripheral cytoplasm, whereas the large lymphocytes have relatively more cytoplasm. The lymphocytes form the second largest group of leucocytes; the small lymphocytes constitute about 25% and large lymphocytes about 3% of the total leucocytes. The size of the small lymphocyte is about 7 μ, whereas it is 10 – 12 μ in the large lymphocytes.

In *monocytes* the cytoplasm is more abundant in relation to the nucleus; the nucleus is ovoid, kidney-shaped or with many indentations. Moreever, characteristically, the nucleus is always placed eccentrically in the cells. The monocytes form 5-7% of total leucocytes. The size ranges from 10-15 μ.

The average life span of leucocytes is two days. They defend the body against invasion by microorganism which are engulfed and digested by them by a process known as *phagocytosis.* Additionally, the lymphocytes form *antibodies* against proteinaceous matter gaining entry into the body. The relative ratio of the leucocytes changes during many diseases. This phenomenon is of great use to pathologists in diagnosis of diseases. For example, the concentration of eosinophils increases significantly during infection by worms, and in allegic conditions such as asthma and hay fever. Another very serious disorder of the body is *leukemia*. This is a form of *blood cancer* and is characterized by an uncontrolled increase in the number of leucocytes (up to 1,000,000/mm^3 of blood), with the apperance of immature forms in the blood stream. Leukemia is accompanied by anemia, and is eventually fatal because the bone marrow cells that normally produce erythrocytes are displaced by proliferation of leucocyte-forming tissue (leukosis).

Thrombocytes or *platelets*: In addition to the erythrocytes and leucocytes, thrombocytes or platelets are another group of formed elements observed in the blood. These are small fragments (2-5 μ) of cells which lack nuclei, and function in blood clotting. They are oval in shape and have the capability of amoeboid movements. There are 250,000 platelets/mm^3 of blood. The life span of the thrombocytes is also very short - perhaps only a few days in duration.

HEMOPOIESIS

Hemopoiesis is the formation of the formed elements of blood. In mammals it takes place in the *bone marrow,* the Kupffer's cells of the liver *splenic sinuses,* and *lymph nodes.* These together are referred to as the *reticuloendothelial system.* Of the various hemopoietic centres, the bone marrow gives rise to all the three types of blood cells, while others give rise to lymphocytes.

Hemopoietic Bone Marrow

Only the *red bone marrow* is hemopoietic. Red marrow is found in the flat bones (such as ribs or sternum) and in the extremities of limb bones. In the red marrow, it is not possible to differentiate between precursor cells of erythrocytes, leucocytes, or platelets. There are two theories of hemopoiesis: *unitarian,* which states that all types of blood cells are formed from only one type of *stem cell* (mother cells): *pluralistic,* which states that different types of blood cells arise from different stem cells. The unitarian theory is more acceptable (Fig. 14.2). There are reticular mother cells in the red marrow of the bone which give rise to cells known as *hemocytoblasts.* The latter are non-circulating undifferentiated cells which lie free in the network of bone marrow reticulum. Hemocytoblasts are the precursor cells (the suffix *blast* describes a precursor cell) of erythrocytes, granulocytes, agranulocytes, and the platelets. All blood cells originate in the reticuloendothelial system, and return there when old or diseased to be destroyed.

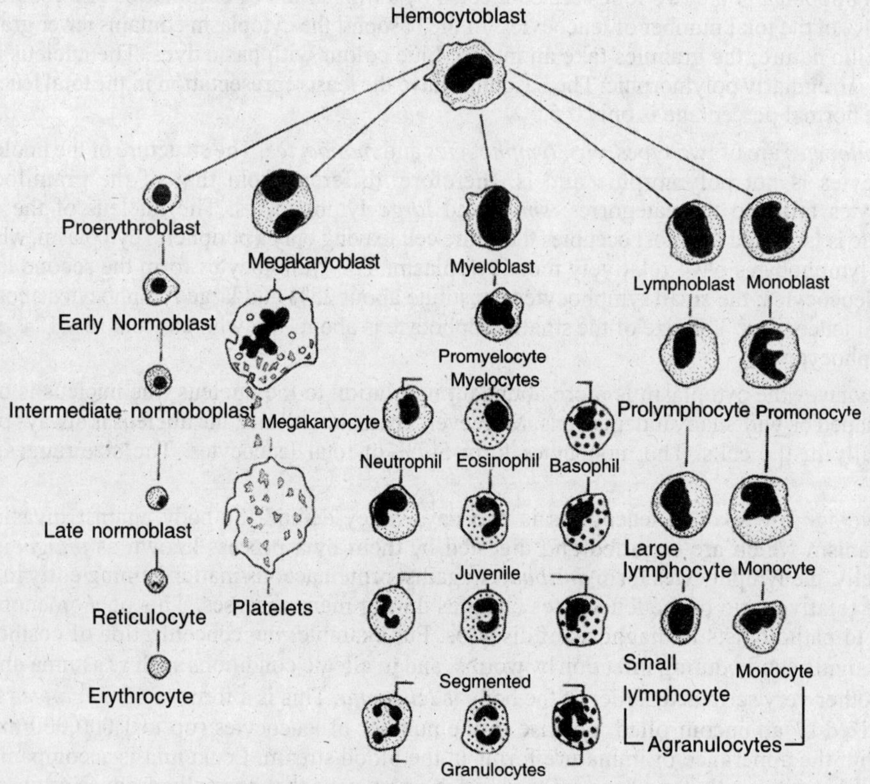

Fig. 14.2. Hemopoiesis of the formed elements of blood.

Erythropoiesis

Mature erythrocytes in the circulating blood develop from a pool of hemocytoblasts in the red bone marrow. The development and the maturation of the erythrocyte from the hemocytoblast is called *erythropoiesis.* During erythropoiesis the hemocytoblasts pass through several stages each of which has been given specific names; in a sequential order these stages are referred to as *proerythroblast* (E_1), *large early normoblast* (E_2), *small normoblast* (E_3), *intermediate normoblast* (E_4), *late non-dividing normoblast* (E_5), *reticulocyte* (E_6), followed by the *mature erythrocyte* (E_7). From stages E_1 to E_2, the cell undergoes 3 or 4 mitotic divisions. However, at the intermediate normoblast (E_4) the cell stops further divisions; in the E_5 stage the *nucleus is squeezed out and the mature erythrocytes are formed.* The time for erythropoiesis is about 120 hours.

Vitamin B_{12} and *folic acid* are two important factors for the normal development of erythrocytes. In their absence, the proerythroblasts (E_1) of bone marrow, instead of maturing into normoblasts, give rise to immature cells called *megaloblasts.* The latter lose their nuclei and appear in the circulation as large *macrocytes.*

The life span of erythrocytes is 120 days. The aging erythrocytes return to the liver and spleen where they are temporarily stored. The *liver* and *spleen,* therefore, are also referred to as *zones of erythrocyte retention.* From these organs the erythrocytes are transported to the reticuloendothelial system where they are destroyed. Some red blood cells are also destroyed by the spleen, and by lytic agents while circulating through the capillaries.

The destruction of old erythrocytes is compensated by the production of new cells to keep constancy in the red blood count in a normal individual. This is controlled by a feedback mechanism mediated by a hormone *erythropoietin,* produced in the kidneys by the *juxtaglomerular cells.* There are several other humoral factors such as *erythrocytosis stimulating factor* (ESF) and a special lipid which also governs erythropoiesis. Erythropoietin is absent in fetus, and it is not yet known what substance causes erythropoiesis in the fetus.

It may be mentioned here that sites of erythropoiesis gradually change with development. In the early embryo erythropoiesis occurs in blood vessels in the yolk-sac membrane, in the fetus in the liver and spleen, in the infant throughout the skeleton, and in the adult in the ribs, sternum, and the extremities of the long bones.

Anemia: Anemia is described as a condition in which the red blood cell count or the amount of hemoglobin per cell is less than normal. Loss of blood from the body due to hemorrhage, excessive bleeding during the menstrual cycle, injury, etc., causes *hemorrhagic anemia.* A dietary deficiency of iron can also cause a characteristic anemia known as *hypochromic anemia;* this may be *normocytic* or *macrocytic.* These types of anemias develop by abnormal erythropoiesis when immature and fragile erythrocytes enter the blood stream. Other nutritional factors also adversely affect the formation of hemoglobin and normal development of the red blood cells. These factors, already described, are certain vitamins such as folic acid, ascorbic acid, niacin, riboflavin, and vitamin B_{12}. The deficiency of the latter causes *pernicious anemia* which is macrocytic and hypochromic.

Another form of anemia, called *sickle cell anemia* (*sicklemia*), is linked with heredity. It is prevalent in people of negroid orgin. It is caused by an abnormal hemoglobin known as *hemoglobin* S (Hb^S). The latter is formed under the influence of a particular pair of abnormal genes (H^S). Erythrocytes containing Hb^S are normal so long they get sufficient supply of oxygen. However, when such erythrocytes lose oxygen to the tissues (*via* capillaries), the erythrocytes become crescent or sickle shaped. These sickled cells clog the blood vessels and are subsequently destroyed. This results in sickle-cell anemia. However, this type of anemia is associated with a person possessing *homozygous* condition of the gene. In a heterozygous condition an individual, with only one gene for Hb^S and another for normal gene, becomes ill only at high altitudes. A group of human anemias due to imbalance in the ratio of alpha and/or beta hemoglobin units, occuring chiefly in children of Mediterranean origin, is referred to as *thalassemia.* It is a genetically determined defect of hemoglobin synthesis.

Leucopoiesis

All categories of leucocytes are formed in the red bone marrow. However, most of the lymphocytes originate in the lymph tissue.

Hemocytoblasts of bone marrow give rise to *myeloblasts;* these form *promyelocytes.* The latter give rise to three types of myeloctytes, viz. *Neutrophil, eosinophil,* and *basophil myelocytes.* The myelocytes mature in the circulating blood and form *metamyelocytes.* These metamyelocytes mature into the corresponding neutrophils, eosinophils, and basophils. This takes place in two stages-*juvenile, metamyelocyte* and *segmented metamyelocyte* stage.

The *lymphocytes* are formed in the bone marrow from hemocytoblasts. They pass through two stages, viz., the *lymphoblast* and *prolymphocyte* stage in that order. The latter finally differentiates into the *large* and *small* lymphocytes of the circulating blood.

As stated earlier, lymphocytes originate from the lymph tissue; in the fetal stages the thymus gland (Chapter 11) is the chief source of lymphocyte production. In the early stages of fetal development lymphocytes producing cells are "blue-printed" from the thymus and transferred to lymph tissues. The thymus gradually loses its lymphopoietic activity so that the major source of lymphocyte production in the adult is the lymph tissue.

Lymphocytes of *thymic origin* are concerned with *cellular immunity,* whereas lymphocytes "processed" in the lymph tissue of alimentary canal, tonsils, lymph nodes, and appendix are concerned with *humoral immunity.* When exposed to foreign *antigens,* these lymphocytes are activated (*immunoblasts*) and differentiate into *plasma cells* or *immunocytes.* These cells secrete the *immune globulins* (antibodies).

Monocytes, too, develop from hemocytoblasts, and pass through *monoblasts* and *promonocyte* stage before appearing in the blood as mature monocytes. Monocytes may also arise in spleen. Some hematologists believe that the monocytes are "free" reticuloendothelial cells (referred to as *histicytes*) which have migrated into the circulating blood.

Thrombocytopoiesis

Platelets are formed in red bone marrow, lungs, and other tissues. In the bone marrow the hemocytoblasts give rise to *megakaryoblasts* which develop into large cells (40-150 μ) known as *megakaryocytes* (150-300 μ). The latter are *polyploid* cells containing about four times the diploid number of chromosomes in a large nucleus. Each of these cells produces about a thousand *anucleate* platelets. Platelets are formed by the division of the cytoplasm of the megarkaryocyte. The cell membrane of the platelet is formed by the fusion of small vesicles in the cytoplasm. The megarkaryocytes migrate towards capillaries which they partially penetrate; the cell membrane disappears allowing the platelets to enter the circulation. In addition, megakaryocytes also throw out pseudopodia which are nipped off to become platelets. The production of thrombocytes or platelets is governed by a humoral factor known as *thrombopoietin* which acts in a manner similar to erythropoietin during erythropoiesis.

COAGULATION OF BLOOD

When a blood vessel is cut or ruptured, the blood coming out of it soon changes into a red gelatinous mass known as the *clot*. This mass slowly contracts leaving a yellowish fluid known as *serum. The serum should not be confused with plasma*; serum is the fluid left behind after the formation of the clot, while plasma is the blood without its formed elements. The formation of the clot (*coagulation of blood or blood clotting*) is a complicated process.

It the clot is examined under the microscope, it shows *agglutinated* blood cells entrapped in a fine meshwork of thin *fibrillar* structures. The fibrils which form the network in the clot are composed of *fibrin.* Fibrin is produced from a plasma protein, *fibrinogen,* which represents about 5% of the total plasma proteins. Clotting is, therefore, synoymous with the synthesis of fibrin from fibrinogen. Fibrinogen is converted to fibrin by the action of *thrombin* which exists in the plasma in the form of *prothrombin.* Prothrombin is converted into thrombin by the action of *thromboplastin* and Ca^{++}. A relatively simple scheme of blood clotting is shown in under:

$$\text{Prothrombin} \xrightarrow{\text{thromboplastin} + Ca^{++}} \text{thrombin}$$
$$\downarrow$$
$$\text{Fibrinogen} \rightarrow \text{Fibrin}$$

The conagulation of blood is not as simple as shown in the flow diagram above, because the synthesis of thromboplastin requires the participation of several *factors.* Scheme of coagulation system is given in Fig. 14.3.

The *coagulation* of blood requires altogether the interaction of *thirteen* distinct chemically identifiable substances which have been called *blood clotting factors*. These are designated in Roman numerals according to an International Convention. *Their normal order indicates the sequence of their discovery and not the sequence of their interaction.* These factors are:

Factor I (*Fibrinogen*): a soluble plasma protein.

Factor II (*Prothrombin*): the precursor of thrombin formed in the liver; its synthesis depends upon the presence of vitamin K.

Factor III (*Plasma Thromboplastin factor-PTC*): a protein found in plasma, and tissue extracts; this converts prothrombin into thrombin in the presence of several *coagulating factors* and phospholipid. In case phospholipids are absent, these are provided by blood platelets in the form of *cephalin.*

Factor IV (Ca^{++})

Factor V (*Labile Factor*): found in plasma but is consumed during clotting and not present in serum.

Factor VII (*Stable Factor* or *Autoprothrombin* I): found in both the plasma and serum.

Factor VIII (*Antihemophilic factor*- AHF): is a macromolecular complex composed of the AHF and von Willebrand factor (vWF); the latter accours for 99% of the mass of factor VIII which is a glycoprotein synthesized by megakaryocytes and is consumed during clotting and absent in serum.

Factor IX (*Christmas Factor*): found in plasma; it is activated during clotting, hence its activity in serum is much greater than in plasma.

Factor X (*Stuart-Power Factor*): present in plasma and serum.

Factor XI (*Plasma Thromboplastin Antecedent*—PTA): found in plasma and serum.

Factor XII (*Hageman Factor*): this factor is found in plasma; it was found to be absent in Hageman family, hence the name.

Factor XIII (*Fibrin Stabilizing Factor*): found in plasma which converts soluble fibrin into insoluble fibrin.

There is no such factor VI; the thirteenth factor is the phospholipids produced by the blood platelets. *Factors* V to XII participate in the formation of thromboplastin which is essential for the conversion of prothrombin to thrombin.

Thromboplastin: Thromboplastin is really the end product of several enzymatic reactions and it is active in the conversion of prothrombin to thrombin. However, because it was found very late that thromboplastin does not exist in the active form, two substances with thromboplastic activity were erroneously designated as thromboplastin. These are *intrinsic thromboplastin* found in the plasma and *extrinsic thromboplastin* found in tissue. These are in fact thromboplastin precursors, and not active thromboplastins. Because of customary usage, these are still called thromboplastins.

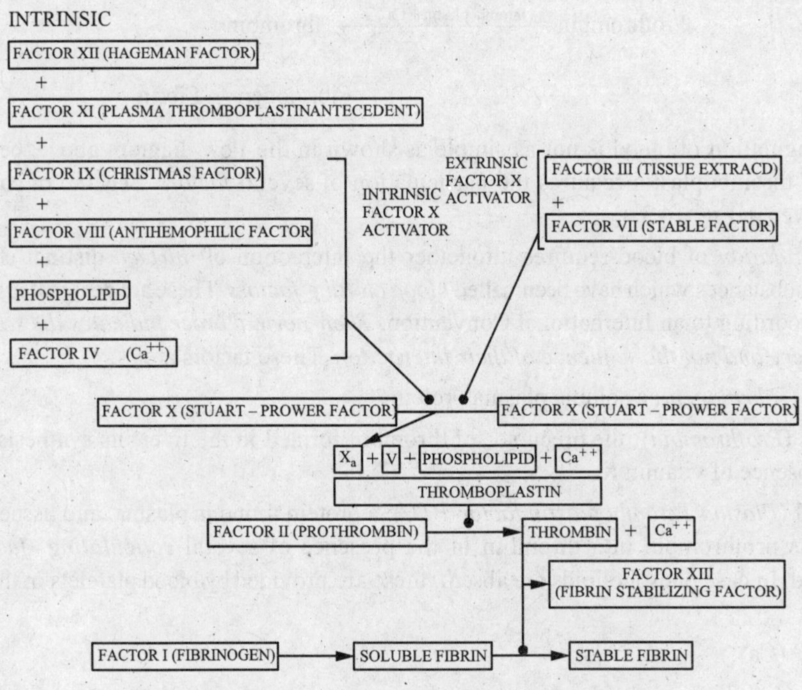

Fig. 14.3. Flow-diagram showing the coagulation of blood.

Intrinsic Thromboplastin. Before intrinsic thromboplastin becomes active it has to react with calcium (*Factor* IV), *Factor* V, *Factor* VIII, *Factor* IX, *Factor* X, *Factor* XI, *Factor* XII, and *phospholipids* found in the platelets.

Extrinsic Thromboplastin. Before extrinsic thromboplastin can become active, it has to react with Ca^{++} (*Factor* IV), *Factor* V, *Factor* VII, and *Factor* X.

In the formation of both these thromboplastins, the key reaction is the conversion of *Factor* X to its active form X*a* (*a* = active). This requires extrinsic thromboplastin *Factor* VII and Ca^{++}. The X*a* then combines with *Factor* V to form thromboplastin. The process is more complicated in the case of intrinsic thromboplastin, and is referred to as *enzyme cascade.*

The thromboplastins, once they are formed, react with prothrombin to form thrombin which in turn hydrolyses the fibrinogen to fibrin monomers.

$$\text{Fibrinogen} \xrightarrow{\text{Thrombin}} \text{Fibrin monomer} + \text{peptides}$$

The fibrin monomers then polymerise to form fibrin polymer which is *soluble* in water.

$$\text{Fibrin monomer} \longrightarrow \text{Fibrin polymer}$$

The fibrin polymer, in the presence of *Factor* XIII (fibrin stabilizing factor present in the plasma) gets converted into an *insoluble fibrin* clot.

Hemophilia. Hemophilia is characterized by severe hemorrhage even after relatively minor injuries by bleeding in the joints. The time of clotting which is usally 3-6 minutes in normal individuals may extend up to several hours in case of hemophilic patients. Hemophilia is a *sex linked* disease and appears exclusively in *males.* It is carried by the absence of *factor* VIII (antihemophilic factor—AHF and/or due to the deficiency of von Willebrand factor (vWF). The gene for hemophilia is carried by X-chromosome and in the female its action is countered by the presence of the other heterochromosome X (♀ = XX). As such hemophilia does not appear in the female but she can transmit it. She is called the *carrier.* In male the action of hemophilia gene is not counteracted (♂ = XY), and the individual exhibits hemophilia. The often quoted example of sex-linked hemophilic patients are the descendants of Queen Victoria of England.

Anticoagulants. Anticoagulants are substances that are used to prevent clotting. There are five classes of anticoagulants: (1) *decalcifying agents, e.g.,* oxalates and citrates, (2) *antiprothrombins,* notably heparin, dicoumarol, (3) *antithromboplastins,* such as protamines and histones, (4) *antithrombins, e.g.,* heparin, hirudin and thiol-derivatives, and (5) *antifibrinogens* such as neutral salts and $KMnO_4$. However, the most well-known, least toxic and most physiological anticoagulant is heparin. It is a polysaccharide esterified by SO_4 groups. It is found mainly in the liver but quantities are also present in lung tissue.

BLOOD GROUPS

In order to understand *blood groups,* it is essential to define *antigen* and *antibody.* An antigen is a macromolecular compound, generally a protein, which when administered to mammals stimulates the synthesis of a specific compound. This compound combines with the antigen and is called an antibody. The antibodies are γ -globulin (proteins); when an antigen combines with its specific antibody it causes *agglutination.* Agglutination is the phenomenon of collecting into lumps of cells dispersed in a fluid medium. In the case of blood, antigens and antibodies are called *agglutinogens* and *agglutinins,* respectively. *The agglutinogens are present in the erythrocytes and agglutinins in the plasma.*

In the human beings there are two agglutinogens referred to as A and B, and two plasma agglutinins called *anti*-A and *anti*-B. Blood groups are named in accordance with the antigenic substance found in the red blood cells. On this basis, the blood of human beings has been divided into four blood groups: *group* A (having antigen A and antibody B), *group* B (having antigen B and antibody A), *group* AB (having both antigens A and B but no antibodies), and *group* O (having both antibodies and no antigens). The blood groups were first described by Karl Landsteiner in 1900 who formed the so-called Landsteiner law which states that (1) if an antigen is present in the erythrocytes the corresponding antibody must be absent from the plasma, and (2) if the antigen is absent in the erythrocytes the corresponding antibody must be present in the plasma. Although the first law is essential, for if both agglutinogen and agglutinin are present, the blood would agglutinate; the second law, although a fact, is not essential because if the agglutinogen is absent, the presence of absence of the corresponding agglutinin is immaterial. Table 14.1 summarizes the erythrocyte-plasma relationship on the basis of agglutinogen-agglutinin reaction.

Table 14.1 Summary of Erythrocyte-Plasma Relationship on the Basis of Agglutinogen-Agglutinin Reaction

Group	*Agglutinogen (in Erythrocytes)*	*Agglutinin (in plasma)*	*Erythrocytes React with Plasma*
A	A	*anti*-A or agglutinin β	*anti*-A or the Plasma of B and O subjects
B	B	*anti*-B or agglutinin α	*anti*-B or the plasma of A and O subjects
AB	A and B	None	*anti*-A and *anti*-B or all plasma (B, A, O)
O	Neither A or B, i.e., O	*anti*-A or agglutinin β and anti B or agglutinin α	No reaction

The implications involved in the above Table may be briefly explained as follows: If a blood sample is agglutinated by antibody A but not by B, the blood is said to belong to group A. If the blood sample agglutinates by antibody B and not by A, the blood will belong to group B. The blood, if it agglutinates by both antibodies, it belongs to group AB, and if it does not agglutinate at all it is said to be of group O.

For the reasons described above, it is important that during blood transfusions, blood from the correct group should be used. The blood of the *donor* should be such where erythrocytes do not agglutinate by the plasma of the *recipient.* An error in this regard has fatal consequences. Blood of the same group can be transfused into the recipient without any problem of agglutination. Blood from a different group may also be transfused if it is compatible with the blood of the recipient. The following flow diagram describes the phenomenon of donors and recipients in a nut-shell. *The direction of the arrow shows the recipients:*

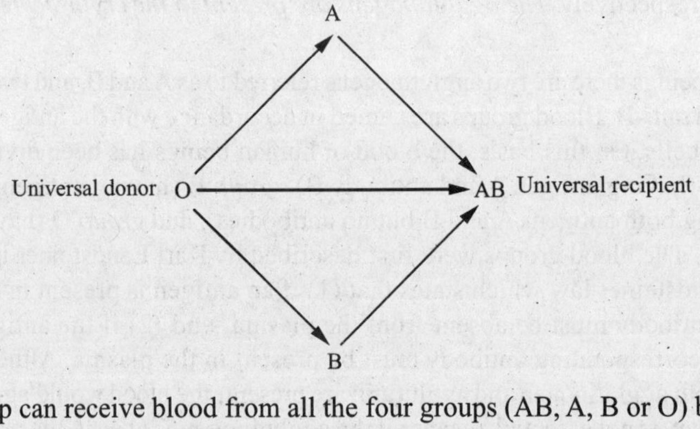

Thus, AB group can receive blood from all the four groups (AB, A, B or O) because it does not have any agglutinins. These are, therefore, called *universal recipients. Group O,* on the other hand, cannot receive blood from any group except O as its plasma contains agglutinins for both A and B agglutinogens. However, since the erythrocytes of O group do not contain any agglutinogens, it can be given to recipients of all the four groups (O, A, B, or AB). Hence, group O is known as the *universal donor.* Although the agglutinins present in the plasma of group O donors contain *anti*-A and *anti*-B, they are not very important. The agglutinins present in the plasma of donors can be usually ignored as they are sufficiently diluted in the blood of the recipient; in addition, these are also neutralized by the large amount of antiglutinogens present in the body fluids of the recipient.

The approximate distribution of blood groups amongst Hindus in India is as follows: A, 24%, B, 38%: AB, 8%; O, 30% Curiously, the incidence of agglutinogen B increases statistically from western Europe and Americas to the East (Turks, Arabs, Indians, and Chinese). *A very interesting example of blood groups in human populations is natives of Peru; they are cent per cent (100%) of group O.*

A word of warning, however, may be given here. Human erythrocytes can be agglutinated by numerous animal sera and this is independent of blood groups. These have been referred to as "*anti-man agglutinins*". Further, it has been reported that blood group A has two additional agglutinins— α_1 and α_2. But since they react below body temperatures, they have no importance in blood transfusions. In addition, agglutinin α_2 can also agglutinate red blood cells of group O. The name *anit-*O *agglutinin* is, therefore, given to α_2 and it follows that even group O erythrocytes contain a reactive substance – *agglutinogen* O.

Rh FACTORS

Human blood contains another factor which in independent of blood groups described above. On the basis of this factor, human blood can be divided into two groups Rh-*positive* and Rh-*negative*. This factor is also taken into consideration during blood transfusion. The Rh *system* can be explained as follows: The symbol Rh is derived from *rhesus.* When rabbits are injected with erythrocytes of the monkey *Macaccus rhesus* an *anti-Rhesus* rabbit serum (or *anti*-Rh) is formed. The latter strongly agglutinates erythrocytes of the monkey, and is also capable of agglutinating blood from about 85% of the human population. These persons are, therefore, called Rh-positive (Rh^+), whereas those in which the anti Rh does not cause agglutination are called Rh-*negative* (Rh^-). Persons who are Rh^+ possess what is known as the D *antigen*. The occurrence of this D antigen is governed by heredity. The presence of Rh^+ factor is caused by a *dominant gene.* First transfustion of an Rh^+ individual with Rh^- blood causes the production of antibodies so that a subsequent transfustion of Rh^- blood may even produce a reaction severe enough to be fatal.

Rh factor sometimes becomes very important when the parents belongs to different Rh groups. There is no problem if both parents have same Rh groups. However, if parents have *different* Rh groups, reaction between the mother and the fetus sometimes kills the latter. If the father is Rh^- and the mother Rh^+, the fetus is always Rh^+ since this factor is governed by a dominant gene. However, even, if the father is Rh^+ and the mother Rh^- the child will always be Rh^+. During the first pregnancy, the D antigen in the erythrocytes of the Rh^+ fetus may cross the placenta and cause the production of antibodies in the Rh^- mother. These antibodies usually do not affect the first child. During subsequent pregnancies, the antibodies already present in the mother may cross the placenta and gain access in the fetal circulation, and destroy fetal erythrocytes. The extent of damage to the fetal erythrocytes is variable. In some cases, the destruction is so severe that it causes the death of the fetus. In most cases, however, the reaction is not so severe. In such cases, if the new born shows signs of hemolytic disease, its blood is completely replaced by compatible blood and the newborn can be saved. It is a relief provided by the nature that even in incompatible parents, however, it is uncommon for the mother to develop antibodies after repeated pregnancies.

LYMPH

Lymph is an *extracellular fluid* drained by *lymph vessels.* It is formed from the intercellular fluid surrounding all cells. *Plasma, interstitial fluid* and lymph are all extracellular fluids with distinct identity (Table 14.2) and should not be confused with each other. Plasma is the fluid part of the blood and is confined only to the blood vessels; interstitial fluid, surrounding all cells, is a mixture of the ultrafiltrate of blood and substances coming out from cells; lymph is a yellow coloured fluid which is formed by the selective absorption of the interstitial fluid in the lymphatic capillaries (Fig. 14.4). Lymph constitutes about 25% of the total weight of all mammals.

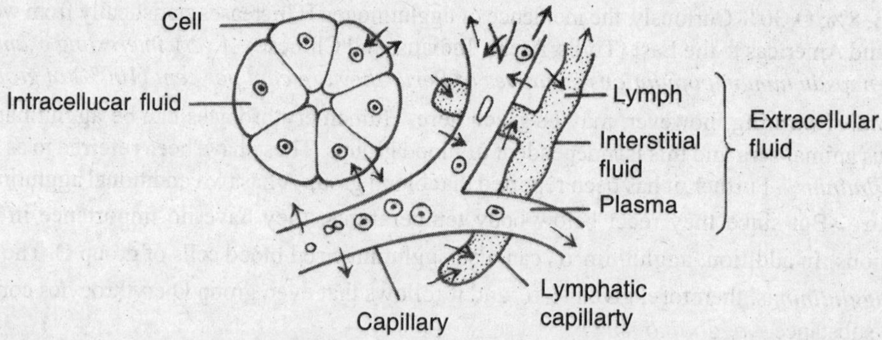

Fig. 14.4. Diagrammatic representation of exchange between intracellular fluid, lymph and blood.

Lymph differs from plasma in having about 40% of the protein concentration of the latter and slightly higher chloride; other components have the same composition as in plasma (Table 14.2). Fibrinogen content of the lymph is low; therefore, coagulation of lymph is very slow. Lymph has about the same leucocyte count (8,000 mm^3) as in blood. Part of these leucocytes escapes through capillary walls into the lymph by a process known as *diapedesis;* most of the leucocytes are, however, lymphocytes which are produced in the lymphatic system itself.

Table 14.2 Chemical Composition of Plasma and Lymph

	Plasma	*Lymph*
Protein (g/100ml)	6.85	2.61
Chloride (mequiv/1)	392	413
Calcium (mequiv/1)	10.4	9.2
Urea (mg/100 ml)	22	23.5
Sugar (mg/100 ml)	123	124
Non-protein nitrogen (mg/100 ml)	27.2	27
Amino acids (mg N/100 ml)	4.9	4.8

The *lymph vessels* are found in all parts of the body and range in size from that of capillaries to larger veins. The structure of lymph vessels is similar to that of capillaries and veins; like the veins, the larger lymph vessels are also provided with valves which prevent backflow.

Most of the blood which passes through arterial capillaries enters the venous capillaries, and is then returned to the heart. However, some of it, filtered through arterial capillaries, enters into interstitial spaces. The ultrafiltrate, in addition having inorganic salts, sugar, urea, etc., also contains some protein. The protien content of the interstitial fluid is less than the plasma. Part of the ultrafiltrate is reabsorbed by the capillaries; however, only small molecules are reabsorbed while the macromolecules cannot enter the capillaries once they have come out. The interstitial fluid is circulated back to the blood through a system of vessels known as the *lymphatic system.* The protein in the interstitial spaces, although unable to enter the blood spaces is able to enter into the fine capillaries known as *lymphatics,* probably through pinocytosis.

The *lymphatics* are present nearly in all parts of the body. The lymph system commences from lymphatics which unite to form larger and larger lymph vessels, which finally open into two large trunks in the upper portion of the body (the *right lymphatic duct* and the *thoracic duct*). The thoracic duct receives lymph from most of the body, including the whole of the intestine, and opens into the left subclavian vein. The right lymphatic duct receives lymph from the right side, head neck, right arm, right side of the heart, and opens into the right subclavian vein.

The movement of the lymph is brought about by breathing movement and contraction and relaxation of skeletal muscles of the body. The flow of lymph from intestinal lymphatics (*lacteals*) is brought about by the rhythmic wave-like motion of millions of villi of the intestine.

In lower vertebrates, muscular "lymph hearts" propel the lymph by their pulsating action. But lymph hearts are absent in mammals.

At many points in the body, where the smaller lymph vessels join the larger ones, masses of cells aggregate to form *lymph nodes.* These produce lymphocytes and antibodies, and filter out bacteria and foreign particles.

Some of the important functions of the lymph are as follows. The lymph returns the ultrafiltrate of the blood from the interstitial spaces back to the blood stream. The role played by capillaries in this direction is negligible. Finer lymph vessels are more permeable to larger molcules such as protein and fats as compared to the capillaries. Thus, the lymphatic system serves to return these molecules and transport them to the blood stream. It is in this manner that most of the products of fat digestion are absorbed from the intestinal wall by the lymphatic network in the villi. Lymph plays a major role in body's defence against disease by the production of lymphocytes and antibodies, and by the filtering action of the lymph nodes. Filtration by the lymph nodes removes microorganisms, foreign particles (for example, dust and soot), cancer cells which have detached from cancerous tissues. These are destroyed by the phagocyte action of lymphocytes. In some cases, bacteria trapped in lymph nodes cause infection and swelling of the nodes.

CIRCULATION

The blood is kept in constant circulation through the cardio-vascular system. The cardiovascular system consists of two units- a pumping organ called the *heart,* and a system of canals consisting of *arteries, veins,* and *capillaries.* Arteries carry blood from the heart, branching and growing smaller and thinner until they break up into a network of capillaries. The capillaries, in turn, fuse into larger and larger vessels, the veins, which convey the blood back to the heart. The interchange of substances between the blood and the various cells of the body occurs during the passage of the blood through the capillaries. The organization of the circulatory system is such that the blood circulates only in one direction. This was first discovered by William Harvey in 1628, and is considered to be the first major discovery of physiology.

STRUCTURE OF THE HEART

The *heart* is a muscular organ. It lies in the central portion of the chest cavity directly behind the sternum and is slightly tilted to the left so that the heartbeat is felt on the left side. The weight of heart in the adult human male and female is about 0.43 and 0.4 per cent of the body weight, respectively.

The heart is surrounded by a double-walled structure-the *pericardium.* This consists of on outer fibrous and an inner serous *pericardial membranes*. The space between the two pericardial membranes is filled with a thin sheet of fluid-the *pericardial fluid.* The pericardium cushions the heart against shocks, and also minimizes the friction on the outer surface of the beating heart.

The human heart, in fact all mammalian and avian hearts, are made up of four chambers consisting of *left* and *right atria* (auricles) and *left* and *right ventricles.* The function of the atria is to receive blood from the various organs of the body and pump it to the ventricles; the function of the ventricles is to pump the blood to the various ograns of the body. The atria and the ventricles are completely separated from each other by a fibrous structure known as the *fibrotendinous ring.* Most of the heart muscles are attached to this ring.

Internally the two atria are separated by a *thin interatrial septum,* and the two ventricles by a thick *interventricular septum.* The atrium and the ventricle of the left side are, therefore, completely partitioned off from the atrium and the ventricle of the right side. Because of this the human heart, although a composite four-chambered structure, is made up of *two relatively independent pumps;* left atrium and left ventricle forming one pump, and the right atrium and the right ventricle forming the other pump.

The *right pump* receives blood returning from all parts of the body (except the lungs) *via* the *superior and inferior vena cava.* The blood is then sent by the *pulmonary artery* to the lungs where carbon dioxide is released and oxygen taken up. The freshly aerated blood returns from the lungs *via* the *pulmonary vein* to the *left pump.* The latter then propels this blood out through the largest artery of the circulatory system-the *aorta,* to all other parts of the body (Fig. 14.5). The right side of the heart is, therefore concerned with blood circulation to the lungs, known as *pulmonary circulation,* whereas, the left side of the heart pumps blood to all the other system and is concerned with *systemic circulation.*

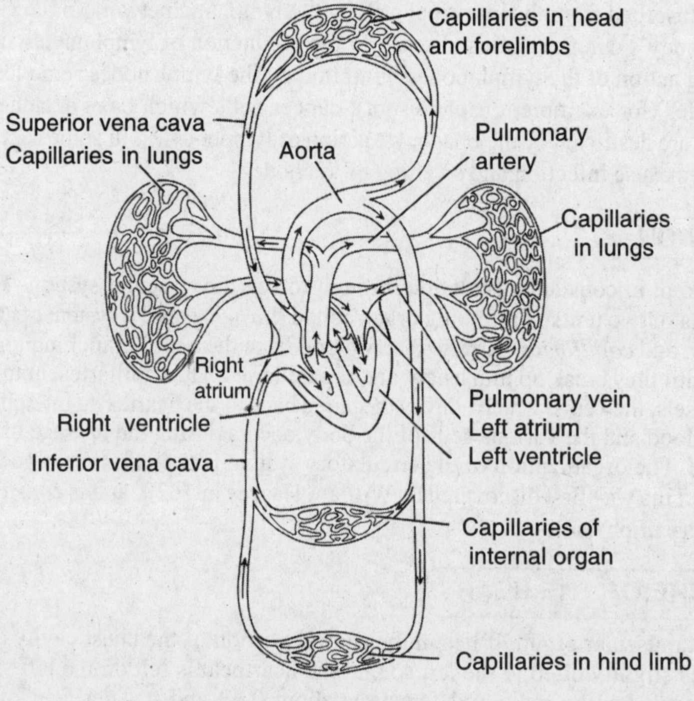

Fig. 14.5. Diagrammatic representation of the general plan of the circulatory system.

The heart consists of three layers: an outer epithelial lining called *epicardium,* middle layer-*myocardium,* and an inner endothelial lining known as *endocardium.* The thickness of the myocardium varies in different parts of the heart: the atrial myocardium is thin as they have to do small amount of work in pumping the blood to the ventricles; the ventricular myocardium is thick as the ventricles do most of the work of the heart for pumping the blood throughout the body by way of the blood vessels. Moreover, the myocardium of both atria and ventricles is a complicated structure. In the ventricles, it sends out projections in the ventricular cavity, which are known as *papillary muscles* (Fig. 14.6).

Heart Valves

Each atrium opens into the ventricle of its own side through *atrioventricular orifices.* The *right atrioventricular orifice* is guarded by the *tricuspid valve* (consisting of three membranous flaps), and the *left* by the *mitral valve* or *bicuspid valve,* consisting of two membranous flaps (Fig. 14.6). These atrioventricular valves are *inlet valves* permitting the blood to flow from the atria to the ventricles, called the *inflow tract,* but not in the reverse direction. Part of each valve is attached to the fibrotendinous ring, which surrounds the two atrioventricular orifices (atrioventricular rings). The inner edge of both the atrio-ventricular valves is free. However, the free margins are attached to the papillary muscles of the ventricles by strong fibrous cords known as *chordae tendineae.*

The base of the aorta and the pulmonary artery are closed by *semilunar valves* (Fig. 14.6). Each semilunar valve is made up of three cusps of very thin fibrous tissue. These are *outlet valves* opening away from the heart - the *outflow tract.* Both atrioventricular and semilunar valves are fibrous structures covered on both sides by the endocardium.

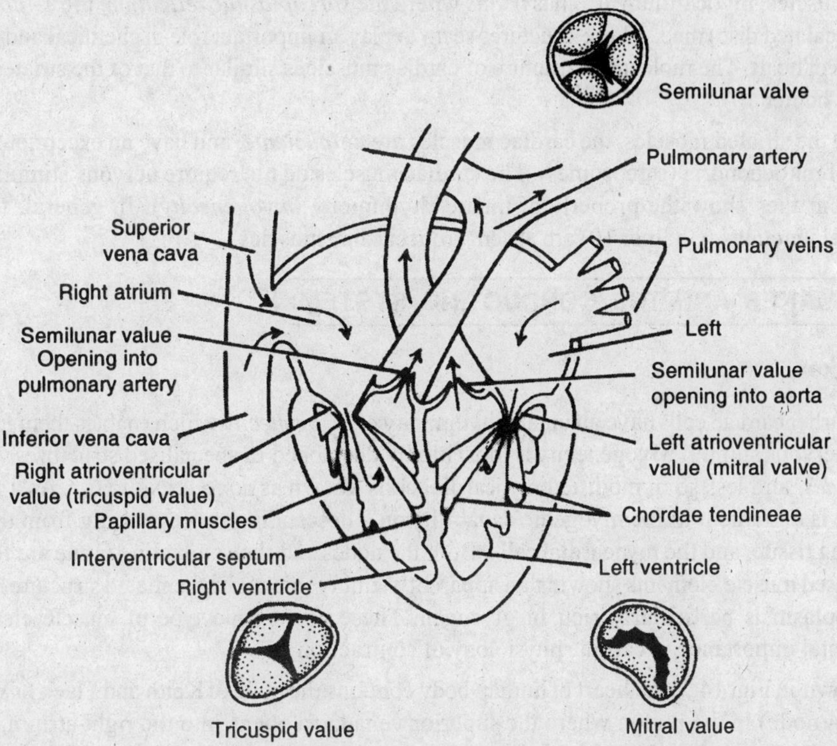

Fig. 14.6. Cut-away view of the heart showing its various chambers and valves; small insets at the corners show the detailed structure of the valves.

When the ventricles fill with blood they begin to contract. The pressure thus produced in the ventricles causes the inlet valves (tricuspid and mitral valves) to snap shut. Almost simultaneously the outlet valves (semilunar valves) open, permitting the discharge of the blood from the right and left ventricles to the pulmonary artery and aorta, respectively. The subsequent relaxation of the heart and the consequent fall in pressure within the ventricles are followed by the closure of the outlet valves (semilunar valves) and the opening of the inlet valves. This permits the blood to enter from the atria into the ventricles. The opening-closing action of the heart valves produces two principle basic *sounds.* The first is the strongest and corresponds to closure of the atrioventricular valves; the second, separated from the first sound by a short period of time, corresponds to the closure of the semilunar valves. The *first* and the *second heart sounds* are traditionally represented by "lub" and "dupp", respectively. Between the second and the first sound of the following cycle there is a period of silence called the "great silence". In addition to the above, two more sounds are heard; the *third-heart sound* can be heard in children and young adults and is set up when the left ventricle is changing from a state of active relaxation to that of passive distension. *The fourth heart sound* or the atrial heart sound is inaudible in normal persons, and caused by contraction of the atria when the blood rushes through the atrioventricular valves. The determination of anomalies in these sounds has been the basis of diagnosing valvular defects. Defective or "leaky valves", upon closing, produce additional sounds or "murmurs" which may furnish important clues regarding the nature of valvular defect.

Cardiac Muscle

The structure of the cardiac muscle has already been given (Chapter 10). However, the following remarks may be added. These are essentially striated muscles in which the *actin* fibres are attached to the *intercalated discs* which correspond to the Z *line* of striated muscles (Fig. 10.7). Like the striated muscles, myocardium also has *triads* where the *sarcoplasmic reticulum,* the T-*tube*, and Z-*line* (intercalated disc) meet. These structures seem to play an important role in chemical and electrical activation of heart. The molecular structure of cardiac muscle is similar to that of the striated skeletal muscle (Chapter 10).

Unlike the striated muscles, the cardiac muscles are *uninucleate,* and have an exceptionally large number of mitochondria (sarcosomes). The cardiac muscles do not require nervous stimuli for their contraction; they show the property of innate rhythmicity (*myogenecity*). In general, therefore, myoneural junctions (Chatper 10) are absent from cardiac muscles.

PACEMAKER AND THE CONDUCTING SYSTEM

Pacemaker

As said earlier cardiac cells have an innate rhythmicity or *myogenecity* which enables them to contract without nervous stimuli. Myogenecity is most highly developed in specialised structures which are called *nodes,* and less so in modified myocardial cells known as *conducting tissue.* As a matter of fact there is a gradient in the myogenic capability, in a descending order, starting from the nodes, conducting tissue, and the myocardial cells. Both the nodes and the conducting tissue are formed by anastomosed muscle elements showing an apparently embryonic undifferentiated structure, of which the sarcoplasm is particularly rich in glycogen. These are unique type of muscle elements of fundamental importance in cardiac physiology of contraction.

As shown in Fig. 14.7, the heart of human body contains the node of Keith and Flack or *sinoatrial node* (S-A node) in the region where the superior vena cava opens into the right atrium. There is another nodal tissue, the node of Aschoff and Tawara or *atrioventicular node* (A-V node), situated above the point of insertion of the tricuspid valve. Both these nodes are capable of generating spontaneous electrical currents at regular intervals. The S-A node has a faster intrinsic rate of generating impulses as compared to the A-V node. The conducting system carries the impulses from the node to the entire heart and causes the myocardial cells to contract. Since the S-A node sends impulses at a much higher rate than the A-V node, it controls the rate (*pace*) of the heart beat. The S-A node is, therefore, known as the *pacemaker* of the heart. If the S-A node is experimentally destroyed the A-V node takes over the role of a pacemaker, and the heart beats at a rate at which the A-V node fires its impulses. Since the pacemaker originates from muscle cells, it is known as a *myogenic pacemaker* in contrast to neurogenic pacemaker which is in the form of nerve cells present in the heart. Neurogenic pacemakers are found in all adult invertebrates except molluscs.

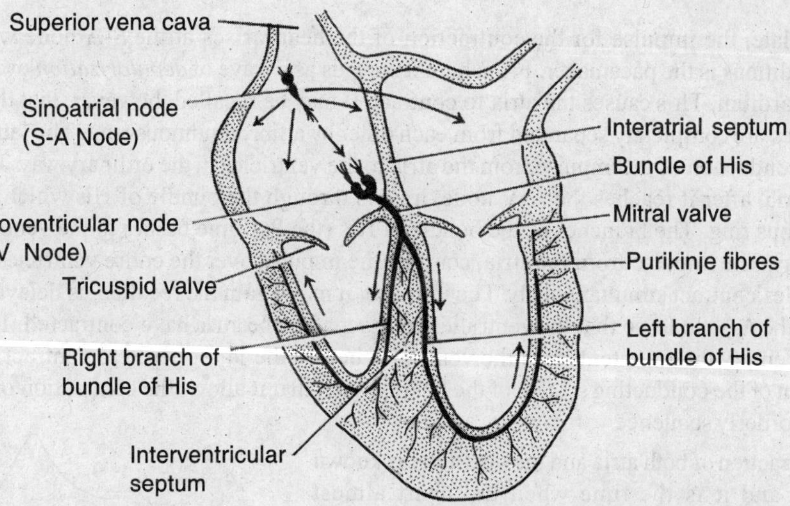

Fig. 14.7. Cut-away view of the heart showing the nodes and the Purkinje system.

Conducting System

The conducting tissue is organized in the form of well defined tract which together form the *conducting system* of the heart. The impulses generated at the S-A node spread over the sheet of the tissue comprising the two atria. In doing so, the impulses excite a succession of muscle fibres which together produce contraction of the atria. There does not appear to be any definite pathway in the atria and the impulse proceeds in a diffuse manner through them. However, certain histologists have described bundles passing from the S-A node to the left atrium (*bundle* of Bachmann) and to the right atrium (*bundle of* Thorel). However, the existence and function of these bundles are not fully explained. In the human atrium the impulse is transmitted at a rate of 900 mm/sec so that the atria are completely activated within 0.08 to 0.12 sec. The impulse, thus spread in the atria, rapidly reaches the A-V node. However, when the impulse reaches the A-V node, there is a time lag such that the ventricular mass beats about 1/10 sec after the atria. This delay is caused by a group of cells situated in the centre of the A-V node.

The impulse from the A-V node is carried to the ventricle by specialized conducting tissue called the *bundle* of His. The ultra-structure of the conduction tissue in the bundle of His has been shown to consist of small fusi form cells, poor in myofibrils and containing cytoplasmic granules of 0.2 to 0.6 μ in diameter. The bundle of His, after travelling a short distance, enters the interventricular septum where it divides into two main branches known as the main *left* and *right branches.* These travel in the septum nearly up to the apex of the ventricle, and then divide into large number of branches which travel upwards as Purkinje *fibres.* The bundle of His and Purkinje fibres are together known as Purkinje *tissue* (Fig. 14.7).

Conduction in the ventricle is thus carried out by the Purkinje tissue which results in a coordinated and almost simultaneous contraction of the ventricles. Relaxation of the atrial walls begins to take place at about the time the ventricles have started contracting. Normally the human ventricles are completely activated in 0.08 to 0.1 sec. This can be accounted in the following manner: 0.02 sec in the interventricular septum, 0.04 sec in the ventricle walls, and 0.02 sec across the myocardium. The latter is referred to as the *transmural conduction.*

In spite of the fact that the S-A node generates the impulse, if the S-A node is destroyed or suppressed the heart continues to beat but at a slower rhythm called the *septal rhythm* initiated by the A-V node. When the A-V node in suppressed the heart beats with a slow irregular rhythm called the *ideoventricular rhythm.* This is generated by the bundle of His.

Electrical Excitation of the Heart

To recapitulate, the impulse for the contraction of the heart arises at the S-A node which under normal conditions is the pacemaker. From here it spreads as a wave of *depolarization* over the entire atrial myocardium. This causes the atria to contract. It may be recalled, however, that the atria and the ventricles are completely separated from each other by a fibrotendinous ring. The latter does not permit the conduction of the impulse from the atria to the ventricles in the ordinary way. The impulse from the atria, after it reaches the A-V node, travels through the bundle of His which pierces the fibrotendinous ring. The branches of the bundle of His (the Purkinje tissue) to the ventricles in the ordinary way. The impulse from the atria, conduct the impulse over the entire ventricles so that the two ventricles contract simultaneously. The impulse, it may again the recalled, is delayed for about 1/10 sec at the A-V node, so that the ventricles contract after the atria have contracted. This permits the atria to empty their contents into the ventricles before the latter begin to contract. The entire organization of the conducting system of the heart is such that it allows the contraction of its various parts in an orderly sequence.

The contraction of both atria and the ventricles is known as *systole,* and it is the time when the heart almost completely empties the blood into aorta and the pulmonary artery. The relaxation of the heart is known as *diastole;* during this period the hart gets refilled with blood returning from the superior and inferior vena cava.

It has been stated earlier that the heart tissue has the property of innate rhythmicity or myogenecity which is most highly developed in the S-A node. The latter is capable of generating electrical impulses at regular intervals. These do not require any stimuli from the nervous system. The exact biochemical cause of the generation of pacemaker impulse is not known, except that it is a process of rhythmic repetitive oxidation.

The ionic basis of *resting* and *action potentials* has been explained in Chapter 9. The mechanism of the generation of action potential in the pacemaker is the same as in the nerve cells. The cardiac cells maintain a resting potential of 90 mV with the outside positive with respect to inside. When the pacemaker is excited then the magnitude of action potential is about 30 mV with the outside negative with respect to inside. The mechanism for the formation of action potential is typical, *i.e.*, the Na^+ permeability rises sharply during spike formation. There are, however, certain differences between the action potential of a nerve and a cardiac muscle cell. In the latter, the action potential lasts for a much longer duration (400-500 msec) in Purkinje fibres where it is the longest than in the nerve fibres. This is perhaps due to slow disapperance of sodium permeability and absence in the rise of K^+ permeability after the spike has been formed. The long duration of the spike is very important as during this period the heart is *refractory* and cannot be stimulated. The duration of the action potential goes on increasing as the

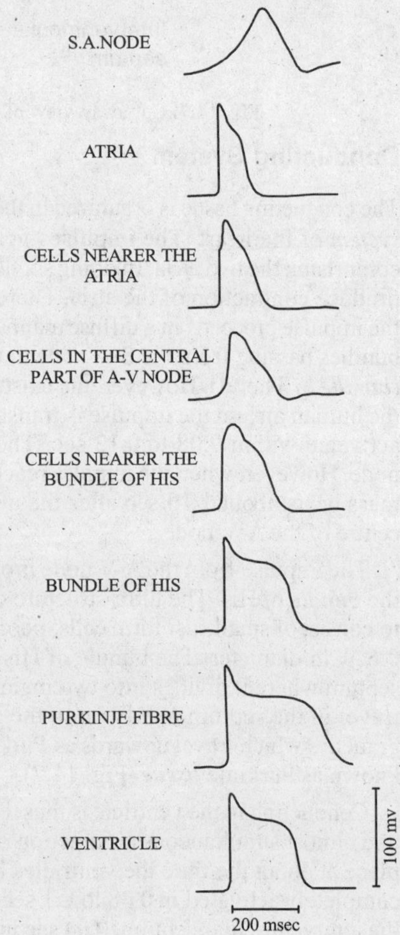

Fig. 14.8. Characteristic shapes of the action potential of the various cells of the heart; note that the duration of the spike progressively increases from the S-A node to the Purkinje fibre.

impulse descends from the S-A node to the Pukinje fibres. The duration of the spike of the working myocardial cell, however, is less than the Purkinje fibres (Fig. 14.8). The threshold for the generation of the impulse at the node and the duration of the diastole of the heart can be altered by the presence of neurotransmitters, and ions like K^+ and Ca^{++}. Once the impulse has reached the myocardial cells, the mechanism of *excitation-contraction coupling* is the same as described for muscles (Chapter 9). The excitatory impulse reaches in the *triads* which allows Ca^{++} to be liberated; the latter activates the myofilaments by binding them to ATP. The energy of contraction for the heart is thus also derived from ATP. However in the case of heart muscle, there is large store of *creatine phosphate;* a transphorylation enables the energy in creatine phosphate to be utilized for the synthesis of ADP. Creatine phosphate thus acts as a reserve of high energy bonds in the myocardium.

The Electrocardiogram

The changes in *electrical potential* associated with the initiation and spread of cardiac impulse can be recorded from the surface of the body by an appropriate instrument. The instrument is usually an extremely sensitive galvanometer known as the *electrocardiograph.* The graphic recordings, made on a moving strip of paper, are known as *electrocardiogram* (ECG or EKG).

It may be recalled that the muscle cells of the heart are connected with each other through *tight junctions* or *nexuses* and as such the impulse travels very rapidly over entire section (atria and ventricles) of the heart so that these behave as a single unit. The amount of current generated when the excitation spreads over the atria and ventricles is quite high; it creates a field of electricity which traverses over the entire body, and can be measured on the entire body surface. The strength of the potential in the field gets feebler as the electrical field moves away from the heart. The tissue between the heart and the body surface acts as a *volume conductor*, for example, like a solution of electrolytes. The current generated by the heart is so much that the other action potentials of the body produce very slight distortions in the recordings of the electrical potential of the heart.

The actual measurements are carried out by a number of electrodes. One of these is always the recording or the *exploring electrode* and the other an *indifferent electrode* which is kept at zero potential. The position of the electrodes is designated by the letter "V". There are four standard leads VR (right arm) VL (left arm), VF (left foot) and VC (chest). Various combination of these leads are used.

Fig. 14.9 shows a typical electrocardiogram taken from the right and left arm. There are three essential components of the ECG. These are P, QRS *complex* and T to which a fourth component U is sometimes added. The letters P, QRS and T were chosen since they have not been previously included in physiological terminology, and thus would not lead to confusion with other systems. In the intact heart, the P *wave* corresponds to *atria activation,* QRS complex (*fast wave*) with the activity of the ventricle during the transmission of the excitation through the Purkinje fibres, and T (*slow wave*) wave with the regression of the impulse from the ventricle. The U *wave* frequently occurs in enlarged hearts of athletes.

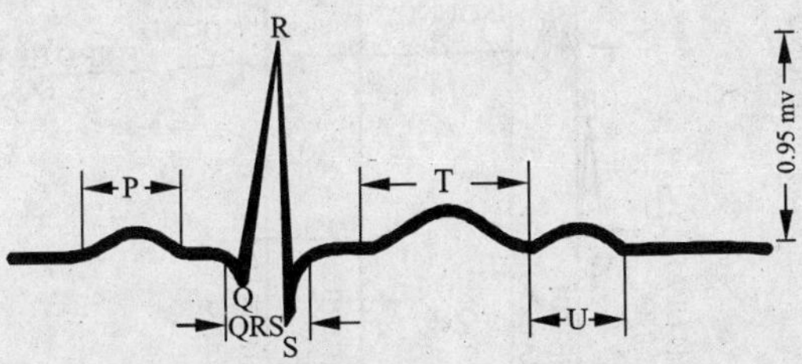

Fig. 14.9. A typical human electrocardiogram recorded from the two arms.

The duration of each electrical event is as follows : the P wave lasts 60—110 msec; 120—200 msec for the PR interval (time for atrioventricular conduction); about 60—110 msec for QRS complex;

and 200 msec for the duration of T. The amplitude of the electrical changes recorded by electrocardiograph in normal man are as follows: 0.1 mV for P, 0.95 mV for R, and 0.30 mV for T. These values, however, must not be considered to be absolute. There are important variations in cardiac potential recorded at different distances from the heart.

Because electrocardiograms reveal facts about the origin and conduction of action potential of the heart, electrocardiograph is a tool which detects abnormalities of the heart.

HEMODYNAMICS

The electrical events described in the preceding section give rise to an orderly sequence of mechanical events, the *cardiac cycle* which result in the characteristic beating of the heart. *The cardiac cycle of the left side of the heart can be described as follows* (Fig. 14.10).

The cardiac cycle is conventionally described from the stage when the atrium is activated, and the ventricle is already full of blood (*ventricular filling*). (The stage of the cardiac cycle at which the description begins is immaterial). This then means that the contraction of the atrium that follows its activation does not add much blood to the ventricle. During the ventricular filling phase, pressure in the atrium is naturally greater than in the ventricles, and blood, therefore, moves from the atrium into the ventricle.

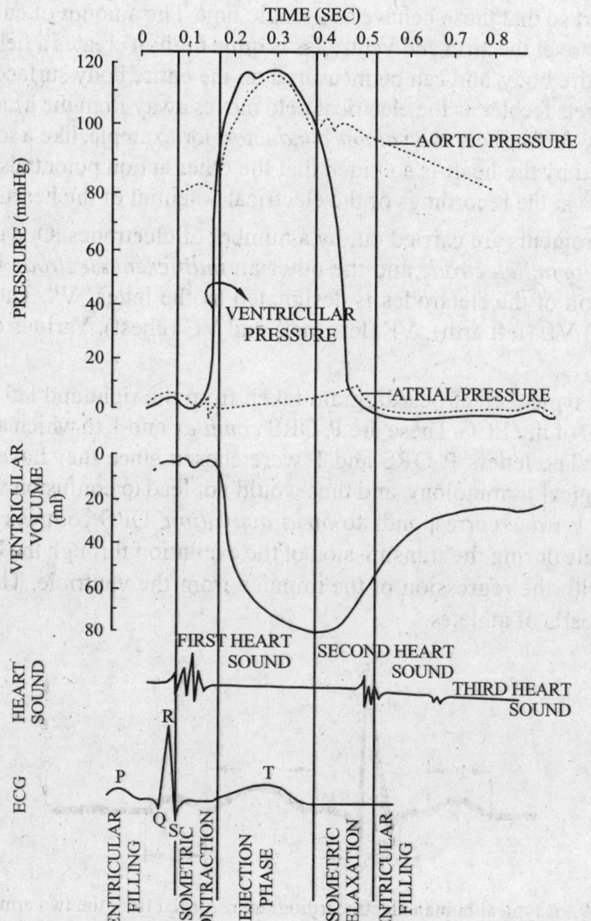

Fig. 14.10. Diagram showing the principle hemodynamic events associated with the cardiac cycle in man.

When the ventricle begins to contract following the QRS complex of the ECG, the pressure in the ventricle rises above that in the atrium and the mitral valve closes. This is the period of isometric contractions. When this pressure created by the *isometric contraction* of the ventricle just increases that in the aorta, the aortic valves open, the *ejection phase* of the blood begins. The ventricular pressure increases until the blood in the ventricle is emptied into the aorta, and the aortic valves close. The ventricle then begins to relax; this is the period of *isometric relaxation;* the pressure within the ventricle falls below that of the aorta, and the aortic valves close. At the end of the ejection phase, the intraventricular pressure is sufficiently low as compared to the atrium. The mitral valve opens and blood begins to fill the ventricle again. This completes the cardiac cycle. The ventricular filling phase of the cardiac cycle is divided into two phases, the fast filling phase (the QRS wave) during which most of the filing occurs, and the phase of *diastasis* (T wave) during which filling of the ventricle is slowly completed. The period of diastasis is greatly reduced when a speeding of the heart rate occurs, and it increases the output (to be discussed later) of the heart.

The cardiac cycle of the left and right sides of the heart is similar. However, the right ventricular pressure and the pulmonary arterial pressure are only about one-sixth (1/6th) of their counter parts in the left heart. This difference signifies the work done by the heart, discussed later.

Cardiac Output

The number of times the heart contracts per minute is the *heart rate,* and the volume of blood ejected by the left ventricle in each beat is the *stroke volume. Cardiac volume,* therefore, is the amount of blood pumped by the left side of the heart/minute, and can be expressed by the following equation:

C.O. = H.R. × S.V., where C.O. is the cardiac output, H.R. is the heart rate, and S.V. is th stroke volume.

In a normal and healthy person the H.R. is 72/minute and the S.V. 70 ml/minute (Fig. 14.10); cardiac output (C.O.), therefore, is approximately 5000 ml/minute.

It is obvious that output of the right and left ventricles must be the same otherwise the entire circulatory system would come to a standstill.

Starling's Law of Heart

The cause of cardiac contractibility is not simply an electrical phenomenon; it is both mechanical and electrical phenomena. Mechanical tension is required for the initiation, maintenance or amplification of the spontaneous contraction of the heart. It was shown by Starling in 1914 that the force of cardiac contraction increased with the resting lengths of the ventricular muscle fibres. By resting length it was meant the length of the fibres at the time of the maximum diastole. In other words, it might be said that the greater is the volume of blood in the ventricle at the moment of maximum diastole (*end diastolic pressure*; EDP), the greater will be the strength of the heartbeat. This observation is known as the "law of heart" or Starling's *law of the heart*. The actual behaviour of the heart at different diastolic pressure is depicted graphically in Fig. 14.11.

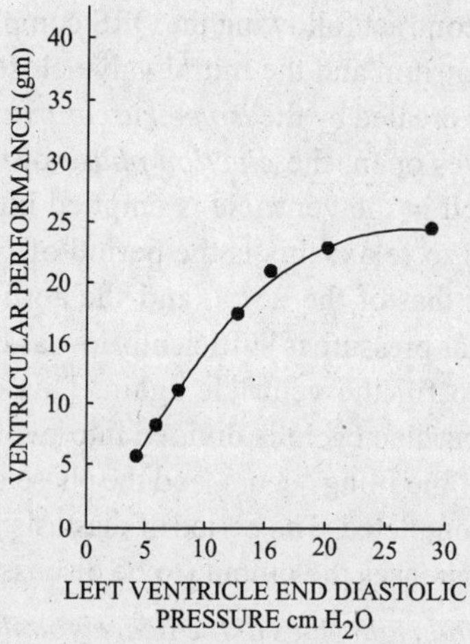

Fig. 14.11. Graph showing the relationship between the end diastolic pressure and the force of contraction of the ventricle.

Starling's law of the heart is very crucial in adapting the heart to its pumping action. When the ventricle contracts, it expels the blood which it contains. However, there is always a residual blood left in the ventricles. This volume of the left-over blood is added to the blood volume flowing into the ventricle during the next cardiac cycle. As a result of the increased EDP, contraction of the heart becomes stronger. Consequently, the residual volume left during the previous beat does not accumulate. This phenomenon is described as *compensation* because the heartbeat becomes stronger to compensate its failure (in expelling all the blood) in the previous beat. *Diastole,* therefore, is a mechanically *active phase* as it determines the strength of the systole. A heart that does not take care of the accumulating blood in the ventricle is described to go into decompensation. Such a phenomenon is characteristic of failing hearts; the heart becomes enlarged than the normal heart due to accumulated blood.

It must not be construed from the above that Starling's law is the only phenomenon which governs compensation in the heart. The nervous system, as will be seen later, also, controls and regulates cardiac output in healthy individuals. It is, however, possible that in disease the output of the ventricle is regulated according to the Starling's law.

Work done by the Heart

In order to function as a pump, the heart performs two functions simultaneously:

In the case of left ventricle:

(1) It pumps certain volume of blood (V) against the pressure (P) in the aorta. The work done (potential energy) during this can be expressed as P × V.

(2) In moving the blood through the blood vessels, it imparts kinetic energy to the blood which can be expressed as ½ mv^2, where *m* is the mass of blood propelled during each beat and *v* is the velocity of blood movement. Therefore, the total work done (*w*) by the left ventricle will be the sum of these two components

$$w = PV + \frac{1}{2}mv^2$$

In the case of right ventricle:

It has already been stated that the right ventricle encounters a pressure 1/6th as compared to the left ventricle, so that the potential energy in this case is 1/6PV. Nevertheless, the kinetic energy (1/2 mv^2) done by both right and left ventricles is the same.

Therefore, the total amount of work done (W) by both ventricles is:

$$W = PV + \frac{1}{2}mv^2 + \frac{1}{6}PV + \frac{1}{2}mv^2$$

or,

$$W = \frac{7}{6}PV + mv^2$$

The final equation thus arrived depicts the work done by the heart/unit of time.

Suppose, aortic pressure P = 100 mm Hg

or,

$$P = 100 \times 13.6 \text{ mm } H_2O$$

or

$$P = \frac{100 \times 13.6}{10} \text{ cm } H_2O$$

A column of water of $\frac{100 \times 13.6}{10}$ cm height and 1.0 cm cross section will, therefore, exert a pressure of

$$P = \frac{100 \times 13.6 \times 981}{10} \times 60 \text{ dynes/cm}^2,$$ where 981 is gravitational force.

If the output of the left ventricle per beat is 60 ml

Then, V = 60

The velocity of blood is 50 cm/sec

Therefore, $\frac{1}{2}mv^2 = \frac{1}{2} \times 60 \times (50)^2$

The work done by the left ventricle will be

$$PV + \frac{1}{2}mv^2 = \frac{100 \times 13.6 \times 981}{10} \times 60 + \frac{1}{2} \times 60 \times (50)^2 \text{ ergs.}$$

CORONARY BLOOD FLOW AND NUTRITION OF THE HEART

The cardiac muscles are nourished through coronary circulation. Blood enters into the *coronary arteries* through openings just ahead of the semilunar valves of the aorta. This opening is not guarded by valves, therefore, blood continuously flows through the coronary system during the entire cardiac cycle. Most of the blood in the coronary anteries flows during diastole; during systole, the blood flow is, however, reduced because the developing tension of the cardiac muscle compresses the walls of coronary blood vessels. The *coronary veins* drain the blood into the right ventricle through *coronary sinus* (Fig. 14.12), and some of the veins even directly open into the ventricle. The total coronary blood flow is about 3 to 10% of cardiac output.

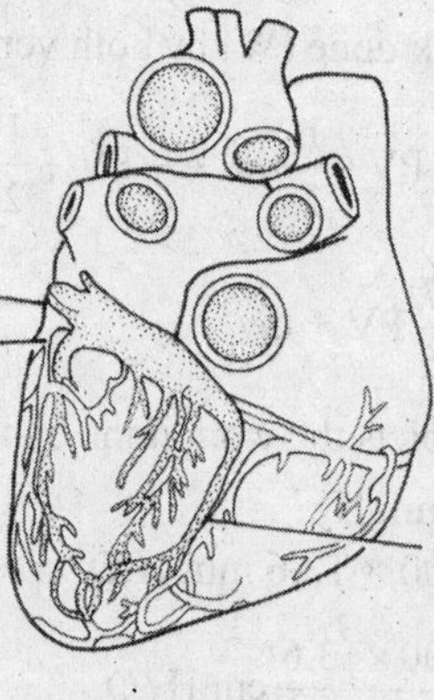

Fig. 14.12. Coronary circulation in man.

Coronary Occlusion. Interruption of circulation in any coronary vessel is called a coronary occlusion. This normally occurs due to deposition of *fatty substances* and *cholesterol* in the *coronary arteries* (arteriosclerosis), or an obstruction caused by a blood clot or *thrombus* (thrombosis). Coronary occlusion deprives the heart of its oxygen and it soon stops functioning. Cardiac muscle has no physiological mechanism to incur oxygen debt, since it does not utilize anaerobic metabolism. When occlusion occurs, the cardiac muscles receiving blood from the affected vessel dies, and a scar is formed and becomes non-functional. This area is called a cardiac *infarction.* If infraction covers a large area, the heart stops, and death occurs by heart failure.

Besides oxygen, other factors such as balanced electrolytes in the cardiac muscles, pH, availability of glucose, and temperature are equally important for the proper functioning of the heart. An excess of Na^+ is toxic for the heart, and it is the reason why intake of table salt is reduced in cardiac patients; excess of K^+ may cause complete stoppage of heart in diastole; and excess Ca^{++} can result in heart failure during systole.

NERVOUS REGULATION OF HEART

The heartbeat is intrinsically generated within the cardiac muscles, but the rate and strength of the beat are to a large extent under the regulation of the *autonomic nervous system.* The centers for the regulation of the heartbeat are located in the medulla and are known as the *medullary cardiovascular centre.*

Medullary Cardiovascular Centre

The medullary cardiovascular centre is located on the floor of the fourth ventricle; it consists of two centres known as the *cardio-accelerator* and *cardio-inhibitor* centres. However, in view of recent researches it is more appropriate to name them as *pressor* and *depressor* centres, respectively. There is however, no anatomical demarcation, and they probably overlap. The presser centre stimulates *sympathetic* discharge which accelerate the heart; it appears that the depressor centre is continuously sending impulses to the pressor cenre in order to initiate sympathetic discharges (Fig. 14.13).

The *efferent* nerves from the *heart* and the *carotid sinus* end in the depressor centre. The latter, in addition to sending stimuli to the sympathetic centre, also sends motor cardio-inhibitory nerves to the heart.

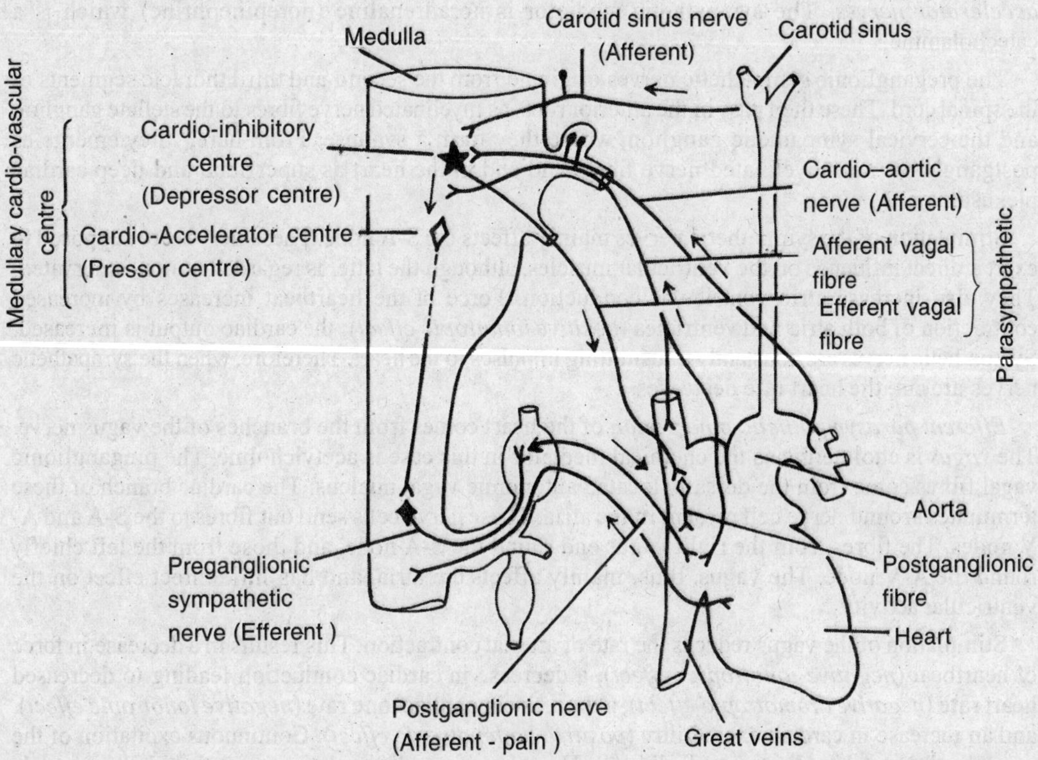

Fig. 14.13. Diagram showing the mechanism of nervous centre of the human heart. The arrows depict the direction in which the impulse is traveling.

Nervous Supply to the Heart

The heart receives both sensory and motor nerves from the autonomic nervous system.

Afferent Nerves: The heart sends out afferent nerves through several routes. A number of these nerves pass through the inferior cervical sympathetic ganglion, and upper four thoracic sympathetic ganglia in the spinal cord. These afferent nerves probably carry *pain sensations* from the heart. The main afferent nerves arise from two places: (1) those which arise form the heart and the aorta (*cardio-aortic nerve*), and (2) from the carotid sinus (*carotid sinus nerve*). One of the branches of the cardio-aortic nerve arises from the aortic arch and left side of the heart, while the other arises from the right side of the heart near the great veins (Fig. 14.13). They travel with the vagus to the medulla. These send out impulses at every systole but the frequency of their discharge increases when the pressure in the left ventricle and aorta is high. Since these nerves are excited by a rise in pressure in the aorta, these are known as *baroreceptors*.

The *carotid sinus nerve* arises in the *carotid sinus* (Fig. 14.13), which is the dilation at the beginning of the internal carotid arteries. The nerve arises from the *stretch receptors* of the sinus. The carotid sinus nerve passes up in a branch of the glossopharyngal nerve to the medulla and ends at the *cardio-inhibitory centre.* The distension of the carotid sinus by an increased blood pressure in it produces a reflex slowing of the heart.

Efferent Nerves: The efferent autonomic nerve supply to the heart consists of both *sympathetic* and *parasympathetic* nerves. Sympathetic innervation of the heart is adrenergic and when stimulated is responsible for *cardio-acceleration.* Sympathetic nerves, therefore, are also called as *cardio-accelerator nerves.* The sympathetic mediator is noradrenaline (norepinephrine) which is a catecholamine.

The preganglionic sympathetic nerves originate from the second and third thoracic segments of the spinal cord. These then pass in the anterior roots as myelinated nerve fibres to the stellate ganglion and the cervical sympathetic ganglion, where they form a synapse. From here, they emerge as postganglionic non-myelinated nerve fibres and end in the heart as superficial and deep cardiac plexuses.

Stimulation of the sympathetic nerves mainly affects the S-A node. These nerves are supposed to exert a direct influence on the ventricular muscles, although the latter is regarded as non-innervated. They also increase atrio-ventricular conduction. Force of the heartbeat increases by increased contraction of both atria and ventricles (*positive ionotropic effect*); the cardiac output is increased. Sympathetic nerves are constantly transmitting impulses to the heart. Therefore, when the sympathetic nerves are cut, the heart rate decreases.

Efferent parasympathetic innervation of the heart comes from the branches of the vagus nerve. The *vagus* is cholinergic as the chemical mediator in this case is acetylcholine. The preganglionic vagal fibres come from the dorsally located autonomic vagal nucleus. The cardiac branch of these terminates around nerve cell present in the atria. These nerve cells send out fibres to the S-A and A-V nodes. The fibres from the right vagus end round the S-A node, and those from the left chiefly round the A-V node. The vagus, thus, mainly affects the atria, and has little direct effect on the ventricular activity.

Stimulation of the vagus reduces the rate of arterial contraction. This results in a decrease in force of heartbeat (*negative ionotropic effect*), a decrease in cardiac conduction leading to decreased heart rate (*negative dromotropic effect*), a decrease in cardiac tone rate (*negative tonotropic effect*), and an increase in cardiac excitability (*positive bathmotropic effect*). Continuous excitation of the vagus leads to total cardiac arrest in diastole. However, if vegal stimulation is continued, the ventricle begins to contract while the atrium remains inactive. This phenomenon is called *ventricular escape* from vagal inhibition. This indicates that the ventricles when deprived of impulses from the atria can beat independently at their own rhythm. This is because the walls of the ventricle are stretched very strongly by the incoming blood which causes them to contract powerfully (see Starling's law of the heart).

The Cardiac Reflex

Stimulation of any of the afferent nerves described above causes a reflex reaction from the heart. Depending upon the nature of the stimulus, the heart may increase or reduce its rate and force of contraction. When the nerve endings of the cardioaortic nerve supplying the aorta and the left side of the heart are stimulated by a rise in the blood pressure in the aorta, the impulse is carried to the depressor centre. Similarly, any rise in arterial blood pressure also stretches the carotid sinus which sends sensory impulses to the depressor centre. Both these cause the inhibition of the heart through the motor branch of the vagus nerve. When the pressure in the great veins and right atria rises due to accumulation of blood, the branch of the cardio-aortic nerve supplying the area carries the information to the depressor centre. This then stimulates the pressor centre and causes an acceleration of the heart which quickly empties the blood.

It may be recalled that the depressor centre continuously stimulates the pressor centre to send out sympathetic discharges which accelerate the heart. However, when the depressor centre receives stimuli which demand an inhibition of the heart, the impulses send to the pressor centre are reduced. It has also been suggested that the activity of the pressor centre is increased by an increase in the hydrogen ion concentration in the interstitial tissue which is mainly influenced by increase in P_{CO_2} level. The medullary cardio-vascular centres also receive afferents from the hypothalamus which may either inhibit or accelerate the heart.

Effect of Drugs on the Heart

The cardiac muscle is peculiar in the sense that it is much more sensitive to drugs applied directly on it as compared to other muscles. The nervous control of the heart in a nutshell is as follows. It is stimulated by sympathetic nerves which are adrenergic and inhibited by parasympathetic nerves which are adrenergic and inhibited by parasympathetic nerves which are cholinergic. All those drugs which mimic the action of adrenergic nerves, *i.e.*, those which have an action similar to noradrenaline, therefore, stimulate the heart, for example, adrenaline and dopamine. It is because of this that adrenaline-like drugs are administered in an emergency to failing hearts. Similarly, all those drugs which have an action similar to acetylcholine inhibit the heart. Muscarine, an alkaloid found in rotten betel-nuts slows or arrests the heart. Atropine, on the other hand, causes and acceleration of the heart as it blocks the cholinergic action of the vagus (parasympathetic) nerve.

PHYSIOLOGY OF BLOOD VESSELS

Vessels circulating blood in the body are of three types, *viz.*, *arteries* which carry blood from the ventricles to the different parts of the body, *veins* which carry the deoxygenated blood from the tissues to the heart, and *capillaries* which from an anastomoses between the arteries and the veins. Ultrafiltration and gaseous exchange take place through the capillaries.

Arteries

Histologically two types of arteries can be distinguished, namely *elastic arteries* and *muscular arteries*. The elastic arteries are the larger arteries such as the aorta, pulmonary artery, etc. These are also referred to as the *central arterial reservoir*. The walls of these arteries are made up of an inner layer (*tunica intima*) consisting of endothelium, a middle layer (*tunica media)* which is thin elastic and contains little smooth muscle, and the outer layer (*tunica externa* or *tunica adventitia*) consisting of large amount of connective tissue, and is fibrous (Fig. 14.14). Muscular arteries are peripheral arteries and are different from elastic arteries in containing more smooth muscle in the tunica media. The very finest of the peripheral arteries, known as *arterioles,* are almost 100% smooth muscle.

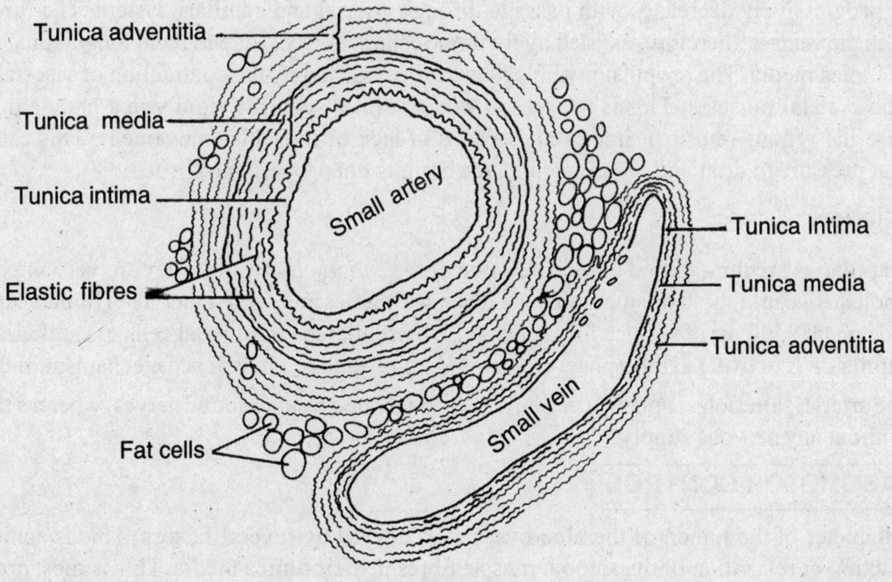

Fig. 14.14. Transverse section showing the structure of the wall of the artery and vein.

Arterial Pressure: It is the pressure of blood against the walls of the main arteries. Usually the term is used to indicate pressure existing in the brachial artery located just above the elbow. Pressure throughout the whole circulation diminishes from arteries next to heart round the veins next to arteries. Arterial blood pressure undergoes a rhythmic change reaching a maximum during ventricular systole when the blood is pumped into the arteries, and decreasing during the ventricular diastole. The two pressures are referred to as *systolic* and *diastolic pressure* respectively. A blood pressure represented as 120/80 represents 120 and 80 as the systolic and diastolic pressures, respectively in mm Hg. Incidentally, 120/80 is considered to be the normal limits in a healthy adult ; pressures above and below these are referred to as *hypertension* (high blood pressure) and *hypotension* (low blood pressure), respectively. However, the blood pressure varies considerably, and is affected by factors such as age, sex, heredity, and physical and emotional states.

Pulse

The pulse is felt over any artery lying close to the skin. However, the usual method is to feel the pulse of the radial artery near the wrist. The pulse is caused by the ventricular systole which sends spurts of blood into the aorta. As a result the elastic wall of the aorta at the site of the thrust of blood distends and contracts. The result is a transmission of a wave of alternate stretching and contraction of all arteries and arterioles. This phenomenon is called the *pulse wave* or the *pulse*. It must be emphasized here that the pulse is not due to the arrival of the blood at the wrist where pulse is usually felt, but due to the transmission along the elastic walls of the impact of blood suddenly entering the aorta. The pulse rate usually varies from 70 to 72 in men, and 78—82 in women. The pulse provides information about the functioning of the heart and circulation in general.

Veins

Veins and *venules* (*a small vein*) are also made up of the same three coats (Fig. 14.14) found in the arteries. But there is less smooth muscle in the veins than in arteries. Most veins contain *semilunar venous valves* which oppose backward flow of blood. If these valves are closed, varicose veins develop.

The propulsive action of the heart has little effect on the venous blood flow since the force of blood progressively decreases with passage through arterial and capillary system. The flow of the blood in the veins is, therefore, assisted by rhythmic contractions of circular and longitudinal muscles in the tunica media. The propulsion of blood is further assisted by the contraction of visceral organs and the skeletal muscles. This is the reason why a soldier standing rigidly in a parade may faint because his venous return in impaired, because of lack of muscular movement. This causes the arterial pressure to drop and, consequently, the brain is not properly nourished.

Capillaries

The capillaries are thin-walled tubes consisting of the *endothelium* only. They are very important in metabolism because these are the sites for exchange of gases and food materials. Human capillaries provide about 2400 m^2 of surface for exchange. Differentiated endothelial cells of capillaries (such as Kupffer *cells* of liver) exhibit phagocytosis and, thus, function in defence mechanism of the body.

The arteries, arterioles, and the capillaries are innervated by autonomic nerves, whereas the veins are without any nervous supply.

VASOMOTOR CONTROL

The diameter of the lumen of the blood vessels is altered by several factors. This is achieved by contraction or relaxation of the smooth muscle fibres in their tunica media. This is most prominent in the arterioles. The smooth muscles are arranged in such a manner that their contraction narrows the lumen of the vessel (*vasoconstriction*) and their relaxation widens their lumen (*vasodilation*). There are several factors which can cause a change in the lumen of arterioles and small arteries. These are controlled by *local*, *nervous* and *hormonal factors.*

The *resistance vessels, viz., small arteries and arterioles,* offer main resistance to blood flow. These are controlled by a *local control* (metabolites, temperature, transmural pressure, P_{O_2} in pulmonary alveoli), *nervous control* (vasoconstrictor nerves, vasodilator nerves, antidromic impulses, and vasomotor reflexes), and *hormonal control* (adrenaline, noradrenaline, angiotensin, kinins, and histamine). These control measures are described as under.

Local Control

The local control is effected by metabolities, temperature, transmural pressure, and P_{O_2} in pulmonary alveoli.

The effect of *metabolites* is usually evident during exercise and temporary occlusion of blood flow in an arteriole. This results in an increased concentration of the products of metabolism in the vessels which causes the vessels to dilate resulting in an increased blood flow. The metabolites are CO_2, hydrogen ion excess, oxygen lack, K^+ excess, increased osmolality of the local tissue fluids, ATP, adenosine phosphates, and bradykinin.

Local temperature of the various parts of the body may cuase vasodilation of vasoconstriction ; the responses are independent of the autonomic nervous system because these phenomena also occur in denervated parts of the body. Warm temperature causes vasodilation of the local blood vessels to enable more blood at the body temperature to perfuse these parts, and bring the temperature down. Similarly, cold temperature causes local vasoconstriction so that loss of heat from the blood to the environment is reduced.

The *transmural pressure* is defined as the difference in pressure between the surrounding tissue and intravascular pressures. Changes in transmural pressure cause local changes in intravascular pressure. If the pressure in the artery, which is about 100 mm Hg, rises, the blood vessel dilates; the blood flow increases slowly until at a high critical pressure when it rises sharply.

When the partial pressure of oxygen falls in the alveoli of lungs, vasodilation occurs. However, fall in P_{O_2} causes vasoconstriction in coronary blood vessels.

Nervous Control

Vasoconstrictor nerves cause blood vessels to constrict which is mediated by the release of *noradrenaline* by sympathetic nerve fibres in the smooth muscle fibres of the blood vessels. The vasoconstricting ability lies in the fourth ventricle of the medulla which forms the *vasoconstrictor centre.* The caudal region of the vasoconstrictor centre can cause vasodialation; therefore, this region is known as the *vasodilator centre.* These two centres together are called the *vasomotor centre.*

The vasoconstrictor impulses pass down the medulla into the spinal cord. The preganglionic vasoconstrictor fibres leave by thoracic and upper lumber nerves, and form synapses with the sympathetic ganglia. Postganglionic fibres arise from the sympathetic ganglia and some of these go straight to the main blood vessels. Others return to the mixed spinal nerve which carry them to the peripheral blood vessels (Fig. 14.15).

Vasodilator nerves transmit impulses the vasodilation in the blood vessels. Vasodilator nerve fibres occur in both *sympathetic* and *parasympathetic* systems. This is why sympathetic systems have both vasodilator and vasoconstrictor abilities.

Vasodilator impulses travel over sensory nerves in a direction opposite to that of ordinary impulses so that the same nerve performs both sensory and motor function. Therefore, these have been termed *antidromic impulses.*

Vasomotor reflexes are those reflexes which override the local requirements, mentioned earlier, of the body.

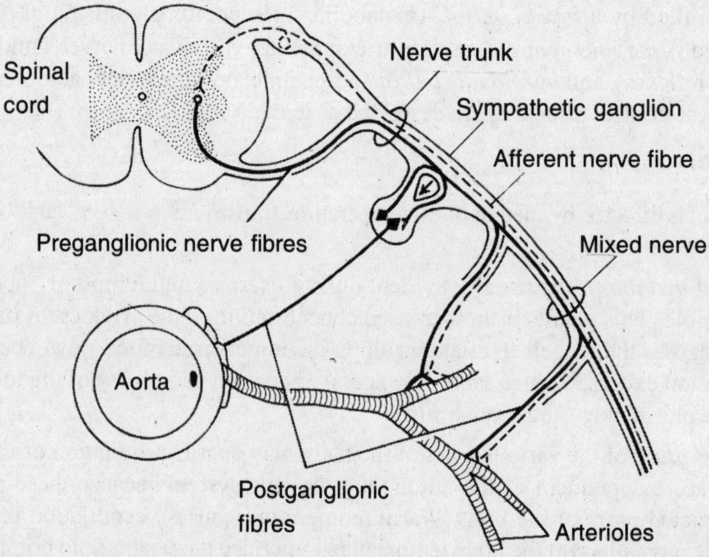

Fig. 14.15. Diagram showing the mechanism of nervous control of the arterioles.

The role of cardio-aortic and the carotid sinus nerves have already been discussed in connection with the nervous regulation of the heart. It will be recalled that these are sensory nerves which transmit impulses to the medullary cardiovascular centre regarding the pressure in the aorta and the carotid sinus, and regulate the heart beat in a reflex manner. In addition to this, these nerves also send afferent branches to the vasomotor centre. When these nerves are stimulated due to increased pressure on the aorta, their impulse to the vasomotor centre causes vasodilation and the blood pressure falls. In summary, therefore, the afferent nerves by their impulses cause two functions, viz., inhibition of the heart and fall of blood pressure. The interrelation between the two is obvious. The interrelation described above between the arterial blood pressure and heart rate is known as Marey's *law* which states that a rise in blood pressure causes fall in heart rate, and vice-verse.

Hormonal Control

Besides nervous regulation of the vessels, hormones also play an important part in the behaviour of the blood vessels. *Adrenaline,* secreted by the adrenal medulla causes sudden temporary fall of arterial pressure, and then at once followed by a sudden rise; *noradrenaline* is more effective, and the effects are similar to adrenaline. The cutaneous vessels are constricted, but the vessels of the skeletal muscles dilate. This results in a rise in systolic pressure. The phenomenon is easy to understand; sympathetic nerves which are adrenergic cause acceleration of the heart and vasoconstriction of the blood vessels. Circulation of adrenaline in blood vessels is, therefore, apparently a *sympathomimetic* action. *Angiotensin* causes vasoconstriction and, therefore, rise in arterial blood pressure. The other details of angiotensin action is discussed in chapters 11 and 15. *Bradykinin,* a peptide, formed during secretion of sweat and salivary glands, causes vasodilation in man. As a matter of fact, it is the most potent vasodilator.

Control of Capillaries: The capillaries are made up only of *tunica intima* and do not have any nerve supply. However, the openings of the arterioles in the capillaries are provided with *precapillary sphincters* which regulate the blood supply to the capillaries. The change in the diameter of the arteriole which is brought about by the nervous system, however, seems to control precapillary sphincters. The vasomotor centre, therefore, exercises an indirect control over the flow of blood through the capillaries. The nervous system, therefore, plays a less important role in the regulation of blood supply to the capillaries. The more important role is played by hormones, such as adrenaline, neurotransmitters (acetylcholine, 5-hydroxytryptamine) and tissue metabolities like CO_2, histamine, and kinins.

15

Chapter

Excretion and Osmoregulation

Excretion and *osmoregulation* are two vitally important functions of the body. These accomplish removal of toxic and useless substances, and balance of water respectively. The two are closely related and are carried out by the *kidneys*.

In a living system, thousands of enzymes are always at work. These enzymes are either engaged in the synthesis of large molecules from smaller molecules or breaking down of larger molecules into smaller ones. The former process is called *anabolism* and the latter *catabolism*. The two processes together are called metabolism. A fully grown animal which has attained constant weight has limited anabolic activity. However, catabolism is a continuous process. During catabolism the enzymes of the body keep on breaking large molecules into smaller and smaller molecules in order to obtain energy (Chapter 13). During this process a stage is reached when the body is no longer able to utilize the energy left in such small molecules. For instance, carbohydrate catabolism, after glycolysis and T.C.A. cycle produced CO_2 and water. A number of similar substances are produced during the catabolism of various substances. Although, these molecules always contain some potential energy the body does not possess an adequate enzyme system to obtain it. Such substances are of no further use to body and are, therefore, called 'waste products'. The act of removal of 'waste products' from the body is called excretion.

Before a detailed discussion of excretion is taken, the following four factors must be clearly understood:

1. That water is a metabolic waste, though not always useless.
2. That most waste products are water soluble and, therefore, tend to raise the osmotic pressure of body fluids.
3. That a majority of waste products are toxic and have to be quickly removed from the body.
4. The waste products, in the case of mammals, are removed in the form of solutions.

It is obvious from the above facts that (1) catabolism should raise the water level in the system (2) excretion should lower the water level since all the excretory products are removed in the form of solutions. The formation and elimination of waste products, therefore, has a profound effect on water content ofd the body. This, however, does not mean that there is an indiscriminate addition or removal of water in the system. This would disturb the osmotic pressure of the body fluids which is really maintained with great precision. The mechanism by which this constancy in osmotic pressure is maintained has been described as *osmoregulation*.

EXCRETION

The Product of Metabolism

It has been mentioned earlier, in this book, that there are three principal groups of substances which come under the category of food: (1) Carbohydrates, (2) Lipids, (3) Proteins. All three groups are catabolised and contribute their share of energy to the body.

Metabolism of Carbohydrates. The metabolism of carbohydrates has been dealt in detail in Chapter 13. At the end of the carbodydrate metabolism CO_2 and H_2O are produced after the carbohydrate has gone through glycolysis and Kreb's cycle. This can be summarised as follows.

$$\underset{\text{Hexose}}{C_6H_{12}O_6} + 6O_2 \longrightarrow \underset{\text{Carbon dioxide + water}}{6CO_2 + 6H_2O}$$

Metabolism of Fat: When fat is required for energy it is transported from the adipose tissue to the site of metabolism in the form of nonesterified fat (glycerol and fatty acid). Glycerol changes to glyceraldehyde 3—phosphate. The latter enetrs the glycolytic chain and Kreb's cycle and the end product is CO_2 and H_2O (Chapter 13). The fatty acids enter the β-oxidation-helix-cycle in which they loose two carbons with a complete turn of the wheel and yield one *acetyl CoA* (Chapter 13). Acetyl CoA acts with oxaloacetic acid, coming from the metabolism of carbohydrates, and forms citrate (Fig. 15.1); the citrate enters the Kreb's cycle. The end product is again CO_2 and H_2O.

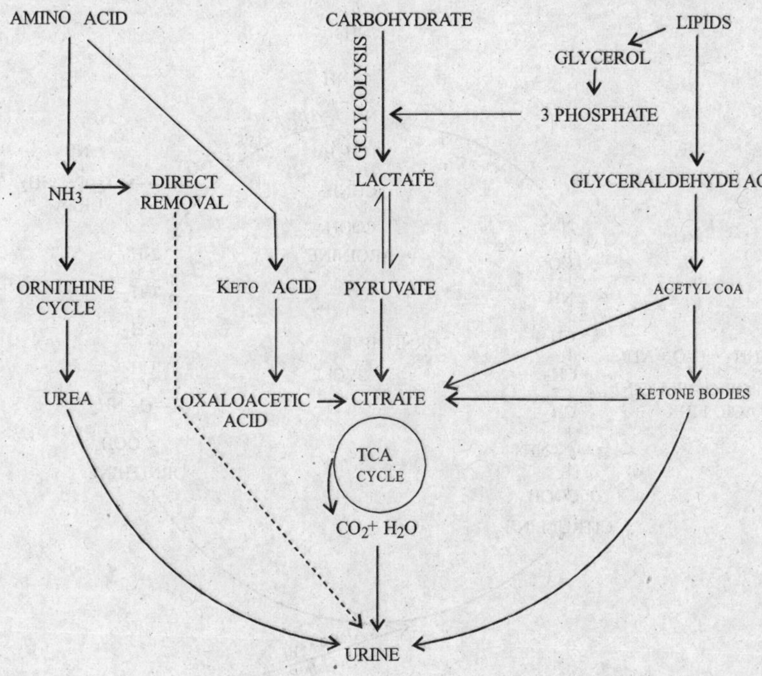

Fig. 15.1. Formation of excretory metabolites during the catabolism of carbohydrates, fats and lipids.

All the acetyl CoA, however, does not get oxidised in this manner. Part of it is metabolised as follows.

Two molecules of acetyl CoA condense to form *acetoacetyl CoA*; this is hydrolysed to acetoacetic acid by the help of β-*hydroxybutyric* dehydrogenase and reduced NAD.

The three compounds, acetoacetic acid, acetone and β-hydroxybutyric acid are together known as *ketone bodies* or *acetone bodies*. The ketone is hydrolzed by a process called *ketolysis* which yields CO_2 and H_2O. In case the ketone bodies are not oxidised they accumulate in blood. These are toxic and cause the condition known as *ketonaemia*. When the ketones are excreted through the urine the condition is known as *ketonuria*.

Metabolism of proteins: When proteins are catabolised they first break up into their component amino acids. The *amino acids* are converted into keto acids by the removal of NH_3 (deamination).

$$\underset{\text{Amino acid}}{NH_2-\overset{\displaystyle R}{\underset{\displaystyle COOH}{C}}-H} + H_2O \xrightarrow[\text{Oxidase}]{\text{Amino acid}} \underset{\text{Keto acid}}{\overset{\displaystyle R}{\underset{\displaystyle COOH}{C}}=O} + NH_3 + H_2 + NAD$$

The keto acids enter Kerb's cycle to yield energy while the ammonia is removed through excretion in the following manner:

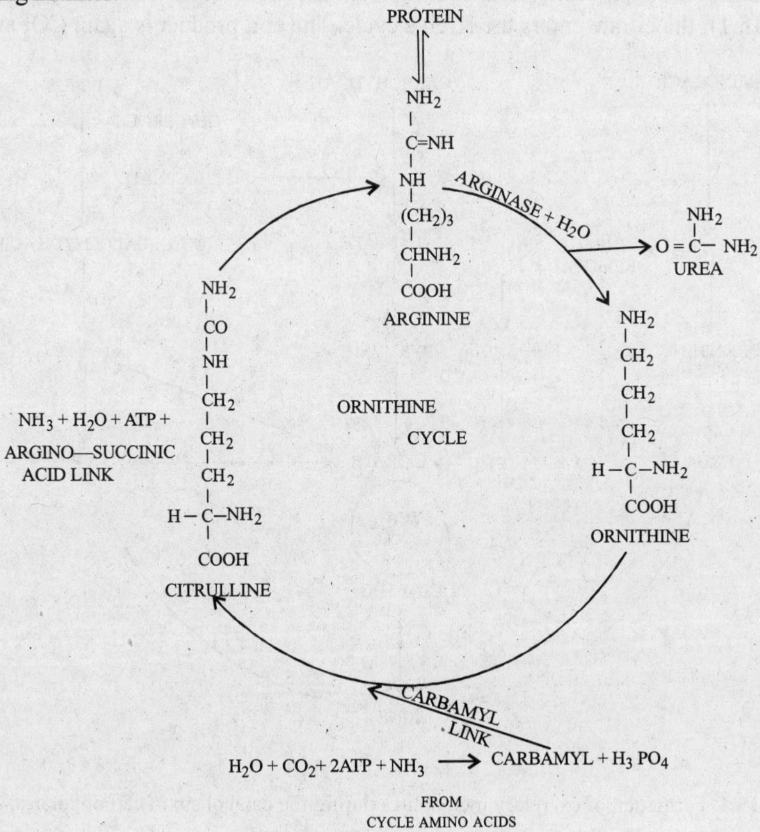

Fig. 15.2. The ornithine cycle

Fate of ammonia: Ammonia, produced during protein breakdown, is highly toxic. Very small quantities of this is excreted as ammonia. Most of the ammonia is converted to urea which is formed by two molecules of NH_3 and one of CO_2 with the elimination of water. This, however, does not mean that the synethesis of urea is a single step process. In fact the synethesis of urea takes in the liver by way of a repeating cycle called *Ornithine cycle* (*Kreb's urea cycle*).

During the breakdown of proteins arginine is obtained. One molecule of arginine is broken down into one molecule of urea and one of ornithine in the presence of enzyme *arginase*. The ammonia produced during protein breakdown is actually utilized during reformation of arginine from ornithine. The cycle proceeds as follows: Ammonia from amino acid deamination, CO_2 from Kreb's cycle and 2 molecules of ATP along with water combine to form a compound *carbamyl phosphate* with the help of the enzyme *carbonyl phosphatase synthetase*.

$$NH_2 + CO_2 + H_2O + 2\,ATP \xrightarrow{\text{Carbamyl Phosphate synthetase}} \underset{\text{Carbamyl Phosphate}}{H_2N{-}\overset{\overset{\displaystyle O}{\|}}{C}{\sim}P} + 2\,ADP + H_3PO_4$$

Carbamyl phosphate combines with ornithine to form citrulline. This reaction is called *Carbamyl phosphate link*. In the second step, asparatic acid from TCA cycle combines with citrulline with the breakdown of one molecule of ATP to form *argino succinic acid*. The latter splits into arginine and fumaric acid. Fumaric acid enters the TCA cycle while arginine again gets ready to release another molecule of urea. The conversion of citrulline to arginine has been called argino-succinic-acid-link. One molecule of urea formation requires 3 ATP for its synthesis. In addition to urea small quantities of *uric acid* and *creatinine* are also excreated. During the formation of uric acid the purines, adenine and guanine are first deaminated to form hypoxanthine and xanthine respectively. These are then oxidised to form uric acid (2, 6, 8 oxypurine). Creatinine is formed when creatine-phosphate, a phosphogen, breaks doen to provide energy for muscle contraction. It may be mentioned that creatine resulting from breakdown of creatine phosphate may again be regenerated to form the phosphogen without the formation of creatinine.

Thus, during the breakdown of carbohydrates, lipids and protenis, CO_2, water, ketones, ammonia, urea and uric acid are formed. Simultaneously, substances like creatine phosphate, nucleotides etc. are also being catabolised. Besides this there is a continuous intake of salt which should be removed. Of the above listed metabolites most of the CO_2 is removed through the lungs. The question of removal or addition of water has been taken up later in this chapter. The remaining substances are selectively picked up by the kidneys and removed as urine. The kidneys have a very specialized structure for this selective removal.

Structure of the Kidneys

All animals possess a pair of kidneys which are present on either side of the vertebrae of the lumbar region. Each kidney is bean-shaped with an indentation on the inner side called the hilum (Fig. 15.3); nerves, ureter, and blood vessels enter the kidneys at the hilum. If a kidney is cut through the middle in two longitudinal halves, two regions are revealed to the naked eye—an outer *cortex* and an inner *medulla* each of which lodges different parts of the kidney (Fig. 15.3).

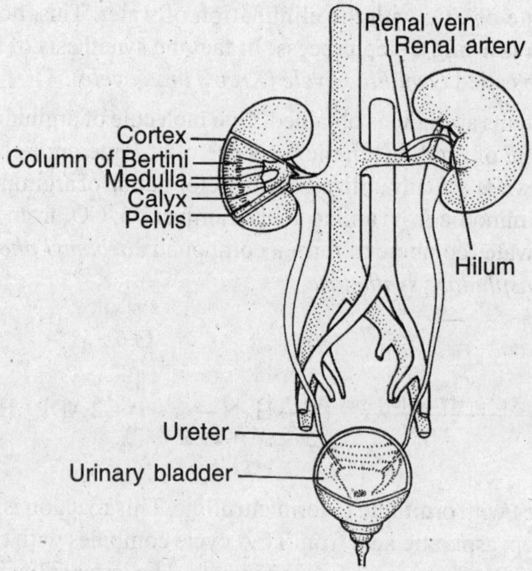

Fig. 15.3. Excretory system of man. A portion of the left kidney been sectioned to show the internal structures.

Each kidney is provided with a million tubular excretory units called *nephrons* (Fig. 15.4). A single nephron is made up of two distinct parts (1) Glomerulus which is made up of a tuft of capillaries carrying blood to the excretory unit called renal tubule (2) Renal tubule which is the excretory part as it removes 'undesirable' constituents of blood and converts these into urine.

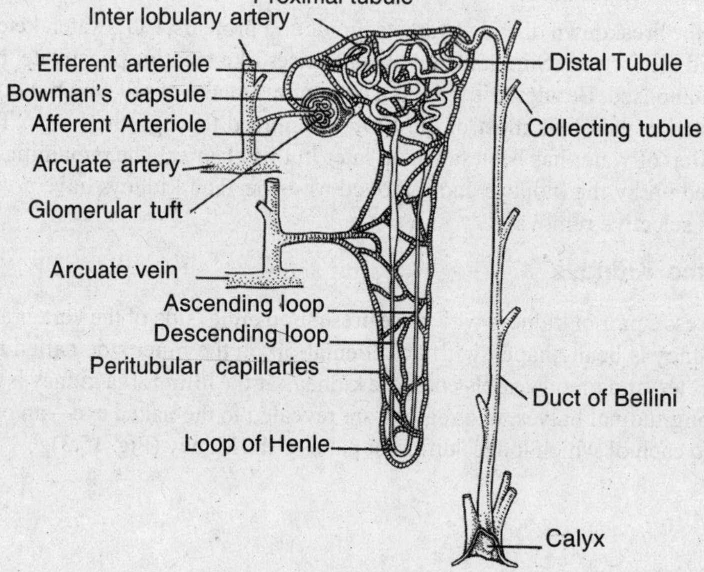

Fig. 15.4. A single excretory unit or nephron along with its blood supply.

Glomerulus: Each glomerulus is made up of about 50 capillaries which are bent into short loops to form a small tuft of blood vessels called *glomerular tuft* (Fig. 15.4). The glomerular tuft is lodged in a cup-shaped structure termed *Bowman's capsule* which forms the beginning of the renal tubule (Fig. 15.4).

Renal tubule: The Bowman's capsule was discovered by Bowman, in 1842, who showed that the tuft of capillaries called glomerulus pushed into the tubule in a manner a first would push into an inflated rubber ball. The invagination thus formed is in the shape of a capsule which is a double walled structure around the glomerulus. The capsule, a blind end of the nephron, evidently does not open into the renal tubule. The Bowman's capsule and the glomerulus together constitute the *Malpighian corpuscle*. Although the cavity of the Bowman's capsule does not open in the renal tubule yet fluid can flow from the into the tubule.

The rest of the renal tubule is divided into three segments. The segment of the tubule which immediately follows the Bowman's capsule is called *proximal tubule* (Fig. 15.4). The proximal tubule is a highly convoluted structure which leads into a U-shaped segment of the renal tubule, called *loop of Henle*. The latter enters straight into the medulla of the kidney, makes a hairpin bend, and comes back to the cortex. The parts of the tubule leading to and going away from the bend are respectively called the *descending* and *ascending* limbs of the loop of Henle. The ascending limb of the loop of Henle is continued into a *distal convoluted tubule*. The distal convoluted tubule opens into the *collecting tubule*. That part of the distal tubule which makes a junction with the collecting tubule is called *junctional tubule*. Several collecting tubules unite to form a *duct of Bellini*. The ducts of Bellini open into *Calices*, which are funnel shaped structures present in the medulla of the kidney. (Fig. 15.4).

The cortex and medulla of the kidney are structural entities carrying out different functions. The cortex contains the glomeruli and the proximal and distal tubules of the nephrons. The medulla has two structures (1) *Pyramids* whose apices open into the calyces; the calyces are large collecting ducts into which the ducts of Bellini empty their contents. All the calyces unite to form a funnel shaped structure called the *pelvis* of the kidney (Fig. 15.3). (2) In between the calyces there are *columns of Bertini* (Fig. 15.3). The loops of Henle of the various nephrons are present in these columns.

The walls of the Bowman's capsule are made up of a single layer of flatened epithelial cells called podcytes. The podocytes send out small tentacle like processes each of which has small projections called *pedicells*. The latter are in close contact with the cells of the glomerular capillaries.

The cells of the proximal tubules are long and cuboidal. The inner border of each cell gives off about 150 thin microvilli which impart it a brush like appearance (Fig. 15.5). These cells are therfore called brush border cells. These greatly increase the surface area of the proximal tubules. The total area of the brush border cells of each human kidney is about 30 sq. metres.

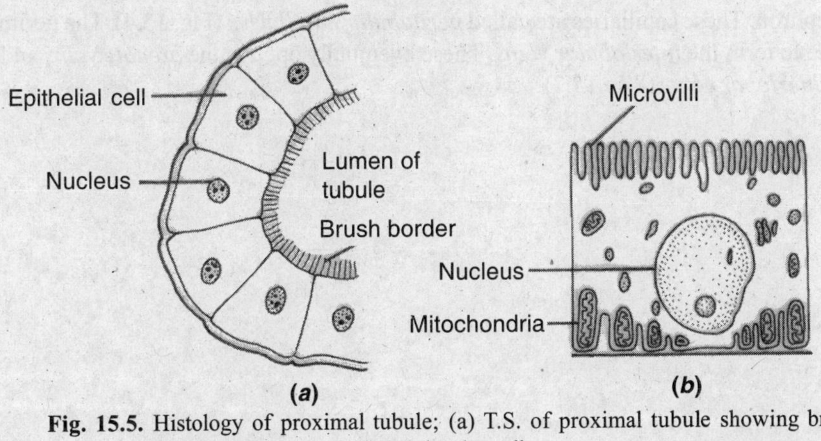

Fig. 15.5. Histology of proximal tubule; (a) T.S. of proximal tubule showing brush-border cells. (b) Ultra structure of a brush-border cell.

The entire descending limb of the loop of Henle, and part of the ascending limb is thin walled and narrower than the rest of the renal tuble.

The cells of the distal tubule are also large and columnar. These have a large number of vesicles (probably secretory) and some microvili (Fig. 15.6).

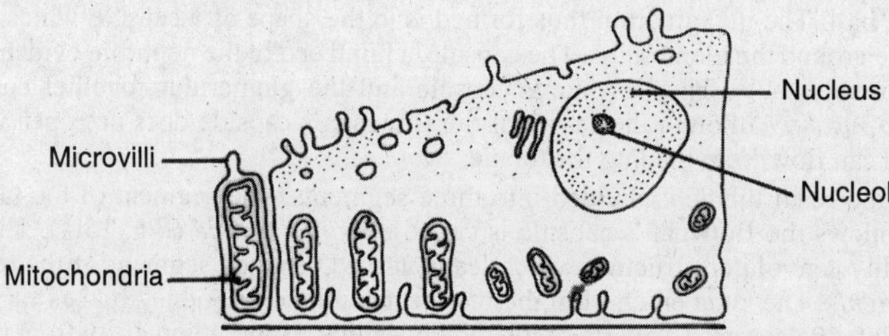

Fig. 15.6. Diagram of ultra-structure of single cells of the distal tube.

Blood Supply of the Kidney

The human kidney weighs only about 0.4 to 0.7% of the total body weight yet it receives more than 20% of the blood pumped by the heart. The kidney receives approximately 1200 ml. of blood per minute which is sent through a pair of *renal arteries*. These are short vessels which arise directly from the dorsal aorta so that the blood pressure in these is quite high. The renal artery enters the kidney at the hilum and gives off branches called *interlobar arteries*. These are present between the renal pyramids and radiate towards the periphery. The interlobar arteries give off *arcuate arteries* which run between the cortex and medulla parallel to the surface of the kidney (Fig. 15.7). The acruate arteries give off numerous fine branches into the cortex called *interlobular arteries*. The interlobular arteries give off fine branches which ultimately divide into arterioles (Fig. 15.7). These arterioles are called afferent arterioles and carry blood to the glomerulus. The afferent arterioles break up into small capillaries which bend on themselves and form the glomerular tuft. Before leaving the glomerulus they again unite to form a second arteriole called the *efferent arteriole* (Fig. 15.4), which carries the blood away from the glomerulus. The diameter of the efferent arterioles is smaller than the afferent arterioles. Due to this, a pressure builds up in the glomerulus. Furhermore, the diameter of both the afferent and efferent arterioles can be controlled. On emerging from the glomerulus, the efferent arterioles again break up into another set of capillaries which surround the tubule of the nephron. These capillaries are called *peritubular capillaries* (Fig. 15.4). The peritubular capillaries unite to form the *interlobular veins*. These eventually open in the *arcuate veins* and from there into the *interlobar veins* (Fig. 15.7).

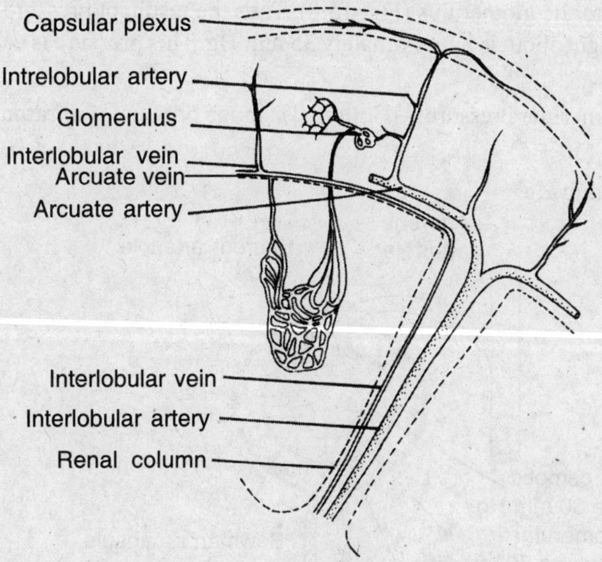

Fig. 15.7. Diagrammatic representation of the main blood vessels of the kidney.

The kidney is innervated both by sympathetic and parasympathetic nerves (Chapter 9), which control the diameter of the blood vessels and consequently the amount of blood entering the kidneys.

PHYSIOLOGY OF EXCRETION

The phenomenon of exeretion in mammals is synonymous with the formation of urine. Urine formation takes place in three steps *viz. filtration, reabsorption and secretion.*

Filtration

The walls of the capillaries of glomerulus are very thin and have minute pores which range from 200—900 Å in size. Molecules up to 10,000 molecular weight pass freely through these pores. Large molecules, above 50,000 molecular weight, cannot to through them at all.

The diameter of the afferent arterioles is more than the efferent arterioles, therefore, more blood enters the glomerulus than can get out of it. This builds up a pressure within the glomerulus and the blood within the glomerulus seeks other channels to get out. As a consequence to this, water along with dissolved electrolytes and other small molecules is forcibly filtered out from the blood into the Bowman's capsule. However, larger molecules like most proteins cannot pass through these pores and are left behind in the blood. This process is called *ultra-filtration*.

Filtration Pressure of Glomerulus

The filtration through the glomerulus depends on three factors (1) The pressure in the glomerulus (*glomerular pressure*) which is about 70 mm Hg. This pressur teneds to force the fluid of the glomerulus into the Bowman's capsule (Fig. 15.8). (2) the walls of the Bowman's capsule as well as the pressure in the renal tubules (*intrarenal pressure*) cause a back pressure which tries to prevent the filtrate from entering the capsule. This pressure is normally about 5 mm Hg. (3) The blood has a certain colloidal osmotic pressure which tends to bring in the fluid from the Bowman's capsule into the glomerulus. The colloidal osmotic pressure is approximately 30 mm Hg.

Thus the glomerular pressure of 70 mm Hg tends to push the fluid out into the capsule. The intrarenal pressure of 5 mm Hg and the colloidal osmotic pressure of 30 mm Hg try to drive the fluid of the Bowman's capsule into the glomerulus (Fig. 15.8). Thus the resultant pressure by which the glomerular filtration is brought about is approximately 35 mm Hg. This pressure is called *filtration pressure*.

Filtration pressure = Glomerular pressure—(Colloidal osmotic pressure + intrarenal pressure)

= 70 – (30 + 5)

Filtration pressure = 35 mm Hg.

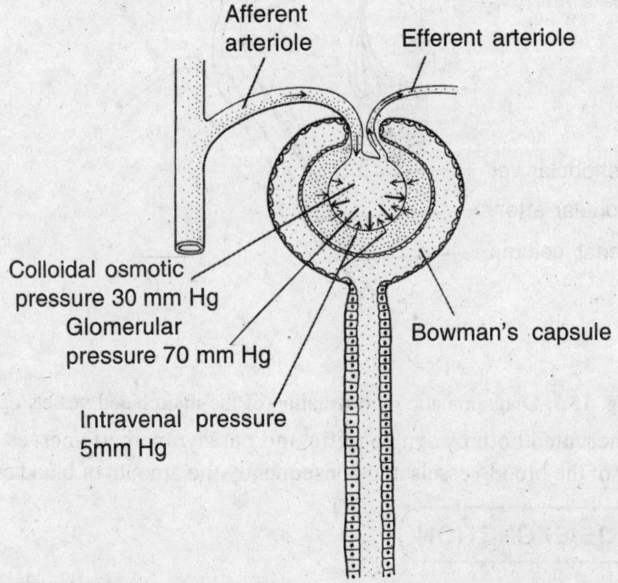

Fig. 15.8. Diagram showing ultrafiltration in the Bowman's capsule. A pressure builds up in the glomerulus due to difference in the diameter between afferent and efferent arterioles. Since the diameter of the efferent arteriole is small only a part of the blood entering the glomerulus can leave it. The remaining blood presses against the walls of the glomerular capillaries at a pressure of 70 mm of Hg (glomerular pressure). This has been depicted by thick arrows. The outflow of blood is checked by two factors: the colloidal osmotic pressure due to the plasma proteins and the intrarenal pressure. This has been depicted by thin arrows. These two together amount to about 35 mm Hg. The blood is thus filtered at a pressure of 35 mm Hg.

It must be remembered that the figures of pressure given above are 'normal' pressures which are subject to a great deal of variation. The diameters of the afferent and efferent arterioles are under the influence of the nervous system. An increase in the diameter of the afferent arteriole, and decrease in the diameter of the efferent arteriole increases glomerular pressure. This causes an increase in the quantity of the ultrafiltrate. A decrease in the diameter of the afferent arteriole and increase in the diameter of efferent arteriole causes fall in glomerular pressure which reduces the quantity of the ultrafiltrate.

The filtrate which enters the Bowman's caspule has a composition similar to that of plasma except that it has very little protein. In other words it is almost deproteinised plasma.

Table 15.1 Approximate concentration of various substances in plasma and urine. It should be noted, however, that even in a normal healthy person these concentrations vary to a considerable extent.

	Plasma 100 ml.	*Urine 100 ml.*	*Ratio* = $\frac{Urine}{Plasma}$
1. Water	90.9	95.0	
2. Proteins, fats etc.	7.9	almost negligible	
3. Glucose	0.1	almost negligible	
4. Nitrogenous waste			
(*a*) Urea	0.03	2.0	66.6
(*b*) Uric acid	0.002	0.03	15
(*c*) Creatinine	0.001	0.125	125
(*d*) Ammonia	0.0001	0.05	500
5. Anions			
(a) Chloride as Cl	0.37	0.6	1.6
(b) Phosphate as P	0.003	0.12	40
(c) Sulphate as S	0.003	0.18	60
6. Cations			
(a) Sodium	0.30	0.60	2
(b) Potassium	0.02	0.15	7
(c) Magnesium	0.0025	0.01	4
(d) Calcium	0.01	0.015	1.5

In spite of this the urine which is voided is remarkably different from plasma (Table 15.1). This is because the filtrate while passing through the tubular part of the nephron loses most of its water and essential substances through *reabsorption*. Moreever, new compounds are added to it through *secretion*.

Tubular reabsorption

Nearly 1750 litres of blood passes through the kidneys every day. Of this nearly 150—180 litres, is filtered out through the glomerulus. However, the total amount of urine passed out by a healthy human being is only 1—1.5 litres, the rest being reabsorbed in the tubules of the nephrons from where it goes to the blood through the *peritubular capillaries*. The greatly coiled tubules with their characteristic epithelial lining provide a large surface area for absorption. The renal tubules, it will be recalled, are closely surrounded by a network of peritubular capillaries which are separated from the tubules only by small interstitial spaces (Fig. 15.9). The blood contained in peritubular capillaries comes from the glomerulus. This blood, therefore, has already lost large amounts of water, and substances of low molecular wight through ultrafiltration. The blood, still, has nearly all the plasma proteins left in it which create a higher colloidal osmotic pressure in the capillaries in comparison to the fluid in the tubule so that the salts and water in the tubule tend to move from the renal tubule to the peritubular capillaries by the simple process of diffusion. All the substances present in the ultra-filtrate, however, do not go back to the blood by this process.

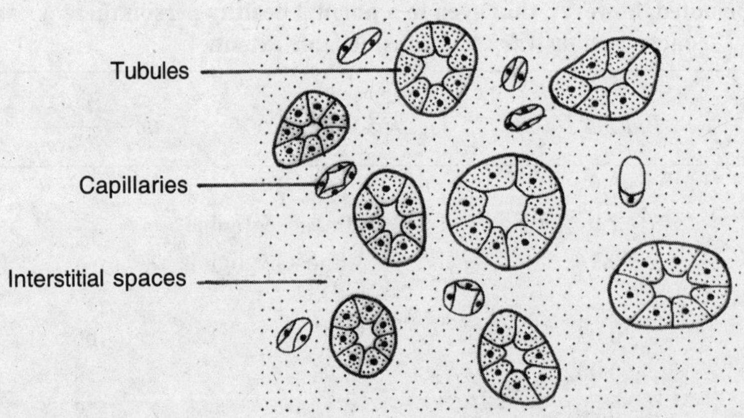

Fig. 15.9. T.S. of tubules and peritubular capillaries. Note that the capillaries are in close proximity of the tubules being separated only by small interstitial spaces.

Some are returned to the blood by *active reabsorption* or *active transport*. Thus the reabsorption in the tubule takes place by two processes—diffusion and active transport.

Reabsorption by diffusion: The walls of the tubules are selectively permeable which readily allow water, sodium chloride, amino acids, glucose, bicarbonates etc. to pass into the blood. Substances which have to be excreted like, uric acid, creatinine, phosphates, sulphates, however, do not pass through the tubular membrane. About 98-99% of the total water in the ultra-filtrate is absorbed. Most of the reabsorption of water (80-90%) takes place in the proximal tubule. About 5% is absorbed in the thin part of loop of Henle and the rest in the distal tubule. The proximal tubule is also responsible for the reabsorption of about 75% of the solids of the ultrafiltrate. All the proteins, glucose and amino acids are absorbed in the part. Part of the phosphate, most of the sodium chloride, bicarbonates and potassium are also absorbed in the proximal tubule. Potassium, however, is resecreated in the distal tubule (Fig. 15.10).

Active reabsoption: Many substances, present in the ultra filtrate, are actively absorbed, instead of passively diffusing through the walls of the tubule. Active absorption, however, is not limited to any selected metabolite. Nearly all electrolytes and even water are actively absorbed. This can be easily explained. Any substance will diffuse back into the blood only up to the point till its concentration in the blood does not reach the level at which it is present in the ultrafiltrate, after which diffusion will stop. However, salts, aminor acids, glucose etc. are not excretory products; their excretion would, therefore, put a needless burden on the body. They are, therefore, actively absorbed even when the concentration of these substances is lower in the blood as compared to the filtrate. Glucose, aminor acids and NaCl are actively absorbed in the proximal tubule while water ans some more NaCl is abosrbed in the distal tubule (Fig. 15.10).

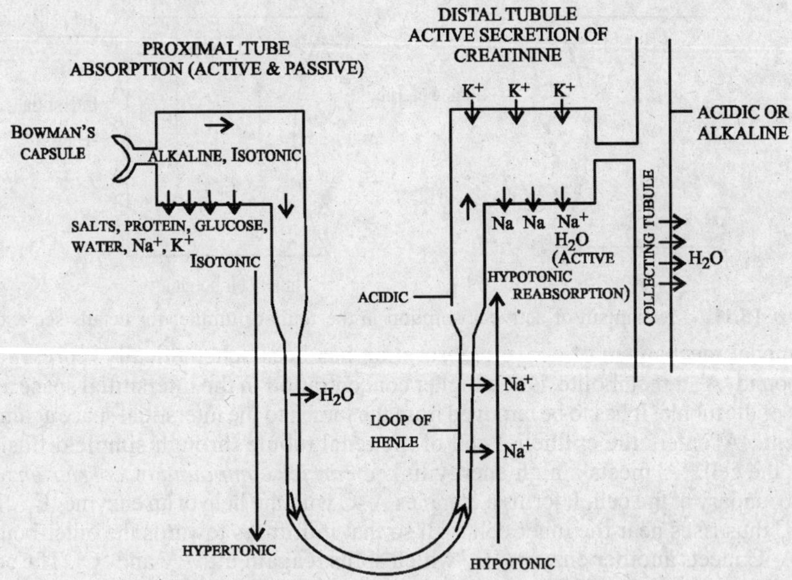

Fig. 15.10. Mechanism of urine formation in a single nephron.

Active reabsorption, also called active transport, takes place against a concentration or electrochemical gradient. This phenomenon is like carrying a load uphill. Diffusion, on the other hand, is like taking a load down hill.

High concentration

Active transport — ↑ ↓ — Diffusion

Low concentration

Carrying a substance against a concentration or electrochemical gradient involves work and therefore energy. The amount of work done for moving a substance against a concentration gradient can be calculated as follows:

Work = K × conc. gradient × quantity

$$= RT\, ln \frac{C_1}{C_2}$$

Where K is a constant

C_1 = higher concentration

C_2 = lower concentration

R = gas constant

T = absolute temperature

ln = natural log

The energy for this work comes from metabolic reactions.

Active transport is, therefore, also called metabolically linked transport. Obviously, active reabsorption stops when energy, is not available.

Mechanism of Active Absorption

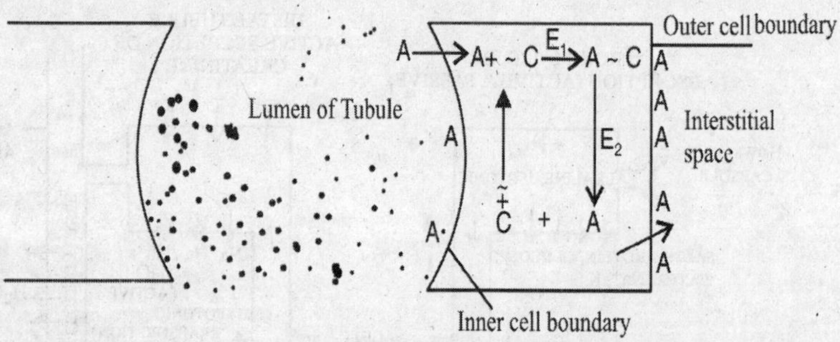

15.11. Mechanism of active absorption in the uninary tubule; for details see text.

A model mechanism of active reabsorption has been schematically represented in Fig. 15.11. Compound 'A', a metabolite, is at a higher concentration in the intersititial space as compared to the lumer of the tubule. It has to be removed from the lumen to the interstitial space against a concentration gradient. 'A' enters the epithelial cell of the renal tubule through simple diffusion. As soon as it enters the cell. 'A' meets a high energy linked *carrier compound* or *carrier enzyme*. 'C' near the inner boundary of the cell. It forms a complex A~C with the help of an enzyme 'E_1'. The concentration of A~C thus rises near the inner cell wall so that it diffuses towards the outer boundary of the cell. Here A~C meets another enzyme 'E_2' which splits it again into 'A' and 'C'. The concentration of A thus goes on rising locally near the outer wall of the cell till it reaches a level higher than the concentration of 'A' in the interstitial space. Due to this 'A' diffuses out into the interstitial space. This entire process requires energy. 'C' again picks up a high energy bond from metabolic processes and the cycle continues. The complex A~C is continuously produced near the inner surface and is continuously split near the outer surface of the epithelial cells of the tubules. The carrier molecule 'C' is visualized as a large molecule, most likely an enzyme, which cannot diffuse out through the cells wall.

Active secretion. All substances excreted through the urine do not come from the glomerular filtrate. Some substances are also brought into the tubules from the blood against a concentration gradient. This phenomenon is called secretion. In human beings substances like creatinine, H ions, para-amino hippuric acid, potassium and anitboitics like penicillin are secreated. The potassium which appears in the urine is actively secreted in the distal tubule because most of the potassium coming from the glomerular filtrate is absorbed in the blood from the proximal tubules (Fig. 15.10). Since both potassium and H ions are catains they complete for the same transport mechanism. If concentration of K^+ is high then less H^+ is secreted, if concentration of H^+ is high, less K^+ is secreted. This plays an important part in pH regulation through the kidney (Chapter 7).

Osmotic pressure of urine

The urine is isotonic to plasma in the proximal tubule as it is an ultrafiltrate of blood. As the filtrate flows into th descending limb of the loop of Henle it losses water and becomes *hypertonic*. In the ascending limb sodium is actively reabsorbed from the filtrate, so that the 'urine' becomes *hypotonic*. In the distal tubule also sodium is again actively reabsorbed in the blood so that the concentration of sodium becomes very high around the distal tubule and forms the so called 'Brine bath' around it. This brine both passively draws out water from the distal tubule due to exosmosis. This is known as passive or *obligatory reabsorption of water*. The unine after having lost water passively, once more becomes isotonic. In the collecting duct more water is reabsorbed so that the urine is made hypertonic. The amount of water absorption in the collecting duct depends on the needs of the body and is called *facultative reabsorption* of water.

From the above discussion it is clear that the entire process of excretion has been designed on very extravagent lines. The kidney produces enormous volumes of filtrate, most of which is reabsorbed. The human kidney, for example, filters and reabsorbs the entire fluid environment of the body 15 times a day.

Of the nearly 1200 g. of sodium chloride filtered every day only 5.10 g. is excreted. Similarly, 500 g. of sodium bicarbonate and 150 g. of glucose are filtered although negligible quantities of these are excreted (Table 15.1). Similar reabsorption of a number of other substances also takes place. Reabsorption recovers the substances valuable to the body and thus maintains a constant level of the internal medium. The relationship of filtration and reabsorption is called *renal clearance*.

The Concept of Renal Capacity and Renal Clearance

As mentioned earlier, a large quantity of blood is fltered through the glomerulus every minute. To measure the glomerular filtration rate—G.F.R. (amount of filtrate/minute), inulin, a poly-saccharide, found in dahlia tubers is injected in the blood. It is noteworthy that, inulin is not metabolisede at all so that the entire inulin injected, is filtered through the glomerulus and passed out into the urine without any reabsorption. If the concentration of inulin in the plasma and urine is determined and, if the urine secreted every minute is measured, then it is possible to calculate the overall filtration rate in the following manner:

Suppose—

Volume of urine passed per minute = V_1

Volume of glomerular filtrate per minute = V_2

Concentration of inulin in urine mg/ml = C_1

Concentration of inulin in plasma mg/ml = C_2

Since the glomerular filtrate is an ultrafiltrate of plasma and inulin is not metabolised by the body, therefore

Concentration of inulin in the filtrate will also be = C_2

Now,

Amount of inulin excreted per minute = $V_1 \times C_1$

Amount of inulin filtered per minute = $V_2 \times C_2$

Since all the inulin filtered is excreted out therefore—

$$V_1 C_1 = V_2 C_2$$

or

$$V_2 = \frac{V_1 C_1}{C_2}$$

or

$$\text{G.F.R.} = \frac{V_1 C_1}{C_2}$$

Example:

The concentration of inulin in plasma is 0.4 mg/ml, concentration of inulin in urine is 50 mg/ml and the volume of urine passed out is 1 ml/minute. Calculate the G.F.R.

From this

$$V_1 = 1 \text{ ml}$$

$$C_1 = 50 \text{ mg}$$

$$C_2 = 0.4 \text{ mg}$$

$$V_2 = \frac{1 \times 50}{0.4} = 125 \text{ ml/min or (180 litres/day)},$$

This is called plasma clearance or *renal clearance* for inulin, since it is the volume of plasma which is cleared of inulin every minute. The clearance for creatinine also is 125 ml because like inulin it also does not get reabsorbed. All substances, however, are not completely cleared as part of these are selectively absorbed in the renal tubule. As the amount of reabsorption is different for each substance, therefore, their plasma clearance is also different. For example, plasma clearance for urea is 60 ml/min.

The kidney fails to function in case of certain diseases or due to poisoning with heavy metals like gold, cadmium etc. Under these conditions the removal of excretory metabolities is hampered and their level in blood rises. A common manifestation of kidney disorder is rise in the level of blood urea—a condition known as *uremia*. During kidney disorders the concentration of acids also rises in the blood causing acidosis and the patient goes into comma. If the patient continues to drink water the body swells (edema) because of the failure of the kidney to get rid of the ingested water. If kidney failure is complete a person dies within one or two weeks.

Certain Properties of Formed Urine

Urine which is finally voided after filtration, absorption and secretion has a composition given in Table 15.1. Normally 1—1.5 litres of urine is discharged every day; however, its quantity varies according to the water requirements of the body. The specific gravity of urine is higher than plasma (specific gravity of plasma is 1.010) ranging from 1.010 to 1.025. In case of kidney failure, due to lack of reabsorption, the specific gravity of urine gets fixed at the specific gravity of plasma—a condition known as *isothenuria*. Urine has a faint yellow colour due to the pigment urochrome, which is formed from the breakdown of proteins. The concentration of urine varies according to the time of day, food and water intake, temperature etc. Even in the normal urine a few leucocytes and some nonsquamous epithelial cells are present. In spite of the fact that urine creates a general distaste and revulsion, it is a completely clean and sterile solution.

Voiding of Urine (Micturition)

The urine formed in the tubules of the kidney is collected into a funnel shaped structure called the *pelvis* of the kidney (Fig. 15.3). The pelvis leads into a short and narrow duct called the *ureter*. The ureters from two kidneys open into a sac like structure, *the urinary bladder* (Fig. 15.3). The ureters are made up of smooth muscle. They show perstaltic movements similar to the alimentary canal.

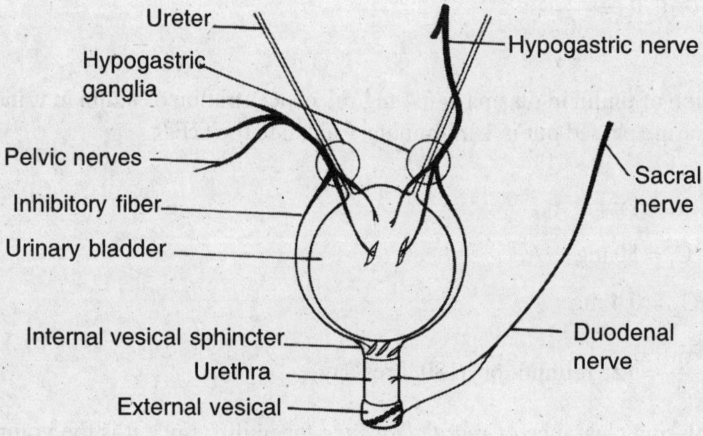

Fig. 15.12. Diagram showing the innervation of the urinary bladder and urethra

Each wave of peristalsis of the ureter takes 12—60 second to complete which sends little spurts of urine into the bladder. The bladder opens outside through a duct called the urethra (Fig. 15.12). The opening of the ureters and urethra make off a triangular area called *trigone*. Proximally where the urethra opens in the urinary bladder its opening is gurded by a sphincter muscle called the *internal vesical sphincter*. The urethra has another sphincter at its distal end called the *external vesical sphincter.* In males the urethra is long and the external sphincter is powerful. It is able to control the opening of the urthra even when the internal sphincter is damaged.

The walls of the bladder, except for the trigone, are made up of special muscles called *detrusor muscles*. The muscles are innervated both by the sympathetic (*hypogastric*) and the parasympathetic (*Pelivic*) nerves. Both these nerves pass through the hypogastic ganglia. The internal sphincter is also innevated by the hypogastric nerves (Fig. 15.12). The bladder expands as it gradually fills up. There is very little change in pressure the bladder fills up almost passively. Further entry of urine into the bladder causes it to stretch and the bladder pressure rises sharply. When about 600 ml of urine has collected in the bladder the pressure becomes very high (Fig. 15.13).

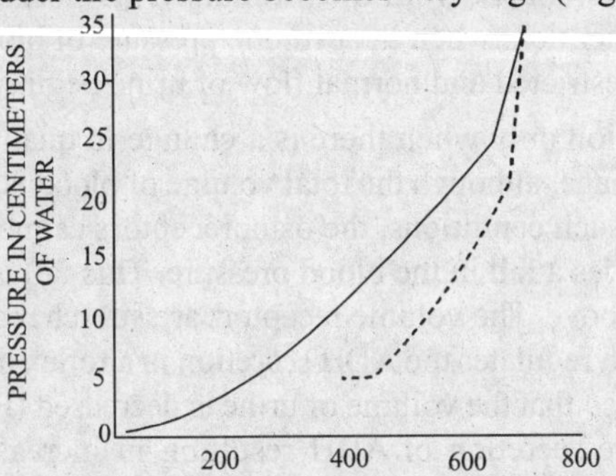

Fig. 15.13. Graph showing the rise in pressure as the urinary bladder gradually fills up with urine. Note that during the first 300 ml of urine the rise in pressure is small; later when there is around 600 ml urine in the bladder the pressure rises sharply. Solid line shows rise in pressure; dashed line fall in pressure after micturiation.

High pressure in the bladder stimulates the stretch receptors which send impulses to the spinal cord through the pelvic nerves. This stimulates the bladder in a reflex manner; the internal sphincter opens and powerful contractions, initiated in the bladder, prepare if for emptying.

The opening and closing of the external vesical sphincter of the urethra is controlled by nerves. The centres for these nerves lie in the conscious portion of the brain, namely midbrain and pons. The sphincter in adults opens only when it is convenient to urinate, otherwise, it remains closed even when the bladder pressure is high.

The control of micturition is, therefore, both voluntary and involuntary. The nervous mechanism which exercises a voluntary control on micturition develops late. This is why babies urinate whenever the bladder is full. Later on babies can exercise control over micturition during the day but they fail to do so at night. Bed wetting at night called *noctural enuresis* is common till a child is 3—4 years old. In imbecile adults and in those where the connection of the sphincter with brain and spinal cord has been damaged involuntary micturition takes place.

OSMOREGULATION OR FLUID CONTROL IN THE BODY

The total amount of body water is maintained with great accuracy. The human body has 50-70% water. The percentage of water in the body depends on age and the quanitity of adipose tissue. The latter is very crucial as it differs from one individual to another and has very little water. The quantity of water in fat free tissue is approximately 90% and is kept at a constant level.

Of the total body water 70% is present within the cells (intracellular) and 30% as extracellular fluid. Plasma forms about 8% of the total body water. There is free movement of water from cells to intercellular fluid and from there to blood and *vice-versa*.

Fluid is lost from the body through sweating, respiration, urination and defecation. Of these, evaporation through lungs and skin is inevitable as the former is related to respiration and the latter with thermoregulation (See Chapter 17).

The major source of fluid loss is urine. The volume of urine can be varied depending upon the amount of water the body can afford to loose. The fluid volume in the body is controlled thorugh two devices (1) Controlled urination (2) Controlled drinking.

(1) *Control of urination:* The volume of urine is monitored by three kinds of receptors (1) Osmoreceptors present in the hypothalamus (2) Volume receptors present in the auricles and, large veins (3) Pressoreceptors present in the large arteries. The latter two are in fact stretch receptors. When water level in the body is low the osmotic pressure of blood rises (*hemoconcentration*) and the osmoreceptors of the hypothalamus are stimulated. This causes the release of vasopressin (Chapter 11), an antidiuretic hormone (ADH), from the pituitary. This hormone causes a decrease in the quantity of urine by water retention; when the osmotic pressure of blood comes back to normal the release of vasopressin gets restricted and normal flow of urine begins.

The osmoreceptors function only when there is a change in quantity of water in blood. When blood is lost due to haemorrhage, although the total volume of blood is reduced there is no change in its osmotic pressure. Under such conditions, the osmoreceptors cannot play any part. Reduction in blood volume, however, causes a fall in the blood pressure. This fall is detected by the volume and pressoreceptors mentioned above. The volume receptors are stretch receptors and a change in blood volume stimulates then which regulates the ADH secretion in a reflex manner. Fall in blood pressure increases the output of ADH so that the volume of urine is decreased (Fig. 15.14). Similarly a rise in blood pressure decreases the secretion of ADH resulting in increased urine output. The above description, however, does not completely explain the mechanism of the control of urine output. Aldosterone, a *mineralocorticoid*, produced by the adrenal cortex, is a very important hormone regulating the Na^+ and K^+ balance in the distal tubule. Increase of aldosterone promotes the retentionof Na^+ and excretion of K^+ and H^+. However, this is not all; it has been found that when there is an increased Na^+ loss, then the volume of urine is also high. Aldosterone, therefore, indirectly effects the output of urine. Secretion of aldosterone is controlled in a complex manner. There are special receptors in the renal blood vessels which when stimulated, cause the secretion of an enzyme called renin from the kidney. The secretion of renin decreases when blood volume is low and there is need to conserve water. Renin catalyses the transformation of plasma protein *angiotensinogen*, present in the blood to form angiotensin I and angiotensin II. When the angiotensins reach the adrenal cortex they cause release of aldosterone. Aldosterone on reaching the kidney reduces the quantity of sodium excretion and consequently reduces water loss (Fig. 15.14).

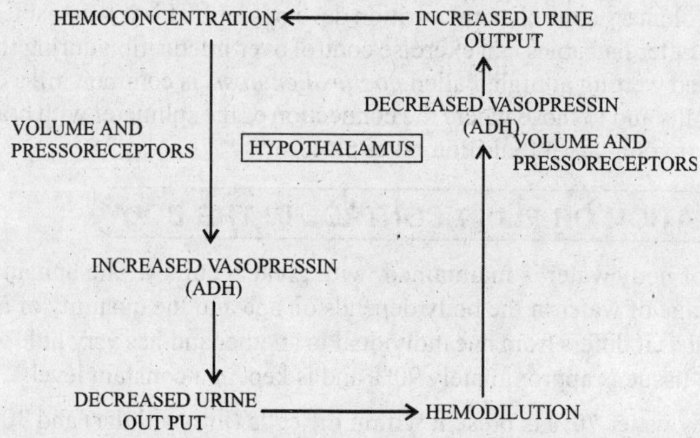

Fig. 15.14. Diagram showing the regulationof urine output in relation to water requirements of the body: (For the details see text).

(2) *Control of drinking*. Control of urine output alone is not responsible for the fluid regulation of the body. The lost fluids are compensated by regular water intake by drinking.

The hypothalamus contains a drinking centre which is provided with osmoreceptors. Whenever the blood volume gets low the drinking centre is stimulated causing the sensation of thirst. As soon as enough water is taken the sensation of thirst is over and the person stops drinking (Fig. 15.15). The drinking of water does not immediately dilute the blood as the water takes time to get absorbed. In spite of this, a person stops drinking as soon as enough water is taken. It is, therefore, evident that although thirst depends on the concentration of blood quenching does not.

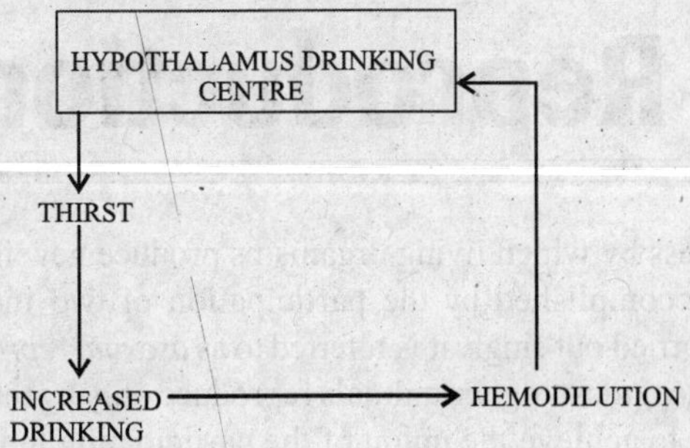

Fig. 15.15. Diagram showing the regulation of water intake. (For details see tex).

Controlled amount of drinking and controlled volume of urine output thus maintain a constant concentration of water in the body.

16
Chapter

Reproduction

Reproduction is the process by which living organisms produce new individuals of their own kind. When reproduction is accomplished by the participation of two individuals it is called *sexual reproduction*; when it is carried out singly it is referred to as *asexual reproduction*. Sexual reproduction is of widespread occurrence; most higher animals reproduce only by this method. The basic demand of sexual reproduction is to achieve the union of the participating individuals at the cellular level. This is accomplished by the union of the nuclei of the *ovum* with the *sperm* to form the *zygote*. This phenomenon is known as *fertilization*. The two participants in sexual reproduction are morphologically and functionally different—the one producing the ovum is called the *female* and the other producing sperms is known as *male*. The reproductive system of nearly all mammals is equipped to bring about the successful development of the zygote into a new individual within the body of the female.

The reproduction system is provided with gonads which produce *sex cells* (*germ cells or gametes*).These are *testes* in the male and *ovaries* in the female. The gonads besides producing the germ cells also secrete a variety of hormones. In addition, the reproductive system is provided with accessory glands, and ducts which transport the gametes. An account of the reproductive system of man follows.

MALE REPRODUCTIVE SYSTEM

The male reproductive system consists of *testes, seminal vesicles, prostate,* and *bulbo-urethral glands,* and the ducts: *epididymis, urethra* along with supporting structures such as *scrotum* and penis (Fig. 16.1).

The *testes* are both an exocrine and endocrine gland and are the principal functional units of the male genital apparatus. The testes are small ovoid glands lodged in a pouch-like structure called the *scrotum*. Each testis is encased by a white fibrous membrane known as *tunica albuginea*. The letter extends into the testis as *septa*, dividing the testis into about 250 *lobules* (Fig. 16.1). Each lobule contains about three seminiferous tubules which are surrounded by numerous *interstitial cells* or *cells of Leydig* (Fig. 11.8). The seminiferous tubule is lined by a three-layered epithelium, consisting of two types of cells—*supporting cells* and *germinal cells.* The germinal cells form the *spermatozoa*. Among the supporting cells there are cells of specialized function called the *cells of Sertoli* (Fig. 11.8). These harbour the heads of the developing sperms and provide them the nourishment. They are stimulated by testosterone and FSH to secrete messengers that stimulate sperm production and maturation. They also secret the protein hormone *inhibin* which inhibits FSH secretion. The release of sperms from the Sertoli cells is known as *spermiation*. All the seminiferous tubules of a testis come together and form a plexus called the *rete testis*. The latter gives rise to about a dozen fine ducts known as *efferent ductules*. These unite to form a single duct which is highly coiled and is known as the *epididymis*. When fully extended the epididymis measures about 20 feet. The epididymis serves as a passage for the sperms, secretes a part of the semen, and stores sperms prior to ejaculation. The sperms which are immobile when they leave the seminiferous tubules, attain motility on reaching the epididymis. The distal end of the epididymis opens into the *vas deferens* (Fig. 16.1).

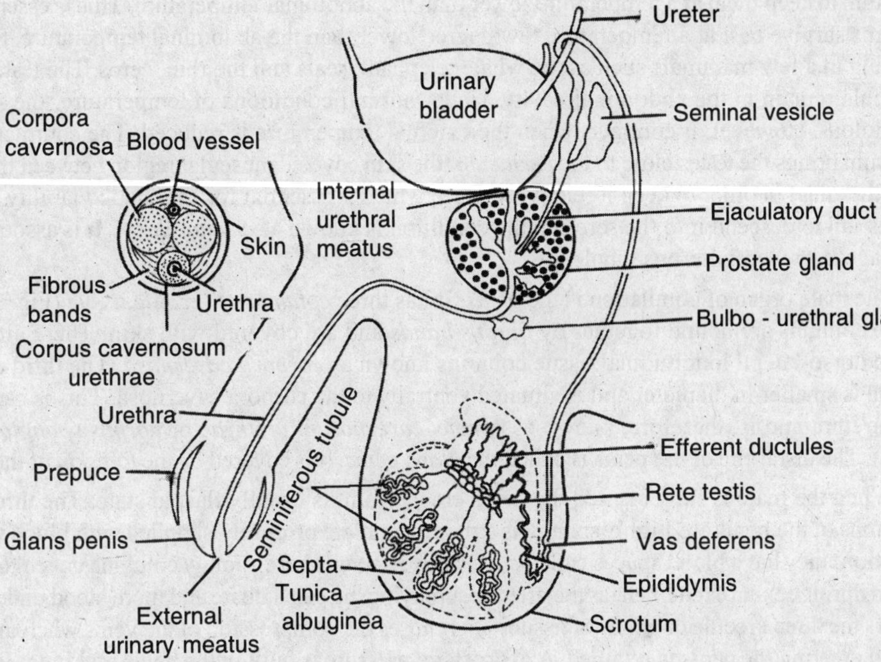

Fig. 16.1. The male sex organs. The inset at left shows cross-section of the penis at the plane AB. It shows the three columns of erectile tissues.

The vas deferens ascends along the posterior border of each testis, passing through the *inguinal canal* which is a narrow passage connecting the abdomen to the scrotum. The part of the vas deferens where it enters the inguinal canal is enclosed in a fibrous cylinder called the *spermatic cord*. The spermatic cord also encloses nerves, lymphatics and blood vessels, and serves to suspend the testis into the scrotum. Within the abdominal cavity the vas deferens is joined by a duct from the *seminal vesicle* (Fig. 16.1). The common duct thus formed is known as the *ejaculatory duct*. The ejaculatory duct from each side open in the *urethra*. The urethra extends from the internal *urethral orifice* of the *urinary bladder* to the *external urethral orifice* (or *urinary meatus*) located at the distal end of the *penis* (Fig. 16.1). The urethra, therefore, transports both urine and semen to the exterior.

The *seminal vesicels* are a pair of pouch-like glands present between the posterior surface of the urinary bladder and the rectum. They secrete a viscous fluid which contains nutrients helpful in spermatic metabolism. Just below the bladder and surrounding the proximal end of the urethra is the *prostate gland* (Fig. 16.1). The prostatic fluid is alkaline in nature so that it protects the sperms from the acidity of the male urethra and female vagina. It also provides a fluid medium for the motility of the sperms. There are a pair of *bulbo-urethral* or *Cowper's gland* (Fig. 16.1) located on either side of the urethra below the prostate gland. These glands discharge a slimy mucoid secretion which lubricate the urethra during coitus or sexual intercourse.

Semen *is a whitish fluid which contains spermatozoa and a mixture of the secretions from the seminal vesicles, prostate, and bulbourethral (Cowper's) glands. Each drop of semen contains about 300,000,000 sperms. Only one sperm, however, fertilizes the ovum, but the presence of millions of sperms appear to be necessary to insure fertilization.*

The *scrotum* is internally divided into two halves by a *septum*; each half contains a testis, the corresponding epididymis and the proximal part of the vas deferens. The testes are lodged in the scrotum to keep them at a temperature lower than the abdominal temperature. This is essential as the sperms survive best at a temperature few degree lower than the abdominal temperature. Scrotum is lacking in a few mammals such as the whale, elephant, seals and the rhinoceros. The testes in these animals remain in the abdominal cavity. Under normal conditions of temperature, the scrotum is pendulous. However, it contracts when the external temperature is reduced. The contraction of the scrotum brings the testes close to the *perineum* (the skin covered muscular region between the scrotum and the anus) in order to keep it relatively warm which is essential for spermatic viability. When the testes fail to descend into the scrotum the condition is known as *cryptorchism.* It is associated with damage to the seminiferous tubules.

The male organ of copulation is the *penis*. It has three *columns of erectile tissue* (Fig. 16.1). The three columns are bound together by *fibrous bands* and are covered with skin. There are a pair of large dorso-lateral longitudinal tissue columns known as *corpora cavernosa.* The third column of tissue is smaller in diameter and is situated ventrally to the corpora cavernosa. This is canalized by the urethra and is, therefore, known as *corpus cavernosum urethrae* or *corpus spongiosum* (Fig. 16.1). The distal end of the penis is called the *glans penis*. It is covered by the *foreskin* or the *prepuce.*

When the male is fully excited the penis erects from its usually flaccid state. The three erectile columns of the penis are highly spongy is structure and are profusely supplied with blood vessels. In addition, they have blood spaces called *cavernous spaces*. These spongy columns get engorged with blood during sexual excitement as the arteries supplying the penis dilate and more blood under pressure enters the three erectile columns of tissue. This causes the compression of the veins where most of the blood entering the penis is retained. A hydrostatic pressure is built in the penis making it long, hard, and erect. This takes place in the preparation of the insertion of the penis deep into the vagina for the safe deposition of the sperms. After the sperms have been released (*orgasm*) the arteries constrict and the blood returns from the spongy tissue so that the penis once again becomes flaccid.

The testes are both exocrine and endocrine in function. Spermatozoa are produced by the seminiferous tubules and the male sex hormones by the interstitial cells of Leydig. *Spermatogenesis* of the germ cells lining the tubules lead to the formation of the spermatozoa. Mitotic division of the germ cells or *spermatogonia* give rise to *spermatocytes* (Fig. 11.9). These undergo meiotic division; each spermatocyte finally producing four haploid cells known as the *spermatozoa* or sperms. Human sperm averages about 0.05 mm in length. Its structure comprises a *head, neck, middle piece,* and a tail (Fig. 16.2). The head is composed mainly of the nucleus of the cell with only a thin cytoplasmic sheath and the cell membrane around it. On the front of the head is a caplike structure, the *acrosome*, which contains lysosomes and associated enzymes. The neck contains the centrioles of the sperm cell. The middle piece is closely packed with mitochondria which provide energy, in the form of ATP, for the locomotion of the sperm. The tail which is covered by a long extension of the cellular membrane possesses most of the cytoplasm of the sperm. It also has the *contractile element* of the sperm, responsible for the whiplike undulations of the tail. The contractile element is in the form of an *axial filament* that extends the entire length of the middle piece and the tail (Fig. 16.2). The filament, surrounded by an *axial sheath,* is made up of a number of fine fibres.

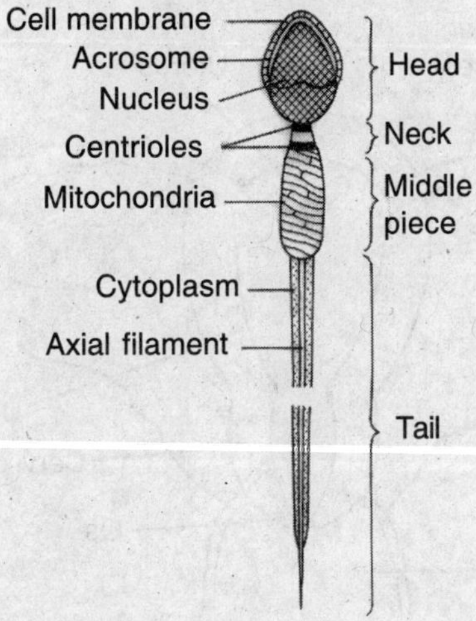

Fig. 16.2. The human sperm.

The *semen* which is a viscous fluid provides a suitable medium for the swimming of the sperms. In addition, it contains nutritive materials which keep the sperms viable. Fresh mammalian semen primarily contains polypeptides, phosphorylcholine, inositol, ergothioneine, citric and, fructose, an important enzyme hyaluronidase, and inorganic ions, viz. Na^+, K^+, Ca^{++}, Mg^{++}, Cl^-, and PO_4^{--}. *The presence of zinc in the semen is perhaps the highest among all the body fluids*. Its concentration is 14 mg/100 ml of semen. Fructose in semen forms an important source of energy and help in sperm motility. It is anaerobically converted into lactic acid through the Embden-Meyerhof pathway (Chapter 13). During this process ATP is formed which is necessary for sperm motility.

The semen shows a tendency to *agglutinate* when exposed to air; the mechanism is very similar to the clotting of blood, and involves the eventual conversion of fibrin to fibrinogen. This would hamper the free movement of the sperms. In order to prevent this both the males and females produce *antaglutinins*. In the males antaglutinins are produced in the prostate gland, and in the females in ovarian follicles, fallopian tubes and the mucosa of the cervix. Antaglutinins of males contain sugar, sulphuric acid residues acid a vitamin E derivative. The antaglutinins of females contain sugars but no sulphur.

Vitamins are important for sexual activity. It is known that the post-natal development of the testes is retarded during biotin deficiency. Pyriodoxine and B_1 deficiency retards weight increases and spermatogenesis in testes, and inhibits the development of the interstitital cells of Leydig. Ascorbic acid protects against B_1 avitataminosis. Deficiency in vitamins E, A, B_1, B_2, B_{12}, C, M, FF, and H affects the whole genital tract.

The *testicular hormones* are known as *androgens*. Besides these, the semen also contains a hormone-like substance known as *prostaglandins*. A description of these has been given in Chapter 11 on Hormones and Endocrine Glands. It is interesting to note that there is a correlation between total semen prostaglandin concentration and the degree of fertility.

FEMALE REPRODUCTIVE SYSTEM

The female reproductive system consists of the *ovaries, the fallopian tubes (uterine tubes), the uterus, vagina*, and the *vulva* (Figs. 16.3 and 16.4).

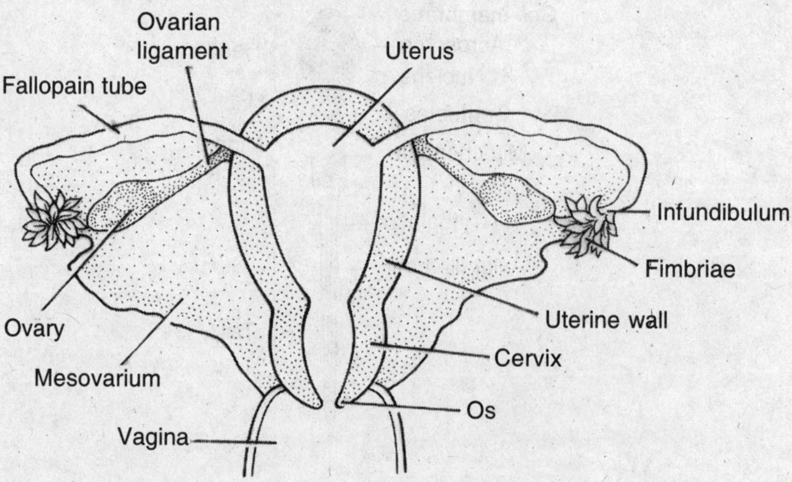

Fig. 16.3. Main features of the internal genital organs of the female.

The *ovary* is the principal functional unit of the female genital system. As in the case of testes, the ovary is both an exocrine (producing the *egg cell* or the *ovum*) and an endocrine gland (producing female sex hormones—the *estrogens*). The ovaries are two almond-shaped structures. These are located on the sides of the lower abdominal cavity, one on either side of the uterus and are attached by the *mesovaria* and the *ovarian ligaments* (Fig. 16.3). The surface of the ovary consists of columns of epithelial cells which at intervals penetrate deep into the *stroma* or *ovarian connective tissue* (Fig. 11.10). During the development of the female small groups of cells separate from these columns and become arranged into definite structures known as *primitive* or *primordial follicles*. Each follicle contains a large single cell, the *oocyte,* which is surrounded by a single layer of cells. At the time of birth the newborn female infant contains several thousand primordial follicles in each ovary. As soon as the female becomes sexually mature (*puberty or menarche* which is reached by the age of 12 to 15 years), a single follicle (primary follicle) undergoes maturation under the influence of FSH of the anterior pituitary gland (Chapter 11). The process of maturation of the follicle takes place in the following manner.

The *primary follicle* consists of an *oocyte* surrounded by a single layer of flattened epithelial cells (Fig. 11.10). As the primary follicle matures it moves deeper into the stroma of the ovary, the oocyte increases in size, and the follicular epithelial cells proliferate mitotically. This forms a multilayered stratified eipithelium, called the *granulosa layer,* around the oocyte. The follicle at this stage is described as *secondary follicle* (Fig. 11.10). The cell membrane of the oocyte is now called the *zona pellucida*. The growing follicle is soon encapsulated by a sheath of tissue (*theca*) derived from the stroma. The theca, later, gets differentiated into an outer *theca externa* and a vascularized inner *theca interna.* The theca interna is separated from the granulosa layer by a thin *membrana propria.* Later, spaces appear in the granulosa which join together to form a cavity known as the *antrum*. The antrum is filled with fluid (liquor folliculi). The follicle is now known as *vesicular* or *Graafian follicle.* The granulosa layer surrounding the ovum is known as corona radiata; whereas that attaching the avum to the antrum is referred to as *cumulus oophorus* (Fig. 11.10). When the Graafian follicle ruptures, it releases the ovum with the zona radiata intact around it. The phenomenon of the release of the ovum from the Graafian follicle is described as *ovulation.* The follicle wall usuallly ruptures at a site opposite the cumulus oophorus. After ovulation the collapsed follicle transforms into a new endocrine structure, the *corpus luteum.* Successive maturation of other individual follicles is followed periodically, about once every 28 days, till about the age of 45 years after which production of Graafian follicles ceases (*menopause*). The details of the process of ovulation and the accompanying phenomenon of menstruation are discussed in the section on menstrual cycle. The whole process is largely controlled by various hormones.

In summary, the process of maturation of the primary follicle into the ovum is as follows. The *oogonia,* corresponding to the spermatogonia, give rise to *oocytes.* These, unlike spermatocytes, virtually complete their differentiation as ova before they undergo reduction division. Unlike sperm formation, each oocyte gives rise to *one ovum* only. The three other very small cells known as the *polar bodies,* which result from unequal division of the oocytes are discarded. The process of the formation of the ovum from the oogonium is known as *oogenesis.* The mature human ovum or egg cell is perhaps the largest cell produced by the body, and it measures about 0.14 mm in diameter.

Mention may be made here of a phenomenon known as *atresia.* This is a degenerative process whereby eggs are lost from the ovary without ovulation. The ova in *atretic follicles* (Fig. 11.10) give off polar bodies and undergo mitotic or amitotic divisions producing several cells within the zona pellucida. However, these are eventually destroyed. Some workers have considered the multiplication of the ovum to be an attempt to undergo parthenogenetic development.

A duct called the *oviduct* or *fallopian tube* (Fig. 16.3) lies in close proximity of each ovary. The proximal end of each fallopian tube opens into the abdominal cavity through a funnel-like structure known as the *infundibulum.* The outer margins of the infundibulum have tentacle-like structures called the *fimbriae* (Fig. 16.3). When the Graafian follicle ruptures it releases the ovum in the abdominal cavity. This is captured by the fimbriae and drawn inside the fallopian tube. The fallopian tubes are lined by ciliated epithelial cells which together with the peristaltic contraction of the tube bring about the downward passage of the ovum in the tube. The fallopian tubes open into the upper part of the *uterus* or womb (Fig. 16.3). Fertilization of the ovum usually occurs in the fallopian tubes. As the fallopian tubes are not actually connected to the ovaries, the fimbriae sometimes fail to bring the ovum in the tubes. If such an ovum gets fertilized in the abdomen, it results in a serious condition known as *ectopic pregnancy.*

The *uterus* or the *womb* is a hollow muscular organ about the size of the clenched fist in the adult non-pregnant woman. It is pear-shaped structure located between the bladder and the rectum, and has a Y-shaped internal cavity. The two arms of the 'Y' are continuous with the fallopian tubes. The lower end of the uterus is smaller and narrow than the main body (*fundus*), and is known as the *cervix.* The latter opens into the upper end of the vagina through its mouth or *os* (Fig. 16.3). The cervix functions as a sphincter, and is normally kept closed at all times except during *parturition* (childbirth) when it softens and undergoes marked dilation. The uterine wall is made up of three layers: (1) an outer serous layer (*perimetrium*), (2) a middle thick smooth muscle-cell layer (*myometrium*), and (3) an inner layer of glandulo-epithelial cells (*endometrium*). During pregnancy the uterus houses the embryo and provides it with nourishment through the *placenta.* During childbirth, the powerful contraction of the myometrium serves to push the child out through the vagina. At the time of *menstruation* the endometrium sloughs off resulting in characteristic menstrual bleeding.

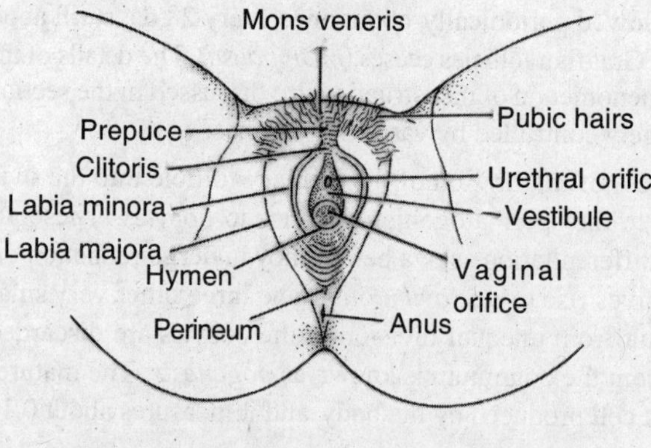

Fig. 16.4. Vulva or the external female genitalia.

The *vagina* (Figs. 16.3 and 16.4) is the copulatory organ of the female. It receives the erect penis and also the ejaculated seminal fluid of the male during sexual intercourse. The vagina is about 4 inches in depth and is dilatable. The lining of vagina consists of stratified squamous epithelial cells devoid of glands. It also fulfills the function of serving as the lower part of the birth canal through which the fetus passes enroute to become a newborn infant. In addition, the vagina also serves as an excretory passage for uterine secretions and menstrual discharge. In females who have not had sexual intercourse (a virgin), the external end of the vagina is closed or bordered around by a fold of mucous membrane called the *hymen* (Fig. 16.4). The hymen is generally ruptured during the first act of sexual intercourse. However, this is not always so; the hymen may be ruptured accidentaly during exercise as in women gymnasts, or even by bicycling.

The female external genital organ is called the *vulva* (Fig. 16.4). It consists of the *vaginal vestibule* in which the external openings of the *urethra* and the vagina are located. The opening of the urethra lies above the *vaginal opening.* The vestibule is enclosed by two pairs of lip-like structures: the outer and large hair-covered *labia majora,* and the inner smaller hairless *labia minora.* The lips of each pair are joined at the anterior and posterior ends of the vestibule. Occupying the apex of the vaginal vestibule is a small erectile organ called the *clitoris* (Fig. 16.4). The clitoris is homologous to the male penis, but unlike the latter it does not have the urethral opening. It becomes engorged with blood during sexual excitement and functions as an excitatory organ. The clitoris is covered by a *prepuce* or *foreskin* which is formed by the labia minora. The top part of the vulva is a triangular fatty pad covered with *public hair.* This is known as the *mons veneris* (Fig. 16.4). A pair of bean-shaped glands (*Bartholin's glands*) are present on either side of the *vaginal orifice* deep in the perineal tissue. These glands open by ducts into the space between the hymen and labia minora, and are homologous to the bulbo-urethral (Cowper's) glands in the male. Bartholin's glands secrete a pre-coital lubricating fluid.

Another important structure which should be described here is the *perineum* (Fig. 16.4). It is the skin covered muscular region between the vaginal orifice and the anus. The perineum usually ruptures during childbirth. In order to avoid this an incision (*episiotomy*) is made in the perineal tissue just before the passage of the fetus down the vagina. This provides more room for the fetus, and minimizes damage to the perineum and anal sphincters.

The biochemistry and mechanism of action of the ovarian hormones (estrogen, progesterone, ovarian androgens, and relaxin, and the endocrine control of the female reproductive system by the anterior gonadotrophic hormones have already been described in the chapter on Hormones and Endocrine Glands.

REPRODUCTIVE CYCLES IN FEMALES

Most mammalian females show seasonal and cyclic changes in *ovulation* (release of ovum from the ovary) and reproductive activities. Breeding is usually restricted to a particular time of the year depending upon the species. This forms a part of natural selection because the babies are usually born in spring which is favourable for the birth and care of the young. The brief period during which, and *at no other time,* the female copulates with the male is known as the "*heat*" or *estrus*. The species in which the onset of the estrus coincides with the release of the ovum are referred to as *spontaneous ovulators,* e.g., rats, mice, hamsters, guinea pigs, dogs, etc. In certain species such as rabbits, however, ova are not released till after coitus; these are known as *induced ovulators. Estrus cycle* may be defined as a reproductive cycle of short duration (usually 5 to 60 days according to species) which is associated with structural and physiological changes in the female. This occurs periodically in non-pregnant females.

The various changes associated with estrus cycle are as follows. In the beginning the thin endometrium of the uterus thickens, its blood supply increaes, and its uterine glands deepen as follicular maturation proceeds. The Graafian follicle ruptures, releases the ovum and collapses. The space created by the collapse of the follicle is invaded by the proliferation of the cells of the granulosa and theca interna as well as the cells of the ovarian stroma. The structure thus formed is an endocrine unit known as the *corpus luteum.* It secretes the female sex hormone *progesterone.* With the release of the ovum and the subsequent appearane of the corpus luteum, the endometrium continues to thicken. If fertilization occurs, the *fertilized egg* (zygote) implants itself in the endometrium where it develops into a new individual. The corpus luteum is maintained during most of the pregnancy. However, if fertilization does not occur the ovum degenerates, the corpus luteum begins to disappear, the endometrium regresses and returns to its original thin state. The endometrium does not build up until the beginning of the next estrus cycle. If fertilization does not take place, the estrus cycle is repeated regularly according to the mating pattern of the species. Depending upon the species a mature female may experience one or several estrus cycle within a breeding season. For example, in cats and dogs two estrus periods occur per year, whereas in the mouse and rat the female has an estrus cycle about every 4 or 5 days.

MENSTRUAL CYCLE

In the human females as well as in higher primates (anthropoid apes) the estrus cycle is not very distinct and exists in a modified form known as the *menstrual cycle* (Fig. 16.5). Sexual activities in these are not limited to any particular "heat" period, but is a regular process because more often than not it is carried out for pleasure rather than procreation. The onset of the menstrual cycle is signalled by a rapid regression and sloughing off the superficial layer of endometrium. This results in the rupture of blood vessels which causes bleeding through the vagina. As the endometrium is highly vascular (Fig. 16.5), the bleeding is rather substantial and lasts for 4 to 7 days. The entire phenomenon is known as *menstruation.* Menstrual phase is regarded as the beginning of the primate reproductive cycle. This is because it corresponds with the formation of new follicles in the ovaries, and is the easiest period of the cycle to recognize. *Menstrual bleeding is absent in the non-primates.*

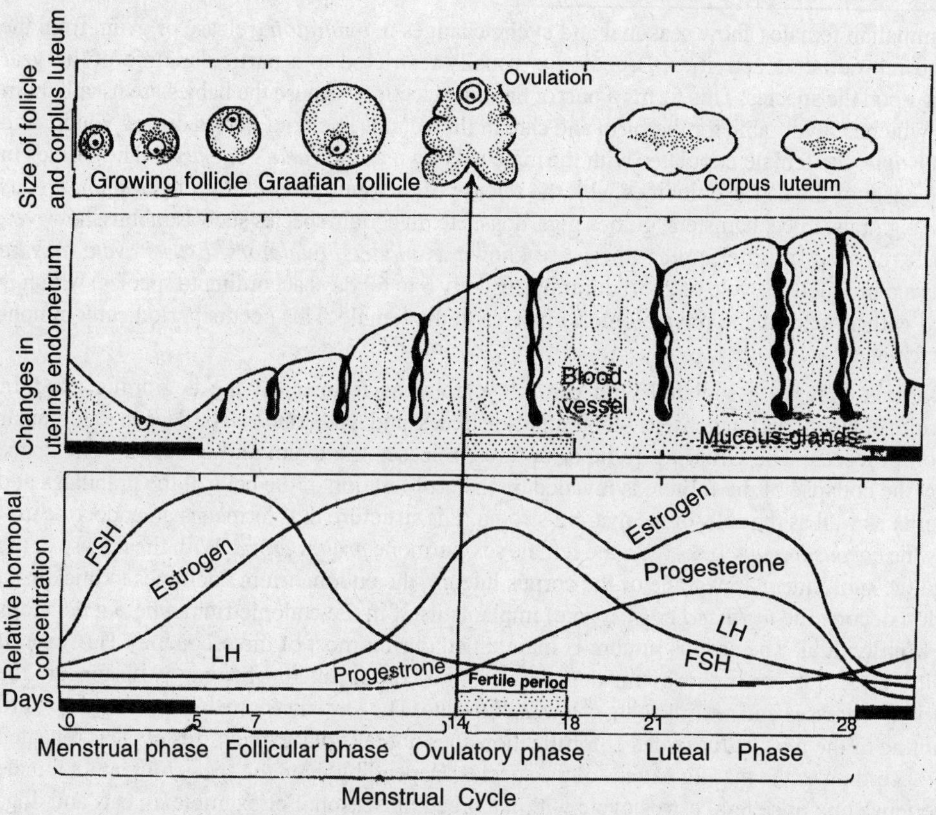

Fig. 16.5. Scheme of events in a typical menstrual cycle. The diagram illustrates growth of the ovarian follicle, the changes in the uterine endometrium, and the relative concentration of circularing hormones. (For details see text).

The menstrual cycle is usually of 28 days duration (Fig. 16.5). It consists of four phases: (1) *follicular (proliferative),* (2) *ovulatory,* (3) *luteal (progestational),* and (4) *menstrual.* The *follicular phase* is characterized by the progressive development and maturation of an ovarian follicle in one of the ovaries (Fig. 11.10 and 16.5). It is controlled by the secretion of FSH from the anterior pituitary gland. The follicle, as it develops, secretes increased doses of the female hormone *estrogen.* The latter promotes the redevelopment of the uterine endometrium for the reception of a fertilized egg. The follicular phase ends and the ovulatory phase (Fig. 16.5) begins with the rupture of the follicle and the release of the ovum. This takes place 10 days after the cessation of menstrual flow (or 14 days after the start of the menstrual flow). The *luteal phase* (Fig. 16.5) starts immediately after ovulation when the cavity of the ruptured follicle (under the influence of LH from the anterior pituitary gland) begins to fill with a mass of developing yellow cell—the *corpus luteum.* The latter produces the hormone *progesterone.* Like estrogen, progesterone also promotes the development of the uterine endometrium for the implantation of the embryo. Progesterone simultaneously inhibits the development of any other ovarian follicles. It also stimulates the development of the mammary glands to keep them in readiness for providing milk to the new-born infant. *Prolactin* secreted by the adenohypophysis also participates together with LH in maintaing the corpus luteum and, therefore, progesterone secretion. The fertilized ovum implants itself in the endometrium where it develops into a fetus. The corpus luteum is maintained throughout pregnancy.

However, if the ovum is not fertilized and implanted in the uterus by the 27th or 28th day after start of the previous menstrual flow, the corpus luteum undergoes degeneration. This results in cessation of progesterone secretion (Fig. 16.5), and is accompanied by a decreased gonadotrophic hormones by the anterior pituitary gland. Because of the lack of progesterone and estrogen the thickened endometrium begins to degenerate and sloughs off, and there is bleeding. This makes the beginning of menstruation and another menstrual cycle. This is called the menstrual phase (Fig. 16.5). The anatomical and physiological changes during menses cause emotional irritability and tension in the menstruating female. This is partly psychological and partly due to the low level of sex hormones in the body at this time.

As already said, ovulation occurs about 14th day of the menstrual cycle. The released egg can survive for 24 to 96 hours and, therefore, 14th to 18th day is called as the *fertile period* (Fig. 16.5) when fertilization is likely to take place.

The menstrual cycle begins at the onset of puberty when there is sudden rise in the production of ovarian hormones. The female remains fertile as long as the menstrual cycle is successively repeated. In most cases menstruation stops (*menopause*) between the ages of 45 to 50 years. *Although the female becomes infertile after menopause, sexual desires being psycho-somatic may persist throughout life.*

The menstrual cycle in humans average about 28 days from start to finish before the next cycle begins. Repair of the uterine endometrium follows, and with the next cycle the endometrium begins to thicken progressively once more. If fertilization takes place, endometrium is maintained and shows further changes. These changes will be discussed in the section on embryonic development.

COITUS

In all mammals, including human beings, there is *internal fertilization*. This requires the introduction of the erect penis into the vagina (*sexual intercourse*) and the ejection of the seminal fluid into it. The stimulation of sensory nerve endings in the penis as a result of friction against the lubricated walls of the vagina culminates in a short period of intense sexual excitement called the *orgasm*. The orgasm is accompanied by *ejaculation*—the forceful ejection of a jet of semen from the penis. The volume of the semen ejaculated during coitus depends upon the interval between two coituses. The average volume ejaculated is 3 to 5 ml and contains about 300×10^6 sperms. *Orgasm in females is accompanied by rhythmic contraction of the vaginal and uterine walls, but no special discharge or ejaculation of fluid occurs, except that there is a secretion of the many glands lining the reproductive tract as a result of sexual excitement.*

FERTILIZATION

During coitus, the sperms are usually released near the cervix. They progress in the female genital tract by swimming with the help of their tails. Contractions of the vagina, cervix and uterus speed up the ascendancy of sperms into the fallopian tubes [Fig. 16.6 (a)] at a rate of about 2 mm/minute. Fertilization takes place in upper third of the fallopian tube [Fig. 16.6 (b)].

Fertilization is the union of the nuclei of sperm and ovum which results in the formation of the zygote. The ovum is surrounded by a layer of follicle called corona radiata which must be dissolved and penetrated before fertilization can take place. The corona radiata cells are held together by a mucopolysaccharide of hyaluronic acid polymers. The semen contains a hydrolysing enzyme called hyaluronidase which dissolves the cementing material of the corona radiata when the sperm carrying the enzyme comes in contact with it. Once the corona radiata has been dissolved the sperm gains access to the ovum. The sperm penetrates the plasma membrane of the ovum, and its head and middle piece enter the ovum leaving the lashing tail behind. Immediately after this the ovum secretes a membrane around it known as the *fertilization membrane*. This protects the ovum from being penetrated by another sperm. The successive mitotic divisions of the zygote produce the embryo which moves down the fallopian tube. Within 10 days the embryo consisting of several hundred cells is implanted in the uterus [Fig. 16.6 (c)], where it is completely covered by the rich mucosa lining the endometrium.

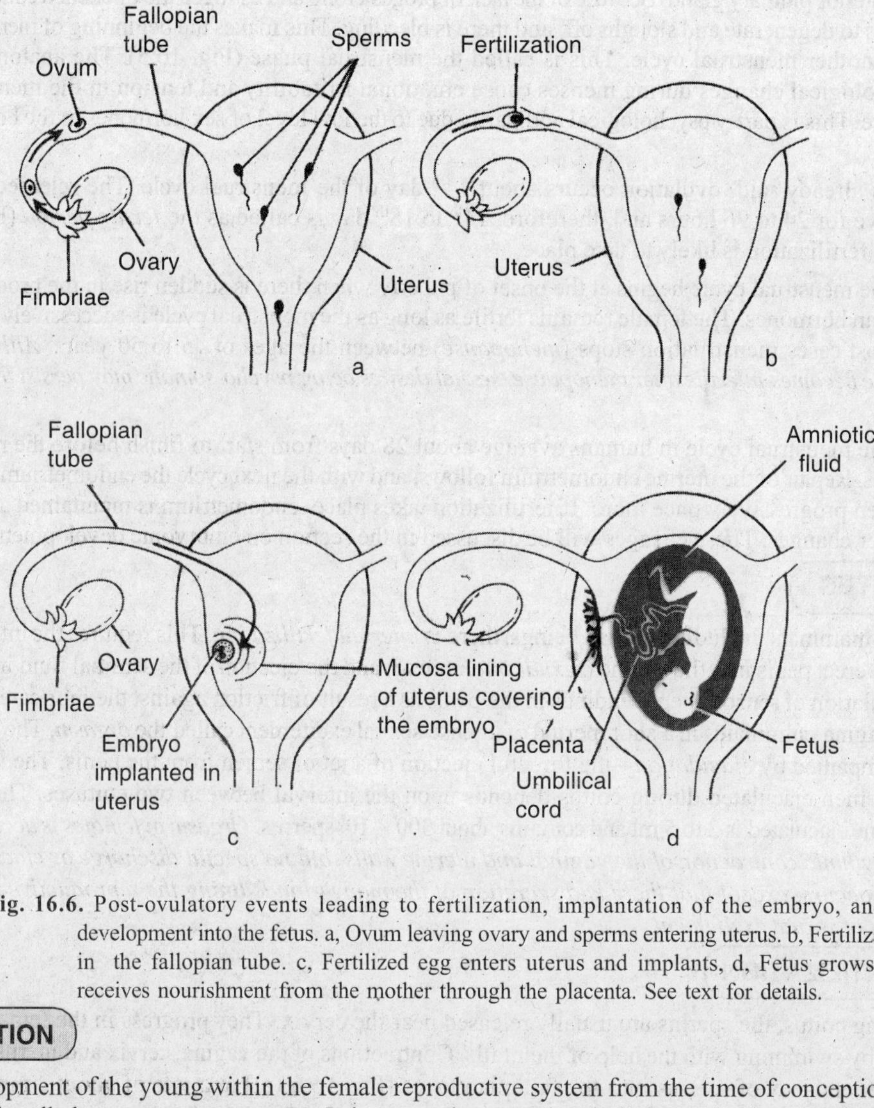

Fig. 16.6. Post-ovulatory events leading to fertilization, implantation of the embryo, and its development into the fetus. a, Ovum leaving ovary and sperms entering uterus. b, Fertilization in the fallopian tube. c, Fertilized egg enters uterus and implants, d, Fetus grows and receives nourishment from the mother through the placenta. See text for details.

GESTATION

The development of the young within the female reproductive system from the time of conception to childbirth is called *pregnancy or gestation*. In humans, the normal period of gestation is approximately 266 days, whereas it is 16 days in the hamster, 355 days in horse, and 22 months in the elephant.

During the course of its development the embryo gets attached to the maternal uterus by means of a specialized structure known as *placenta* [Fig. 16.6 (d)]. The latter contains both the embryonic and maternal tissue in close association. There is exchange of small molecules between maternal and embryonic blood in placenta; oxygen and food substances enter the embryo and urea and carbon dioxide leave it; only minute amount of proteins are exchanged. The placenta also produces estrogen and progesteron which keep the uterus adapted for pregnancy. The embryo is connected to the placenta by means of a cord called *umbilical cord* [Fig. 16.6 (d)].

Eight to ten days after fertilization, the human embryo is in the *blastocyst stage* consisting of the primary germ layers and several extra embryonic structures which include the *trophoblast, amnion*, and the *yolk sac* which form part of the placenta. By the end of the first month the embryo is half an inch long and has the beginnings of the brain, eyes, stomach, kidneys, and the heart. The heart beats approximately 60 times/minutes.

Between the 5th and 8th weeks the limbs begin to appear as buds. By the end of the second month most of the internal organs of the future adult are formed, and the embryo is referred to as the *fetus*. During the third month the nails begin to appear and the external sex organs differentiate so that the male and the female can be distinguished. The fetus is about 3 inches in length.

In the 4th month hair appears on the head and the body. In the 5th month the fetus is about a foot long and blood formations in the bone marrow commences. By the 6th month the fetus start moving in the uterus and its presence is clearly felt by the mother. The fetus at this stage is fully formed.

In the final two or three months of gestation the fetus only gains in weight, and is able to survive if born prematurely. At this time most of the antibodics from the mother are transmitted to the fetus. The antibodies are needed for protection from infectious diseases for the first six months after the infant is born.

During the course of gestation the uterus increases in size approximately 20-fold. The human fetus towards the latter part of gestation normally assumes a position with its head directed downward [Fig. 16.6 (d)] towards the vagina in preparation for childbirth.

PARTURITION

The process by which the fetus is brought out of the mother's womb is called child-birth or parturition. Birth begins with the involuntary contraction of the uterine walls called *labour*. The eventual effect of the hormones is to increase the contractibility and irritability of the uterus. In the later stages, the mother voluntarily contracts the abdominal muscles also to aid the labour. The contraction of the uterus forces the head of the fetus against the cervix, causing it to dilate. The amniotic sac bulges into the vagina and ruptures, liberating the amniotic fluid through the vagina. The fetus is finally expelled out by vigorous contractions of the uterus and the abdominal walls.

The umbilical cord which connects the newborn to the uterus by way of the placenta is cut close to the baby's body. Within the next 15-30 minutes further contractions of the uterine wall expel the placenta and other extra embryonic structures. The duration of birth process or labour may last about 16-20 hours at the time of first childbirth. During childbirth the relative concentration of estrogens and progesterone changes. There is an increase in the concentration of the former which initiates the uterine contraction essential for childbirth.

LACTATION

Mammary Glands

All mammals are provided with milk-secreting glands of cutaneous origin called the *mammary glands*. These are *modified sebaceous glands*. The glands are found in both sexes but are functional only in the female. The immature glands consists of 16 to 25 *lobes* each of which has a *lactiferous duct* that extends up to the *nipple* Fig. 16.7 (a and b). The nipple is surrounded by a pigmented circular area called the *areola.* In the *prepuberal* (pubescent), *puberal* and *adult virgin* female, growth and branching of the lobes occur Fig. 16.7 c under the influence of the increasing output of the ovarian hormones. At *puberty* and in the adult virgin a rapid and extensive deposition of fat occures in mammary glands resulting into the developmjent of breasts. The mammary glands develop further during pregnancy. The areola darkens, and the lactiferous duct of each lobe dilates into a *sinus,* again constricts, and opens at the tip of the nipple. There is a further arborization of the lactiferous ducts and numerous *ductules* are formed Fig. 16.7d. The tips of the ductules give rise to end buds which develop into the *alveoli* Fig. 16.7a and e. The entire development of the breasts is termed *mammogeneis.* The basic structure of mammary gland, therefore, is that of a typical exocrine gland; it consists of alveolar cells which secrete milk and excretory ducts opening outside on the nipple.

It is during the later part of gestation that the mammary glands becomes active in order to produce milk (*lactogenesis*). This takes place in two stages; the *colostrogenic stage,* and the *lactogenic stage.* These stages are associated with the secretion of *colostrum* and *milk,* respectively. Colostrum is a yellow viscous fluid secreted by the alveoli towards the end of pregnancy. It contains inorganic salts, lactose, casein, and fat droplets, leucocytes, numerous desquamated epithelial cells, vitamins, and *granular bodies of* Donne (or *colostrum bodies*) filled with fatty droplets. After delivery when the baby is put to the breast, this protein-rich fluid is all that the infant gets until about the third to fifth day, when the milk Fig. 16.7 (f) comes in (*lactogenic stage*).

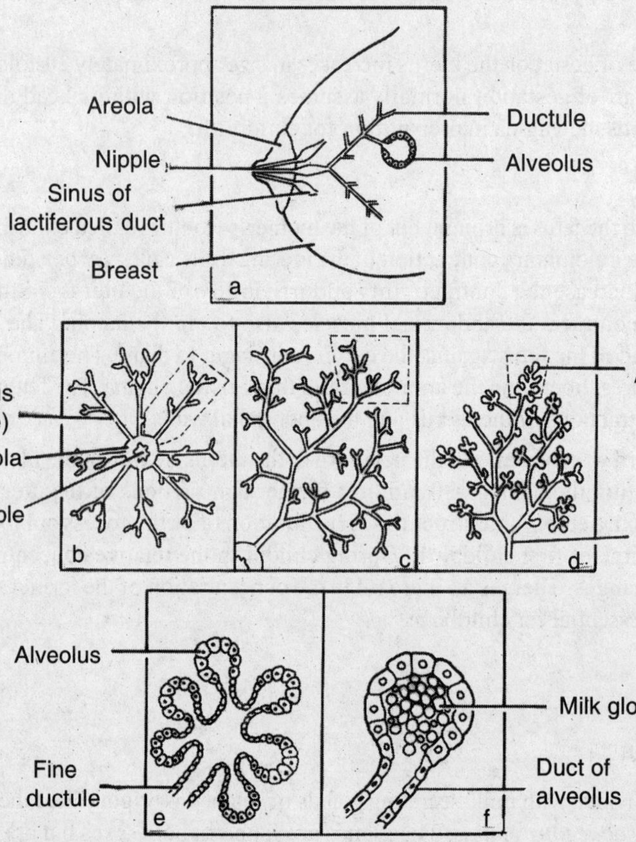

Fig. 16.7. Diagrams illustrating mammogenesis. a, Typical structure of the breast during pregnancy. b, Immature gland. c, Magnified segment of the boxed area in 'b' to show the mammary gland in the adult virgin. d, Boxed area of 'c' enlarged again to show the development of alveoli during pregnancy. e, Structure of several alveoli. f, Accumulation of milk globules within an alveolus in lactating female.

Human milk contains 88% water, 1.2% of three essential proteins (casein, β -lactoglobulin, and α -lactoglobulin), 6.5% of lactose and gynolactose (a mixture of at least 14 sugars some of which contain nitrogen), 3.5% fat, and 0-2% inorganic materials. It also contains sialic acid, cellular and nuclear fragments, urea, vitamins (mainly A, D, E, D, B_1, and B_{12}), and enzymes (lactase, amylase, lipase, phosphomonoesterases, xanthine oxidase, etc). *Donne bodies and leucocytes which are characteristically present in the colostrum are absent in the milk.*

With the cessation of breast-feeding (*weaning*), gradually the colostrum stage reappears followed by regression of the alveoli, and the breasts become similar in structure to *adult virginal state.* After menopause only the excretory ducts remain, and in many cases the connective tissue is substituted by fats. This leads to an increased size of the post-menopause mammary glands. During old age (*senescence*) fats, however, disappear, the connective tissue become a fibrous system, the breasts shrivel and become small.

HORMONAL CONTROL OF LACTATION

Initiation and maintenance of milk secretion and its delivery to the infant is called *lactation*. This is controlled by a variety of hormones, viz., estrogen, progesterone, adrenal corticoids, prolactin, FSH, LH, and oxytocin.

The following important events cause lactation. During the later stage of pregnancy, progesterone secreted by the placenta aids in the development of the mammary glands for lactation. The high level of estrogen also secreted by the placenta, however, inhibits the secretion of prolactin essential for milk production. The expulsion of the placenta at child birth decreases the estrogen level in the mother resulting in an increased secretion of prolactin and subsequent synthesis of milk for the child. The suckling action of the newborn at the nipples of the mother sends afferent nerve impulses to the hypothalamus, resulting in the release of oxytocin from the posterior pituitary gland. The oxytocin stimulates the ejection of milk from the mammary glands. Oxytocin is, therefore, also known as the *milk-let-down-ejection-factor.* Suckling is necessary for the continued formation of milk by the breasts. Optimum milk secretion lasts for about a period of 8 to 9 months, after which it gradually diminishes, and finally ceases.

17

Chapter

Temperature and Thermoregulation

Animals can function efficiently only within a very narrow range of temperature called the *biokinetic zone*. The biokinetic zone for most animals lies within 10—45°C. This imposition of a temperature limit, for the optimal functioning of the body, is because most enzymes can function efficiently, only within this range. Below and above this range enzymatic activity falls rapidly. When the temperature is far removed from the biokinetic zone, other damages also take place. However, since the environmental temperature range is quite wide (–79 to 95°C), animals have developed two distinct regulatory mechanisms to counter the temperature stress: (1) In lower animals, which are *thermoconformers*, although the body temperature changes with the environmental temperature, the physiological activities are adapted to function in spite of the temperature stress. This phenomenon is known as *piokilothermy* and animals showing this are known as *poikilotherms*, or *cold blooded animals*. (2) Higher animals, which are *thermoregulators*, maintain a constant body temperature, in spite of the fluctuations in environmental temperature. This phenomenon is known as *homeothermy*; the animals (birds and mammals) possessing this ability are known as *homeotherms* or *warm blooded animals*.

At this stage, two additional factors are necessary in understanding temperature regulation and its implication on the adaptability of the body. These are *acclimatization* and *acclimation*. The former deals with the compensatory changes in the organism necessitated by the changing climatic, physical, and biotic factors in nature. Acclimation, on the other hand, pertains to compensatory change in the organism in response to changes brought about under experimental conditions in the laboratory.

PHYSIOLOGICAL EFFECTS OF HEAT AND COLD

Heat

High temperature is injurious to life; extremely high temperatures are lethal to all forms of life. Although some organisms called *thermophilic* can tolerate extremely high temperatures, most animals however, survive only in the biokinetic zone; these are known as *mesophilic*.

Heat destroys living cells in various ways:

(1) Most proteins coagulate at high temperatures and become relatively inactive. Heat deaths are, therefore, mostly due to thermal inactivation (coagulation) of proteins. Enzymes, which are also proteins, are reversibly changed by mild rise in temperature. However, when the temperature is high these are irreversibly altered. The duration of exposure to heat is also crucial. Although many enzymes can stand short exposures to high temperatures, they become inactive if exposed to even 45°C for longer periods. In many cases heat causes the dissociation of enzymes into their constituent polypeptides. When the constituents reassemble the resulting protein is either catalytically less active or totally inactive.

(2) It has been suggested that heat also causes the production of a coagulating enzyme in the cells. Exposure to high temperature releases calcium at the outer boundary of the cytoplasm. This in turn causes the liberation of the coagulating enzyme which causes an irreversible gelation of the cell.

(3) Heat causes the melting of the lipid layer of the tripartite cell membrane which results in the disruption of the membrane. It has been demonstrated that animals acclimatized to high temperatures contain lipids of higher melting point as compared to control individuals.

Cold

Low temperature reduces the overall rate of thermochemical reactions. Most enzyme-catalysed reactions come to a stand still at 0°C. Most forms of life, therefore, cannot survive freezing. Organisms which can tolerate subzero temperatures are called *cryophilic*. Besides the suspension of enzymatic activities, cold death is caused by several other factors also:

(1) When the water around the cell starts freezing into ice, the concentration of solutes in the unfrozen water becomes very high. This makes the extracellular fluid hyperosmotic to the cytoplasm causing withdrawal of water from the cell. Consequently, the concentration of electrolytes within the cell becomes very high and the cell dies. Ability to survive sub zero temperature is, therefore, the ability to withstand high concentrations of solutes in the cytoplasm.

(2) At subzero temperatures ice formation takes place in the cytoplasm also. Needle like ice crystals mechanically destroy the cell membrane as well as the fine cytoplasmic organelles like the cytoplasmic reticulum, mitochondria etc.

(3) Freezing also removes lipids from the cell memberance and makes it fragile. On getting warm the membrane just disintegrates.

(4) Ice formation, as already said, increases the concentration of solutes in the cytoplasm, which alters its electrolyte balance and pH. This causes the breaking of enzymes into their component polypeptide units, rendering them inactive.

HOMEOTHERMY

In the animal kingdom, only birds and mammlas are abe to maintain a constant body tempeature. This is kept between 36°C and 42°C. Moreover, each species has its own 'normal' body temperature, which is maintained quite accurately even under extreme climatic conditions. Appreciable departure from the 'normal' temperature invariably leads to death. The phenomenon of the maintenance of a constant body temperature is called *homeothermy*. Homeotherms have the distinct advantage over poikilotherms in being above to function at a relatively constant level of efficiency throughout the year. Homeothermy has been one of the major steps in evolution, as it has enabled animals to avoid a compulsory period of inactivity (hibernation/aestivation) which poikilotherms have to undergo. Besides this, piokilotherms has to evolve an elaborate means involving regulation of metabolic pathways, endocrines, behaviour etc. to cope with the changes in the environment. Homeothermy bypasses these elaborations by maintaining a constant body temperature. According to Barcroft (1934), "...Here then is a very fine issue, the cold blooded animal successfully adopting ingenious mechanisms, first biochemical, then physiological, in order to adapt..." The warm blooded predecessor has laboriously beaten out, invoking the nervous system to reverse the normal biochemical relationship and gaining a new freedom by adjusting, not itself to the internal environment but the internal environment to itself.

The temperature of man is measured in the mouth, rectum and axilla (armpit); the most common is the oral method. The oral temperature is maintained around 37°C (98.6°F) with only small fluctuations. In order to maintain the body temperature accurately, the heat gained by the body should be equal to the heat lost by it.

In addition to heat gained and lost from the environment, the body actively generates and loses heat. The various sources of heat gain and heat loss are given in Table 17.1.

Table 17.1 Showing the various mechanisms of heat gain and heat loss by the body

Heat gain	*Heat loss*
1. From environment	1. To environment
(a) convection	(a) convection
(b) conduction	(b) conduction
(c) radiation	(c) radiation
2. Production of heat by metabolism—*thermogenesis*	2. Heat loss due to evaporation
(a) basal heat production	(a) basal heat loss
(*i*) specific dynamic action of food	(*i*) insensible perspiration
(*ii*) unconscious tensing of muscles	(*ii*) water evaporation through lungs
(*iii*) disease	(*iii*) warming of respired air
(*iv*)	(*iv*) through urine and faeces
(b) active heat production	(b) active heat loss
(*i*) exercise	(*i*) sweating
(*ii*) shivering	(*ii*) panting

Heat Gain

Body gains heat from two sources (1) environment (2) heat obtained from metabolism or *thermogenesis*.

Heat gain from environment: When the environmental temperature is higher than body, the latter gains heat from the surroundings through *convection* and *conduction*. Besides this, the body also gains heat by absorbing radiations thorugh the skin. This is independent of the ambient (surrounding) temperature. Another factor which decides heat gain is the pigmentation of the skin. A dark skin, which absorbs more radiations, gains more heat as compared to a light skin. In summers, light coloured clothing, which reflects heat is, therefore, better, as compared to a dark naked body.

Heat production by the body: All the energy liberated during metabolism is not converted into ATP. Only about 40% gets converted into ATP, the rest is released as heat. This is known as *thermogenesis*. During standard resting conditions (basal conditions—conditions under which the individual is under complete physical and psychial rest, 12—18 hours after meal and in an equable environmental temperature) the heat production is approximately 1700 Cals in males and about 1500 Cals in females per day. Males and females, indulging in moderate physical activity, require about 3000 Cals and 2500 Cals respectively per day. This rises further in individual who do heavy physical work.

The heat produced by the various organs of the body depends on the degree of their activity. Although the rate of heat production by heart and brain does not vary very much, the heat produced by the skeletal muscles, on the other hand, varies a great deal; it is very low when muscles are resting and very high when muscles are active. Active muscles in fact, are the most important source of thermogenesis.

Specific dynamic action of food: The rate of metabolism is influenced by the nature of the diet. In general, whenever food is taken, the rate of metabolism rises. All classes of food, however, do not stimulate metabolism to the same extent. The extent to which a particular food influences the rate of metabolism is referred to as its *specific dynamic action*. Proteins have a higher specific dynamic action; these raise the metabolic rate by about 30%; carbohydrates and lipids only about 4.5% and a mixed diet about 10%. The increase caused by a protein diet reaches its peak 3—5 hours after ingestion. In cold weather, therefore, protein, owing to its high specific dynamic action, gives valuable assistance in warming the body. High protein diet, therefore, is not suitable for a hot climate, as the greater heat production caused by protein intake, puts an additional burden on the cooling mechanism of the body.

The popular belief in India, that high protein foods like egg, meat, etc. have a 'hot' property, while carbohydrate containing foods like rice, sugar etc. have a 'cool' property is, perhaps, based upon the specific dynamic action of these foods.

When the body temperature falls, body produces additional heat by active mechanism like shivering, increased production of hormones, etc., the details of which have been given later.

Heat Loss

Body losses heat basically by two means (1) to environment (2) water evaporation from the body.

Loss to environment: The body continuously emits heat (infrared) radiations which are absorbed by the surrounding objects. The amount of heat lost by radiation depends on the surface area of the body and the skin temperature. When the environmental temperature is below body temperature heat is lost to the surrounding air by convection. The amount of heat lost in this manner depends on the temperature gradient between the body and the environment—the higher the difference the faster is the heat loss. For instance, the amount of heat lost froma bare head may amount to half the resting heat in man, when the outside temperature is 4°C. Heat loss by the body due to *conduction* is quite low as air is a very bad conductor of heat.

Heat loss due to evaporation: Water evaporates from the body through various sources (Table 17.1). Evaporation of water takes place from the lungs during respiration. This amounts to about 300 ml/day. The skin loses about 600 ml/day of water as a result of *insensible perspiration*. The latter is caused by diffusion of water to the surface of the skin; it cannot be seen or felt and is not a secretion of the sweat glands. As the latent heat of water is 0.58 Cals/ml the heat loss due to the evaporation of insensible perspiration is about 400 Cals. The heat loss due to respiration is about 200 Cals. In addition to this, about 80 Cals are spent in warming the respired air in cold weather and another 50 Cals are lost through urine and faeces.

When the body temperature rises, it loses heat by active means, such as sweating, panting, etc., the details of which have been given later.

Thermoregulation

The maintenance of a constant body temperature in the face of fluctuations in environmental temperature, is known as thermo-regulation. True homotherms can maintain their body temperature in spite of variations in the environmental temperature, whereas poikilotherms cannot . A lower mammal like *Echidna*, which is not a true homeotherm, maintains its temperature with in accuracy (Fig. 17.1).

Heat production curve: The relationship between body temperature and ambient temperature is graphically represented by a *heat production curve* (Fig. 17.2).

There is a lower limit of ambient temperature below which the production of the heat begins to rise. This is known as *lower critical temperature* and it is 25-27°C in man. Similarly, when the temperature rises above a certain point—*upper critical temperature* (about 35°C in man) then also heat production rises (Fig. 17.2). The temperature between the upper and lower critical temperatures (between C and D in Fig. 17.2) is called *zone of thermal neutrality*. Heat production in this zone is minimum and should be regarded as the *basal heat production* if all the other requirements of a standard resting state are met.

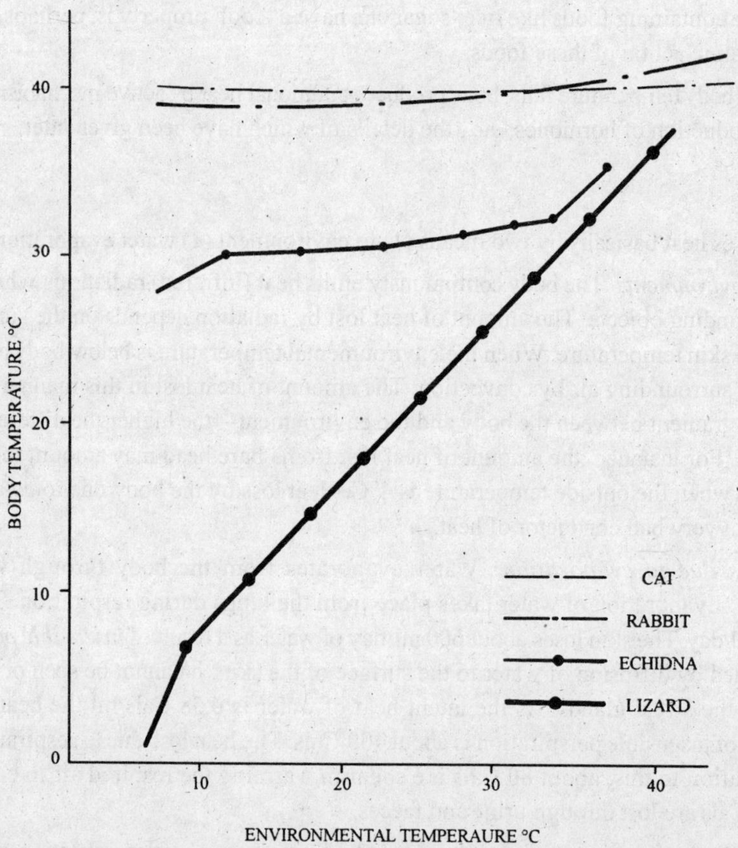

Fig. 17.1. Figure showing the variation in body temperature with change in environmental temperature in true mammal (cat and rabbit), primitive mammal (*Echidna*) and a poikilotherm (lizard).

Temperature gradient of the body: The temperature is human beings is maintained around 37°C. There is a diurnal variation of body temperature; the lowest temperature is recorded early in the morning (35.8°C or 96.5°F) and the highest in the evening (37.3°C or 99.2°F). The rise of temperature during the day is due to increased heat production from muscular activity and metabolism of food. 37°C, described as normal temperature is the temperature of the blood as recorded in the mouth. In the armpit it is about 0.5°C lower and in the rectum it is about 0.5°C higher than the oral temperature. The temperature of all parts of the body is thus, not maintained uniformly at 37°C. In cold climates, when the environmental temperature is low, there is a gradient of temperature from the skin to the blood, Table 17.2.

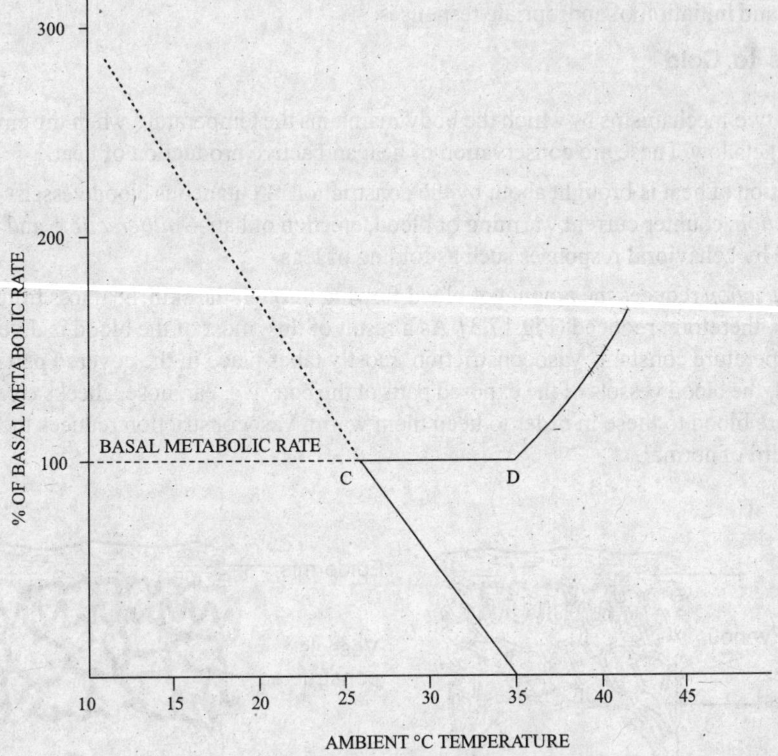

Fig. 17.2. Heat production curve of man, showing the rate of production of heat in response to change in ambient temperature. C—lower critical temperature; D—upper critical temperature. There is incresed heat production below C and above D; between C and D—zone of thermal neutrality, heat production is at the basal metabolic rate.

Table 17.2 Temperature of different parts of the body with an ambient temperature of 18°C

	Average temperature °C
Skin	33.0
Subcutaneous tissue	33.6
Deep muscle	36.2
Mouth (arterial temperature)	37.0

Control of Body Temperature

In order to detect environmental temperature, the body has temperature sensing devices or *thermal receptors*. There are both peripheral and central receptors; the former are present in the skin and the latter in the *heat loss centre* of the anterior hypothalamus. The skin has two sets of thermal receptors—one sensitive to cold and the other sensitive to heat. These two are present in a definite pattern over the surface of the skin. These two are present in a definite pattern over the surface of the skin. These receptors are so sensitive that they can even detect changes of a few thousandth of a centigrade. The heat receptors are active between 25—45°C and the cold ones between 10—35°C. The impulses from the receptors are sent by afferent nerves to the heat maintenance centre in the posterior hypothalamus, which initiates reflex responses in the body. The thermal receptors of the hypothalamus record the temperature of the blood. The hypothalamus, thus, has two function *viz.*, sensing of temperature and initiation of appropriate responses.

Responses to Cold

There are two mechanisms by which the body maintains the temperature when the environmental temperature gets low. These are conservation of heat and active production of heat.

Conservation of heat is brought about by the constriction of cutaneous blood vessels—*peripheral vasoconstriction*, counter current warming of blood, erection of hair—*piloerection*, and reduction of body surface by behavioral responses such as folding of legs.

Vasoconstriction reduces the amount of blood flowing through the skin; heat loss from the surface of the skin is, therefore, reduced (Fig.17.3). As a result of this, most of the blood is diverted to keep the core temperature constant. Vasoconstriction mostly takes place in the covered parts of body. In extreme cold, the blood vessels of the exposed parts of the body *e.g.* ear, nose, cheeks etc., dilate, thus directing more blood to these in order to keep them warm. Vasoconstriction reduces the heat loss to about one-third of normal.

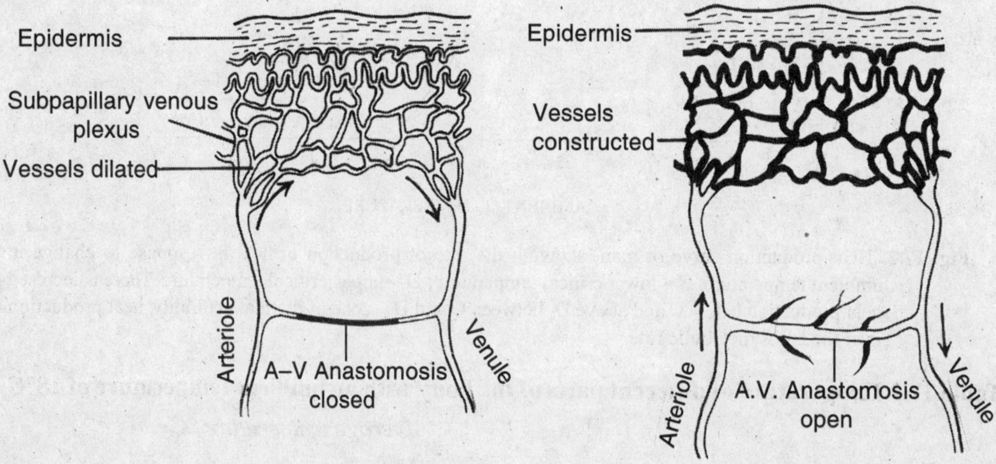

Fig. 17.3. Figure showing (a) peripheral vasodilation and (b) peripheral vasoconstriction. When cutaneous blood vessels dilate, the artero-venous (A-V) anastomoses close and when the cutaneous blood vessels constrict, the anastomoses open.

Counter-current warming of blood is another mechanism which keeps the blood ward. In cold weather, there is considerable loss of heat through the long extremities of the body *viz.* the arms and legs. When the outside temperature is near freezing, the temperature of the blood in the legs may get as low as 20°C. The return of blood to the heat at 20°C is injurious. In order to avoid this there is an effective system of warming the blood before it re-enters the heart. As the blood leave the aorta, its temperature is 27°C, which progressively falls and reaches 20°C in the leg capillaries (Fig. 17.4).

The chilled blood enters the veins and goes back towards the heart. As the arteries and veins lie in close proximity with each other, the chilled venous blood gets warmedby the warm arterial blood. This process is so effective that the blood nearly reaches a temperature of 36°C before entering the heart. This mechanism has been called *counter-current system of temperature control*.

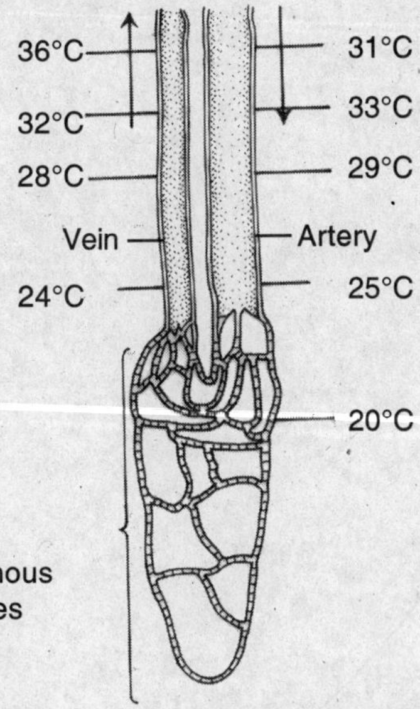

Fig. 17.4. Counter-current warming of blood in the extremities. The blood gets cold, when it reaches a long extremity during cold weather. When the blood returns through the veins, it gets rewarmed by the adjacent arterial blood.

Piloerection protects the body from the vagaries of extreme cold. All homeotherms are provided with hairs or feathers. When the weather is warm the hairs lie flat and the animals are said to wear a thin fur coat. However, when temperature falls the hairs erect and the animals appear thickly coated with fur. The thickened fur coat acts as an insulator; the air trapped in the hairs is stilled so that it gets warmed by the body. This decreases heat loss by convection. The erection of hairs is carried out by small muscles at the root of the hairs, which are controlled by the central nervous system in response to the messages received from the thermal receptors of the skin.

As the temperature falls further, the skeletal muscles start contracting involuntarily. The movement is at first irregular, then assumes the form of quick involuntary movements called shivering. During shivering, small groups of muscles contract asynchronously. Due to the asynchronous movement they do not move the parts associated with them in a coordinated manner. Heat produced by shivering may be as high as five times the basal level (Fig. 17.5).

Shivering is partly controlled by the nervous system through the hypothalamus and partly through the secretion of adrenaline from the adrenal medulla. Adrenaline production raises the metalolic rate (Chatper 11). Babies do not shiver, because their body is effectively insulated by brown fat. The latter comprises about 6% of the body weigth in a healthy baby.

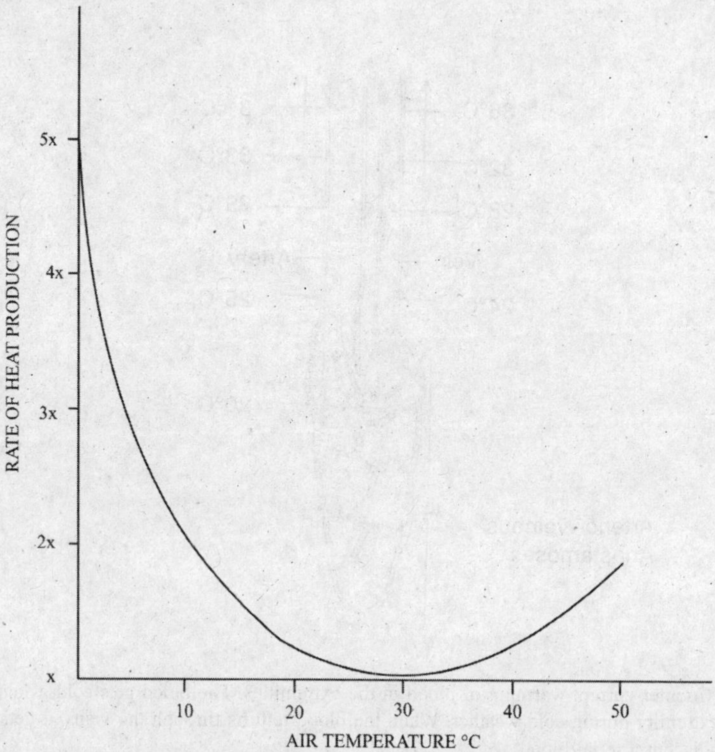

Fig. 17.5. Change in the rate of heat production with change in temperature. The heat production at 30°C is minimum and may be taken as basal. Note that there is a fivefold increase from 30 to 0°C.

When the body is cooled still further, the adrenal cortex is also stimulated by ACTH; this causes increase in the tissue ascorbic acid which decreases susceptibility to cold.

Responses to Heat

When the surrounding temperature is high, the body makes efforts to lose heat. The heat receptors are stimulated, which trigger off a series of reactions. To begin with, the mechanisms mentioned above are reversed; shivering stops, vasodilation occurs so that there is increased blood flow in the skin, piloerection is abolished so that thermal conduction of skin is increased 5—6 times, and the secretions of the adrenal medulla are reduced. These processes are only passive and are effective only up to 30°C. Above 30°C, however, the active processes for losing heat are initiated.

As discussed earlier, most of the loss of the body heat is due to evaporation of water in the form of *insensible perspiration* and *sweating*. As a first step towards cooling, the amount of insensible perspiration increases. If this is not enough, the temperature of the skin rises, which causes sweating in a reflex manner. The threshold of skin temperature in man for sweating is different for different parts of the body e.g., the axilla sweats before the forearm. Usually the threshold for sweating lies between 30 and 32°C.

Under basal conditions, heat loss due to evaporation is about 20% of the total heat loss of the body. Above the upper critical temperature, heat loss due to evaporation becomes very high (Fig. 17.6). Normally, a man loses about 1 litre of water *per day* when the weather is cool. As compared to this, he may lose 15 litres of water *per hour* while doing hard work in a desert. Sweat output increases at the rate of about 20 ml/hour for every 1°C rise in temperature. Sweat is not pure water, it contains amongst other things salts and aminor acids. During profuse sweating man loses large quantities of salt. Death caused by heat-stroke is often due to salt imbalance, rather than rise in body temperature. Dogs and other animals, which do not have sweat glands, vaporize water through the respiratory tract by panting.

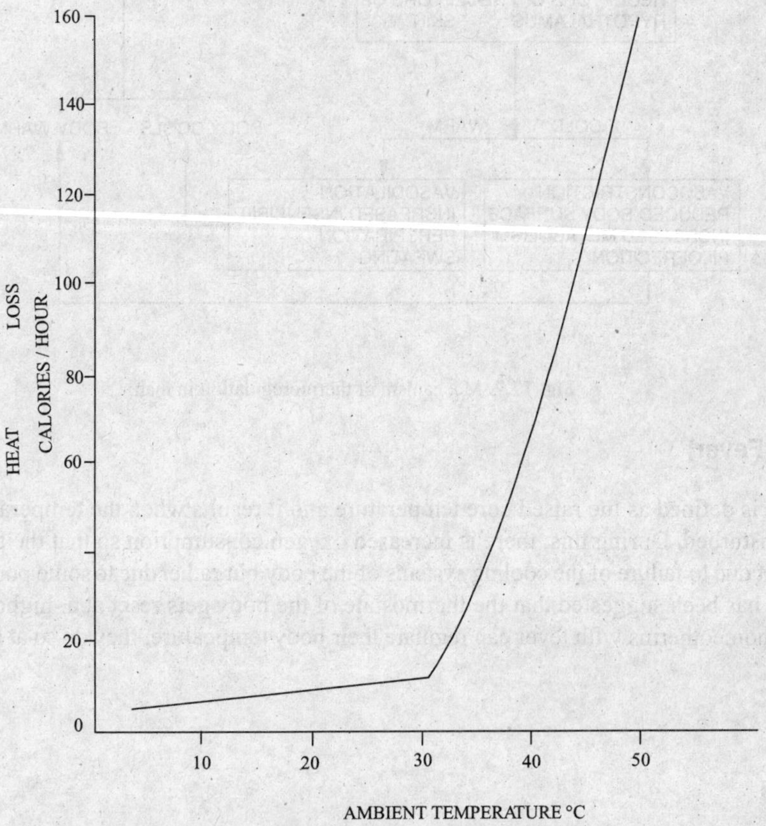

Fig. 17.6. Figure showing increase in the rate of heat loss with rise in ambient temperature.

Individuals who have been acclimatized to high temperatures develop higher capacity of work at these temperatures with lesser discomfort. In such individuals sweating starts at lower temperatures, the sweat contains less salt and the efficiency of the heart increases. As long as a homeotherm can sweat and the sweat can evaporate he can go on cooling his body and stand temperatures as high as 115°C. However, at such high temperatures, individuals get dehydrated very quickly and die. Sweating and vasodilation, like shivering and vasoconstriction are controlled by the central nervous system.

In brief, when the environment is cold, the hypothalamus triggers off a sequence of vasoconstriction, increased metabolism and shivering; when it is hot, it triggers off a sequence of vasodilation increased moisture production and sweating (Fig. 17.7).

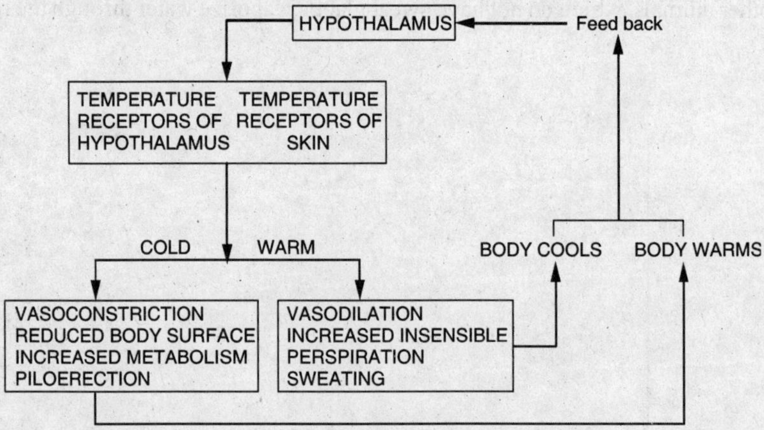

Fig. 17.7. Mechanism of thermoregulation in man.

Pyrexia (Fever)

Pyrexia is defined as the raised core temperature and it results when the temperature regulating centre is disturbed. During this, there is increased oxygen consumption so that the heart rate rises. Fever is not due to failure of the cooling systems of the body but rather due to some poorly understood reasons. It has been suggested that the thermostate of the body gets reset at a higher temperature. Although homeotherms with fever can regulate their body tempeature, they do so at a higher level.

18
Chapter
The Special Senses

Every living organism is provided with specialized structure which in coopeation with the nervous system detect what goes on outside or inside the body. Such specialized structures are known as *receptors* or *sense organs*. In the unicellular animals, the entire body can perceive environmental changes and give appropriate responses. In higher animals, onthe other hand, most of the cells of the body are far removed from each other as weel as from the outside world. Therefore, the means by which they can percieve changes in environment is through sensory organs. During the course of evolution many cells have appeared which have specialized in the reception of external stimuli. Usually, a receptor cell is sensitive to only one kind of stimulus like light, temperature, sound etc. All receptor cells are basically transducers as they convert various forms of enery of the stimulus into electrical energy of the nerve impulse. The receptor may occur as a single specialized cell or as an organization of many similar cells in the form of an organ. In lesser lower animals (protozoa to coelenterata) the receptors are usually single-celled structures, whereas in the more evolved animals, they may be multicellular and complex structures.

In man, there are special receptors for perception of *light, sound, smell, taste and touch*. In addition to these, there are equally important receptors present within the body. For example, the receptor organs of muscles (known as muscle spindles) which provides a measure for muscle stretch; receptor present in the carotid sinus which is sensitive to blood pressure; and, receptors present in the hypothalamus sensitive to the temperature of blood.

Some of the receptors are phenomenally sensitive, *e.g.*, the human ear can differentiate sounds just higher than those generated by thermal motion of molecules in the air. Similarly, the human eye can be stimulated by even less than ten individual photons of light. The sense of smell has also evolved tremendously, that in a moth species, the male is able to detect the presence of female from a distance of even three miles.

GENERAL CONSIDERATIONS

Every receptor in the body, regardless of its function, works through the central nervous system. The nerve impulses sent from all the receptors are waves of depolarization and are, therefore, similar in nature (Chapter 9). The high precision and efficiency of the central nervous system is, however, able to distinguish the impulses arising from specific receptors. The central nervous system recognises the quality of the impulse by identifying the receptor from which the impulse is arising. It recognises the magnitude of the impulse from: (1) number of fibres firing at a time and (2) number of spike fired every second.

A receptor, like any other cell, can be stimulated by any kind of energy provided it is off sufficient strength. However, the receptors are specialized for a specific stimulus for which the threshold is very low. The retinal cells of the eye, for example, have a very low threshold for light and may respond even to strong mechanical, thermal and electrical stimuli.

A factor common to all receptors is that the nature of sensation perceived by the brain is dependent on the nature of the receptor stimulated, rather than the nature of the stimulus. Regardless of the fact whether the stimulus is light, temperature, or electricity, the eye, for example, will always transmit senszation of light, and the nose that of smell. It follows from this that the same stimulus can evoke different sensations, but the nature of the latter depends upon the nature of the receptor organ which perceives it. This is called *Muller's Law of Specific Irritability* or 'Law of specific energies'. The nature of the sensations which are produced by receptor is known as its *modality*. For example, the skin has four modalities, *viz.*, touch, warmth, of cold and pain. Within each modality there may be sub-divisions of sensations, *e.g.*, with sound there may be tone, loudness etc.

Another characteristic common to all receptors is that the height of spike does not depend on the strength of stimulus. The reason being that sensory nerves, like all other nerves, follows the all or none law (Chapter 9) so that the stimulus, whether weak, mode-rate or strong, produces spike of same amplitude. The characteristic which distinguish a weak stimulus from a strong one is the frequency of the discharge of the nerve fibre. A weak stimulus generates fewer spikes as compared to a strong stimulus (Fig. 18.1).

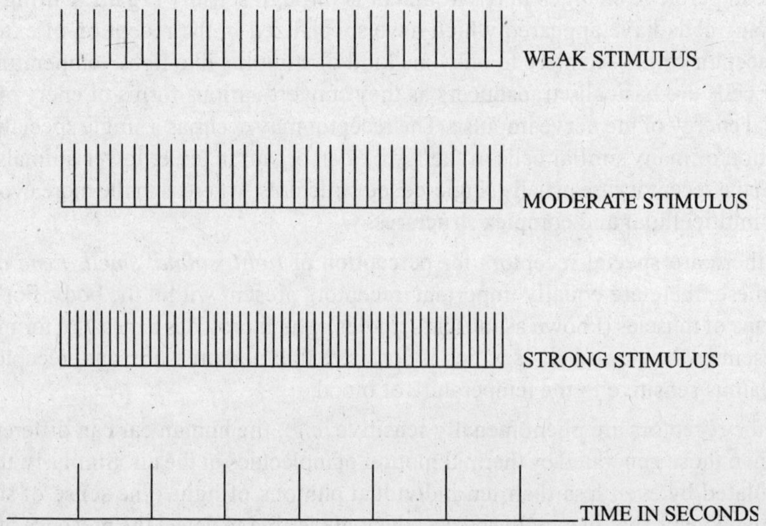

Fig. 18.1. Diagrammatic representation of the discharge of a sensory nerve generated by stimuli of varying strength.

The actual sensation produced by a receptor is a logarithmic and *not* a linear function of the stimulus provided. It mainly depends upon the background stimulus, which is a constant. Thus, a person sitting at a quiet place can hear very low voice but if he is moving in a busy and noisy street, his communicant may have to shout very loudly to be heard by him. The relationship of sensation to a stimulus is expressed by the *Weber-Fechner Law* which is expressed as:

$$\text{Sensation} = K \log I \times C$$

Where C is a constant of background intensity at which the sensation measured.

I is the intensity of the stimulus and

K is also a constant peculiar to the modality.

But this law it is clear that if the intensity is increased tenfold, the amount of sensation becomes double.

An additional characteristic of some receptors is the capacity of their adaptation to a stimulus. For instance, if one has kept a hat on his head, after some time he may not be aware of it. This is called the adaptation of receptors. The touch receptors are adapted most quickly.

Connections of Sensory Units with Central Nervous System

A general organization of sensory nerve tracts is presented in Fig. 18.2.

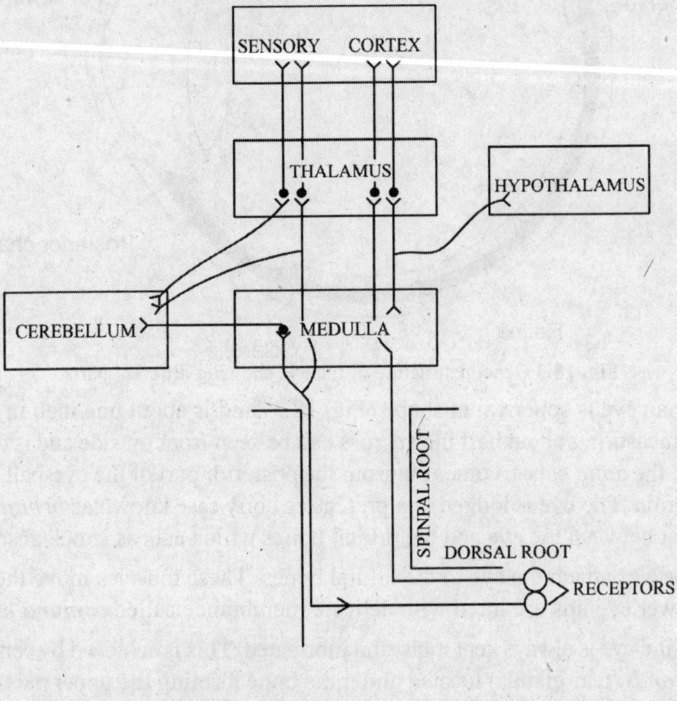

Fig. 18.2. Diagrammatic representation of general organization of sensory portion of nervous system.

Sensory fibres from some of the receptors enter the spinal cord via dorsal roots. Some of the fibres make synaptic connections immediately after entering the spinal cord; other traverse the entire spinal cord and synapse in the medulla; still others travel right up to the thalamus where they make synaptic connections. The thalamus is a major sensory co-ordinating region of vertebrate fore-brain and most of the fibres synapse in the thalamus. The tracts in the spinal cord which carry such fibres are known as *spinothalamic tracts*. Similarly there are other tracts *e.g., spinocerebellar tracts* which have the fibres going to cerebellum.

The sensory fibres from head region are carried through the cranial nerves. Thus, the V or *trigeminal* nerve includes sensory fibres associated with the sense of pain, touch, temperature and pressure. Similarly, VII or *facial*, IX or *glossopharyngeal* and X of *vagus* nerves carry sensory fibres associated with different sense. There are three cranial nerves which are purely sensory. These are optic, olfactory and auditory which are concerned with the eye, nose and ear, respectively.

Most of the sensory imputs go to the thalamus. But, the thalamus acts only as a way station for sensory connections. Its major connection with other brain parts include tracts to the verebral cortex, the hypothalamus and the cerebellum.

ORGANS OF SPECIAL SENSE

Vision

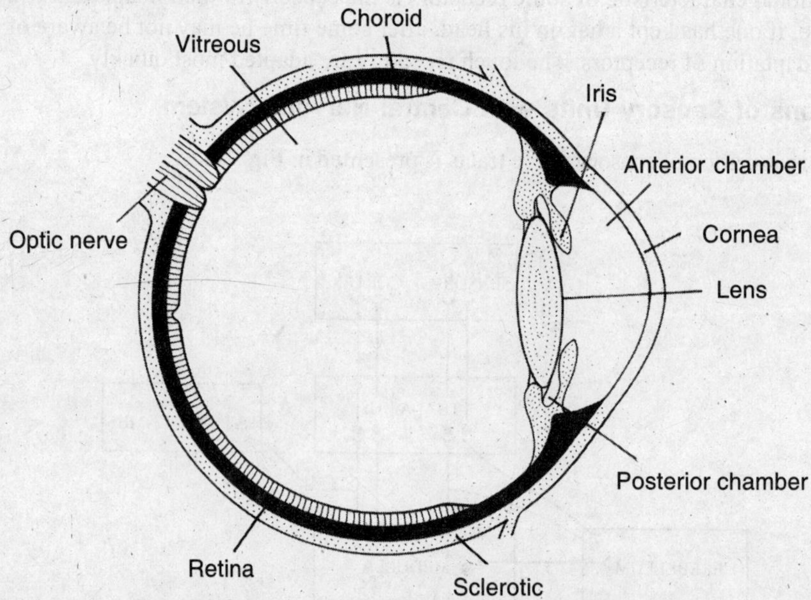

Fig. 18.3 General structure of the eye showing different parts.

Eye: The human eye is spherical in shape (Fig. 18.3) and is about one inch in diameter. Only a circular window measuring about half inch across can be seen from outside and is termed as *cornea*. A compact nerve, the *optic nerve*, comes out from the posterior part of the eye-ball and carries nerve impulses to the brain. The eye is lodged in a protective bony case known as *orbital cavity*. There is a thick layer of fat between the eye and the orbital bones which acts as shock absorber.

A set of *six* muscles attach the eye to the orbital bones. These muscles move the eye in the orbit. The upper and lower eye lids are lined with delicate membrance called *conjunctiva*.

The surface of the eye is always kept moist and lubricated. This is achieved by continuous secretion of the *lacrimal glands* (tear glands) located under the bone forming the upper part of the eye socket. The lacrimal fluid (tears) is drained by the *lacrimal ducts* which open in the nose through the naso-lacrimal duct (Fig. 18.4).

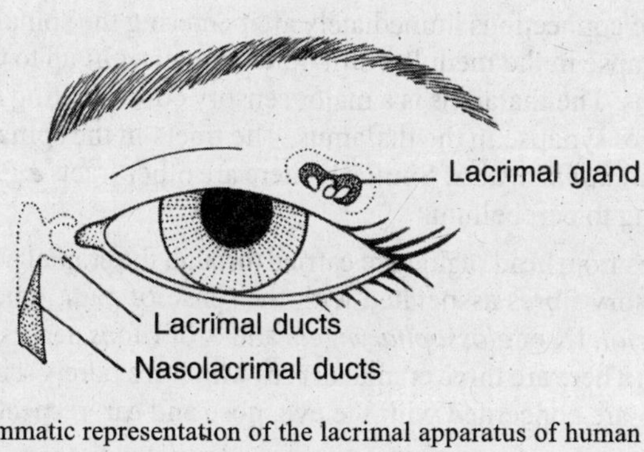

Fig. 18.4. Diagrammatic representation of the lacrimal apparatus of human beings.

If the secretion of lacrimal fluid is hampered, the delicate membrane, covering the eye soon dries up and the eye gets inflamed and finally damaged.

The eye is composed of three layers; they are: an outer or *sclerotic*, a middle or *choroid* and an inner or *retina*.

The *sclerotic* is made up of hard coat of fibrous material. It gives shape to the eye and prevents the inner parts from minor injuries. The part of the sclerotic visible from outside is known as *white of the eye*. The sclera is made up of several rows of flat cells cemented together. In the centre of the white of the eye, there is a window known as cornea.

The *choroid* is richly supplied with fine blood vessels. It is present in the entire eye ball except for a small circular window— like opening known as *pupil*. The circular band of choroid surrounding the pupil is termed as the *iris*. The colour of the iris varies from race to race giving the eye its characteristic hue—black, brown, blue or hazel eyes.

The retina is the light sensitive layer of the eye. It converts the light energy into nerve impulses which are transmitted to the brain through the optic nerves. The eye is provided with an ovoid *crystalline lens* (Fig. 18.3) suspended through the *suspensory ligaments* and located approximately 2 mm behind the cornea. It is biconvex but its rear surface is more convex than the front. The lens is perfectly transparent. However, due to disease or abnormality, it may become opaque or semi-transparent and thus cause blindness. Such a condition is known as *cataract*. The lens divides the eye into two fluid filled chambers. The front chamber is filled with fluid known as *aqueous humor* which is an extracellular fluid. The chamber behind the lens contains a protein matrix that from a gelatinous but clear fluid called the *vitreous humor*.

As the distance between cornea and lens, and lens and retina has to be kept constant, the eye acts like a camera. This is maintained by the two fluids present in the eye. This is obtained by maintaining the pressure of theh internal fluid at a near constant pressure (intra ocular pressure) of 20—30 mm Hg.

The *aqueous humor* enters the eye through the ciliary body. The ciliary boides contain large number of arterial capillaries. The pressure in these capillaries is very high so that the flud always leaks into the front chamber of the blood. Part of the fluid diffuses into the vitreous humor, but the larger portion flows around the lens.

After the aqueous humor has passed around the lens, it passes through the *spaces of Fontana* into a venous sinus around the eye known as the *canal of Schlemm* (Fig. 18.5). The latter opens into the ocular vein. This keeps a continuous circulatin of aqueous humor around the lens.

Blockade of the canal of Schlemm results into high intra-ocular pressure; the pressure rises to the extent that the eye ball on palpation feels like a nut. This condition is known as *Glaucoma*. When the pressure becomes abnormally high, Glaucoma may even cause blindness.

Retina: A microscopic structure of retina that it is made up of five distinct layers from outside to inside *viz.*, (1) *pigmented cells* (2) *rods and cones* (3) *nuclear* (inner and outer) *layers* (4) *ganglion cells* and (5) *nerve fibres*.

The layer of pigment cells: The pigmented layer consists of a single layer of irregular cells cemented together containing granules of a dark pigment. Their main function is to absorb light and prevent blurring of the image which may occur due to reflection.

Hence, it perform the same function as the black paint inside a camera.

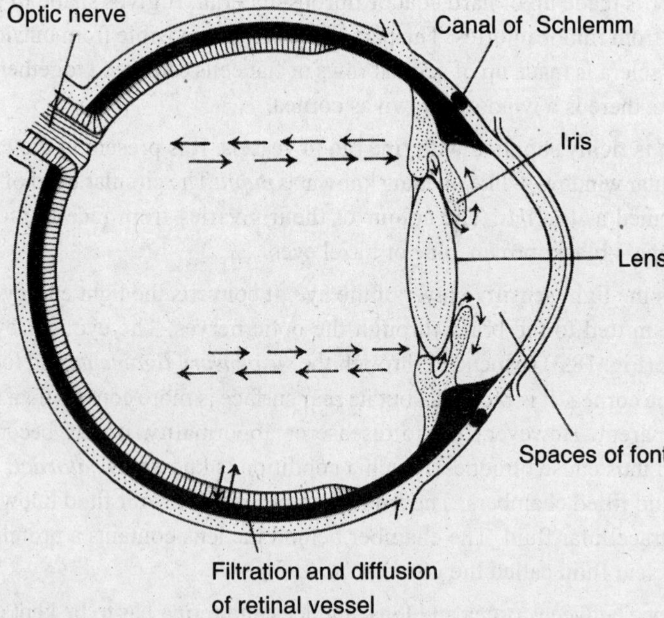

Fig. 18.5 Diagrammatic representation of the fluid system of eye.

The layer of rods and cones: The layer of rods and cones comprises cells which are rod-like and cone-like in structure respectively (Fig. 18.7 A and B). This layer lies in close contact with the pigmented layer. The rods and cones are modified nerve cells and are the visual receptors of the eye. When these are stimulated by light, they set up impulses in the nerve cells which are transmitted to the brain through the *optic nerve*. In most animals, the cones are found clustered together in a depression on retina, the *fovea centralis* (Fig. 18.3). The fovea is concerned with the acute vision *i.e.*, the perception of finer details. The does not mean that cones are not present on other parts of retina. They are present through out the retina intermingled with rods but their number progressively decreases as the distance from fovea increases. The rods are responsible for night and day vision and the cones for colour vision. Both the rods and cones are provided with nuclei at their inner ends which form a layer known as the *outer nuclear layer*. These nuclei given off axons.

Nuclear (inner and outer) layers: Lying adjacent to layer of rods and cones there is a layer of nerve cells known as the inner *nuclear layer of first order neurones*. The dendrites of the cells of the inner nuclear layer synapse with the axons arising from the nuclei of the rods and cones forming the outer plexiform layer (Fig. 18.6). The axons of the inner nuclear layer, on the other hand, synapse with the dendrites arising out of the second layer of nerve cells known as the ganglionic layer. The synapses thus formed are collectively known as *inner plexiform layer*.

Layer of ganglion cells: As mentioned above, the inner nuclear layer synapse with the *ganglionic layer*. The latter is formed by large plump of nerve cells and thus forms a *second layer of neurones*.

Layer of nerve fibres: The axons of the ganglion cells form the last layer of the retina, known as the *nerve fibres layer* of *the second order neurones*. These nerve fibres travel forward for a short distance and then turn horizontally forming a felted structure. There are about 500,000 fibres. These converge at the back of the eye ball and collect to form the thick *optic nerve*.

It may be mentioned at this point that only one cone (or at most two or three) is innervated by a single fibre of the optic nerve whereas 10—100 rods may shere a single fibre. This provides the cones with acute vision (Fig. 18.6).

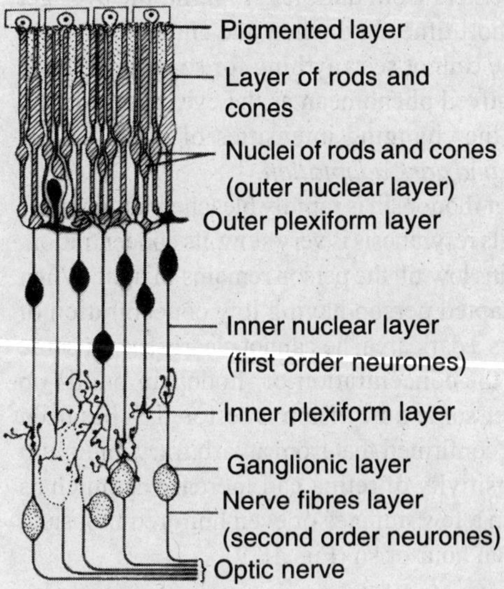

Fig. 18.6. Simplified diagram of retina to show its chief layers.

Chemistry of Vision: The rods and cones change the light energy into nerve impulses. Vitamin A (Chapter 6) plays a very important role in this transormation and is the basic chemical which is utilized by both rods and cones for synthesizing substances which are sensitive to light. In the rods vitamin A is converted into a substance known as *retinine*. The retinine then combines with a protein *opsin* present in the rods to form a light sensitive pigment—*rhodopsin* or *visual purple.* When the rod is exposed to light the rhodopsin get bleached into a substance known as *lumi-rhodopsin* or *transient orange*. The latter is very unstable and in about 1/10th of a second gets converted into another substance known as *metarhodopsin* or *indicator yellow.* This too is unstable and finally breaks to form retinine and opsin. The change to lumirhodopsin is a light reaction whereas all the other reactions are dark or thermal. The lumi-rhodopsin initiates a chemical reaction, independent of light, which gives rise to a nerve impulse. The impulse is transferred from the rods into the optic nerve and from there into the brain. The rate of resynthesis of rhodopsin is very high in darkness, with the result its concentration in the rods gradually builds up.

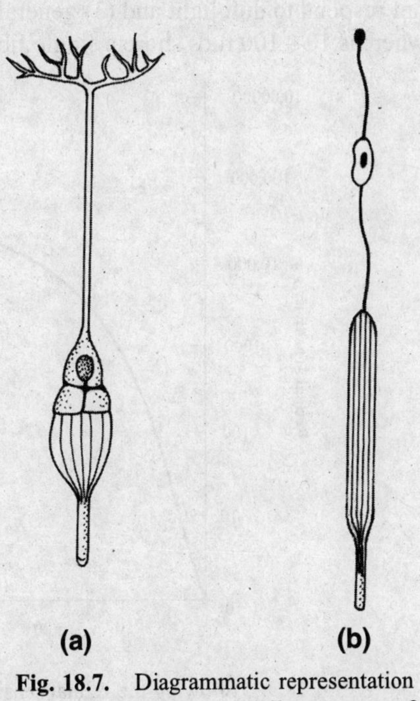

Fig. 18.7. Diagrammatic representation of the two different types of cells found in retina. (a) - a cone and **(b) - a rod.**

Light and Dark Adaptation. It is a common experience that when one enters from darkness to light, the eyes get dazzled for a short time. And when one enters from light to darkness, one cannot see anything for some time. Both these are short-lived phenomean as the eye can gradually adapt itself to the changing intensities of light. This is known as light *and dark adaptation*.

In bright light rhodopsin is rapidly bleached into retinine and opsin; and its resynthesis is very slow. Its concentration, therefore, remains low till the person remains in light. When such a light adapted person having low concentration of rhodopsin enters a dark area, he cannot clearly see for some time. In dark, the concentration of rhodopsin builds up rapidly and after some time vision is restored. It has been experimentally confirmed that from an arbitrary minimum level of the sensitivity of retina can increase as much as thousand fold in a few minutes or even hundred thousand times in about an hour or so (Fig. 18.9).

But when such a dark adapted person reenters a bright area, light adaptation occurs and the retinal sensitivity falls down abruptly within a few minutes (Fig. 18.9).

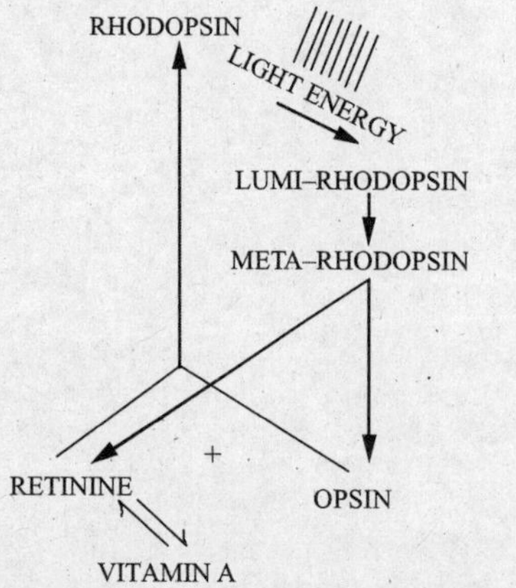

Fig. 18.8. Schematic representation of the retinine—opsin in chemical cycle responsible for light sensitivity of the retinal rods.

Colour Vision : The *cones* are responsible for colour vision. They are different from rods in many respect, *e.g.*, (1) cones respond selectively to colours (2) they are less sensitive to light and hence do not respond to dull light and (3) generally only one cone is innervated by a single optic nerve fibre whereas 10—100 rods share a single fibre (Fig. 18.6).

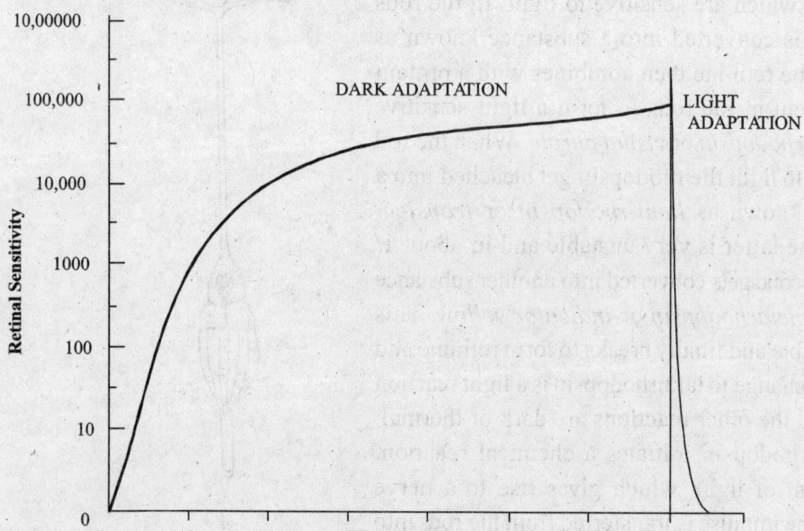

Fig. 18.9. Showing the retinal sensitivity increasing from an arbitrary value of 1 up to even 100,000 in about an hour or so in darkness. Further, it shows that when a person enters lighted area from a dark area, within about 10 minutes, light adaptation occurs and the retinal sensitivity falls from 100,000 to 1 in about 10—15 minutes.

The primary function of cones is to perceive and appreciate colours. For the appreciation of different colours, different types of cones are present in the eye. In lower animals, it has been seen that there are seven types of cones which respond to the seven colours of the spectrum. In human beings, there are only three types of cones, showing maximum sensitivity to blue, green and red. Since colour perception is the ability to distinguish between different wave-lengths of light, the human eye is sensitive to the entire visible spectrum. One reason being that other colours besides blue, green, and red have a wavelength falling within the range of these colours. The spectral sensitivities of blue, green and red colours are shown in Fig. 18.10.

It is obvious from the figure that the cones responsible for perception of blue colour are capable of responding to any wave-length between 400-500 mμ. This means that these cones can also appreciate violet, and green colours, although with lesser sensitivity than blue colour.

The cones for green are sensitive to light between wavelength 450-675mμ. This means that the green cones can respond to blue, green, yellow, orange and red light. However, they respond to green light with greater intensity than any other colour.

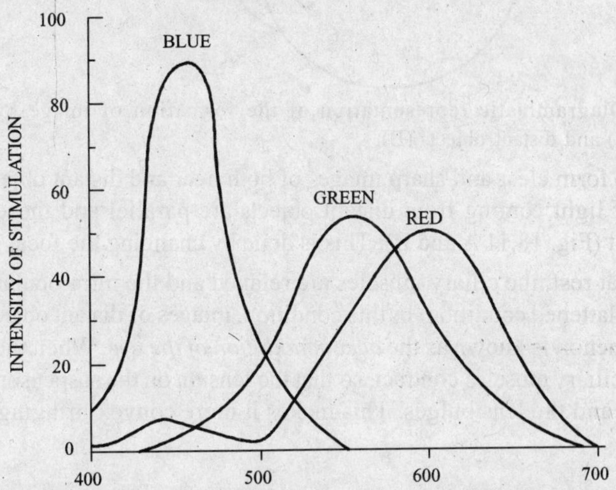

Fig. 18.10. Graph showing the spectral sensitivity curves for blue, green and red lights

The red cones are sensitive to light of all wavelengths, but they respond to the raddish-orange light more strongly than any other colour.

Colour Blindness: Colour blindness is a sex-linked recessive character found in about 8% males. This is due to a recessive gene for which the male does not have a dominant allele. It is the inability to distinguish between the various colours. It is caused by the absence of any of the three types of cones. If a person is deficient in red cones, he can see green, yellow, orange and red colours but he cannot differentiate between these colours. Similarly, if a person lacks green cones, he is not able to distinguish between green, yellow, orange and red colours. This type of colour blindness is termed as *red-green colour blindness.*

In a very rare case people lack blue cones. Such type of people are unable to differentiate between violet, blue, and green colours. This type of blindness is commonly called *blue weakness*.

Functioning of the Eye: The eye functions as a camera. The light coming from an object passes through the cornea and the lens and forms a small inverted image upon the retina. The iris of the eye functions like the diaphragm of the camera and controls the amount of light entering the eye.

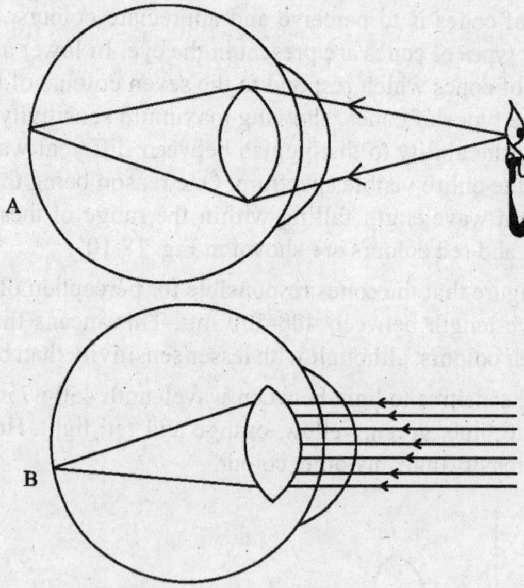

Fig. 18.11. Diagrammatic representation of the formation of image by the eye from near (A) and distant objects (B).

The eye is able to form clear and sharp images of both near and distant objects regardless of the fact that the rays of light coming from distant objects are parallel and those coming from near objects are divergent (Fig. 18.11 A and B). This is done by changing the focal length of the lens.

When the eye is at rest, the ciliary muscles are relaxed and the intra-ocular pressure keeps the lens in a somewhat flattened condition. In this condition, images of distant objects are in focus (Fig. 18.12). This phenomenon is known as the *accommodation of the eye*. When the eye is to be focused at a near object, the ciliary muscles contract, so that the tension on the suspensory ligaments holding the lens is reduced, and the lens bulges. This makes it more convex bringing nearer objects into focus (Fig. 18.12).

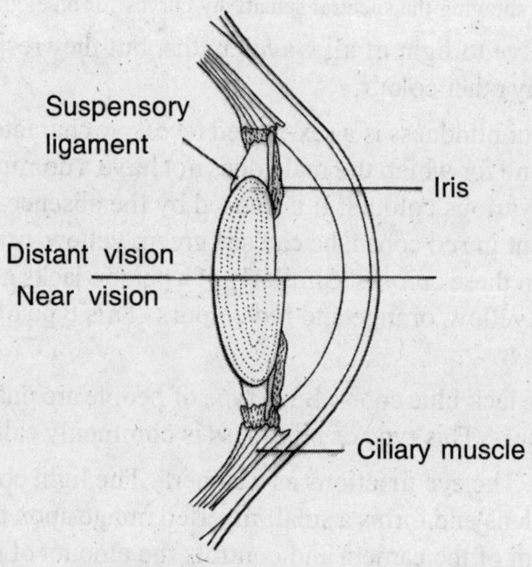

Fig. 18.12. Diagrammatic representation of the shape of lens for distant and near vision respectively.

The elasticity of lens gets reduced with age. This is the reason that elderly people have difficulty in accommodating their sight for near objects.

Optical Defects

A healthy eye focuses the parallel light waves exactly on the retina. This is called as emmetropia (Fig. 18.13 A). However, three common abnormalities may occur due to which the light rays are not focused exactly on retina. These are *myopia* (near-sightedness), *hypermetropia* (far-sightedness) and *astigmatism*.

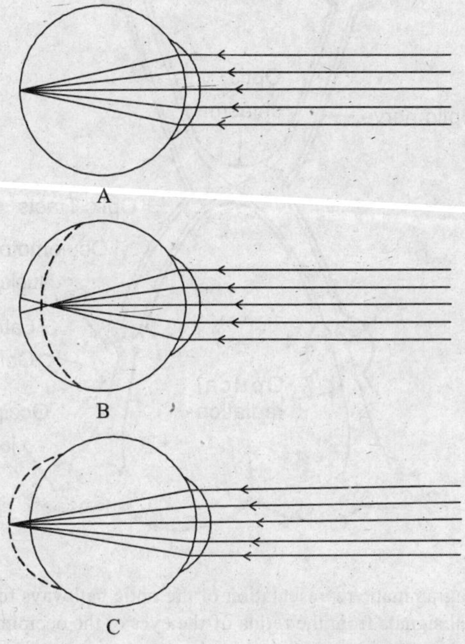

Fig. 18.13. Diagrammatic representation of the focusing of light rays by emmetropic eye (A); myopic eye (B); and hypermetropic eye (C).

Myopia is a condition in which the lens refracts the light rays to the same degree as the normal eye does. But in this case the eye-ball becomes longer so that the distance between the lens and retina increases. Hence the rays are focused a bit in front of the retina (Fig. 18.13 B). Although myopic persons can see near objects they have difficulty in seeing distant objects clearly. Near-sightedness can be corrected by wearing concave lenses.

Hypermetropia or far-sightedness is a condition in which the distance between retina and lens is reduced due to flattening of the back of the eye-ball. In such cases, the lens is unable to converged the rays upon the retina and a blurred image is formed (Fig. 18.13 C). Convex lenses are used to correct this distance.

Astigmatism is a common structural defect in the lens; the rays of light entering the eye instead of being brought to a common focus are focused at different points on the retina. This arises from unequal refraction of the lens.

Neuronal Connections of the Retina with the Brain

Nerve fibres coming from the retina through the optic nerves travel posteriorly under the brain and join to form *optic chiasma*. The chiasma then splits and enters the brain mass. Shortly after entering the brain, *optic tracts* enter the *lateral geniculate bodies* of the thalamus. At this level connections are made with the accommodation of the eye. Finally, from the lateral geniculate bodies the nerve fibres go to the occipital lobe of the cerebral cortex (Fig. 18.14).

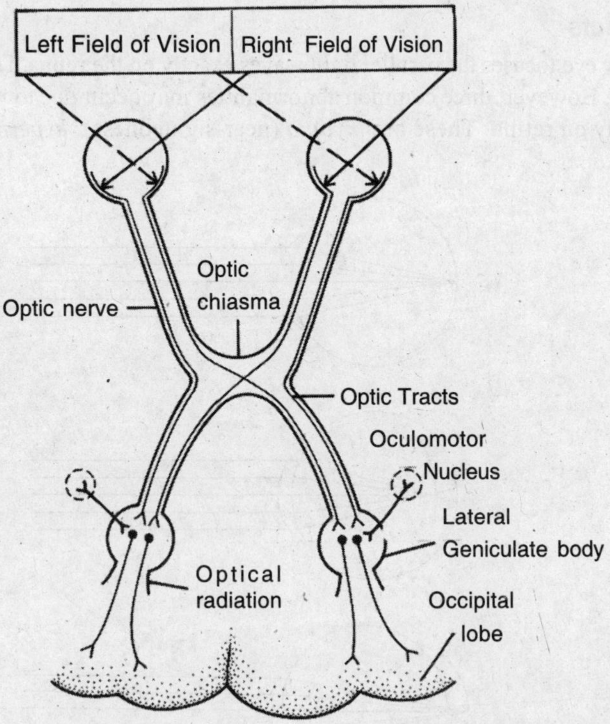

Fig. 18.14. Diagrammatic representation of the optic pathways for transmission of visual signals from the retina of the eyes of the occipital lobe.

Fibres from the right side of left eye cross with the fibres of the left side of right eye at the optic chiasma. Due to this, the nerve impulses from the right side of the field of vision go to the left occipital cortex and those from the left side of the field of vision go to the right occipital cortex.

HEARING

Ear: *The ear* consists of three different compartments, each of the which plays different role in the mechanism of hearing. The different regions of the ear are termed as the outer (external), middle and inner (internal) ears (Fig. 18.15).

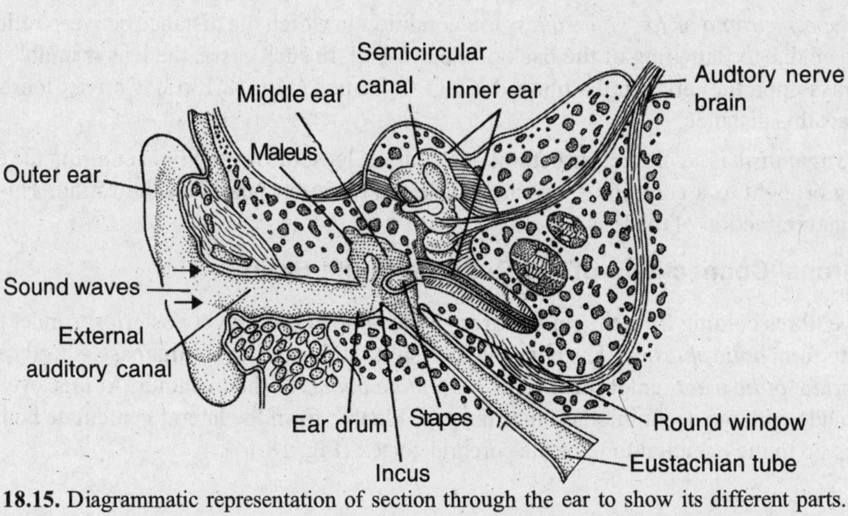

Fig. 18.15. Diagrammatic representation of section through the ear to show its different parts.

The Outer (External) Ear : The external ear consists of two parts: (1) the irregularly shaped shell of skin and cartilage, projecting from the side of the head called the *pinna* or *auricle* (2) a short funnel shaped canal called the external *auditory canal*. The inner end of this canal is blind, being closed by a thin membrane, called the *tympanic* or *drum membrane* which separates it from middle ear.

The Middle Ear or Tympanum: The middle ear lies on the inner side of tympanic membrane. It is surrounded by temporal bones from all sides except for the outer wall which is formed by tympanic membrane. A chain of three bones or *ossicles* form the middle ear. These are the outermost hammer or club-shaped *malleus*; the middle bicuspid tooth-like *incus* and the innermost stirrup-like *stapes* (Figs 18.15 and 18.16).

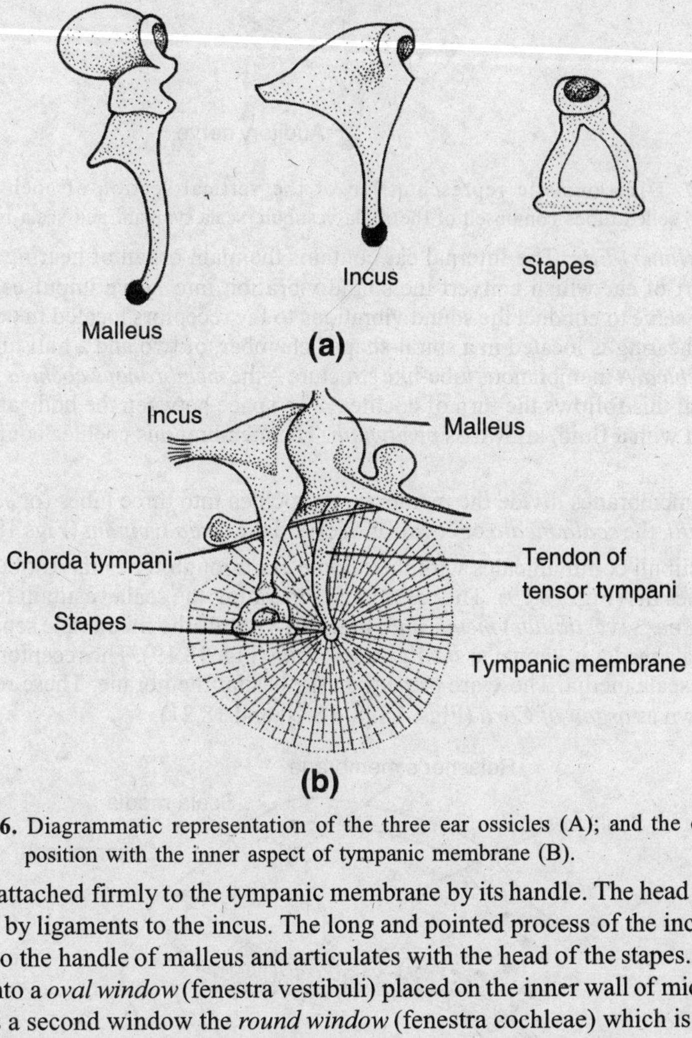

Fig. 18.16. Diagrammatic representation of the three ear ossicles (A); and the ossicles in normal position with the inner aspect of tympanic membrane (B).

The malleus is attached firmly to the tympanic membrane by its handle. The head of the malleus is firmly attached by ligaments to the incus. The long and pointed process of the incus passes downwards parallel to the handle of malleus and articulates with the head of the stapes. The oval base of the stapes fits into a *oval window* (fenestra vestibuli) placed on the inner wall of middle ear. Situated a little below is a second window the *round window* (fenestra cochleae) which is closed by a thin membrane.

The tympanic membrane is furnished with a muscle—*tensor tympani* (or drum stretcher) which keeps the tympanum (ear drum) stretched for perception of high-pitched sounds. There is yet an other muscle, the stapedius, which stretches from the wall of tympanum to the stapes. It prevents the oval base of stapes from entering into the fenestra vestibuli.

A small tube, the *Eustachian* or *auditory* tube connects the cavity of the middle ear with the pharynx.

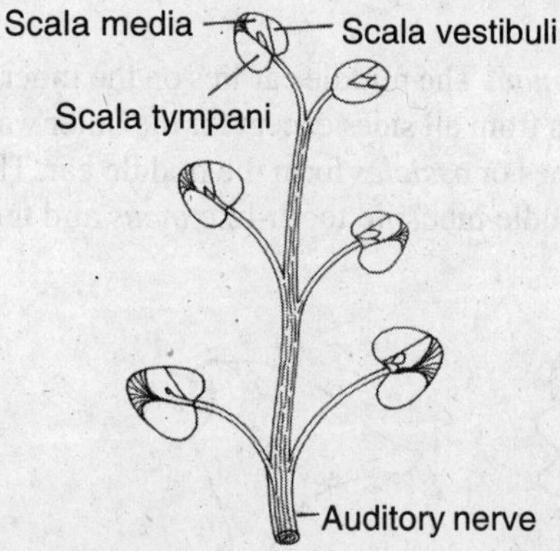

Fig. 18.17. Diagrammatic representati]on of the vertical section of cochlea showing the coiled tubes composed of the scala vestibuli, scala tympani and scala media.

The Internal (Inner) Ear: The internal ear contains the main organ of hearing . Receptors are found in this part of ear which convert the sound vibration into nerve impulses. The outer and middle ears only serve to conduct the sound vibrations to the receptors located in the inner ear. The actual organ for hearing is located in a spiral-shaped chamber, of two and a half turns, of temporal bones termed *cochlea.* A membranous tube-like structure—the *membranous cochlea* is present within the bony cochlea; this follows the turn of cochlea. The space between the body and membranous structure is filled with a fluid, known as *perilymph*. The membranous cochlea contains a fluid, the *endolymph.*

Two delicate membranes divide the membranous cochlea into three tubes (or *scala*). These are the *scala vestibuli,* the *scala media* or cochlear duct and the *scala tympani* (Figs 18.17 and 18.18).

The scala vestibuli communicates with the middle ear through fenestra vestibuli in which, the foot-plate of stapes fits (Fig. 18.19). The membrane separating the scala vestibuli from scala media is known as *Reissner's* (*vestibular*) *membrane* (Fig. 18.18) while the membrane separating the scala tympani from scala media is known as *basilar membrane* (Fig. 18.19). The receptor cells of hearing are found in the scala media. These are present on the basilar membrane. These receptor cells are collectively known as *organ of Corti* (Figs. 18.18, 18.20 and 18.21).

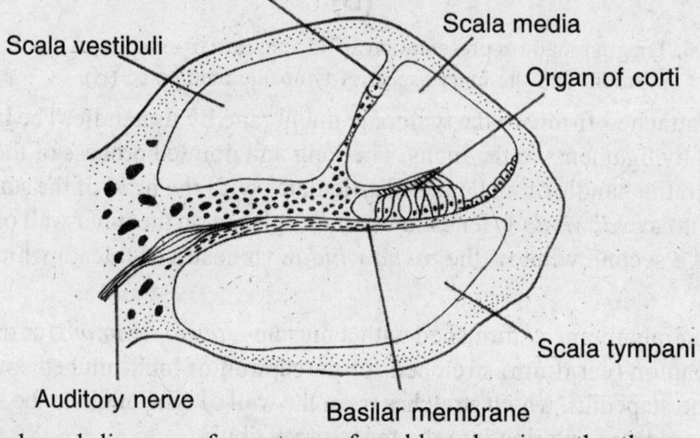

Fig. 18.18. Enlarged diagram of one turn of cochlea showing the three scala, Reissner's and Basilar membrane. The organ of Corti is present at the Basilar membrane.

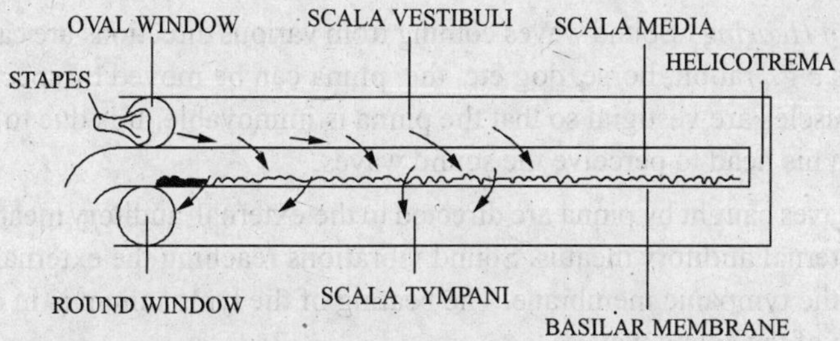

Fig. 18.19. Diagrammatic representation of the cochlea to show the manner in which the vibrations are transmitted from oval window to round window.

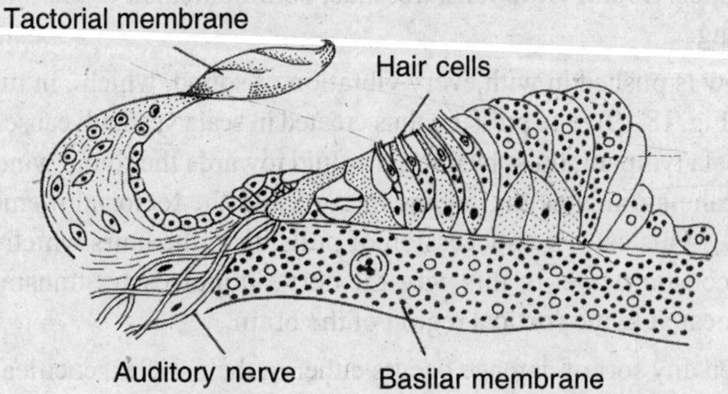

Fig. 18.20. Diagrammatic representation of the microscopic appearance of the organ of Corti which rests on the basilar membrane and converts the sound vibrations into nerve impulses.

The cells of the organ of Corti are long and narrow. Some are provided with hair and are, therefore, known as *hair-cells*. Over-lying the hairs is a delicate membrane which is fastened at one end but free at the other end. This membrane is known as *tectorial membrane.* The tectorial membrane move with every vibration of sound.

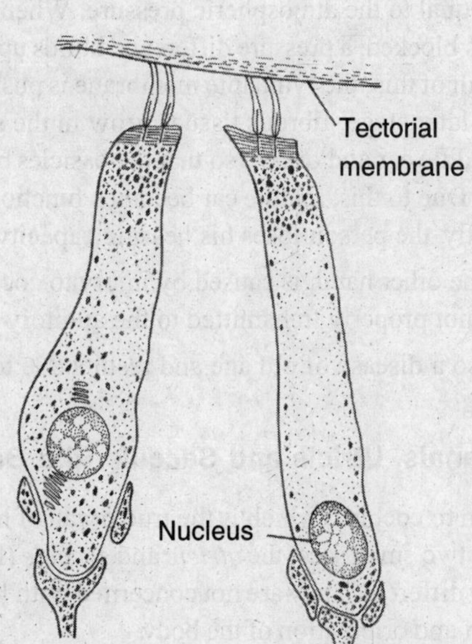

Fig. 18.21. Highly magnified diagram of the cells of the organ of Corti.

Mechanism of Hearing : Sound waves coming from various directions are caught by the pinna. In many mammals *e.g.*, rabbit, horse, dog etc. the pinna can be moved by special muscles. In man, however, the muscles are vestigial so that the pinna is immovable. It is due to this reason that man often has to turn his head to perceive the sound waves.

The sound waves caught by pinna are directed to the external auditory meatus. Sound vibrations reaching the external auditory meatus. Sound vibrations reaching the external auditory meatus set up vibration in the tympanic membrane. The beating of the tympannum is in correspondence with that of vibration of the sound waves.

The vibrations are conveyed through the bones of middle ear to the oval window through the vibration of the stapes. Sound vibrations are, thus, communicated to the internal ear, which is the true organ of hearing.

The oval window is pushed in with every vibration of sound, which , in turn, pushes the fluid of the scala vestibuli (Fig. 18.19). The pressure thus created in scala vestibuli causes the basilar membrane to bulge into the scala tympani, thus, pushing the fluid towards the round window. The movement of the basilar membrane stimulates the organs of corti and the tectorial membrane starts vibrating swiftly (Fig. 18.20). This causes a succession of pulls upon the hairs which are provided with the terminals of the acoustic nerve. In this way, the nerve impulses are transmitted to the centre for hearing which is located in the cortical region of the brain.

Deafness : When any sort of damage occurs either in the ossicles, cochlea, tympanic membrane or in the nerve, it causes partial or total deafness. Deafness is of two types: (1) *Conduction deafness* and (2) *Nerve deafness*.

Conduction deafness may be caused by injury to the ear ossicles or tympanic membrane so that the sound waves are not properly transmitted from the tympanum to the cochlea.

Conduction deafness is also caused by the blockade of the auditory tube. This tube connects the middle ear with the external atmosphere through the nose. The function of this tube is to maintain a pressure inside the ear equal to the atmospheric pressure. When, due to any reason, *e.g.*, allergy or injury or cold, the tube is blocked, a pressure difference builds up between the external atomosphere and that of ear. As a result of this, the tympanic membrane is pushed inwards and the formation of a serous fluid starts. At a later stage, fibrous tissues grow in the serous fluid. This tissue may grow between the cavity of middle ear and ossicle so that the ossicles become firmly bound to the walls of the cavity of middle ear. Due to this, middle ear becomes functionless and sound waves fail to reach the cochlea. Consequently, the person loses his hearing capacity and becomes *deaf*.

Nerve deafness, on the other hand, is caused by injury to cochlea itself or to the auditory nerve. Hence the impulses are not properly transmitted to the auditory cortex.

Nerve deafness is also a disease of old age and results due to the partial degeneration of nerve and cochlea due to aging.

The Semicircular Canals, Utricle and Saccule (the Balancing Organs)

The inner ear in addition to cochlea (which is the true organ of hearing) is also provided with three semicircular canals, and two small sacs, the *utricle* and *saccule* (Fig. 18.22 A). The function of these canals and sacs is totally different. They are not concerned with hearing, but are the chief organs for maintaining equilibrium and orientation of the body.

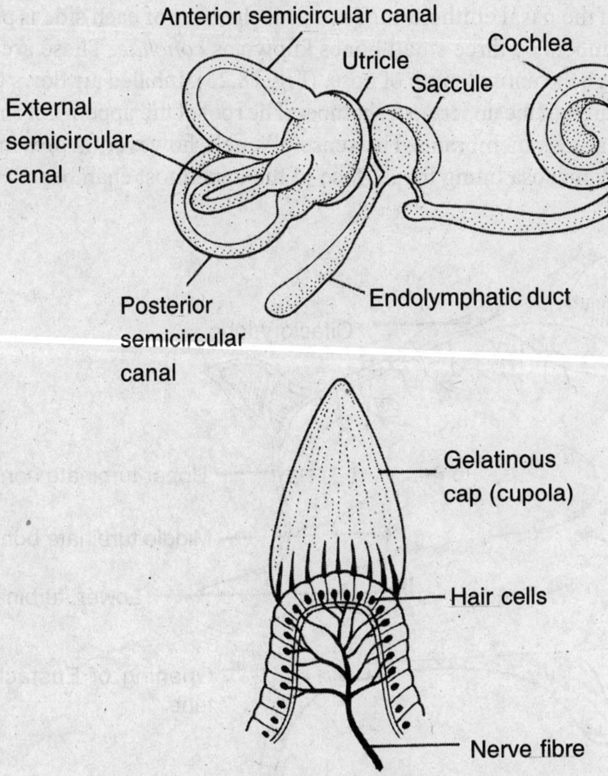

Fig. 18.22. The internal ear showing the semicircular canals, utricle,cochlea etc. (A) and (B) Section of receptor present on the ampulla of semicircular canals showing cupola, hair cells and fibres of vestibular nerve.

There are three semicircular canals which are placed at right angles to each other. These are named according to their position: *viz.*, anterior, posterior and horizontal semicircular canals. All the semicircular canals open in the utricle. One end of each of these canals is swollen into an ampulla which contains receptors for equilibrium. The receptor has sensitive hair cells (Fig. 18.22 B) embedded in a gelatinous cap called the *cupola*. The receptors are supplied by the fibres of vestibular nerve which convey impulses to the medulla and thence to various other parts of the brain.

The receptors of the semicircular canals are stimulated by the movement of the body. The movements cause the endolymph to bend the sensitive hair cells. The brain interprets this and sets up reflexes which cause movements of the limbs and neck to counter this. This is turn, maintains the equilibrium of the body.

Smell and Taste

The taste and smell receptor are *chemoreceptors* because they are excited by chemical stimuli. It implies, therefore, that the characteristic taste and smell depends upon the chemical nature of the material.

SMELL

The sense of taste and smell is not as well developed in man as in some other mammals. However, this deficiency is supplemented by the highly developed vision of man.

The sense of smell is located in the nasal epithelium. The nasal chamber of each side is partially divided into four incomplete chambers by three small bones known as *conchae*. These are placed one above the other running through the entire length of nose. (Fig. 18.23). Inhaled air flows through the lower three chambers but not through the uppermost chamber. The roof of the uppermost chambers of the nasal passage are lined by mucus membrane. The sense of smell, however, is present in the form of yellowish brown olfactory mucosa lining the surface of the uppermost chamber.

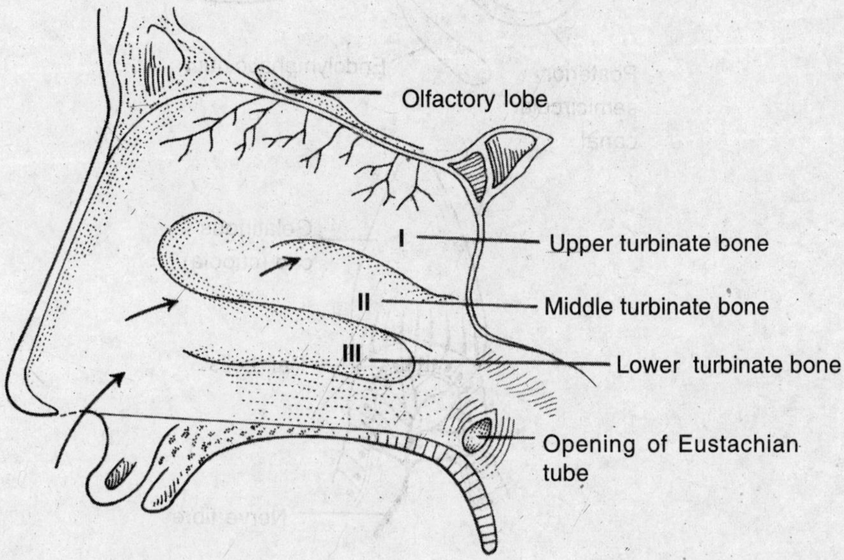

Fig. 18.23. The internal structure of nasal chamber of one side. Figure shows that due to the turbinal bones, the nasal cavity has incompletely divided into four chambers. The arrow indicate the course of inspired air. The area of mucosa above the upper turbinal contains the receptors of smell

Each olfactory cell has about a dozen cilia on its surface; the cilia are modified dendrites. Each cell itself is modified bipolar nerve cell. There are supporting cells present between the olfactory cells which secrete mucus.

The total olfactory region of the nose has an area of 5 cm^2. Each area contains about 50 million sensory cells which lie among supporting cells. The sensory cells give off non-medullated nerve fibres which synapse and finally pass through the cribriform plate to make contact with the olfactory lobes of the brain.

The path taken by inspired air is interesting. The air flows only in the lower three chambers but not through the uppermost chamber. Thus, substances which have to stimulate the olfactory cells must be carried in gaseous form to the uppermost chamber. Experiments have proved that they are carried upwards by *eddies* from the lower chambers during breathing. The eddies are formed by the mixing of the cool air from outside and warm air coming from the lungs. The upward moving eddies take the odourous substances to the sensory cells which are stimulated to give olfactory sensations. It is this reason, that in order to smell more acutely, we take sniffs which allows more eddies to be formed.

An injury to the olfactory mucosa or the olfactory path may totally abolish the sense of smell (anosmia) or alter or change its character (paraosmia). It has been observed that adrenal insufficiency enhances sense of smell; however, a reduction of the sense of smell (hyposmia) occurs during deficiencies of gonadal hormones and vitamin A.

TASTE

The organs of taste are the *taste buds situated* on small projections called *taste papillae* and are found mostly on the surface of the tongue. There are about 9,000 taste buds in adult. Some taste buds are also present in the mucus membrane of soft palate, and a few on *tonsils* and *epiglottis*. In children, these buds are found only on the dorsal surface of the tongue and inside of cheeks.

The taste buds which are embedded in the stratified epithelium of the tongue contain fusiform taste cells surrounded by supporting cells (Fig. 18.24). Each taste cell is provided with a fine hair-like structure. The complete taste bud opens on the surface of the papilla through a pore known as *taste pore*.

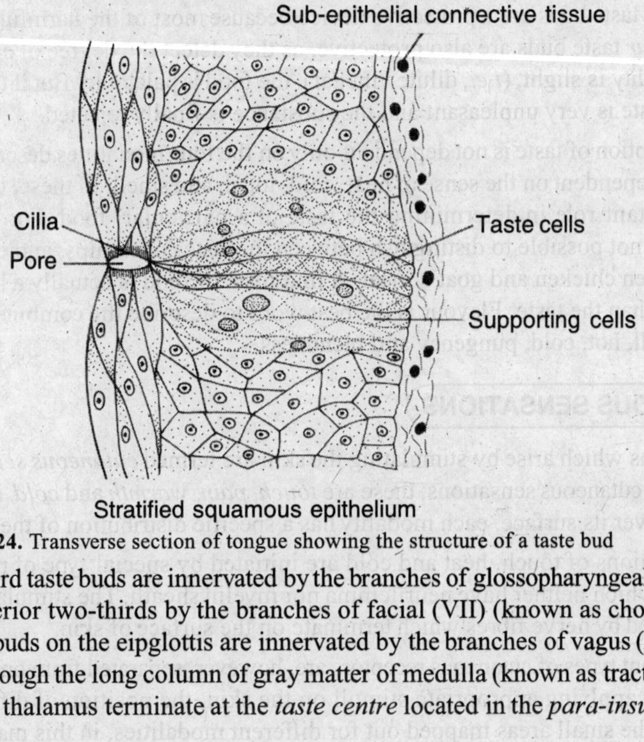

Fig. 18.24. Transverse section of tongue showing the structure of a taste bud

The posterior one-third taste buds are innervated by the branches of glossopharyngeal (IX) (known as lingual) and the anterior two-thirds by the branches of facial (VII) (known as chorda tympani) (Fig. 18.25). The taste buds on the eipglottis are innervated by the branches of vagus (X) nerve. All the nerve fibres pass through the long column of gray matter of medulla (known as tractus solitarius) and finally through the thalamus terminate at the *taste centre* located in the *para-insular* region of the brain.

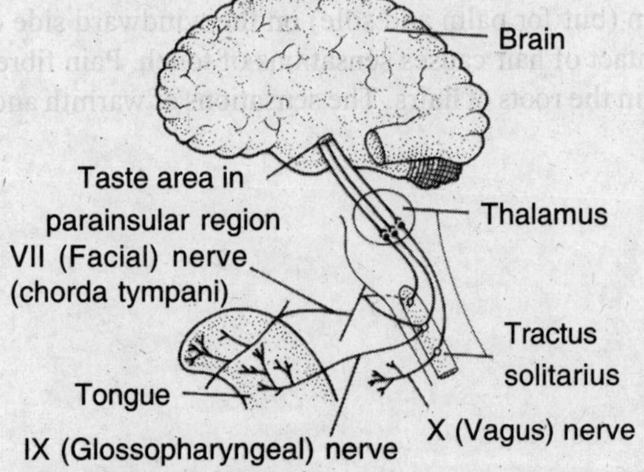

Fig. 18.25. Showing the innervation of tongue taste buds and other taste buds.

A subtance can only be tasted in the form of solution; insoluble substances have no taste. When a solid substance is taken into the mouth it is first dissolved by saliva and tasted thereafter. Substances in solution state enter the pores of the taste buds and stimulate the sensory hairs of the taste cells. With the help of nerve supply, the information is transmitted to the brain where the taste is interpreted.

The taste buds, although smilar in structure, are able to perceive different taste sensations. There are four modalities of taste found in human beings. These are *sweet, salt, bitter* and *sour*.

The *sweet* taste buds discriminate the degree of sweetness of the food. This is in accordance with the amount of sugar present in a particular substance.

The *salt* buds determine the proportion of salt and other electrolyte present in the food.

The *bitter* taste buds are a protective device because most of the harmful substances are bitter in taste. The *sour* taste buds are also protective, as they detect the degree of acidity in the food. If the degree of acidity is slight, (*i.e.*, dilute vinegar), the food is palatable. But if the substance is strongly acidic, the taste is very unpleasant and the substance is finally rejected.

The perception of taste is not dependent only on the four modalities described above. The taste of food is also dependent on the sense of heat, cold, touch and smell, of these, the olfactory sense plays a very important role in determining the taste of a particular food. For example, if the nose is pinched, it is not possible to distinguish between the tastes of turnips, apples and onions or for that matter between chicken and goat's meat. A find plate of food is actually a high developed sense of smell rather than the taste. Flavour is the best word to describe the combined sensations producted by taste, smell, hot, cold, pungency and blendness.

CUTANEOUS SENSATIONS

The sensations which arise by stimulating the skin are termed *cutaneous sensations*. There are *four* modalities of cutaneous sensations; these are *touch, pain, warmth* and *cold.* Although the skin looks uniform all over its surface, each modality has a specific distribution of the receptor.

The sensations of touch, heat and cold are initiated by special type of receptors provided with nerve fibres which neither have neurilemma nor myelin sheath. The stimulations of pain, however, are transmitted by nerve fibres which terminate on the surface of skin.

The different types of cutaneous receptors are, however, separated from one another by measurable distances. By applying appropriate stimuli on the skin, the position of different receptors can be determined; the small areas mapped out for different modalities, in this manner, are referred to as *spots*.

The touch receptors are abundantly found over the tip of the fingers and tongue. They are also found throughout the skin (but for palm and sole) on the windward side of the hairs. This is the reason that even light contact of hair causes sensations of touch. Pain fibres form a network in the skin. They are also found in the roots of hairs. The sensations of warmth and cold have already been dealt.

SELECTED READINGS

Bargamann, W. and Scharrer, B. (eds.) : *Aspects of Neuroendocrinology*. Springer-Verlag, New York. 1990.

Barrington, E.J.W. and Jorgenson, C.B. (Eds.) : *Perspectives in Endocrinology*. Academic Press, New York, 1990.

Bell, G.H., Davidson J.N. and Emslie-Smith, D. : T*extbook of Physiology and Biochemistry*. E.S. Livingstone, Edinburgh. 2001.

Bennet, T.P and Frieden, E. : *Modern Topics in Biochemistry*. Macmillan, New York. 2005.

Brown, G. I. : *Introduction to Physical Chemistry*. Longmans, Green London. 2000.

Davidson, J.N. : *Biochemistry of the Nucleic Acids*. Methuen London, 2003.

Davidson, H. : *Textbook of General Physiology*. J. & A. Churchill Ltd., London. 2005.

Dill, D.B. Adolf., E. F., and Wilber, C.G. (Eds) : *Handbook of Physiology*. American Physiological Society, Washington. 1995.

Eccles, J.C. : *The Synapse*. Scienctific American. January 1965.

Giese, A.C. : *Cell Physiology*. Saunders, Philadelphia, 2000.

Green, D. E. and Baum, H. : *Energy and the Mitochondria*. Academic Press, New York. 2001.

Harris, H. : *Nucleus and Cytoplasm*. Oxford University Press. 1990.

Hoar, W.S. : *General and Comparative Physiology*. Prentice-Hall, Inc. 1986.

Huxley, H.E. : *The Mechanism of Muscular Contraction*. Scientific American. September 1961.

Jellinck, P.H. : *The Cellular Role of Macromolecules*. Glenview, III : Scott, Foresman 2001.

Loewy, A.G. and Siekewitz, P. : *Cell Structure and Function*. Holt, Finehart, and Winston, New York. 2004.

Mcelroy, W.D. : *Cell Physiology and Biochemistry*. Engiewood Cliffs, N.J. : Prentice-Hall, Inc. 1995.

Nalbandov, A.V. : *Reproductive Physiology*. W.H. Freeman, San Francisco & Co. 2003.

Prosser C.L. and Brown, F.A. : *Comparative Animal Physiology*. Saunders, Philadelphia. 1971.

Samson Wright's. *Applied Physiology*. Oxford University Press. 1991.

Tedder, J.M. and Nechvatal. A. : *Basic Organic Chemistry*. John Wiley. New York. 1996.

Watson, J.D. : *Molecular Biology of the Gene*. W.A. Benjamin, New York. 1998.

White., A., Handler, P., and Smith, E.L. : *Principles of Biochemistry*. McGraw-Hill, New York. 2004.

APPENDIX

Units and measures used in physiology and biochemistry based on SI units (*Systime international D'unit's*); an international system of units derived form M.K.S. (*meter-kilogram-second*) system, now in use for all scientific purposes.

Prefixed applied to symbols for various units :

These are never used alone.

T	=	tera-	$= 10^{12}$	G	=	giga-	$= 10^{9}$
M	=	mega-	$= 10^{6}$	k	=	kilo-	$= 10^{3}$
h	=	hecto-	$= 10^{2}$	da	=	deca-	= 10 ;
d	=	deci-	$= 10^{-1}$	c	=	centi-	$= 10^{-2}$
m	=	milli-	$= 10^{-8}$	μ	=	micro-	$= 10^{-6}$
n	=	nano-	$= 10^{-9}$	p	=	pico-	$= 10^{-12}$
f	=	femto	$= 10^{-15}$	a	=	atto-	$= 10^{-18}$

TIME

Hour	= hr	= h Second	= Sec	= s
minute	= min	millisecond	= msec	
1 hr	= 60 min	= 3600 sec		
1sec	= 100 msec	= 1,000,000 μ sec		

MASS

gram	g
kilogram	kg
milligram	mg
microgram	μg
millimicrogram or nanogram	ng
micromicrogram or picogram	pg

1 kg = 1000 g

1 g = 1000 mg = 1,000,000 μg

1 μg = 1000 ng = 1,000,000 pg

1 kg = 2.20 lb

1 lb = 453.36 g

1 tonne (t) = 1000 kg = 0.984 tons

1 ton = 1016 kg

1 g = 0.035 oz

1 oz = 28.4 g

LENGTH

meter	m	micron	μm	yard	yd
centimeter	cm	millimicron	nm	food (feet)	ft
millimeter	mm	Angstrom unit	Å	inch	in

1 m = 100 cm = 1000 mm = 10^{6} μm = 10^{9} mμ = 10^{10}; Å

1 Å = 10^{-10} m = 0.1 nm

1 μm = 10^{-6}m = 1,000 nm = 10.000 Å

1nm = 10^{-9} m = 10 Å

1 cm = 0.394 in

1 km = 3281 ft = 0.6214 mile

1 yd = 0.9144 m

1m = 39.37 in = 1.094 yd

1 in = 2.64 cm = 0.0254 m

1 mile = 1760 yd = 5280 ft

= 1.609 km

1000ft = 0.3048

AREA

The unit of area is the square meter, m^2

1sq in	=	645.2 mm^2
1 sq ft	=	0.093 sq yd
1 m^2	=	1.196 sq yd = 10.764 sq ft
1 sq yd	=	0.836 m^2

VOLUME

The unit of volume is the cubic meter, m^3

cubic millimeter	mm^3	milliliter	ml
liter	1	microliter	μl
cubic centimeter	c.c.		

The volume of 1 kg of water at the temperature of maximum density (4°C)is 1*l*. The millilitre (ml) is one thousandth part of litre. The cubic centimeter (c.c.) is based on the standard of length litre. The cbic centimeter (c.c.) is based on the standard of length that is the metre, and is very nearly but not quite the same as the ml.

1*l* = 1000 ml	1ml = 1000 μl
1 ml = $\frac{1}{1000}$l = 1.000028 c.c.	1 oz = 28.4 ml
1 *l* = 35.196 fluid ounce (oz)	1 gallon = 4.551
1 *l* = 0.061 in^3	1 gallon = 0.00455 m^3
	1 cu in = 1.639 × 10^{-5} m^3
	1 cu ft = 28.32*l*

DENSITY

The unit of density is the kilogram, per cubic metre, (kg/m^3)

1 lb/in^3 = 2.768 × 10^4 kg/m^2

1 *lb/ft*^{3} = 16.023 × kg/m^3

PRESSURE

unit = mm Hg = torr	=	1.33.322 newtons/m^2
1 normal atmosphere	=	760 mm Hg = 760 torr
	=	1033 g/cm^3
	=	1013 millibars
	=	1013250 dyn/cm^2

HEAT

The calorie (cal), also known as the small calorie or gram calorie, is the amount of heat required to raise the temperature of 1g of water from 14.5°C to 15.5° C.

The large calorie (kcal or Cal), also known as the kilogram calorie or kilocalorie, is the amount of heat required to raise the temperature of 1 kg of water from 14.5°C to 15.5°CF

1.000 millicalorie (mcal) = 1 cal

1 kcal = 1000 cal = 4,186.8 joules (J).
In British 'Thermal Unit (B.T.U.) 1 B.T.U. = 0.252 kcal

TEMPERATURE

Temperature (*t*) is customarily expressed in °C – degree Celsius, (degree Centrigrade).

Thermodynamic temperature is measured from absolute zero, *i.e.*, – 273.15°C. At present absolute temperature is measured in degree Kelvin (°K). The unit of Kelvin and Celsius temperature are identical, i.e., 1°C = 1K. However, temperature in Kelvin is measured from absolute zero so that temperature measured in °C is equal to temperature in Kelvin +273.15.

Conversion from Fahrenheit to Centigrade scale of temperature or vice-versa may be made by using the formual:

$$9/5°C = °F - 32$$

ELECTRICAL UNITS

ampere, A	Volt, V
milliampere, mA	millivolt, mV
watt, W	ohm , Ω
farad, F	megohm, M Ω
microfarad, μF	joule, J
coulomb, C	

One ampere (A) is a flow of 6×10^{18} electrons per second past a given point. If I ampere is flowing in a wire, the quantity (Q) of electricity passing a point in the wire in *t* sec is given by Q = I*t* coulombs.

Ohm's Law. The ratio of the potential difference (V) between the ends of conductor to the current (I) flowing in it is a constant, called the resistance (R). R = V/I.

The heat in watts (W) produced in such a resistance is given by

$$W = I^2 R \text{ or } W = I.V$$

PHOTOMETRIC UNITS

Candela (cd). The unit of luminous intensity. A full radiator (black body) at the melting point of platinum emits 60 candelas per square centimetre. The candela is approximately equal to the absolute 'standard candle'.

Lumen (lm). The unit of luminous flux. It is the flux emitted into unit solid angle by a unifrom point source of one candela.

Lux or Metre-candle (lx). The unit of illumination, one lumen per square metre. It is approximately the illumination of a surface placed one metre from a standard candle.

COMMON SYMBOLS

>	greater than/whence/from which is derived
<	less than/from/derived from
∝	infinity
:	ratio ; "is to"
: :	equality between ratios ; "as"
*	birth
+	death
♀	female
♂	male
=	equal to

$\rightleftharpoons$ reversible reaction
+ plus/excess/acid reaction/positive
– minus/deficiency/alkaline reaction/negative
± plus or minus/either positive or negative/indefinite

MISCELLANEOUS

Atomic weight : Atomic weight of an element is defined as the average weight of atoms of specified isotopes relative to the weight of Carbon atom taken as 12.

Molecular weight : The sum of the atomic weights of all the atoms which comprise a molecule.

Gram Atom. The atomic weight of an element expressed in grams, *e.g.*, 12 g of carbon = 1 gram atom of C.

Gram Equivalent : The weight of substance in grams which will combine with or replace 1 gram of hydrogen or 8 grams of oxygen.

Gram Molecule : The molecular weight of a substance expressed in grams **(mole or mol)**.

Molar Solution : A solution containing 1 gram molecule (1mol) of a substance made up to 1 litre of the solution.

Molar Solution : Number of moles of a solute dissolved in 1kg of solvent.

Normal solution : A solution containing 1 gram equivalent of solute per litre of solution. Thus a solution containing 2 gram equivalents per litre is expressed as 2N solution. (This is now an obsolete expression).

Index

A

D

N

O

P